AF326846

Advances in Pattern Recognition

For other titles published in this series, go to
http://www.springer.com/series/4205

Dr. Riad I. Hammoud

(Ed.)

Augmented Vision Perception in Infrared

Algorithms and Applied Systems

 Springer

Editor
Dr. Riad I. Hammoud
Delphi Electronics & Safety
Kokomo, Indiana, USA
riad.hammoud@delphi.com
http://sites.google.com/site/riadhammoud/

Series editor
Professor Sameer Singh, PhD
Research School of Informatics
Loughborough University
Loughborough, UK

ISBN 978-1-84800-276-0   e-ISBN 978-1-84800-277-7
DOI 10.1007/978-1-84800-277-7

Advances in Pattern Recognition Series ISSN 1617-7916

British Library Cataloguing in Publication Data
A catalogue record for this book is available from the British Library

Library of Congress Control Number: 2008939138

Printed on acid-free paper

9 8 7 6 5 4 3 2 1

Springer Science+Business Media
springer.com

Preface

Throughout much of machine vision's early years the infrared imagery has suffered from return on investment despite its advantages over visual counterparts. Recently, the fiscal momentum has switched in favor of both manufacturers and practitioners of infrared technology as a result of today's rising security and safety challenges and advances in thermographic sensors and their continuous drop in costs. This yielded a great impetus in achieving ever better performance in remote surveillance, object recognition, guidance, noncontact medical measurements, and more. The purpose of this book is to draw attention to recent successful efforts made on merging computer vision applications (nonmilitary only) and nonvisual imagery, as well as to fill in the need in the literature for an up-to-date convenient reference on machine vision and infrared technologies.

Augmented Perception in Infrared provides a comprehensive review of recent deployment of infrared sensors in modern applications of computer vision, along with in-depth description of the world's best machine vision algorithms and intelligent analytics. Its topics encompass many disciplines of machine vision, including remote sensing, automatic target detection and recognition, background modeling and image segmentation, object tracking, face and facial expression recognition, invariant shape characterization, disparate sensors fusion, noncontact physiological measurements, night vision, and target classification. Its application scope includes homeland security, public transportation, surveillance, medical, and military. Moreover, this book emphasizes the merging of the aforementioned machine perception applications and nonvisual imaging in intensified, near infrared, thermal infrared, laser, polarimetric, and hyperspectral bands.

This book contains eighteen chapters organized into seven distinctive parts. Part I presents advanced techniques for identifying unique infrared signatures and classifying small-resolution objects above and under the soil. Chapter 1 addresses the challenge of metallic and nonmetallic land mine detection using infrared thermography instead of popular metal detectors. It describes a novel approach for classifying shallowly buried objects in terms of geometric and thermal signatures. Chapters 2 and 3 address the classification problem of vehicles including M35 trucks vs. T72 tanks, using uncooled-infrared and passive polarimetric sensors within a distance up to 12 km.

Part II focuses on describing successful noncontact thermal video analysis methodologies to compute vital sign measurements on the human face, neck, and breast. Chapter 4 presents a wavelet-based signal analysis approach for accurate measurement of arterial pulse based on the thermal infrared imaging of arterial pulse propagation along the superficial arteries of the human body. Chapter 5 describes a coalitional tracking algorithm used in free-contact deception, lie, and breathing-during-sleep monitoring systems. The last chapter of this part (Chapter 6) provides a survey of recent research on thermal infrared imaging in early breast cancer detection.

Part III highlights the recent advances in sensor technology development through combining conventional imaging and spectroscopy (hyperspectral imaging) and their deployment in medical diagnostics, remote sensing, and military target recognition. Two examples of hyperspectral bands are explored in acquiring critical spatial information from an object: visible fluorescence and near-infrared spectrum. Chapter 7 presents a noninvasive skin tumor detection technique based on the analysis of spectral signatures measured from hyperspectral visible fluorescence imaging at a fixed ten-nm spectra resolution. The last chapter of this part (Chapter 8) proposes a methodology to automatically select relevant spectral bands (spectral screening) along with a novel class of target detection methods.

Part IV highlights the use of intensified and thermal sensors for facial recognition. Chapter 9 tackles the problem of face recognition in low-light environment by matching probe and gallery recorded with disparate I2 and thermal sensors. Chapter 10 proposes a novel technique to recognize facial expressions in thermal imagery.

Part V deals with automatic moving object detection in airborne low-resolution near-infrared and thermal videos. Chapter 11 addresses the issue of safe navigation of both manned and unmanned aircrafts and presents an enhanced vision system to locate a runway and detect obstacles on it during approaching and landing operations using an infrared sensor mounted on an aircraft nose. Chapter 12 presents a faster technique than multiscale optical flow for precise localization of boundaries of moving objects in low-resolution thermal videos. The basic principle is combining forward and backward motion history images.

Part VI focuses on fusion and registration techniques of multiple disparate multichannel visible and thermal sensors for accurate image segmentation and pedestrian contour detection (Chapter 13), tracking occluded objects using mutual information (Chapter 14), and pedestrian tracking using up to four cameras (Chapters 15 and 16). Chapter 16 presents a pedestrian detection system based on the simultaneous use of two optical and thermal stereo systems.

The last part of this book (Part VII) focuses on multitarget tracking using laser and infrared sensors. Chapter 17 presents a successful algorithm to track multiple people in a crowded scene using laser imagery. The last chapter (Chapter 18) of this book reports the experimental results of a comparative study of both boosted and adaptive particle filters for affine-invariant target tracking in infrared imagery.

This practical reference offers a thorough understanding of theory and experimental field operational characteristics of key algorithmic building blocks of computer vision systems using nonvisual infrared imagery. It contains enough material to fill a two-semester upper-division or advanced graduate course on topics related to machine vision and its applications, augmented vision, infrared imagery, pattern recognition, remote sensing, and information fusion. Scientists and teachers will find in-depth coverage of recent state-of-the-art work on augmented machine vision in infrared imagery. Moreover, the book helps readers of all levels understand the motivations, activities, trends, and directions of researchers and engineers in the machine vision industry in today's market and offers them a view of the future of this rapidly evolving technological area. It should be noted that while this book provides a brief background review of computer vision and infrared spectrum, it is highly recommended for nonexperts in this area to first read through some popular introductory references on image understanding, remote sensing, and electromagnetic spectrum.

This effort could not have been achieved without the valuable support of my colleagues from several active communities, including IEEE OTCBVS workshop and SPIE Defense and Security Symposium series. I am so grateful for their permission to include their excellent work here; their expertise, contributions, feedback, and review added significant value to this groundbreaking resource.

I would like to extend thanks to all folks at Springer-Verlag, and in particular to Catherine Brett for her valuable support.

Kokomo, Indiana, USA   *Riad Ibrahim Hammoud*
February 12, 2008

Contents

Part II Thermal Imagery & Vital Sign Detection

Part IV Face and Facial Expression Recognition in Intensified and Thermal Imagery

9 Face Recognition in Low-Light Environments Using Fusion of Thermal Infrared and Intensified Imagery

Part VI   Multimodal Imagery Fusion for Human Localization and Tracking

13   Feature-Level Fusion for Object Segmentation Using Mutual Information

14   Registering Multimodal Imagery with Occluding Objects Using Mutual Information: Application to Stereo Tracking of Humans

Contributors

Massimo Bertozzi Vision Laboratory, University of Parma, Italy, bertozzi@vislab.it

Aruni Bhatnagar Division of Cardiology, Department of Medicine, University of Louisville, Louisville, Kentucky, USA

Alberto Broggi Vision Laboratory, University of Parma, Italy, broggi@vislab.it

Sergey Y. Chekmenev Computer Vision and Image Processing Laboratory, Department of Electrical and Computer Engineering, University of Louisville, Louisville, Kentucky, USA

Robert Collins Department of Computer Science and Engineering, The Pennsylvania State University, University Park, PA 16802, rcollins@cse.psu.edu

J. Cui Key Laboratory of Machine Perception (MoE), Peking University, China, cjs@cis.pku.edu.cn

James W. Davis Dept. of Computer Science and Engineering, Ohio State University, Columbus OH 43210, USA, jwdavis@cse.ohio-state.edu

Nicholas A. Diakides Advanced Concepts Analysis, Inc. 6353 Crosswoods Drive, Falls Church, VA 22044-1209, USA, diakides@cox.net

Jonathan Dowdall Iconic Research, LLC, Houston, TX, USA, jonathan.dowdall@iconicresearch.com

Edward A. Essock Department of Psychological and Brain Sciences, University of Louisville, Louisville, Kentucky, USA

Guoliang Fan Oklahoma State University, Stillwater, OK 74078

Aly A. Farag Computer Vision and Image Processing Laboratory, Department of Electrical and Computer Engineering, University of Louisville, Louisville, Kentucky, USA

Mirko Felisa Vision Laboratory, University of Parma, Italy, felisa@vislab.it

Stefano Ghidoni Vision Laboratory, University of Parma, Italy, ghidoni@vislab.it

Cristina Hilario Gómez Universidad Carlos III de Madrid, Spain, chilario@ing.uc3m.es

Paolo Grisleri Vision Laboratory, University of Parma, Italy, grisleri@vislab.it

Riad Hammoud Delphi Corporation, Delphi Electronics & Safety, Kokomo, IN 46901-9005, USA, riad.hammoud@delphi.com

Rida Hamza Honeywell International, Aerospace, Minneapolis, MN 55418, USA

Dinh Nho Hào Vrije Universiteit Brussel, Department of Electronics and Informatics, Pleinlaan 2, 1050 Brussels, Belgium, hao@etro.vub.ac.be
and
Hanoi Institute of Mathematics, 18 Hoang Quoc Viet Road, 10307 Hanoi, Vietnam, hao@math.ac.vn

Joseph P. Havlicek University of Oklahoma, Norman, OK 73019

Benjamín Hernández Centro de Investigación Científica y de Educación Superior de Ensenada, Km. 107, Carretera Tijuana-Ensenada, 22860, Ensenada, BC, México
and
Instituto de Astronomía, Universidad Nacional Autónoma de México, Km. 103, Carretera Tijuana-Ensenada, 22830, Ensenada, BC, México

Mohamed Ibrahim M Honeywell International, Aerospace, Minneapolis, MN 55418, USA

Mark W. Koch Sensor Exploitation and Applications Dept., Sandia National Laboratories, PO Box 5800, MS 1163, Albuquerque, NM 87185-1163, mwkoch@sandia.gov

Seong G. Kong Department of Electrical and Computer Engineering, Temple University, Philadelphia, PA 19122

Stephen Krotosky Computer Vision and Robotics Research Laboratory, University of California, San Diego, 9500 Gilman Dr 0434, La Jolla, CA 92093-0434, krotosky@ucsd.edu

Alex Leykin Indiana University, Department of Computer Science, Bloomington, IN, oleykin@indiana.edu

Kevin T. Malone Systems Engineering Dept., Sandia National Laboratories, PO Box 5800, MS 978, Albuquerque, NM 87185-0978, ktmalon@sandia.gov

William M. Miller Computer Vision and Image Processing Laboratory, Department of Electrical and Computer Engineering, University of Louisville, Louisville, Kentucky, USA

Gustavo Olague Centro de Investigación Científica y de Educación Superior de Ensenada, Km. 107, Carretera Tijuana-Ensenada, 22860, Ensenada, BC, México, olague@cicese.mx

Lae-Jeong Park Department of Electrical and Computer Engineering, Temple University, Philadelphia, PA 19122

Ioannis Pavlidis Computational Physiology Lab, Department of Computer Science, University of Houston, Houston, TX, USA, ipavlidis@uh.edu

Hairong Qi Electrical Engineering and Computer Science Department, University of Tennessee, Knoxville, TN 37996, USA, hqi@utk.edu

Dinesh Ramegowda Honeywell International, Aerospace, Minneapolis, MN 55418, USA

Yang Ran Center for Automation Research, University of Maryland, College Park, Maryland, rany@cfar.umd.edu

Venkatagiri Rao Honeywell International, Aerospace, Minneapolis, MN 55418, USA

Stefan A. Robila Montclair State University, RI 301, Montclair, NJ 07043, USA, robilas@mail.montclair.edu

Eva Romero Centro de Investigación Científica y de Educación Superior de Ensenada, Km. 107, Carretera Tijuana-Ensenada, 22860, Ensenada, BC, México

Mike Del Rose U.S.Army TARDEC, Warren, MI, U.S.A., mike.delrose@us.army.mil

Firooz Sadjadi Lockheed Martin, Saint Paul, Minnesota, firooz.sadjadi@ieee.org

Farzad Sadjadi School of Physics & Astronomy, University of Minnesota, Minneapolis, Minnesota, sadjadi@physics.umn.edu

Hichem Sahli Vrije Universiteit Brussel, Department of Electronics and Informatics, Pleinlaan 2, 1050 Brussels, Belgium, hsahli@etro.vub.ac.be

Vinay Sharma Dept. of Computer Science and Engineering, Ohio State University, Columbus OH 43210, USA, sharmav@cse.ohio-state.edu

R. Shibasaki Center for Spatial Information Science, The University of Tokyo, Japan, shiba@csis.utokyo. ac.jp

Diego A. Socolinsky Equinox Corporation, 207 East Redwood Street, Baltimore, MD, 21202, diego002@equinoxsensors.com

X. Song Key Laboratory of Machine Perception (MoE), Peking University, China, songxuan@cis.pku.edu.cn

Li Tang Duke University, Durham, NC 27708

Nguyen Trung Thành Vrije Universiteit Brussel, Department of Electronics and Informatics, Pleinlaan 2, 1050 Brussels, Belgium, ntthanh@etro.vub.ac.be

Mohan Trivedi Computer Vision and Robotics Research Laboratory, University of California, San Diego, 9500 Gilman Dr 0434, La Jolla, CA 92093-0434, mtrivedi@ucsd.edu

Leonardo Trujillo Centro de Investigación Científica y de Educación Superior de Ensenada, Km. 107, Carretera Tijuana-Ensenada, 22860, Ensenada, BC, México

Panagiotis Tsiamyrtzis Department of Statistics, Athens University of Economics and Business Athens, Greece, pt@aueb.gr

Vijay Venkataraman Oklahoma State University, Stillwater, OK 74078

Guido Vezzoni Vision Laboratory, University of Parma, Italy, vezzoni@vislab.it

Lawrence B. Wolff Equinox Corporation, 9 West 57th Street, New York, NY, 10019, wolff@equinoxsensors.com

Zhaozheng Yin Department of Computer Science and Engineering, The Pennsylvania State University, University Park, PA 16802, zyin@cse.psu.edu

H. Zha Key Laboratory of Machine Perception (MoE), Peking University, China, zha@cis.pku.edu.cn

H. Zhao Key Laboratory of Machine Perception (MoE), Peking University, China, zhaohj@cis.pku.edu.cn

Part I
Infrared Signatures and Classification

Chapter 1
Infrared Thermography for Land Mine Detection

Nguyen Trung Thành, Dinh Nho Hào, and Hichem Sahli

Abstract This chapter introduces the application of infrared (IR) thermography in land mine detection. IR thermography in general and for remotely detecting buried land mines in particular, seems to be a promising diagnostic tool. Due to the difference in thermophysical properties of mines and the soil (mines retain or release heat at a rate different from the soil), soil-surface thermal contrasts above the mines are formed. These contrasts are captured by IR cameras to show the changes in temperature over the mines, which can be used for detecting them. Clearly, the degree of success of such detection technology depends on the factors that affect the formation of the thermal contrasts (signatures), such as the depth of burial; soil properties and attributes, including mine properties (size); as well as the time of day during which the measurement is carried out. Another important factor that strongly influences the viability of IR detection method is the rate of false alarms. Indeed, IR sensors may detect any thermally transmitting objects, not only land mines. It is therefore necessary to develop parameter estimation and decision-making tools that enable the IR technology to distinguish signals resulting from a land mine and unrelated clutter signals. This chapter consists of four sections. The first section introduces the physical principles of IR thermography and gives an overview of its literature. The flowchart of the technique is also given in this section. In the second one, we summarize a thermal model of the soil with the presence of shallowly buried objects. This model is used for studying the influence of soil and land mine properties on the temperature distribution of the soil, especially on its surface. The third section aims at detecting possible anomalies in the soil using IR images and classifying them as mines or nonmine objects. The classification is based on the estimation of the thermal and geometric properties of the detected anomalies. In the fourth section, the performance, in terms of probability of detection and false alarm rates, of the proposed approach for an experimental data set is presented. The processing chain of IR thermography, including data acquisition, data preprocessing, anomaly detection, and estimation of thermal and geometric properties of the detected anomalies, is presented and illustrated using an experimental data set measured in an outdoor minefield. Finally, conclusions on the statistical reliability of the IR technique are drawn.

R.I. Hammoud (ed.), *Augmented Vision Perception in Infrared: Algorithms and Applied Systems,* Advances in Pattern Recognition, DOI 10.1007/978-1-84800-277-7_1, © Springer-Verlag London Limited 2009

1.1 Introduction

Detection and clearance of abandoned land mines have been receiving the special attention of many international communities. Several demining techniques have been investigated and applied, such as manual demining, mechanically assisted demining, and metal detectors. Recently, some new technologies (e.g., ground penetrating radar and infrared [IR] thermography) have been developed. Reviews on currently used land mine detection techniques can be found in [8, 13, 29, 32].

So far, using metal detectors is the most popular technique in demining. The main advantage of this technique is that it can detect very small metallic structures usually present in metallic mines. However, more and more nonmetallic mines (plastic mines, wooden mines) are used. It is extremely difficult or even impossible to detect these mines using metal detectors. As a complement to metal detectors, thermal IR techniques seem to be promising for detecting shallowly buried nonmetallic land mines.

The IR detection relies on the difference of the thermal characteristics between the soil and buried objects. Indeed, the presence of a buried object affects the heat conduction inside the soil under natural heating conditions. Consequently, the ground temperature above the object is often different from that of unperturbed areas. This temperature contrast can be measured by an IR imaging system placed above the soil area used to detect the object.

Several works have been devoted to the study of the underlying physical phenomenology that determines thermal signatures of the soil under natural heating conditions since the well-known paper of Watson [46] published in 1975 (see [1, 3, 4, 10, 15, 20, 21, 24, 25, 27, 34, 35, 38] and the references therein). Among them, Leschack and Del Grande [23] described the effect of the emissivity on thermal signatures of the soil using a dual-band IR system. Janssen et al. [19] made a study in which they used several sensors to determine the thermal evolution of surface-laid and buried mines. Different types of mines and different depths of burial were considered that led to interesting conclusions about the evolution of the thermal contrast measured by the IR sensors. A related study is that of Maksymomko et al. [26], who measured the temperature of real and surrogated mines through several diurnal cycles. Similar analyses of measured thermal signatures of the soil were also carried out in [1, 40]. In [31, 37], the authors analyzed the limits of the IR systems for land mine detection in conjunction with other sensor systems in a remotely operated vehicle. They focused on establishing a relation between the temperature contrast of the soil surface and the temperature gradient inside the soil. The thermal evolution of recently laid mines was also investigated. Finally, the application of polarimetric IR thermography for surface-laid mine detection has also been studied [5, 10].

From measured IR images, it is possible to detect the presence of buried anomalies using image processing techniques such as RX algorithm [16], neuron network [24], or mathematical morphology [6]. However, to classify these anomalies, one has to estimate their physical characteristics (thermal diffusivity), size, and shape from the thermal IR measurements. Such a problem is often solved in two steps. The first step, referred to as *thermal modeling*, aims at predicting the temporal behavior of the soil temperature with the presence of buried objects under natural heating conditions using a mathematical model (*forward thermal model*). The second step, referred to as *inverse problem setting for buried object detection*, consists of using the acquired IR images and the forward thermal model to infer the characteristics of the detected anomalies [42].

Thermal modeling helps understanding the effect of land mines (or generally, buried objects) on the temperature distribution of the soil, especially on its surface. So far, most of the works in IR thermography for land mine detection have focused on defining forward thermal models for surface-laid and buried land mines. Among the existing models, Sendur and Baertlein [36] proposed a three-dimensional thermal model for a homogeneous soil containing a buried antitank mine modeled as trinitrotoluene (TNT). The authors considered smooth and Gaussian distributed rough soil surfaces. They proposed a radiometric model for thermal signatures released from the soil surface and from other sources. The considered radiometric model allows approximating the real temperature of the soil surface from measured IR images. This helps improve the comprehension and interpretation of thermal IR imagery. The same radiometric model was also studied by Pregowski et al. [30]. A three-dimensional model for homogeneous soil containing a buried land mine and considering both the explosive (TNT) and the casing of the land mine was introduced by Khanafer and Vafai [21]. In their paper, the authors analyzed the effect of the casing on the diurnal behavior of the soil temperature distribution taking into account soil-surface roughness. In another work [22], the authors analyzed the effect of a thin metal outer case and air gap on the simulated soil temperature at different depths, including the soil surface and the top and the bottom surfaces of the mine. Finally, in [39], a one-dimensional thermal model taking into account the soil moisture variation was introduced. Although different thermal models have been proposed, their validation, by comparing the simulations with experimental data, has been considered by just a few authors using indoor experiments [28, 39] or outdoor experiments [24, 25, 41–45].

The objective of the inverse problem is to use the measured thermal images and a validated thermal model to detect and characterize buried objects by estimating their thermal and geometric properties. Based on the estimated properties, we conclude if the buried objects are land mines or not. In land mine detection and clearance processes, this step helps reduce the number of false alarms and hence speeds up the clearance process. Although mathematical theory of inverse problems in heat conduction have been intensively studied (see, e.g., [2, 7, 14, 17] and the references therein), to our knowledge its application in land mine detection was just considered for the first time in [24] and improved in [42].

The flowchart of IR thermography for land mine detection can be divided into five steps. First, IR cameras are placed in the minefield to acquire data (IR images) for a period of time. Some auxiliary data (such as soil temperature, weather data) are also acquired. Then several preprocessing steps, including radiometric calibration, image temporal coregistration, atmospheric correction, apparent temperature conversion, and ground projection, are applied to the acquired IR images. The preprocessed IR images represent the soil temperature in the minefield at different time instants. After that, the preprocessed IR images and the auxiliary data are used to estimate soil thermal parameters. These parameters play important roles in thermal modeling in which we simulate the diurnal evolution of the soil temperature. Finally, the IR images and the thermal model are used for the detection of possible buried land mines (objects). This step includes anomaly detection and the estimation of thermal and geometric parameters of the detected anomalies. The flowchart is summarized in Fig. 1.1.

As a summary of our work [42], this chapter is structured as follows. In Section 1.2, we establish a thermal model of the soil with the presence of shallowly buried objects. The inverse problem setting for buried object detection is described in Section 1.3. In Section 1.4, the full processing chain of IR thermography, including data acquisition, data preprocessing, and processing steps are presented and illustrated using an experimental data set measured in an outdoor minefield.

Due to lack of space, we do not demonstrate theoretical aspects and numerical methods for the thermal model and the inverse problem in this work. For these topics, we refer to [42].

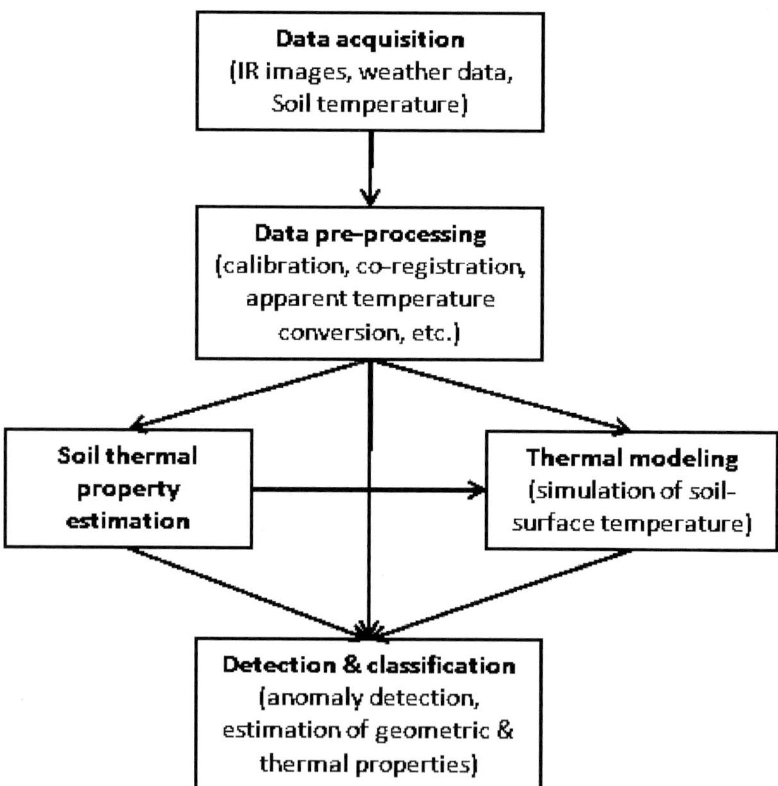

Fig. 1.1 Flowchart of IR thermography for land mine detection

1.2 Thermal Modeling of the Soil, Including Shallowly Buried Objects

In this section, we establish a mathematical model that describes heat transfer processes inside the soil containing buried objects and on the air-soil interface under natural heating conditions. Moreover, we discuss its applicability in practical situations. The model is described as a mixed initial boundary value problem of a partial differential equation of parabolic type.

Throughout this work, for simplicity, we make use of the assumption that the soil and the buried objects under investigation are isotropic and homogeneous that is, their thermal properties are constant (the word *soil* is therefore referred to as a homogeneous soil). Moreover, we also assume that the air-soil interface (the soil surface) is flat, and the variation of the soil moisture content within the investigated volume is negligible during the period of analysis. We do not take into account the presence of vegetation or grass in the soil area under investigation.

1.2.1 Mathematical Formulation

Consider an open rectangular parallelepiped Ω of the soil volume containing a buried object as shown in Fig. 1.2 (although we assume, in the following, that there is only one object buried within the domain Ω, the thermal model we propose below is still valid for multiple objects provided that the boundary conditions are satisfied). We associate the soil volume with an orthonormal Cartesian coordinate system in which the coordinate of a point is denoted by $x = (x_1, x_2, x_3)$. Without loss of generality, we assume that $\Omega = \{x : 0 < x_i < L_i,\ i = 1, 2, 3\}$. We denote by Γ the boundary of Ω and $\Gamma_i^1 = \{x \in \Gamma : x_i = 0\}$, $\Gamma_i^2 = \{x \in \Gamma : x_i = L_i\}$, $i = \overline{1,3}$. We note that Γ_3^1 is the air-soil interface, the only portion of the soil volume

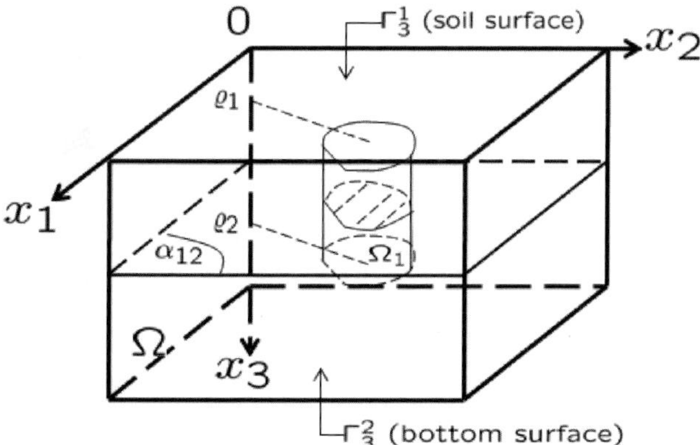

Fig. 1.2 A soil volume including a buried object

accessible to thermal IR measurements, and Γ_3^2 is the bottom of the soil volume. For simplicity of notation, the union of the vertical boundaries of Ω is denoted by Γ_v [$\Gamma_v = (\Gamma \setminus \Gamma_3^1) \setminus \Gamma_3^2$]. Duration of analysis is denoted by $(0, t_e)$ and $S_{i,t_e}^j := \Gamma_i^j \times (0, t_e)$, $i = 1, 2, 3$, $j = 1, 2$; $S_{v,t_e} := \Gamma_v \times (0, t_e)$. Then the temperature distribution $T(x,t)$, $(x,t) \in Q_{t_e} := \Omega \times (0, t_e)$ in the considered domain satisfies the following partial differential equation [9, 33]:

$$\frac{C(x)\partial T(x,t)}{\partial t} = \sum_{i=1}^{3} \frac{\partial}{\partial x_i}\left(\kappa(x)\frac{\partial T(x,t)}{\partial x_i}\right), \ (x,t) \in Q_{t_e}, \tag{1.1}$$

where $C(x)$ (J/m^3/K) is the volumetric heat capacity and $\kappa(x)$ (W/m/K) is the thermal conductivity (of the soil and the buried object) in the domain Ω. These parameters are piecewise constant.

We note that in the setup of the inverse problem for buried object detection presented in the next section, both coefficients $C(x)$ and $\kappa(x)$ are unknown (at least in the location of the buried object). In order to characterize the buried object, these parameters must be simultaneously estimated. Unfortunately, there is not enough information in measured data for obtaining reliable estimates of both coefficients concurrently. To reduce the ill-posedness of the inverse problem, we only consider the problem of estimating the thermal diffusivity $\alpha(x) = C(x)/\kappa(x)$ (m^2/s). Hence, in the following, we approximate the soil temperature distribution by the following equation instead of (1.1):

$$\frac{\partial T(x,t)}{\partial t} = \sum_{i=1}^{3} \frac{\partial}{\partial x_i}\left(\alpha(x)\frac{\partial T(x,t)}{\partial x_i}\right), \ (x,t) \in Q_{t_e}. \tag{1.2}$$

Since the coefficients are piecewise constant, the above equations are only different on the interface between the object and the soil.

We note that the soil temperature distribution is not only governed by (1.2) but also controlled by the initial temperature distribution as well as necessary boundary conditions, such as the prescribed soil temperature or incoming heat flux on the boundary Γ of the domain. These conditions are described in the following.

1. The initial condition expresses the temperature distribution in the domain Ω at the starting time $t = 0$

$$T(x,0) = g(x), \ x \in \Omega. \tag{1.3}$$

It should be remarked that in practice, the initial soil temperature distribution is not given. There are different methods for approximating it (see discussions in [42]). In this work, the initial condition is approximated by interpolations of in situ measured soil temperature at different depths at a given position and time instant, assuming that at that moment the thermal equilibrium between the soil and the buried object takes place, that is, the temperature is constant in horizontal planes [25, 45]. Both experiments and simulations have indicated that the time instant should be chosen around sunrise or sunset.

2. The soil-surface heat flux represents the incoming heat flux q_{net} through the air-soil interface Γ_3^1:

$$-\kappa_s \frac{\partial T}{\partial x_3}(x,t) = q_{net}(x,t), \quad (x,t) \in S_{3,t_e}^1, \tag{1.4}$$

with κ_s (W/m/K) the soil thermal conductivity at the surface layer. Here, we have assumed that the object is fully buried, hence the air-soil interface consists of only soil. This boundary condition controls the temperature distribution inside the soil, so it plays the most critical role in thermal modeling.

In practice, the heat flux on the soil surface consists of different processes that depend on several factors, such as weather conditions, soil type, and moisture content of the soil. Thus it is impossible to be measured. In this work, we assume that the moisture evaporation/condensation on the soil surface is negligible during the period of analysis. Under this assumption, the soil-surface heat flux (1.4) can be approximated as follows [18, 46]:

$$q_{net}(x,t) = q_{sun}(t) + q_{sky}(t) + q_{conv}(x,t) - q_{emis}(x,t), \tag{1.5}$$

where

$q_{sun} = \varepsilon_{sun}E_{sun}$ (W/m^2) is the solar irradiance absorbed by the soil with ε_{sun} being the *solar absorption coefficient* (*solar absorptivity*) of the soil and E_{sun} (W/m^2) the *solar irradiance* on the earth's surface.

$q_{sky} = \varepsilon_{sky}E_{sky}$ (W/m^2) is the sky irradiance absorbed by the soil with ε_{sky} being the *sky absorption coefficient* (*sky absorptivity*) of the soil and E_{sky} (W/m^2) the *sky irradiance* on the earth's surface.

q_{conv} (W/m^2) is the heat transfer by convection between the soil and the air. It refers to the transport of heat between the soil surface and the atmosphere by motion of the air.

q_{emis} (W/m^2) is the thermal emittance of the soil. It corresponds to the thermal radiation emitted by the soil surface.

In (1.5), the solar irradiance E_{sun} and the sky irradiance E_{sky} can be approximated by mathematical formulas or measured by a weather station (see a detailed discussion in [42]). The soil thermal emittance is given by Stefan–Boltzmann's law:

$$q_{emis}(x,t) = \varepsilon_{soil}\sigma T^4(x,t), \quad (x,t) \in S_{3,t_e}^1, \tag{1.6}$$

with ε_{soil} the *soil thermal emissivity*, σ Stefan–Boltzmann's constant, and $T(x,t)$ the soil-surface temperature.

The convection term in (1.5) is usually approximated by Newton's law [9]:

$$q_{conv}(x,t) = h_{conv}[T_{air}(t) - T(x,t)], \quad (x,t) \in S_{3,t_e}^1, \tag{1.7}$$

where h_{conv} (W/m^2/K) is the *convective heat transfer coefficient*. In general, the convective heat transfer coefficient depends on wind speed [18]. Consequently, it varies in time and is difficult to be accurately measured or approximated. In this work, for simplicity, we assume that the model is only applicable in the absence of strong wind, so the coefficient h_{conv} can be considered as a constant.

3. The "sufficient depth condition" assumes that the soil temperature at a sufficiently deep depth does not depend either on diurnal heat transfer processes or on the buried object, so it is approximated by the measured soil temperature at a given location. This assumption is reasonable in practice since we only consider a shallowly buried object that does not affect the soil temperature at the deep depth. Hence the condition on the bottom surface Γ_3^2 of the domain Ω is given by

$$T(x,t) = T_\infty, \ (x,t) \in S_{3,t_e}^2. \tag{1.8}$$

The sufficiently deep depth can be approximated using Angström's method, assuming that the soil temperature is steady periodic (see p. 136 in [9]) or taken from in situ measurements. In practice, the sufficiently deep depth L_3 is usually set not greater than 0.5 m because the measured soil temperature at this depth is almost invariant in diurnal cycles under common soil and weather conditions [42].

4. Incoming or outgoing heat flows on the vertical boundaries S_{v,t_e} are assumed to be negligible. This hypothesis is reasonable under the assumption of homogeneous soil and if the considered domain Ω is so large that we can neglect the effect of the buried object on these boundaries. In practice, given a measured IR image sequence of a soil area, we can roughly estimate the location of a buried object by using anomaly detection procedures (Section 1.4). Therefore, we can select the domain Ω such that the buried object is far from its vertical boundaries. In this case, the heat balance takes place on these vertical boundaries during the period of analysis. This condition is described by

$$\frac{\partial T}{\partial n}(x,t) = 0 \text{ for } (x,t) \in S_{v,t_e}, \tag{1.9}$$

where n is the outward unit normal vector to Γ.

Equation (1.2) with conditions (1.3)–(1.9) is considered as a thermal model of the soil including a shallowly buried object. It should be noted, as mentioned, that the thermal model is also valid for multiple buried objects provided that the hypothesis on the vertical boundaries of the soil domain is still valid.

We note that the thermal model (1.2)–(1.9) is nonlinear due to the nonlinearity of the boundary condition (1.4) on the air-soil interface. Therefore, solving this problem numerically is really time consuming. To overcome this difficulty, the nonlinear condition is usually linearized. In the literature, the linearization has been performed using the modeled sky irradiance [28, 46]. In this work, the linearization is performed by approximating the soil thermal emittance term (1.6) by the following form:

$$q_{emis}(x,t) = \varepsilon_{soil}\sigma T^4(x,t) \approx \varepsilon_{soil}\sigma T_0^4 + 4\varepsilon_{soil}\sigma T_0^3[T(x,t) - T_0], \tag{1.10}$$

if $[T(x,t) - T_0]/T_0 \ll 1$. The value T_0 may be chosen as the mean value of the soil-surface temperature measured by, for example, a thermocouple. From (1.4)–(1.7) and (1.10), we have

$$-\alpha_s \frac{\partial T(x,t)}{\partial x_3} + pT(x,t) = q(t), \ (x,t) \in S^1_{3,t_e}, \tag{1.11}$$

with α_s (m^2/s) being the thermal diffusivity of the soil and

$$p = \frac{\alpha_s}{\kappa_s}(4\varepsilon_{soil}\sigma T_0^3 + h_{conv}),$$

$$q(t) = \frac{\alpha_s}{\kappa_s}[\varepsilon_{sun}E_{sun}(t) + \varepsilon_{sky}E_{sky}(t) + 3\varepsilon_{soil}\sigma T_0^4 + h_{conv}T_{air}(t)]. \tag{1.12}$$

Here, E_{sun}, E_{sky}, T_0, and T_{air} are measured values. In (1.11), we have multiplied both sides by α_s/κ_s.

In summary, we have the following linearized forward thermal model of the soil including buried objects:

$$\begin{cases} \frac{\partial T}{\partial t}(x,t) - \sum_{i=1}^{3} \frac{\partial}{\partial x_i}\left(\alpha(x)\frac{\partial T(x,t)}{\partial x_i}\right) = 0, \ (x,t) \in Q_{t_e}, \\ -\alpha(x)\frac{\partial T}{\partial x_3}(x,t) + pT(x,t) = q(t), \ (x,t) \in S^1_{3,t_e}, \\ \frac{\partial T}{\partial n}(x,t) = 0, \ (x,t) \in S_{v,t_e}, \\ T(x,t) = T_\infty, \ (x,t) \in S^2_{3,t_e}, \\ T(x,0) = g(x), \ x \in \Omega, \end{cases} \tag{1.13}$$

In the formulation of the thermal model (1.13), we have assumed that the soil is homogeneous and the soil surface is flat. Moreover, the moisture content variation and the evaporation/condensation are negligible during the period of analysis. These assumptions, of course, are not accurate under arbitrary soil and weather conditions. However, in our application, we assume that the thermal model can only be applied under reasonably good weather conditions such as dry climates with the absence of strong wind or rain. Under these conditions, the assumptions of the moisture content variation and the evaporation/condensation are acceptable. In addition, we only consider a small soil volume, say, 50 by 50 by 50 cm around each object. Within such a small area, the assumptions on the homogeneity of the soil and flat soil surface may be reasonable. More complicated cases are the topics for future work.

Under the assumption that the soil and the buried object are homogeneous, the thermal diffusivity $\alpha(x)$ is described by a piecewise constant function:

$$\alpha(x) = \begin{cases} \alpha_o, \ x \in \Omega_1, \\ \alpha_s, \ x \in \Omega \setminus \Omega_1. \end{cases}$$

Here α_o is the thermal diffusivity of the object, and Ω_1 is the subdomain of Ω occupied by the object. With this representation and the assumption that the object is fully buried, the second equation of (1.13) is consistent with the soil-surface heat flux equation (1.11).

In practical situations, it should be remarked that the soil thermal diffusivity α_s, thermal conductivity κ_s, and the thermal parameters on the air-soil interface ε_{sun}, ε_{sky}, ε_{soil} and h_{conv} in (1.12) [equivalently, the constant p and the function $q(t)$] are generally not available. Moreover, these parameters are not easily measured in an arbitrary soil area. In our approach, for enhancing the applicability of the thermal model, we estimate these parameters using in situ measurements. The ideas are presented in the next subsections. The validity of the proposed thermal model (1.13) with the estimation of the soil thermal parameters is illustrated in Section 1.4 using an outdoor experimental data set.

1.2.2 Estimation of the Soil Thermal Diffusivity

Soil thermal properties such as thermal conductivity, thermal diffusivity, heat capacity and density depend not only on the soil type but also on the moisture content of the soil. Hence, it may not be accurate if we use given values in the literature. In our opinion, when an experiment is performed, these values should be directly measured or estimated in the soil area under investigation. In this section, we introduce a method to estimate the soil thermal diffusivity from soil temperature profiles measured by thermocouples. The problem setup is described in the following.

Under the assumption of homogeneous soil, its thermal properties do not depend on the location or the depth (to the limit of approximately 50 cm from the soil surface). During an experiment, the soil temperature at different depths can be measured by a *soil temperature profile* (thermocouple) placed at a given position (of homogeneous soil). In the surrounding area of the thermocouple, the heat conduction equation (1.2) can be simplified as a one-dimensional (1D) equation with constant coefficient describing the soil temperature distribution in depth and time. Suppose that the soil temperature is measured at the depths of $0 \leq a < z_1^m < z_2^m < \cdots < z_N^m < b$, then the profiles at the depths of a and b are used as Dirichlet boundary conditions, and the other measurements are considered as measured data for estimating the soil thermal diffusivity. The distribution in depth and time of the soil temperature can be represented as the solution to the following Dirichlet problem for the 1D heat equation:

$$\begin{cases} \frac{\partial T}{\partial t}(z,t) & = \alpha_s \frac{\partial^2 T}{\partial z^2}(z,t),\ a < z < b,\ 0 < t \leq t_e, \\ T(z,0) & = g(z),\ a < z < b, \\ T(a,t) & = f_a(t),\ 0 < t \leq t_e, \\ T(b,t) & = f_b(t),\ 0 < t \leq t_e. \end{cases} \tag{1.14}$$

In (1.14), $f_a(t)$ and $f_b(t)$ are given from the measured soil temperature profiles at the depths of a and b, respectively; the initial condition $g(z)$ is approximated by interpolating the soil temperature measured at different depths at the starting time as discussed. We denote by $T(z,t;\alpha_s)$ the solution to problem (1.14) (to emphasize its dependence on the coefficient α_s) and by $\theta_j(t)$ the measured soil temperature at the depth z_j^m, $j = 1,\ldots,N$. The soil thermal diffusivity α_s in (1.14) is estimated by minimizing the following objective function:

$$\mathcal{F}(\alpha_s) = \frac{1}{2} \sum_{j=1}^{N} \int_0^{t_e} [T(z_j^m,t;\alpha_s) - \theta_j(t)]^2 dt. \tag{1.15}$$

To obtain a physically meaningful value of the soil thermal diffusivity, the following bound constraints are taken into account when solving the above estimation problem:

$$0 < \alpha^l \leq \alpha_s \leq \alpha^u, \tag{1.16}$$

where α^l and α^u are, respectively, the lower and upper bounds of the soil thermal diffusivity.

Since in this problem we seek for only one scalar parameter varying in a given interval, its optimal value can be simply found by the following: We subdivide the interval $[\alpha^l, \alpha^u]$ into small subintervals by the points α_s^k, $k = 0,\ldots,N_s$, $(\alpha^l = \alpha_s^0 < \alpha_s^1 < \cdots < \alpha_s^{N_s} = \alpha^u)$ and calculate the values of the objective function $\mathcal{F}(\alpha_s^k)$ at these points. Then the solution of (1.15) is chosen as the element of the set $\{\alpha_s^k, k = 0,\ldots,N_s\}$ that associates with the minimum value of the set $\{\mathcal{F}(\alpha_s^k)\}$. The length of the subintervals should be chosen based on the required accuracy of the estimation.

1.2.3 Estimation of the Soil-Surface Boundary Condition

As given in (1.11) and (1.12), the soil-surface heat flux depends on the meteorological conditions (E_{sun}, E_{sky}, and T_{air}) and the soil-surface thermal parameters. The meteorological conditions can be measured via a weather station. However, it is not easy to measure the soil-surface thermal parameters in practical situations. Therefore, the soil-surface boundary condition (1.12) is generally unknown. In our approach, this boundary condition is estimated so that the thermal model (1.13) approximates well-measured data.

To estimate the soil-surface boundary condition, let us consider the forward thermal model (1.13) in an area of homogeneous soil. In this area, the model can be reduced to the following 1-D problem describing the soil temperature distribution in depth and time:

$$\begin{cases} \frac{\partial T}{\partial t} = \alpha_s \frac{\partial^2 T}{\partial z^2}, \ 0 < z < L, \ 0 < t \leq t_e, \\ -\alpha_s \frac{\partial T}{\partial z}(0,t) + pT(0,t) = q(t), \ 0 < t \leq t_e, \\ T(L,t) = T_\infty(t), \ 0 < t \leq t_e, \\ T(z,0) = g(z), \ 0 < z < L, \end{cases} \tag{1.17}$$

with α_s being the soil thermal diffusivity estimated as described and L the depth of the soil volume.

Since we want to compare simulations using the thermal model with measured IR images (soil-surface apparent temperature), the soil-surface apparent temperature at a location of homogeneous soil is used as input data for estimating the soil-surface boundary condition. Using the least-squares approach, the problem for estimating the soil-surface boundary condition is equivalent to minimizing the objective function

$$\mathcal{F}(p,q) = \frac{1}{2} \int_0^{t_e} [T(0,t;p,q) - \theta(t)]^2 \, dt, \tag{1.18}$$

with $\theta(t)$ the measured soil-surface apparent temperature, and $T(z,t;p,q)$ is the solution to (1.17). Here we use this notation to emphasize the dependence of the solution on the parameter p and function $q(t)$.

We note that the estimation problem (1.18) is not uniquely solvable if no a priori information is available. Moreover, its solutions may be physically meaningless. To overcome these difficulties, we first represent p and $q(t)$ as functions of the parameters ε_{sun}, ε_{sky}, ε_{soil}, h_{conv} as given in (1.12) and estimate these parameters by minimizing the objective function (1.18) with p and q being replaced by these parameters. Then, the estimated parameters are replaced into (1.12) to obtain an initial guess for estimating the air-soil interface boundary parameter p and function $q(t)$.

In the next section, the thermal model (1.13) is used in the estimation of thermal and geometric parameters of buried objects. Therefore, the accuracy of the estimation depends significantly on the validity of the thermal model. In Section 1.4, the validation of the thermal model (1.13) with the estimated soil diffusivity and the soil-surface boundary condition will be performed by comparing the simulations [using the thermal model (1.13)] to experimental data.

1.3 Inverse Problem Setting for Buried Object Detection

1.3.1 Mathematical Formulation

Given the forward thermal model (1.13) and IR images measured at the air-soil interface, we now formulate the inverse problem for land mine detection. After the detection of buried objects, the main purpose of the inverse problem is to classify them based on the estimation of their thermal as well as geometric properties. Mathematically, it aims at estimating the coefficient $\alpha(x)$ of the domain under consideration. It should be noted that the acquired IR images can be considered as measured soil temperature at the air-soil interface, that is, the boundary Γ_3^1 of the domain Ω. The estimation problem aims at finding the thermal diffusivity $\alpha(x)$ such that the simulated soil-surface temperature using the forward model (1.13) fits the measured data. The most common way to formulate this problem is the least-squares approach, which is equivalent to the following minimization problem:

$$\min_{\alpha(x)} \mathcal{F}(\alpha) := \frac{1}{2} \int_0^{t_e} \int_{\Gamma_3^1} [T(x,t;\alpha) - \theta(x,t)]^2 \, dx_1 dx_2 dt, \qquad (1.19)$$

where $\theta(x,t)$ is the measured soil-surface temperature (IR images). Here we use the notation $T(x,t;\alpha)$ to emphasize the dependence of the solution to the forward model (1.13) on $\alpha(x)$.

We note that, since thermal properties of materials are positive and finite, the following bound constraints must be taken into account in solving the inverse problem (1.19) subject to (1.13):

$$0 < \alpha^l \leq \alpha(x) \leq \alpha^u, \; x \in \Omega, \qquad (1.20)$$

where α^l and α^u indicate the range in which the thermal diffusivity of the object is expected to fall.

1.3.2 Simplification of the Inverse Problem

The inverse problem (1.19) subject to (1.13) and (1.20) for estimating the coefficient $\alpha(x)$ is severely illposed due to lack of spatial information in the measured data. Numerical tests have indicated that it is difficult to obtain reliable estimates unless more constraints or simplifications are used [42]. The constraints or simplifications are chosen based on particular applications. As our objective is to detect land mines and distinguish them from other objects, we assume that land mines are cylinders, but their cross sections are not necessarily circular. Moreover, they are vertically buried. Under these assumptions, a buried object is specified by (1) its depth of burial, (2) its height, (3) its horizontal cross section, and (4) its thermal diffusivity (see Fig. 1.2).

As the object is assumed to be an upright cylinder, we can represent the coefficient $\alpha(x)$ as follows:

$$\alpha(x) = \begin{cases} \alpha_{12}(x_1, x_2), & \text{if } \rho_1 \leq x_3 \leq \rho_2, \\ \alpha_s, & \text{otherwise,} \end{cases} \qquad (1.21)$$

where ρ_1 and ρ_2 are the locations of the top and the bottom surfaces of the object in the soil volume ($0 < \rho_1 < \rho_2 < L_3$), and $\alpha_{12}(x_1, x_2)$, $0 \leq x_i \leq L_i$, $i = 1, 2$, is the coefficient on a horizontal surface of the soil domain across the object as shown in Fig. 1.2. The estimation problem leads to reconstructing the depths ρ_1, ρ_2 and the function $\alpha_{12}(x_1, x_2)$.

In solving the estimation problem, some constraints of the unknown parameters must be taken into account. It is obvious that $\alpha_{12}(x_1,x_2)$ are bounded by α^l and α^u as in (1.20), that is,

$$0 < \alpha^l \leq \alpha_{12}(x_1,x_2) \leq \alpha^u. \tag{1.22}$$

We also remark that, as analyzed in [42] (see also in the next section), the detection can only be possible for shallowly buried objects, say, at most 10-cm deep for common antipersonnel (AP) mines. Hence, the depth of burial ρ_1 should not be too large. Moreover, since we assume that the soil-surface contains only homogeneous soil, the depth of burial must be positive. More precisely, we have

$$0 < \rho_1^l \leq \rho_1 \leq \rho_1^u < L_3, \tag{1.23}$$

where d_1^l is a small positive value that prevents the depth of burial from converging to zero (as the boundary condition (1.11) is only valid for fully buried objects), and d_1^u is the maximum depth of burial at which the object is still detectable.

Concerning the height of the object, we have indicated in [42] (see also the next section) that its effect on the soil-surface thermal contrast is very small when the height exceeds a certain value (approximately 5 cm for common AP mines). Hence, an estimated value of the height is only reliable in this range. Moreover, for simplicity of numerical implementation, we also assume that the height of the object is not less than the discretization grid size in the x_3-direction. Therefore, the following constraints should be added to the estimation problem:

$$h_3 \leq \varsigma \leq \varsigma^u, \tag{1.24}$$

where h_3 is the discretization grid size in the x_3-direction and ς^u is the maximum height of the object at which the estimation is still reliable. Note that this parameter must be chosen so that $\rho_1^u + \varsigma^u < L_3$.

1.3.3 A Two-Step Method for Solving the Simplified Inverse Problem

Note that the simplified inverse problem is still very ill-posed. In our work, we propose a two-step method for solving it. In the first step, we assume that the cross section of the object is given (it can be roughly estimated using anomaly detection procedures described in the next section). Under this assumption, we only have to estimate three parameters: the depth of burial, the height, and the thermal diffusivity. This step helps reducing the ill-posedness of the estimation problem as it reduces the number of unknown parameters. However, its result depends on the accuracy of the cross section being given by the anomaly detection procedures.

To enhance the accuracy of the result of the first step in the second step we use the result of the previous step as an initial guess for estimating the full parameter set, namely, the depth of burial, the height, and the values of the thermal diffusivity on a horizontal plane of the soil domain across the object. The cross section is improved by the estimated values of the thermal diffusivity on the horizontal plane.

Step 1: Cylindrical Object with Given Cross Section

Given the cross section, the object is represented by only three parameters: the thermal diffusivity, the depth of burial, and the height. We denote by $\tilde{\Gamma}_3^1$ the estimated cross section of the buried object. With the estimated cross section and the assumption that the buried object is homogeneous, we can represent α_{12} as follows:

$$\alpha_{12}(x_1,x_2) = \begin{cases} \alpha_o & \text{for } (x_1,x_2) \in \tilde{\Gamma}_3^1, \\ \alpha_s & \text{for } (x_1,x_2) \in \Gamma_3^1 \setminus \tilde{\Gamma}_3^1. \end{cases} \tag{1.25}$$

The unknown parameters need to be estimated in this case are α_o, ρ_1, and ς. To avoid the dependence of the unknown parameters on their units and absolute values, we introduce the following dimensionless variable:

$$v = \left(\frac{\rho_1}{L_3}, \frac{\varsigma}{L_3}, \frac{\alpha_o}{\alpha_s} \right)'.$$

The estimation problem is subject to the constraints of the forms (1.22), (1.23), and (1.24). For convenience, they are rewritten in the vector forms as follows:

$$\left(\frac{\rho_1^l}{L_3}, \frac{h_3}{L_3}, \frac{\alpha^l}{\alpha_s} \right)' \leq v \leq \left(\frac{\rho_1^u}{L_3}, \frac{\varsigma^u}{L_3}, \frac{\alpha^u}{\alpha_s} \right)' \tag{1.26}$$

which mean that each component of v is bounded by the corresponding components of the two other vectors. To make the estimation problem stable, we apply the Tikhonov regularization technique. More precisely, we minimize the following objective function:

$$\mathcal{G}_1(v) = \mathcal{F}(\alpha) + \frac{1}{2}\gamma_1 \|v - v^*\|_{R^3}^2, \tag{1.27}$$

where v^* is an approximation of the desired solution. Here $\|\cdot\|_{R^n}$ denotes the Euclidean norm in R^n. The regularization parameter γ_1 should be properly chosen for each particular problem.

Step 2: Cylindrical Object with the Estimation of the Cross Section

It is clear that results of step 1 depend on the estimation of the object's cross section. To improve these results, in this step we estimate the depth of burial ρ_1, the height ς, and the full function $\alpha_{12}(x_1,x_2)$ in (1.21). Actually, in numerical implementation, this function is replaced by its mean values at discretization grid points of the soil domain. Therefore, the estimation problem is devoted to the reconstruction of the new vector V consisting of the matrix v of the dimensionless mean values of the function α_{12}, ρ_1/L_3 and ς/L_3. The objective function is given as follows:

$$\mathcal{G}_2(V) = \mathcal{F}(\alpha) + \frac{1}{2}\gamma_2\|V - V^*\|^2_{\mathbb{R}^n}. \tag{1.28}$$

Here the dimension n of the space $\mathbb{R}^n$ of the unknown vector V depends on the number of discretization grid points. The problem is also subject to the constraints of the forms (1.22), (1.23) and (1.24).

The above optimization problems are solved by a quasi-Newton algorithm accompanied with an adjoint method for calculating the gradient of the objective functions. The detailed formulation of the gradient of the objective functions can be found in [42].

1.4 Experimental Data and Processing Chain

In this section, we present the flowchart of IR thermography for land mine detection, including data acquisition, pre processing steps, and processing chain (validation of the proposed thermal model, anomaly detection, and classification steps, i.e., the estimation of the thermal diffusivity and geometric properties of buried objects using the two-step method presented in Section 1.3). Some numerical result illustrate the performance of the proposed techniques.

1.4.1 Description of the Minefield and Measurement System

In our work, we test the proposed algorithms for an outdoor experimental data set acquired in the Netherlands in 2001. For a detailed description of the experiment setup and the minefield, refer to [12]. Roughly speaking, the test minefield is a box of sandy soil (sand lane) with the dimensions of $10 \times 3 \times 1.5$ m for length, width, and depth, respectively. The thermal characteristics of the soil in the minefield are given in Table 1.1. There are thirty four mines and nine other objects (time domain reflectometers, test objects) buried in the minefield. Some of the mines are fully buried under the soil; some of them are partly buried. The ground truth of the minefield is described in Fig. 1.3 and Table 1.2. Types of land mines are shown in Fig. 1.4 depicts different types of land mines.

To measure the soil-surface temperature, a multicamera system was fixed on a sky lift placed above the minefield. During the experiment, the measured data were sent to a workstation for processing. The data set consists of a sequence of 479 IR frames acquired by a quantum well infrared photodetector (QWIP), every 5 minutes, from 1600 hours, July 24, 2001, until 0805 hours, July 26, 2001. The temperature resolution of the IR sensor is 0.03 K. Moreover, a weather station was used to measure meteorological data such as solar irradiance, sky irradiance, air temperature, and wind speed during the same period. A soil temperature profile placed at a corner

Table 1.1 Thermal characteristics of some types of soil and TNT

Material	Conductivity [W/(m K)]	Density (kg/m³)	Heat capacity [J/(kg K)]	Diffusivity (m²/s)
Sandy soil	0.75	1650	710	6.402×10^{-7}
Clay soil	1.5062	1500	2959	3.428×10^{-7}
TNT	0.2	1170	1500	1.1396×10^{-7}

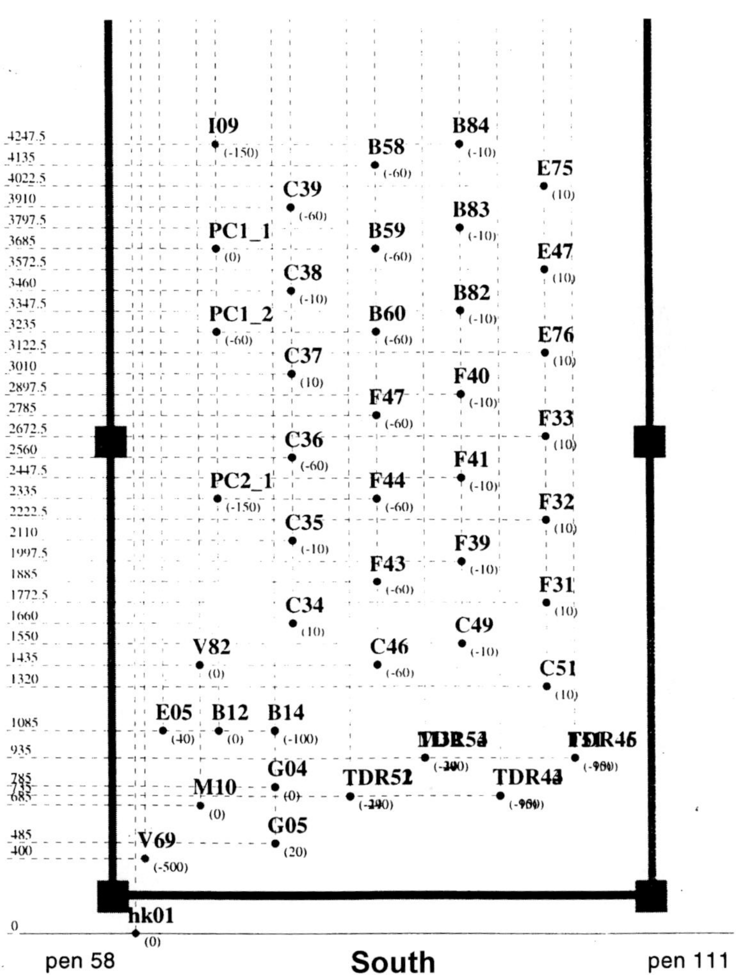

Fig. 1.3 Ground truth of the test minefield

of the test minefield allowed measuring the soil temperature on the surface (0 cm), at 2.5-, 5-, 7.5-, 10-, 20-, 30-, and 50-cm deep. We choose the data (IR images, weather data, soil temperature) from 0000 to 2400 hours, July 25, 2001, to analyze the soil-surface thermal behavior in a diurnal cycle.

Table 1.2 Description of the objects used in the experiment

Code	Type	Shape	Height	Diameter	Case
B**	AP mine (NR22C1)	Cylindrical	53 mm	62 mm	Plastic
C**	AP mine (PMN)	Cylindrical	48 mm	117 mm	PVC
E**	AP mine (NR22C1)	Cylindrical	53 mm	62 mm	Plastic
F**	AP mine (M14)	Cylindrical	42 mm	55 mm	ABS
G**	AP mine	Rectangular	18 mm	NA	PE
I09	AT mine (DM31)	Cylindrical	134 mm	254 mm	Metal
M**	AT mine	Cylindrical	115 mm	300 mm	No casing
PC**	Test objects				
V82	Metal disk				
V69	Thermocouple				
TDR**	TDR probes				

NA not applicable, *PE* polyethylene, *PVC* polyvinyl chloride, *ABS* Acrylonitrile Butadiene Styrene (ABS) Plastic Materials

a NR22C1   **b** PMN   **c** M14

Fig. 1.4 Some plastic land mines

1.4.2 Preprocessing

Before applying the algorithms to the IR data set, a preprocessing chain, consisting of (1) radiometric calibration, (2) temporal coregistration, (3) atmospheric correction, (4) apparent temperature conversion, and (5) inverse perspective (ground) projection, was applied to the acquired IR image sequence.

Radiometric calibration provides images of gray values that relate to the radiance of the surface (W/m^2/sr) using measured temperatures of two blackbodies that are at different temperatures. Temporal coregistration aims at correcting the misalignment of the IR images, which is mainly due to the movement of the camera or platform during the acquisition. Atmospheric correction is used for estimating the attenuation of the radiated thermal energy through the path from the surface to the camera. After the atmospheric correction, the resulting radiance is an estimation of the radiance originating from the surface. This radiance consists of two components. The first component is the radiance due to emission from the objects (having a temperature above 0 K). The second component is the accumulation of all the reflections of the objects in the direction of the camera. If an object has a reflection coefficient that is zero and hence an emission component that is one, the estimated radiance is solely due to emission from the object. Consequently, the temperature of the objects can be estimated using the integral of Planck's equation. If, however, the object is not a blackbody, then its temperature can still be approximated by the so-called apparent temperature. Ground projection maps the IR images onto the surface of the soil, producing a thermal image (with a resolution of 1 cm^2/pixel) sequence of the soil-surface apparent temperature. Detailed descriptions of these steps are given in [11]. For clarity, we recall that the preprocessed IR images used in the following represent the soil-surface temperature.

1.4.3 Estimation of the Soil Thermal Parameters

As discussed in Section 1.2, the input parameters (the soil thermal diffusivity, the soil-surface boundary condition) of the thermal model (1.13) are estimated using in situ soil-temperature measurements. For this data set, the estimated soil thermal diffusivity (6.420×10^{-7} m^2/s) is very close to the reference value (6.402×10^{-7} m^2/s) [24]; hence the latter is used. The estimated value of the parameter p in the air-soil interface boundary condition is $p = 6.76 \times 10^{-6}$, and the function $q(t)$ is depicted in Fig. 1.5.

1.4.4 Validation of the Proposed Thermal Model

As mentioned in Section 1.2, the estimation of the thermal and geometric parameters of buried objects is only reliable if the thermal model (1.13) is valid, that is, it approximates well the temperature evolution of the soil under investigation. To show the validity of the proposed thermal model, we compare the simulated soil-

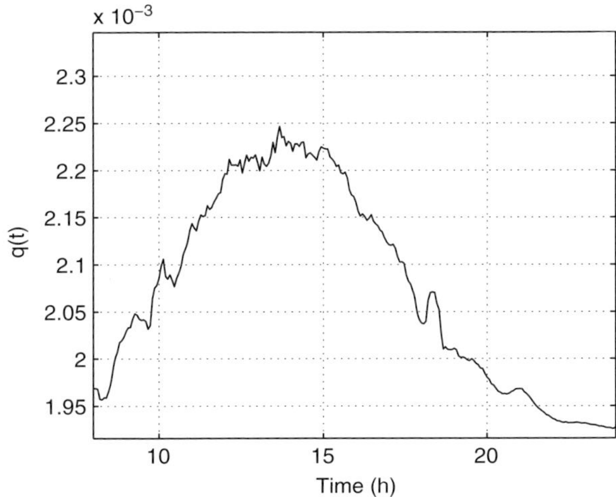

Fig. 1.5 The air-soil interface boundary function q(t)

surface temperature (with the estimated soil thermal diffusivity and the soil-surface boundary condition) to the measured data (soil-surface apparent temperature). For this, we consider three different antipersonnel plastic mines of different sizes and buried at different depths: mine C49 buried at 1 cm, mine C36 buried at 6 cm, and mine B82 buried at 1 cm (see Table 1.2).

In modeling the mines, we note that in [22], the effects of mine insert, top air gap, and thin mental outer case on thermal IR images were analyzed. As noted in [22], the effect of the top air gap is critical on the top and bottom surfaces of the mine, but it is small on the soil surface. Hence, for simplicity, we assume that the mines are fully filled with TNT, whose thermal characteristics are given in Table 1.1.

To simulate the soil temperature, a volume of 40 by 40 by 50 cm around each mine is considered. The shape and the location of the mines are provided by the ground truth. In numerical computations, the soil volume is divided into cells of size of $1 \times 1 \times 1$ cm to make it compatible with the resolution of the measured IR images. The time step size is set to be 60 s.

As analyzed in Section 1.2, we need to start the simulation around sunrise or sunset to have a good approximation of the initial condition. The IR images show that the heat equilibrium happens around 8000 hours, so this time instant is chosen as the starting time. The soil temperature at this time instant is approximated by interpolating the soil temperature at different depths measured by the thermocouples.

Figures 1.6, 1.7, and 1.8 depict the simulated and measured soil-surface temperature above the mines (the left figures), in the homogeneous soil areas (the middle figures), and the thermal contrasts of the mines (the right figures). The figures show that the simulated soil-surface temperatures and thermal contrasts between the mines and the soil follow the same behaviors as those of the measured data. This confirms the validity of the thermal model (1.13) and the advantage of the estimation of the soil-surface boundary condition. However, there are some unexpected peaks around 1000 hours in the measured data. By analyzing the data, we have noticed a shadow in some images around this time, which was probably caused by the cable of the camera system. These peaks can be avoided by either removing those images or smoothing the measured data.

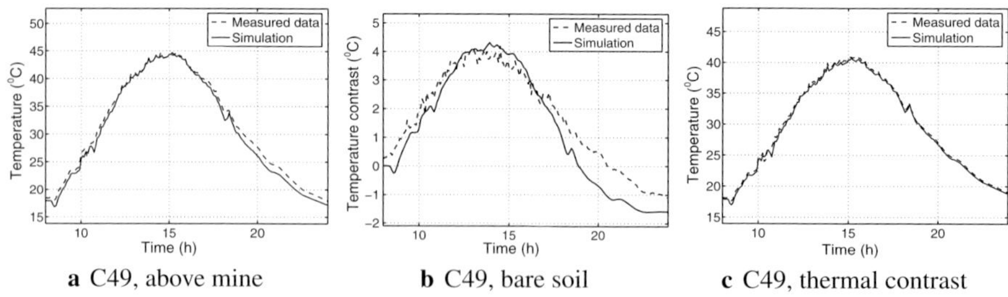

a C49, above mine
b C49, bare soil
c C49, thermal contrast

Fig. 1.6 Simulated and measured thermal behavior and thermal contrast of mine C49: **a** Thermal behavior above the mine; **b** Thermal behavior in the bare soil area; **c** Thermal contrast

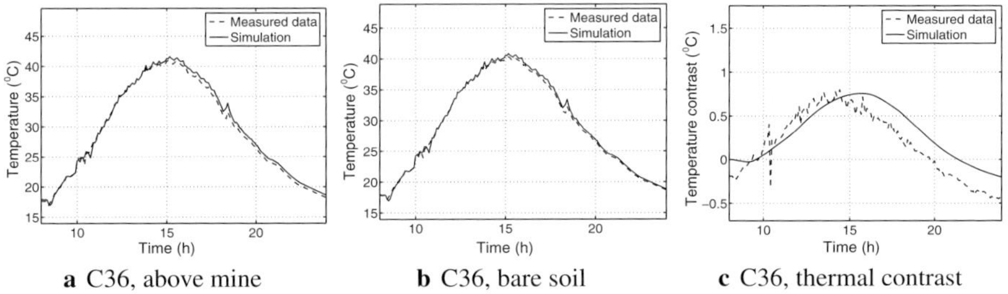

a C36, above mine   **b** C36, bare soil   **c** C36, thermal contrast

Fig. 1.7 Simulated and measured thermal behavior and thermal contrast of mine C36: **a** thermal behavior above the mine; **b** thermal behavior in the bare soil area; **c** thermal contrast

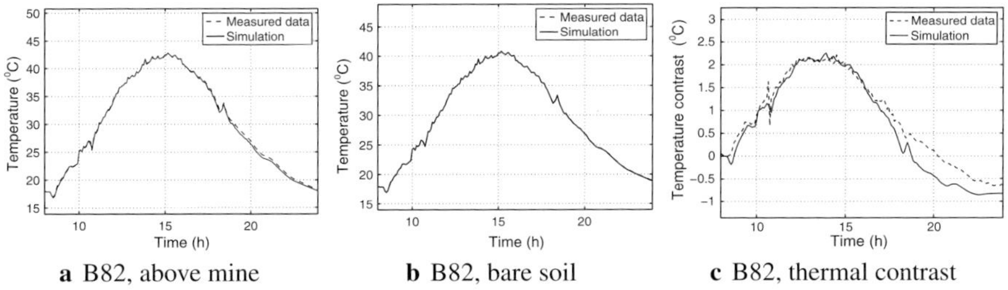

a B82, above mine   **b** B82, bare soil   **c** B82, thermal contrast

Fig. 1.8 Simulated and measured thermal behavior and thermal contrast of mine B82: **a** thermal behavior above the mine; **b** thermal behavior in the bare soil area; **c** thermal contrast

1.4.5 Effect of Mine Properties and Soil Type on the Soil-Surface Thermal Contrast

Having the validated thermal model, we can consider the second step of IR thermography: the inverse problem for the detection and characterization of buried objects. It is evident that the detection and characterization of a buried object depend on the soil-surface thermal contrast evolution caused by the object. This thermal contrast depends not only on weather conditions, but also on several other factors such as the depth of burial, the size of the object, and the difference of the thermal diffusivity between the soil and the object. Logically, a parameter to which the soil-surface thermal contrast is highly sensitive is easy to be estimated in the characterization process and vice versa. In the following, we analyze the effect of these parameters on the soil-surface thermal contrast.

As most land mines are cylindrical in shape, in this section we only consider cylindrical mines that are vertically buried.

1.4.5.1 Effect of the Depth of Burial and the Horizontal Size

It is expected that the deeper a mine is buried or the smaller the mine is, the smaller the thermal contrast on the soil surface is. This behavior is confirmed in

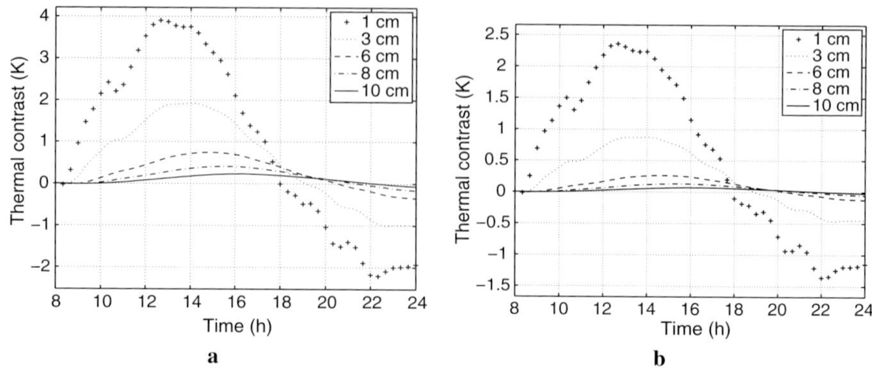

Fig. 1.9 Simulated soil-surface thermal contrast of two mines buried at different depths: **a** mine C49; **b** mine B82

the measured data shown in Figures 1.6, 1.7, and 1.8. Now, we consider the question of how deep the mine should be, or what is the maximum depth of the mine at which the soil-surface thermal contrast above this mine is still noticeable? Of course, this maximum depth depends on the size of the mine. The bigger the mine is, the deeper the depth of burial at which we still can detect the soil-surface thermal contrast.

Having the validated thermal model in mind, we can deal with the question quantitatively using the simulated thermal behavior and thermal contrast of each mine at various depths of burial. Two different mines of the same sizes as C49 and B82 are considered. The mines are simulated at the depths of 1, 3, 6, 8, and 10 (cm). The simulated soil-surface thermal contrast of mines C49 and B82 are plotted, respectively, in Fig. 1.9a, b. From the figure one can notice that the thermal contrasts decrease rapidly as the depth of burial increases from 1 to 10 cm. Moreover, if the measurement noise level is approximately 0.1 K in magnitude (this may happen in practice due to errors coming from various sources, such as the approximation of the model and measurement error), the thermal contrast of the first mine is almost not noticeable at the depth of 10 cm, while it is about 8 cm for the second mine.

Concerning the effect of the horizontal size of mines on the soil-surface thermal contrast, Fig. 1.10 shows that at the same depth of burial, the thermal contrast associated with mine C49 is almost three times higher than that of mine B82, which makes it easier to be detected. It should be noted that the heights of the two mines are almost the same, while the radius of the horizontal cross section of the first mine is two times larger than that of the second one.

1.4.5.2 Effect of the Mine Height

While the soil-surface thermal contrast of a mine is quite sensitive to its depth of burial and radius, our analyses have indicated that the thermal contrast is much less

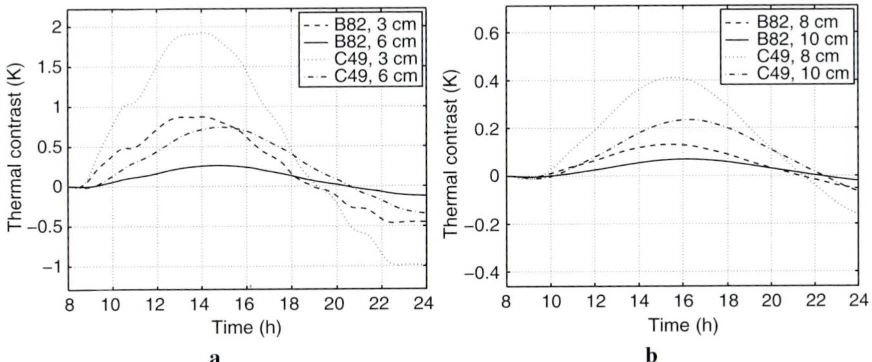

Fig. 1.10 Comparisons of the simulated soil-surface thermal contrast of mines C49 and B82 buried at **a** 3 cm and 6 cm, **b** 8 cm and 10 cm

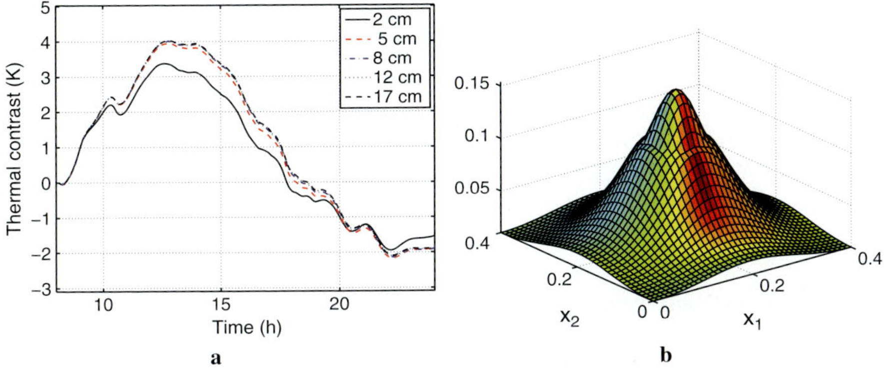

Fig. 1.11 Effect of the mine height on its soil-surface thermal contrast (the depth of burial of 1 cm and the radius of 5.7 cm): **a** thermal contrasts with respect to different heights; **b** the absolute values of the difference between the thermal behaviors at 1400 hours with respect to the heights of 5 and 17 cm

sensitive with respect to the height of the mine. To show this, we consider some simulated mines buried at 1 cm with a radius of 5.7 cm (approximately the same as that of mine C49). The height of the mines is modeled, respectively, as 2, 5, 8, 12, and 17 cm. The simulated soil-surface thermal contrast of these mines are depicted in Fig. 1.11a and the difference of the thermal behaviors at 1400 hours, the time instant of approximately maximum thermal contrast, of the mines associated with the heights of 5 and 17 (cm) is depicted in Fig. 1.11b. The figure shows that the soil-surface thermal contrast of the mines with height greater than or equal to 5 cm are almost identical (if the measurement noise is of magnitude 0.1 K, we can no longer distinguish them). The same behavior is shown in Fig. 1.12 for some mines of radius of 3 cm (which approximates the radius of B82).

If the mines are buried deeper, it is expected that the effect of their heights becomes smaller. From the analysis, we can conclude that the height of a mine has a

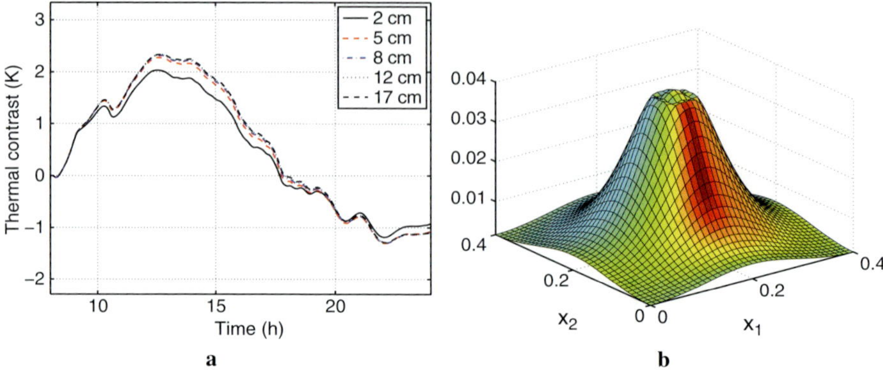

Fig. 1.12 Effect of the mine height on its soil-surface thermal contrast (the depth of burial of 1 cm and the radius of 3 cm): **a** thermal contrasts with respect to different heights; **b** the absolute values of the difference between the thermal behaviors at 1400 hours with respect to the heights of 5 and 17 cm

very small effect on its soil-surface thermal contrast when it exceeds a certain value. This means that it is not easy to have reliable estimates of this parameter in the inverse problem. It may be better to fix this parameter at a certain value (e.g., at the common height of mines) or apply a bound constraint on this parameter (e.g., less than or equal to 5 cm in this case).

1.4.5.3 Effect of Soil Type

The thermal contrast on the soil surface above a buried mine depends not only on the depth of burial and the horizontal size of the mine, but also on the thermal diffusivity of the soil. In the following, we illustrate this effect. To do so, apart from sandy soil, we also consider clayey soil. Its thermal properties are given in Table 1.1. Since the thermal diffusivity of clay is almost two times smaller than that of sand, we expect smaller thermal contrast on the soil surface.

The simulated soil-surface thermal contrast of mines C49 and B82 is depicted in Fig. 1.13a, b, respectively. As shown, the thermal contrast in the clayey soil is actually smaller than that in the sandy soil. Moreover, we can see that the maximum thermal contrast in the clay appears later than that in the sandy soil. This behavior can be explained as follows: At the beginning when the heat starts flowing into the soil, the temperature of the soil is low, and it allows the heat to flow easily in the area of homogeneous soil, but it is blocked above the mines, resulting in increasing soil-surface thermal contrast. After a while, the bottom layers of the soil become warmer and the temperature difference between the top and the bottom layers becomes smaller, resulting in smaller incoming heat flux and, consequently, decreasing thermal contrast. This process happens faster in the sandy soil than in the clayey soil as the thermal diffusivity of sand is higher than that of clay. Hence, the maximum thermal contrast of the mines in the sandy soil appears earlier than that in the clayey soil.

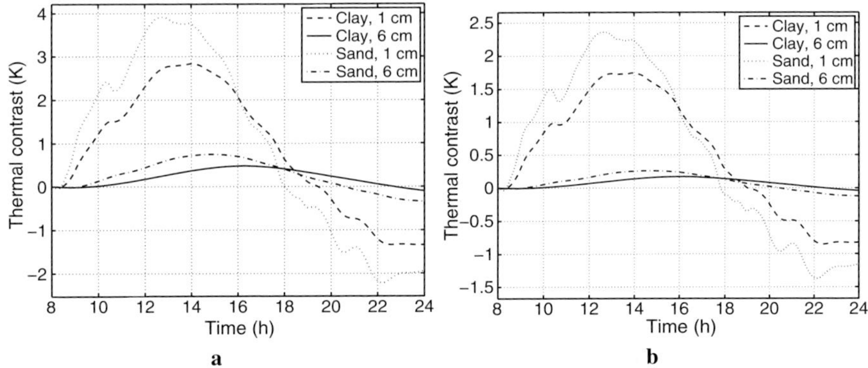

Fig. 1.13 Simulated soil-surface thermal contrast of some mines buried in sandy and clayey soils: **a** mine C49; **b** mine B82

1.4.6 Anomaly Detection and Reduction

As mentioned at the beginning of the chapter, we assume that there is only one buried object in each region under investigation. Hence, in dealing with a full IR image sequence containing several objects, we split it into subimage sequences so that there is only one possible objects in each sequence. To do that, we first apply an anomaly detection procedure to get a rough estimate of the locations of possible anomalies. Then the image sequence is subdivided in such a way that the detected anomalies fall into the middle of the subimages.

In our approach, since estimation results of step 1 depend strongly on the determined cross section of the object, the anomaly detection procedure must not only detect and locate anomalies but also provide rough estimates of their sizes. Our tests have indicated that it is not easy to determine accurately the real sizes of different objects buried at different depths in a single image sequence with usual anomaly detection techniques such as the RX algorithm [16], neuron network [24], or mathematical morphology [6]. We can even lose objects whose thermal signatures are dominated by others [42].

In this work, we consider the anomaly detection and estimation of the size of the detected anomalies separately. That is, we first apply an anomaly detection procedure to the full image sequence to detect and locate possible anomalies. The full image sequence is then divided into subimage sequences in such a way that the anomalies are centered in the subimages. Another procedure is applied to the subimage sequences to estimate the sizes of the detected anomalies.

For the detection step, among the existing anomaly detection techniques, mathematical morphology seems to be preferable due to its high rate of detection and low rate of false alarms [6, 11]. However, when we applied this technique to both simulated and real data, we found that it could not detect some objects buried at approximately 6 cm because their thermal contrasts are low compared to that of others [42]. To obtain better detection results, we propose another method based on the analysis of the thermal contrast of each anomaly with respect to its locally surrounding area. The idea is described next.

Denote by $\{I^n\}$ the measured full IR image sequence. We divide it into overlapped subimage sequences $\{I^n_{i,j}\}$. For each frame $I^n_{i,j}$, we denote by $\sigma^2(I^n_{i,j})$ its variance. The anomaly detection criterion is based on the assumption that IR signatures of anomalies should be different from that of unperturbed soil. Here, the anomaly consists of all pixels in the image satisfying the following condition:

$$\sigma^2(I^n_{i,j}) \geq \delta m_{\sigma^2(I^n_{i,j})}, \tag{1.29}$$

where $m_{\sigma^2(I^n_{i,j})}$ is the mean value of the variance of the image $I^n_{i,j}$, and δ is a threshold parameter. This parameter has been empirically determined. Our tests have shown that $\delta = 1.5$ is a reasonable value for different IR image sequences.

The detection result of the full sequence is the combination of the results of the subimage sequences. The detection results of the test data set using the proposed method are depicted in Fig. 1.14 along with the result of mathematical morphology. It is shown in the figure that the proposed method is able to detect most of the deeply buried objects while they are not detected by mathematical morphology. The number of detected buried land mines of the two methods are, respectively, fourteen and eleven of the total amount of twenty. Although the number of false alarms detected with our method is higher, they may be reduced in the classification step when their thermal diffusivity is estimated.

Considering that our application is mainly detecting buried land mines, one can reduce the number of false alarms based on the phenomenon that the soil-surface thermal contrast should be negative at night and positive during daytime (see [42]). Hence, the anomalies that do not satisfy this property should be classified as nonmine objects and removed from the list of suspected anomalies. Applying this criterion, the reduced detected anomalies are given in Fig. 1.15a. As can be noticed, twenty one of forty six detected anomalies are reduced, of which nine are surface-laid mines. Note that surface-laid objects are beyond the topic of this work

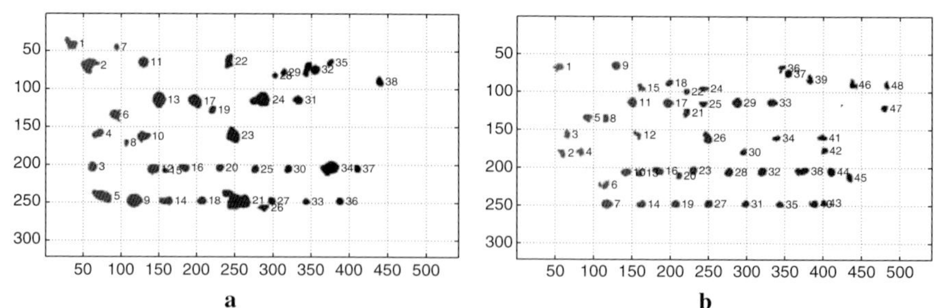

Fig. 1.14 Detected anomalies: **a** using mathematical morphology; **b** using subimage sequences and detection criterion (1.29)

as the thermal model (1.13) is not applied to the case of surface-laid objects, so it is not surprising that we classify the surface-laid mines as nonmine targets. However, in practice these mines can be easily detected using visible cameras. Of the twenty six remaining anomalies, there were fourteen buried mines (of the total number of eighteen buried mines in the minefield), one surface-laid mine, and fifteen other objects and false alarms. Among the detected mines, all the mines buried at 1-cm depth and three of five mines buried at 6-cm depth were accurately located, while the two other mines buried at 6-cm depth were not detected at the same locations as in the ground truth.

The sizes of the detected anomalies generally do not approximate well the real sizes of the objects. To have better approximations, in this work we consider another anomaly detection approach that can estimate rather accurately the sizes of the anomalies regardless of their depths of burial. The idea comes by analyzing simulated soil-surface thermal contrasts of different objects buried at different depths, of discussed next.

Let $\{\theta^n\}$ be the soil-surface temperature at different time instants of the soil domain around a detected anomaly (object). In addition, we assume that the object is buried around the middle of the considered soil domain so that it does not affect the soil temperature at the vertical boundaries of the domain (this can be done by using the result of the detection step). Thus, the soil-surface temperature near the boundaries is assumed to be constant at each time instant. The soil-surface thermal contrasts are then calculated as the difference between the soil-surface temperature of the full soil area and that at the soil-surface boundaries. Denote by $\Delta\theta^n$ the thermal contrast at time instant t_n. Numerical simulations have indicated that except at some time instants when the heat equilibrium between the soil and the object takes place, the thermal contrast $\Delta\theta^n$ depends on both the real size and the depth of the object. However, the larger the object is, the larger the area on the image sequence affected by the object is, regardless of the depth of burial (Fig. 15b). By testing for several objects buried at different depths, we found that the size of the object can be roughly estimated using the following criterion:

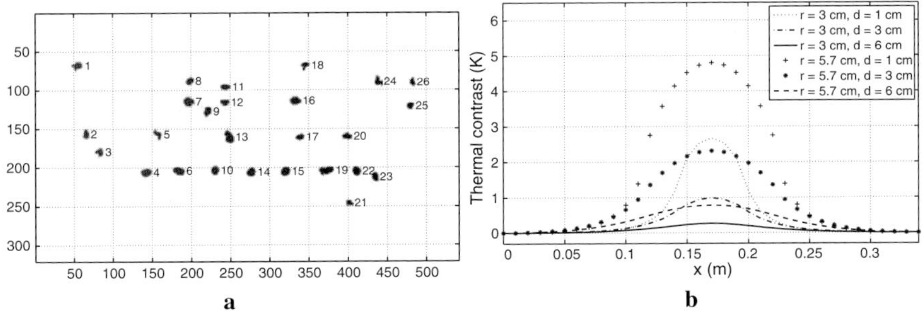

Fig. 1.15 a Reduced detected anomalies of the tested data set; b simulated thermal contrasts at 1300 hours of two different mines buried at different depths. r radius, d depth

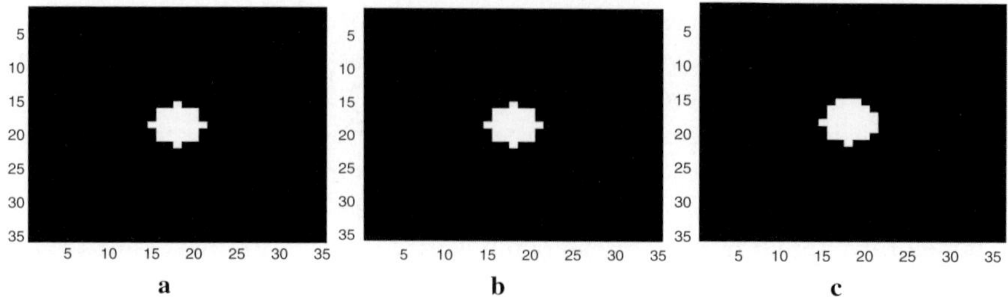

Fig. 1.16 Estimated cross sections of the detected anomalies associated with an object of radius of 3 cm buried at different depths of the simulated data with $\delta_{\theta^n} = 2.7$: **a** 0.01 m, **b** 0.03 m; **c** 0.06 m

$$\Delta\theta^n \geq \delta_{\theta^n}\sqrt{m_{\Delta\theta^n}M_{\Delta\theta^n}}, \qquad (1.30)$$

where $m_{\Delta\theta^n}$ and $M_{\Delta\theta^n}$ are, respectively, the mean value and the maximum value of the thermal contrast $\Delta\theta^n$. The threshold parameter δ_{θ^n} should be appropriately chosen. Our tests have indicated that this parameter can be chosen independent of the depth of burial. Unfortunately, a fixed value is not suitable for different objects of different sizes.

Figure 1.16 shows the estimated cross sections of some simulated anomalies. These anomalies correspond to the same object buried at 0.01-, 0.03-, and 0.06-m depths. For these anomalies, the parameter δ_{θ^n} is chosen to be 2.7. We can see that the estimated cross sections are almost the same (and approximate well the real size) for different depths of burial.

1.4.7 Reconstruction of the Geometric and Thermal Properties

The next step after anomaly detection is the classification of the detected anomalies in terms of geometric and thermal properties. This is done by solving the inverse problems of Section 1.3.

Let us illustrate with a detected anomaly. We consider anomaly 15 in Fig. 15a. This anomaly corresponds to mine B82.

To estimate the cross section of the anomaly, the parameter δ_{θ^n} in (1.30) is chosen to be 2. The estimated cross section is depicted in Fig. 1.17 along with an IR image at 1300 hours and the exact cross section.

In step 1, we start the algorithm with the initial guess of $v^0 = (0.1, 0.075, 0.6)'$. The regularization parameter γ_1 is empirically chosen to be 1 after several numerical tests. As analyzed in [42], the approximation v^* should be reasonably chosen. In this case, it is chosen as $v^* = (0.05, 0.125, 0.25)'$. The bound constraints of the parameters are chosen as

$$\alpha^l = 0.064 \times 10^{-7}, \quad \alpha^u = 6.402 \times 10^{-7}, \quad \rho_1^l = 0.001, \quad \rho_1^u = 0.15, \quad \varsigma^u = 0.06.$$

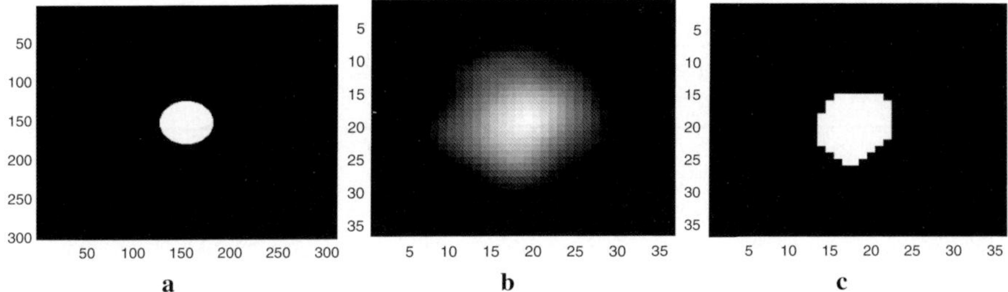

Fig. 1.17 Anomaly 15: **a** exact cross section; **b** IR image at 1300 hours; **c** estimated cross section

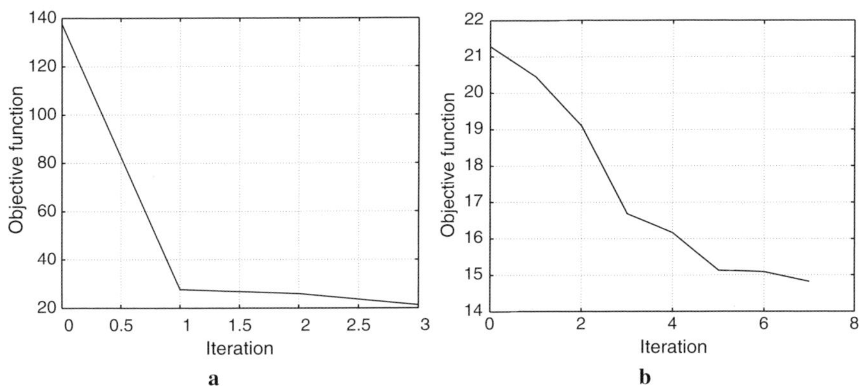

Fig. 1.18 Evolution of the objective functions (1.27) and (1.28): (a) step 1; (b) step 2

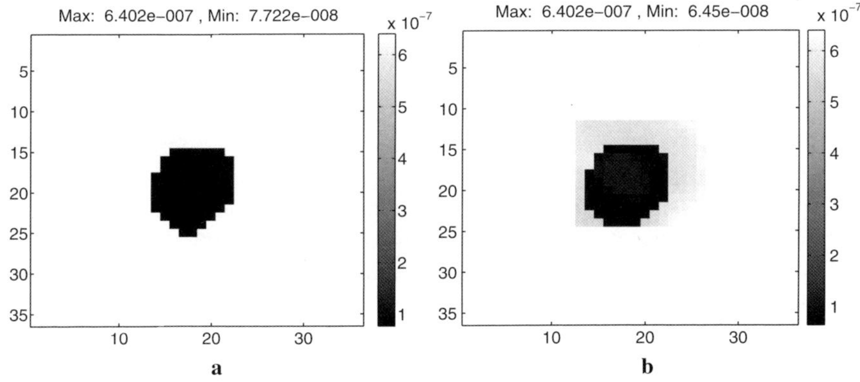

Fig. 1.19 Estimates of the mean values of α_{12}: **a** step 1; **b** step 2

In this example, we stop the algorithm of step 1 after only 3 iterations due to its fast convergence. Step 2 is stopped after 7 iterations when the solution cannot be further improved. The results of the two steps are depicted in Figs. 1.17, 1.18, 1.19 and Tables 1.3 and 1.4.

Table 1.3 Evolution of the solution and the objective function (1.27) of step 1

	Exact	Regularization approximation	Initial guess	Iteration 1	Iteration 2	Iteration 3
Depth	0.010	0.020	0.040	0.021	0.016	0.026
Height	0.053	0.050	0.030	0.045	0.049	0.054
Difference	1.139	1.601	3.841	1.953	1.480	0.772
Objective function			242.664	54.901	38.525	36.650

Table 1.4 Evolution of the solution and the objective function (1.28) of step 2

	Result step 1	Iteration 1	Iteration 2	Iteration 3	Iteration 4	Iteration 6	Iteration 7
Depth	0.026	0.023	0.023	0.022	0.022	0.022	0.022
Height	0.054	0.055	0.055	0.054	0.054	0.054	0.054
Difference	0.772	0.684	0.684	1.010	1.010	1.010	1.010
Objective function	36.650	36.695	36.695	35.602	35.600	35.595	35.592

Table 1.5 Classification result for the tested data set

Confidence interval $\times 10^{-7}$	No. buried mines/ no. anomalies	No. correctly classified buried mines	No. positive false alarms	No. negative false alarms
1.139 ± 0.4	14/26	7	0	7
1.139 ± 0.6	14/26	8	2	6

Figure 1.17 shows that the detected anomaly is larger than the real size of the corresponding mine. Therefore, the estimated parameters of step 1 do not approximate well the real values. However, the thermal diffusivity is indeed improved by step 2.

Tables 1.3 and 1.4 show that the depth of burial of the mine is not really accurately estimated (compared to the reference value). The possible reasons are due to the effect of the surface roughness and the inhomogeneity of the soil, which make the measured IR images not uniform as in simulations. However, for practical applications, this error is acceptable.

1.4.8 Classification of the Detected Anomalies

Although we concurrently estimate the thermal diffusivity, the depth of burial, and the height of the detected anomalies, the classification of the detected anomalies is only based on the estimated thermal diffusivity. The depth of burial and the height are only used as additional information in the clearance process.

To decide if the detected anomalies are mines, we introduce a confidence interval so that the detected anomalies whose thermal diffusivity falls within this interval are classified as mines. Otherwise, they are classified as nonmine objects. Since in our approach we assume that mines are specified as TNT, the confidence interval should be chosen around the thermal diffusivity of TNT, which is given in Table 1.1. The wrongly classified anomalies are divided into two types: positive false alarms and negative false alarms. Positive false alarms consist of all nonmine objects but classified as mines. On the contrary, negative false alarms are the buried mines but classified as nonmine targets.

It is clear that the classification results depend on the choice of the confidence interval. If the interval is large, the number of the buried mines correctly classified is high, and therefore the number of negative alarms is low. However, the number of positive false alarms is high. That means more suspected anomalies must be checked in the clearance process. On the other hand, if the confidence interval is small, the number of positive false alarms is reduced, but the number of negative false alarms is also reduced, that is, some mines may be missed in the classification step, and they can cause danger in the clearance process. Table 1.5 summarizes the classification results of the full data set for two different confidence intervals. The table shows that about 50% of the buried mines are correctly classified in this data set. Compared to the ground truth, we realized that most of the mines buried at 0.01-m depth are detected and correctly classified. However, only one mine buried at 0.06-m depth was correctly classified. This confirms that the IR technique is only useful for shallowly buried objects.

1.5 Conclusions

This chapter introduced the flowchart of IR thermography for land mine detection that consists of several steps from data acquisition, preprocessing, and anomaly detection to classification of the detected anomalies. Although all the steps were mentioned, the main emphasis was devoted to the detection and classification of buried objects in terms of geometric and thermal properties.

The detection and classification of buried objects was divided into two steps: thermal modeling and inverse problem setting. In the first step, a mathematical model was established to approximate the diurnal temperature distribution of the soil with the presence of buried objects. This model helps in understanding the influence of buried objects on the soil-surface temperature, which can be measured by IR cameras. In the second step, the thermal model is used along with measured IR images to estimate the thermal diffusivity and geometric parameters of buried objects. This step helps classify buried objects as mines or not.

The numerical result of an experimental data set showed that this technique helps reduce the number of suspected objects. However, the classification rate is still low (50%) compared to safety requirements of humanitarian demining (almost 100%).

Finally, it should be remarked that no technologies available so far can detect 100% of land mines. So, it is necessary to combine different technologies to enhance the detection rate.

Chapter's References

1. M. Albert, G. Koh, G. Koenig, S. Howington, J. Peters, and A. Trang. Phenomenology of dynamic thermal signatures around surface mines. In R.S. Harmon, J.T. Broach, and J.H. Holloway Jr., editors, *Proceedings of SPIE 5794, Detection and Remediation Technologies for Mine and Minelike Targets X*, pages 846–856, 2005
2. O.M. Alifanov. *Inverse Heat Transfer Problems*. Springer-Verlag, Berlin, 1994
3. S.R. Arridge. The forward and inverse problems in time resolved infrared imaging. *SPIE Medical Optical Tomography Functional Imaging and Monitoring*, SPIE-IS11:35–64, 1993
4. B.A. Baertlein and I.K. Sendur. Role of environmental factors and mine geometry in thermal IR mine signatures. In A.C. Dubey, J.F. Harvey, J.T. Broach, and V. George, editors, *Proceedings of SPIE 4394, Detection and Remediation Technologies for Mines and Minelike Targets VI*, pages 449–460, 2001
5. B.A. Barbour, S. Kordella, M.J. Dorsett, and B.L. Kerstiens. Mine detection using a polarimetric IR sensor. *IEE Conference Publication*, 431:78–82, 1996
6. S. Batman and J. Goutsias. Unsupervised iterative detection of land mines in highly cluttered environments. *IEEE Transactions on Image Processing*, 12(5):509–523, 2003
7. J.V. Beck, B. Blackwell, and S.R. St-Clair Jr. *Inverse Heat Conduction. Ill-Posed Problems*. Wiley, New York, 1995
8. C. Bruschini and B. Gros. A survey of current sensor technology research for the detection of land mines. In *Sustainable Humanitarian Demining: Trends, Techniques and Technologies*, pages 172–187, Mid Valley Press, Verona, VA, December 1998
9. H.S. Carslaw and J.C. Jaeger. *Conduction of Heat in Solids*, 2nd ed, Oxford University Press, Oxford, U.K., 1959
10. F. Cremer. Polarimetric infrared and sensor fusion for the detection of land mines. Ph.D. thesis, TNO Physics and Electronics Laboratory, The Hague, The Netherlands, 2003
11. F. Cremer, N.T. Thành, L. Yang, and H. Sahli. Stand-off thermal IR minefield survey, system concept and experimental results. In R.S. Harmon, J.T. Broach, and J.H. Holloway Jr., editors, *Proceedings of SPIE 5794, Detection and Remediation Technologies for Mine and Minelike Targets X*, pages 209–220, 2005
12. W. de Jong, H.A. Lensen, and Y.H.L. Janssen. Sophisticated test facility to detect land mines. In A.C. Dubey, James F. Harvey, J.T. Broach, and R.E. Dugan, editors, *Proceedings of SPIE 3710, Detection and Remediation Technologies for Mine and Minelike Targets IV*, pages 1409–1418, Orlando, FL, Apr 1999
13. C.P. Gooneratne, S.C. Mukhopahyay, and G. Sen Gupta. A review of sensing technologies for land mine detection: unmanned vehicle based approach. In *Second International Conference on Autonomous Robots and Agents*, pages 401–407, Palmerston North, New Zealand, Dec 2004
14. D.N. Hào. *Methods for Inverse Heat Conduction Problems*. Peter Lang, Frankfurt am Main, Bern, New York, Paris, 1998
15. J. Hermann and I. Chant. Microwave enhancement of thermal land mine signatures. In A.C. Dubey and J.F. Harvey, editors, *Proceedings of SPIE Vol. 3710, Detection and Remediation Technologies for Mines and Minelike Targets IV*, pages 110–114, Orlando, FL, 1999
16. Q.A. Holmes, C.R. Schwartz, J.H. Seldin, J.A. Wright, and L.J. Witter. Adaptive multispectral CFAR detection of land mines. In *Proceedings of SPIE*, Vol. 2496, pages 421–432, 1995
17. V. Isakov. *Inverse Problems for Partial Differential Equations*, vol. 127 of *Applied Mathematical Sciences*, 2nd ed. Springer Science+Business Media, Inc. New York, 2006

18. P.A. Jacobs. *Thermal Infrared Characterization of Ground Targets and Backgrounds*. SPIE Optical Engineering Press, Bellingham, WA, 1996
19. Y.H.L. Janssen, A.N. de Jong, H. Winkel, and F.J.M. van Puten. Detection of surface laid and buried mines with IR and CCD cameras, an evaluation based on measurements. In A.C. Dubey, R.L. Barnard, C.J. Lowe, and J.E. McFee, editors, *Proceedings of SPIE Vol. 2765, Detection and Remediation Technologies for Mines and Minelike Targets*, pages 448–459, 1996
20. A.B. Kahle. A simple thermal model of the earth's surface for geologic mapping by remote sensing. *Journal of Geophysical Research*, 82:1673–1680, 1977
21. K. Khanafer and K. Vafai. Thermal analysis of buried land mines over a diurnal cycle. *IEEE Transactions on Geoscience and Remote Sensing*, 40(2):461–473, 2002
22. K. Khanafer, K. Vafai, and B.A. Baertlein. Effects of thin metal outer case and top air gap on thermal IR images of buried antitank and antipersonnel land mines. *IEEE Transactions on Geoscience and Remote Sensing*, 41(1):123–135, 2003
23. L.A. Leschack and N.K. Del Grande. A dual-wavelength thermal infrared scanner as a potential arborne geophysical exploration tool. *Geophysics*, 41(6):1318–1336, 1976
24. P. López. Detection of landmines from Measured Infrared images using thermal modelling of the soil. Ph.D. thesis, University of Santiago de Compostela, 2003
25. P. López, L. Van Kempen, H. Sahli, and D. C. Ferrer. Improved thermal analysis of buried land mines. *IEEE Transactions on Geoscience and Remote Sensing*, 4(9):1965–1975, 2004
26. G. Maksymomko, B. Ware, and D. Poole. A characterization of diurnal environmental effects on mines and the factors influencing the performance of mine detecting ATR algorithms. In A.C. Dubey, I. Cindrich, J.M. Ralston, and K.A. Rigano, editors, *Proceedings of SPIE Vol. 2496, Detection and Remediation Technologies for Mines and Minelike Targets*, pages 140–151, 1995
27. A. Muscio and M.A. Corticelli. Experiments of thermographic land mine detection with reduced size and compressed time. *Infrared Physics & Technology*, 46(1–2):101–107, 2004
28. A. Muscio and M.A. Corticelli. Land mine detection by infrared thermography: reduction of size and duration of the experiments. *IEEE Transactions on Geoscience and Remote Sensing*, 42(9):1955–1964, 2004
29. J. Paik, C.P. Lee, and M.A. Abidi. Image processing-based mine detection techniques: a review. *Subsurface Sensing Technologies and Applications*, 3(3):153–202, 2002
30. P. Pregowski, W. Swiderski, W.T. Walczak, and B. Usowicz. Role of time and space variability of moisture and density of sand for thermal detection of buried objects—modeling and experiments. In Dennis H. LeMieux and John R. Snell Jr., editors, *Proceedings of SPIE 3700, Thermosense XXI*, pages 444–455, 1999
31. K.L. Russel, J.E. McFee, and W. Sirovyak. Remote performance prediction for infrared imaging of buried mines. In A.C. Dubey and R.L. Barnard, editors, *Proceedings of SPIE Vol. 3079, Detection and Remediation Technologies for Mines and Minelike Targets II*, pages 762–769, 1997
32. H. Sahli, C. Bruschini, and S. Crabbe. *Catalogue of Advanced Technologies and Systems for Humanitarian Demining*. Eudem 2 technology survey report, v.1.3, Dept., of Electron Informatics, Vrije Universiteit Brussel, Brussels, Belgium, 2005.
33. A.A. Samarskii and P. N. Vabishchevich. *Computational Heat Transfer. Volume 1: Mathematical Modelling*. Wiley, Chichester, U.K., 1995
34. M. Schachne, L. Van Kempen, D. Milojevic, H. Sahli, P. van Ham, M. Acheroy, and J. Cornelis. Mine detection by means of dynamic thermography: simulation and experiments. In *The Second International Conference on the Detection of Abandoned Landmines*, pages 124–128, Edinburgh, U.K., October, 12–14, 1998
35. I.K. Sendur and B.A. Baertlein. Techniques for improving buried mine detection in thermal IR imagery. In A.C. Dubey, J.F. Harvey, J.T. Broach, and R.E. Dugan, editors, *Proceedings of SPIE 3710, Detection and Remediation Technologies for Mines and Minelike Targets IV*, pages 1272–1283, 1999
36. I.K. Sendur and B.A. Baertlein. Numerical simulation of thermal signatures of buried mines over a diurnal cycle. In A.C. Dubey, J.F. Harvey, J.T. Broach, and R.E. Dugan, editors, *Proceedings of SPIE 4038, Detection and Remediation Technologies for Mines and Minelike Targets V*, pages 156–167, 2000

37. J.R. Simard. Improved land mine detection capability (ILDC): systematic approach to the detection of buried mines using passive IR imaging. In A.C. Dubey, R.L. Barnard, C.J. Lowe, and J.E. McFee, editors, *Proceedings of SPIE Vol. 2765, Detection and Remediation Technologies for Mines and Minelike Targets*, pages 489–500, 1996

38. S. Sjökvist. Heat transfer modelling and simulation in order to predict thermal signatures — the case of buried land mines. Ph.D. thesis, Linkopings University, Linkopings, Sweden, 1999

39. S. Sjökvist, R. Garcia-Padron, and D. Loyd. Heat transfer modelling of solar radiated soil, including moisture transfer. In *Third Baltic Heat Transfer Conference*, pages 707–714, Gdansk, Poland, September, 22–24, 1999, IFFM Publishers

40. S. Sjökvist, M. Georgson, S. Ringberg, M. Uppsall, and D. Loyd. Thermal effects on solar radiated sand surfaces containing land mines — a heat transfer analysis. In *Fifth International Conference on Advanced Computational Methods in Heat Transfer*, pages 177–187, Cracow, Poland, June, 17–19, 1998. Computational Mechanics Publications

41. S. Sjökvist, A. Linderhed, S. Nyberg, M. Uppsall, and D. Loyd. Land mine detection by IR temporal analysis: physical numerical modeling. In R.S. Harmon, J.T. Broach, and J.H. Holloway Jr., editors, *Proceedings of SPIE 5794, Detection and Remediation Technologies for Mine and Minelike Targets X*, pages 30–41, 2005

42. N.T. Thành. Infrared thermography for the detection and characterization of buried objects. Ph.D. thesis, Vrije Universiteit Brussel, Brussels, Belgium, 2007

43. N.T. Thành, D.N. Hào, P. López, F. Cremer, and H. Sahli. Thermal infrared identification of buried land mines. In R.S. Harmon, J.T. Broach, and J.H. Holloway Jr., editors, *Proceedings of SPIE 5794, Detection and Remediation Technologies for Mine and Minelike Targets X*, pages 198–208, 2005

44. N.T. Thành, D.N. Hào, and H. Sahli. Thermal model for land mine detection: efficient numerical methods and soil parameter estimation. In R.S. Harmon, J.T. Broach, and J.H. Holloway Jr., editors, *Proceedings of SPIE 6217, Detection and Remediation Technologies for Mine and Minelike Targets XI*, pages 517–528, 2006

45. N.T. Thành, H. Sahli, and D.N. Hào. Finite difference methods and validity of a thermal model for land mine detection with soil property estimation. *IEEE Transactions on Geoscience and Remote Sensing*, 45(3):656–674, 2007

46. K. Watson. Geologic applications of thermal infared images. *Proceedings of the IEEE*, 63: 128–137, 1975

Chapter 2
Passive Polarimetric Information Processing for Target Classification

Firooz Sadjadi and Farzad Sadjadi

Abstract Polarimetric sensing is an area of active research in a variety of applications. In particular, the use of polarization diversity has been shown to improve performance in automatic target detection and recognition. Within the diverse scope of polarimetric sensing, the field of passive polarimetric sensing is of particular interest. This chapter presents several new methods for gathering information using such passive techniques. One method extracts three-dimensional (3D) information and surface properties using one or more sensors. Another method extracts scene-specific algebraic expressions that remain unchanged under polarization transformations (such as along the transmission path to the sensor).

2.1 Introduction

Some of the known methods of passive 3D imaging include shape from shading, stereo imaging, and more recently, use of polarization diversity operating in the visible to infrared (IR) range of the electromagnetic spectrum [5, 10, 15, 20, 24]. 3D imaging can be achieved by determining orientation angles of surface normals at every point in the scene. Even though attempts at extracting surface orientations for reflectance polarization for limited sensor placement geometries have been reported [10, 20], the work reported here is novel in its use of emittance polarization for general sensor placement geometries. This is of particular significance for passive sensors.

Polarization signatures, however, are susceptible to distortion due to scattering along the line of sight from source to sensor. When one represents these signatures as Stokes vectors, these distortions can then be represented by a series of linear transformations obeying Mueller calculus. This allows us to extract attributes of the scene that are invariant under such transformations, attributes that are useful in target classification.

We begin in Section 2.2 with a brief overview of electromagnetic waves, polarization, and refraction as well as the particulars of the sensor placement

R.I. Hammoud (ed.), *Augmented Vision Perception in Infrared: Algorithms and Applied Systems*, Advances in Pattern Recognition, DOI 10.1007/978-1-84800-277-7_2,

geometry necessary for successful measurement of the angle of incidence (between the line of sight of the sensor and the surface normal). This angle is key to the measurement of the surface normal vector at a given point and, subsequently, the corresponding index of refraction of the target surface at that same point. To extract attributes of the scene that are invariant under linear transformation, we need to represent polarimetric imagery in terms of a multivariable probability density function. We also discuss how these probability density functions transform via Mueller calculus as viewed in the context of the theory of invariant algebra. We end by deriving seven global invariants that will be useful in later discussions. Section 2.3 then details several applications of this theoretical work in simulations, using single-sensor and dual-sensor examples, as well as laboratory-based experiments. Both 3D information and polarization invariant attributes are extracted for a number of human-made and natural objects.

2.2 Theory

The measurement by passive observation of the surface normal vector of a target and of its index of refraction hinges primarily on two choice pieces of classical physics: Fresnel's transmissivity relations and Snell's law. Fresnel gave us a relationship between the partial polarization of a wave emergent from a surface and the refraction angle. Snell gave us a relationship between the refraction angle and the index of refraction of the surface itself. We begin with a brief review of polarization in the context of traveling electromagnetic waves in linear media. From this, we are able to derive a relationship between the Stokes parameters for the emergent wave (as observed by a remote sensor), the angle of refraction (or angle of incidence from the point of view of the sensor), and the index of refraction at the surface. We then show that by using two sensors to observe a single surface we can use the specifics of the target-sensor geometry and the Stokes parameters as seen by each sensor to measure the angle of refraction at the surface.

As the polarimetric signature of a scene, represented by a set of Stokes vectors, travels in its path to the sensor, it may interact with the intervening media. These interactions could transform the polarization states, making target classifications that rely on the use of undistorted signatures all the more challenging. Hence, it is desirable to look for scene-related attributes that are invariant under these distortions. We begin by representing a polarimetric image in terms of a multivariable probability density function (in terms of Stokes vector parameters) and by treating the distortions as transformations of the Stokes vectors. We then proceed using the rules of Mueller calculus; this allows us to treat these transformations as linear. Casting the derivation of polarimetric invariants in the context of the theory of invariant algebra, in turn allows us to derive algebraic expressions that are unchanged under such linear transformations. These expressions are represented in terms of the coefficients of multivariable polynomials of various order. By using these tools we are able to derive a set of expressions in terms of the polarimetric signature of a scene that will remain invariant under polarization transformations.

2.2.1 Background

We begin by stating Maxwell's equations for a traveling electromagnetic wave in a linear, isotropic, nonconducting, semi-infinite medium:

$$\nabla \cdot \mathbf{B} = 0, \tag{2.1}$$

$$\nabla \cdot \mathbf{E} = 0, \tag{2.2}$$

$$\nabla \times \mathbf{B} = \mu \varepsilon \frac{\partial \mathbf{E}}{\partial t}, \tag{2.3}$$

$$\nabla \times \mathbf{E} = \frac{\partial \mathbf{B}}{\partial t}, \tag{2.4}$$

where $\mathbf{E}$ is the electric field, $\mathbf{B}$ is the magnetic field, μ is the permeability of the medium, and ε is the permittivity [7]. In general, μ and ε are complex functions of ω, the frequency of the fields (where the imaginary components represent the absorption of the material). If we consider a plane wave traveling in the x-axis, $e^{ikx-i\omega t}$, with wave number k and frequency ω, we can relate the wave number to the frequency by

$$k = \omega\sqrt{\mu\varepsilon} \tag{2.5}$$

We can then relate the phase velocity v of the wave in the medium to that in a vacuum by

$$v = \frac{\omega}{k} = \frac{c}{n} = \frac{1}{\sqrt{\mu\varepsilon}}, \tag{2.6}$$

where n is the index of refraction of the medium:

$$n = \sqrt{\frac{\mu}{\mu_0}\frac{\varepsilon}{\varepsilon_0}}, \tag{2.7}$$

where μ_0 and ε_0 are the permeability and permittivity of free space, respectively. Due to its dependence on μ and ε of the medium, n is in general also a complex function of ω.

A plane wave solution to Maxwell's equations must satisfy the Helmholtz wave equation:

$$\left(\nabla^2 + \mu\varepsilon\omega^2\right)\begin{Bmatrix}\mathbf{E}\\\mathbf{B}\end{Bmatrix} = 0 \tag{2.8}$$

We can do so with plane waves of the form

$$\mathbf{E}(\mathbf{x},t) = \mathcal{E}e^{i k\mathbf{n}\cdot\mathbf{x}-i\omega t}, \tag{2.9}$$

$$\mathbf{B}(\mathbf{x},t) = \mathcal{B}e^{i k\mathbf{n}\cdot\mathbf{x}-i\omega t}, \tag{2.10}$$

where $\mathcal{E}$ and $\mathcal{B}$ are perpendicular vector amplitudes (constant in time), and $\mathbf{n}$ is the propagation orientation unit vector (also constant, though possibly complex). These satisfy Helmholtz so long as Eq. (2.5) holds [7]. Furthermore, $\mathcal{E}$ and $\mathcal{B}$ must be such that $c\mathbf{B}$ and $\mathbf{E}$ have the same dimensions and the same magnitude in free space (but differ by n in media).

2.2.1.1 Polarization

The polarization of a plane wave describes the locus of the tip of the electric vector as a function of time in the plane perpendicular to the propagation direction. This locus generally moves in an ellipse. The sense of rotation or *handedness* is denoted right-handed (or left-handed) when rotation is as a right-handed (or left-handed) screw in the propagation direction.

When $\mathbf{n}$ is real valued, we see that $\mathcal{E}$ and $\mathcal{B}$ are in phase. We can then define a pair of orthogonal unit vectors $\hat{\varepsilon}_1$ and $\hat{\varepsilon}_2$ that, along with $\mathbf{n}$, form a basis for defining the polarization of the Electro-Magnetic (EM) wave (see Fig. 2.1). In the new basis, there exist two options for $\mathcal{E}$ and $\mathcal{B}$:

$$\mathcal{E} = \hat{\varepsilon}_1 E_0, \qquad \mathcal{B} = \hat{\varepsilon}_2 \sqrt{\mu\varepsilon} E_0, \tag{2.11}$$

or

$$\mathcal{E} = \hat{\varepsilon}_2 E_0', \qquad \mathcal{B} = -\hat{\varepsilon}_1 \sqrt{\mu\varepsilon} E_0', \tag{2.12}$$

where E_0 and E_0' are constants (generally complex).

Working from this choice, we see that an EM wave built from either choice would be *linearly* polarized [7], such that two solutions present themselves:

$$\mathbf{E}_j = \varepsilon_j E_j e^{i\mathbf{k}\cdot\mathbf{x} - i\omega t}, \tag{2.13}$$

$$\mathbf{B}_j = \sqrt{\mu\varepsilon} \frac{\mathbf{k} \times \mathbf{E}_j}{k}, \qquad j = 1, 2 \tag{2.14}$$

We can combine these two to get a general linearly polarized plane wave with

$$\mathbf{E}(\mathbf{x}, t) = \left(\hat{\varepsilon}_1 E_1 + \hat{\varepsilon}_2 E_2 \right) e^{i\mathbf{k}\cdot\mathbf{x} - i\omega t} \tag{2.15}$$

One way in which the generic state of polarization can be written is as a Stokes vector S, commonly defined [1, 5, 7] as

$$S = \begin{bmatrix} S_0 \\ S_1 \\ S_2 \\ S_3 \end{bmatrix} = \begin{bmatrix} |\hat{\varepsilon}_1 \cdot \mathbf{E}|^2 + |\hat{\varepsilon}_2 \cdot \mathbf{E}|^2 \\ |\hat{\varepsilon}_1 \cdot \mathbf{E}|^2 - |\hat{\varepsilon}_2 \cdot \mathbf{E}|^2 \\ 2\,\mathrm{Re}\left[(\hat{\varepsilon}_1 \cdot \mathbf{E}) * (\hat{\varepsilon}_2 \cdot \mathbf{E}) \right] \\ 2\,\mathrm{Im}\left[(\hat{\varepsilon}_1 \cdot \mathbf{E}) * (\hat{\varepsilon}_2 \cdot \mathbf{E}) \right] \end{bmatrix} = \begin{bmatrix} S_0 \\ S_0 \cos(2\chi)\cos(2\psi) \\ S_0 \cos(2\chi)\sin(2\psi) \\ S_0 \sin(2\chi) \end{bmatrix} \tag{2.16}$$

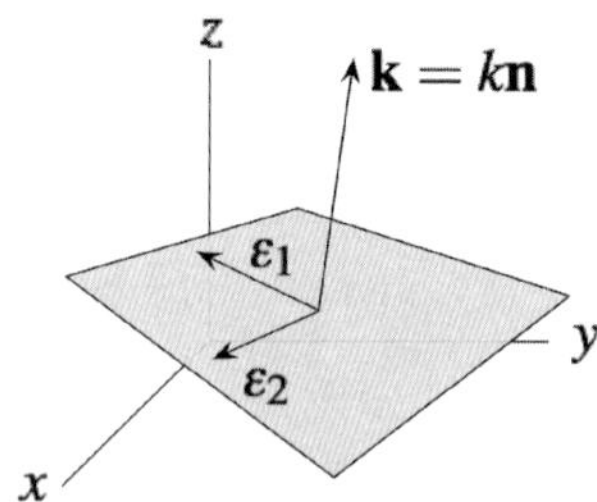

Fig. 2.1 Propagation vector and orthogonal unit polarization vectors [7]

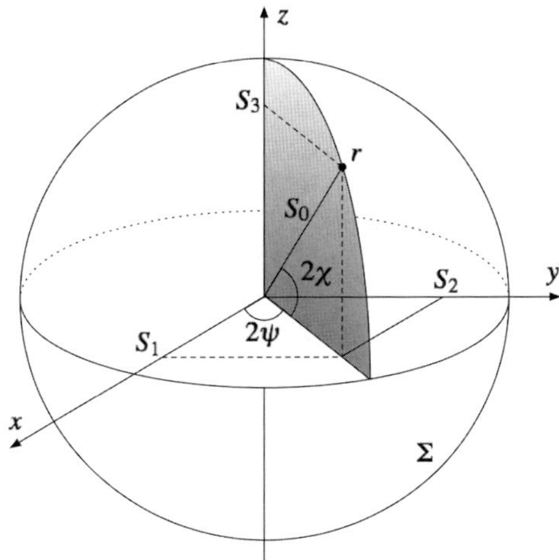

Fig. 2.2 Poincaré representation of polarization, with Stokes parameters as coordinates on a sphere [1]

where S_0 is the signal intensity (or amplitude), χ is the ellipticity angle of the polarization ellipse, and ψ is the angle of linear polarization (sometimes called the azimuth angle). The components of this vector can be visually interpreted using a Poincaré sphere (Fig. 2.2). For a given S_0 amplitude, any possible state of polarization can be mapped uniquely to a point r on the surface of a sphere Σ by the two angles χ and ψ in spherical coordinates or by the values of (S_1, S_2, S_3) in Cartesian coordinates. The two forms are related to each other by

$$\psi = \frac{1}{2}\tan^{-1}\left(\frac{S_2}{S_1}\right), \tag{2.17}$$

$$\chi = \frac{1}{2}\tan^{-1}\left(\frac{S_3}{\sqrt{S_1^2 - S_2^2}}\right). \tag{2.18}$$

Using the scalar value P for the degree of polarization, defined as

$$P = \frac{\sqrt{S_1^2 + S_2^2 + S_3^2}}{S_0}, \tag{2.19}$$

we can further define the degree of linear polarization as a condition on P where the S_3 term (the Stokes parameter that describes circular polarization) goes to zero. In purely linear case, the state of polarization can be written in terms of just two values: the angle of linear polarization ψ and the degree of linear polarization P_{linear}:

$$P_{linear} = \sqrt{S_1^2 + S_2^2}/S_0 \tag{2.20}$$

2.2.1.2 Refraction

When electromagnetic waves transition from one medium to another, the direction
of propagation shifts under the process of refraction. The emergent wave is also
partially polarized in this process. We are considering a special case where thermal
radiation coming from within a target refracts at the surface-air boundary and then
is observed by an IR sensor.

As observed by Fresnel, an unpolarized electromagnetic wave incident on its
surface-air boundary both reflects and refracts. The refracted wave (the emergent
wave) is polarized in a direction perpendicular to the plane of incidence, and the
reflected wave is polarized in a direction parallel to the plane of incidence. These
two components are given by the (Fresnel) transmissivity equations:

$$I_\| = \frac{sin(2\phi)sin(2\varphi)}{sin^2(\phi + \varphi)cos^2(\phi - \varphi)} \tag{2.21a}$$

$$I_\perp = \frac{sin(2\phi)sin(2\varphi)}{sin^2(\phi + \varphi)} \tag{2.21b}$$

where φ is the angle of incidence, and ϕ is the angle between the surface normal
and the line of sight of the sensor (see Fig. 2.3). It has been shown [1] that the degree
of polarization P is related to the above parallel and perpendicular components by

$$P = (I_\| - I_\perp)/(I_\| + I_\perp) \tag{2.22}$$

Inserting Eq. (2.21) into Eq. (2.22) and simplifying gives us

$$P = \frac{1 - cos^2(\varphi - \phi)}{1 + cos^2(\varphi - \phi)}. \tag{2.23}$$

By making use of a few trigonometric identities,

$$\cos(\alpha - \beta) = \cos\alpha\cos\beta + \sin\alpha\sin\beta, \tag{2.24}$$

$$\cos^2\alpha = 1 - \sin^2\alpha, \tag{2.25}$$

and Snell's law ($n\sin\varphi = \sin\phi$, where n is the index of refraction of the surface
material), we can eliminate φ, leading to the following solutions for ϕ [15] (see
Fig. 2.4):

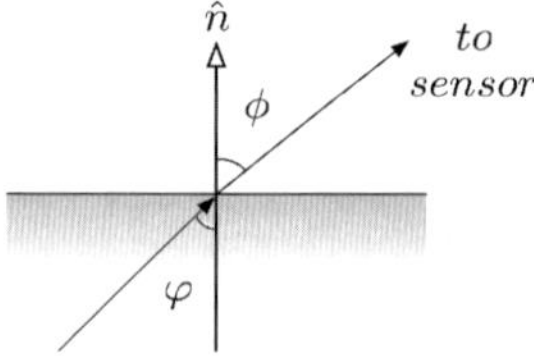

Fig. 2.3 Emergent EM
wave refracts at surface-air
boundary

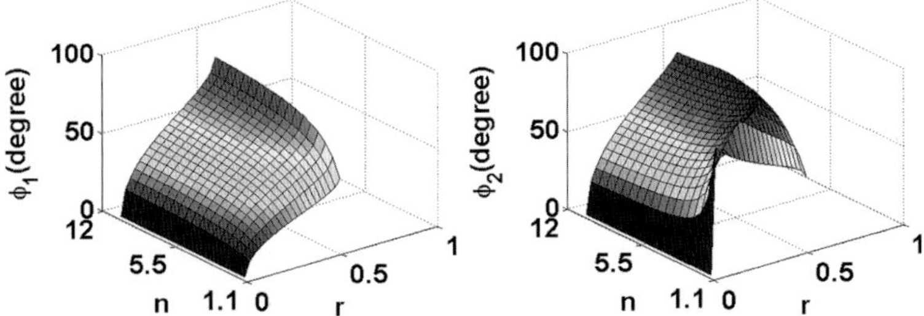

Fig. 2.4 Plots of two solutions *for* ϕ in terms of r (degree of linear polarization) and n (index of refraction). ϕ_1 corresponds to choice of $+, +$; ϕ_2 corresponds choice of $+, -$

$$\phi = \sin^{-1}\left[\pm\left(\frac{\frac{1-P}{1+P}-1}{\pm\frac{2}{n}\left(\frac{1-P}{1+P}\right)^{\frac{1}{2}}-\frac{1}{n^2}-1}\right)^{\frac{1}{2}}\right], \tag{2.26}$$

where four choices come from the two roots, each offering two sign values ($\pm$). The range of P is from 0 to 1. The index of refraction n is in general a complex number, the imaginary part of which is called the *extinction coefficient*. This coefficient indicates the amount of absorption loss when the electromagnetic wave propagates through the material and is typically larger than 1 for transparent materials (such as the ones studied here). Rewriting Eq. (2.26) yields an equation for n:

$$n = \left[\pm\sqrt{\frac{1-P}{1+P}} \pm \frac{1}{\sin\phi}\sqrt{\sin^2\phi\,\frac{1-P}{1+P} - \sin^2\phi - \frac{1-P}{1+P} + 1}\right]^{-1} \tag{2.27}$$

2.2.2 Surface Normal from Geometry

To use Eq. (2.27) to specify the index of refraction of the surface material, one needs to first specify the degree of polarization P and the depression angle ϕ. P can be determined from the Stokes vector [in general, Eq. (2.19), but for our situation, the linear form in Eq. (2.20)], which is given directly by the sensor for each pixel; ϕ can be determined from the sensor-target geometry.

To determine the angle ϕ for use in Eq. (2.27), we need to determine the relationship between the line of sight of the observer (the sensor) and the unit vector normal to the surface at the point on the target the observer is looking. Consider a point O on the target. A line connecting O to an observer A is the line of sight ($\overrightarrow{\rho}_A$) for that observer. Assume that this line forms a depression angle $\phi = \alpha$ with the surface normal at O (see Fig. 2.5). Similarly, the angle of line sight $\overrightarrow{\rho}_B$ of another

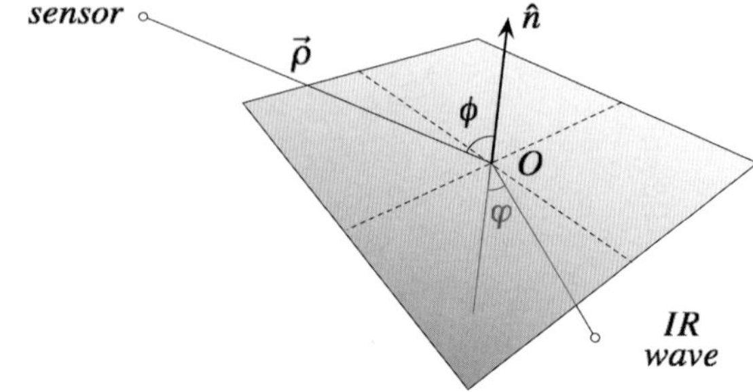

Fig. 2.5 Sensor-target geometry

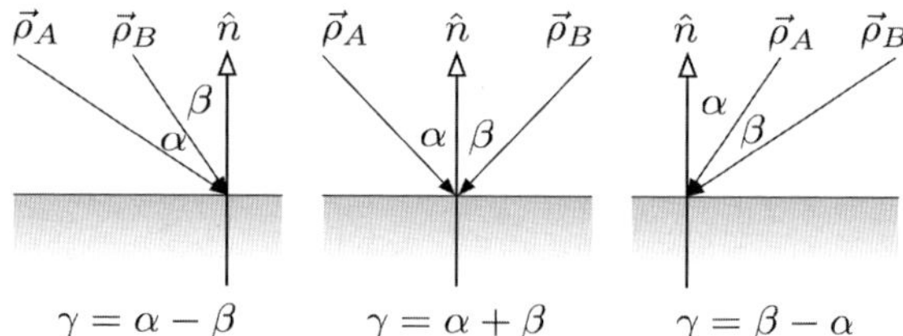

Fig. 2.6 Special two-sensor geometry showing the three possible combinations of depression angles

observer B forms a depression angle $\phi = \beta$ with the surface normal at O. We also label the angle between the two lines of sight $\vec{\rho}_A$ and $\vec{\rho}_B$ as γ.

There are two situations to be addressed, a special case where the surface normal lies in a plane with the two lines of sight $\vec{\rho}_A$ and $\vec{\rho}_B$ and a general case for all other configurations (other than that dealt with in the special case). A third geometry, where the two sensors image the scene from the same point of view (along the same line of sight), is dealt with in a later is dealt with in Section 2.2.2.2.

2.2.2.1 Special Case

There exists a special case where the surface normal $\hat{n}$ lies in the plane defined by the two observers A and B and the surface point O. In this case,

$$\gamma = \begin{cases} \alpha - \beta \\ \alpha + \beta \\ -\alpha + \beta \end{cases} \tag{2.28}$$

where the three cases arise from the three possible relative orientations of the two observers relative to the normal in the plane (see Fig. 2.6).

In principle, one can define α in terms of γ,

$$\gamma = cos^{-1}\left(\frac{\vec{\rho}_A \cdot \vec{\rho}_B}{|\vec{\rho}_A||\vec{\rho}_B|} \right) \tag{2.29}$$

and the degrees of linear polarization for both observers (r_A and r_B). From Eq. (2.26), one can find α and β:

$$\alpha = \cos^{-1}\left(\frac{k_A^2 + x^2 \mp 2xk_A}{1 + x^2 \mp 2xk_A}\right)^{\frac{1}{2}} \tag{2.30a}$$

$$\beta = \cos^{-1}\left(\frac{k_B^2 + x^2 \mp 2xk_B}{1 + x^2 \mp 2xk_B}\right)^{\frac{1}{2}} \tag{2.30b}$$

where $x = 1/n$ and $k_i^2 = \sqrt{(1 - P_i)/(1 + P_i)}$ for $i = A, B$.

Summing both forms of Eq. (2.30) according to the appropriate form of Eq. (2.28) and taking γ as defined in Eq. (2.29) gives a relation for the index of refraction of the surface in terms of the two lines of sight and the two degrees of linear polarization. The surface normal can be easily defined by reconstructing either of the depression angles α or β.

2.2.2.2 General Case

For a more general sensor placement (see Fig. 2.7), one needs to construct azimuth vectors ($\hat{\xi}_A$ and $\hat{\xi}_B$) from the two sensors. Label ψ_A as the azimuth angle of the plane of incidence for observer A and ψ_B for observer B. An *azimuth vector* is defined as a unit vector perpendicular to the plane that contains surface normal and line of sight. This vector then lies in the plane of the surface of the target at point O such that if one takes the cross product of two such vectors (as defined by two observers) one can reconstruct the surface normal.

$$\hat{n} = \hat{\xi}_A \times \hat{\xi}_B \tag{2.31}$$

The individual azimuth vectors, as represented in the image plane of the sensor, are defined by the azimuth angle ψ rotated from the image's vertical axis. In this notation, the special case corresponds to when the two azimuth vectors are parallel and cannot be used to define a unique normal vector.

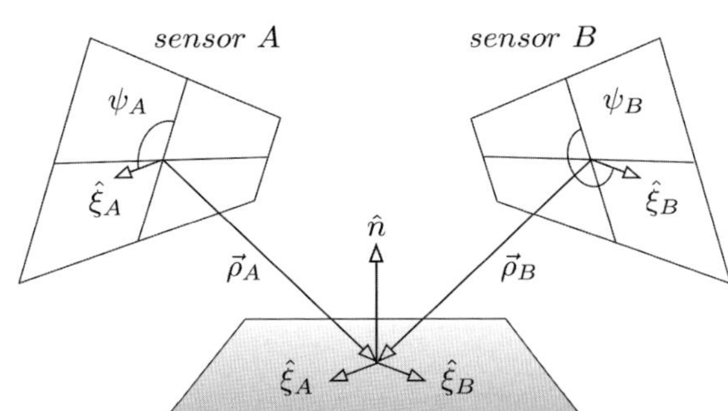

Fig. 2.7 General two-sensor geometry showing surface normal and azimuth vectors

The components of the azimuth vectors can be written in terms of the polar angles (θ_i, η_i) that define the line of sight $\vec{\rho}_i$ in terms of the azimuth angle ψ_i as

$$\xi_i = \begin{pmatrix} sin\theta_i sin\psi_i - cos\theta_i cos\eta_i cos\psi_i \\ -cos\theta_i sin\psi_i - sin\theta_i cos\eta_i cos\psi_i \\ sin\eta_i cos\psi_i \end{pmatrix} \tag{2.32}$$

where $i = A, B$. From the normal vector $\hat{n}$ and either of the lines of sight, one can easily compute the appropriate depression angle using a dot-product:

$$\alpha = cos^{-1}\left(\frac{\vec{\rho}_A \cdot \hat{n}}{|\vec{\rho}_A||\hat{n}|} \right) \tag{2.33a}$$

$$\beta = cos^{-1}\left(\frac{\vec{\rho}_B \cdot \hat{n}}{|\vec{\rho}_B||\hat{n}|} \right) \tag{2.33b}$$

Using Eq. (2.30), one can then solve for the index of refraction for the surface in terms of the two lines of sight, the two degrees of linear polarization, and the two azimuth angles.

Colocal Sensors with Different Spectral Bands

Up to this point we have dealt with sensing using two identical sensors, operating at the same frequency and with all other parameters matched. (In practice, this might be achieved by using the same sensor, moving it from place to place.) If, however, one were to use two sensors that operated at different frequencies, several new possibilities open up. For instance, one can avoid positioning and geometrical concerns by taking images from both sensors along the same line of sight ($\vec{\rho}_A = \vec{\rho}_B$, with $\gamma = 0$). Registration between the two sensors' images would then be much easier.

Dealing with multifrequency sensing leads us to introduce a frequency dependent n: $n(\omega)$, where the dependence on frequency is given by the Sellmeier dispersion formula [8],

$$n^2 = 1 + \sum_i \frac{A_i \lambda^2}{\lambda^2 - B_i^2} \tag{2.34}$$

where the sum has more or fewer terms depending on the desired precision (one or two iterations are not uncommon). The coefficient A_i can be interpreted as the fraction of the molecules in the target material resonant at a given frequency where B_i is the corresponding resonance frequency. Using two sensors and therefore two different observation frequencies, we can only achieve one iteration ($i = 1$, which is to assume that only one resonant frequency is dominant in the material).

Using two instances of Eq. (2.26), with two unknown values of n (n_1 and n_2) and a single line of sight ($\phi_1 = \phi_2$), we can deduce the value of A_1 and B_1 and therefore arrive at a specific value of n for each observed frequency. All that remains is to find the surface normal: Given n (and P from the Stokes vector), we can again use Eq. (2.26) to find ϕ. This, in addition to the azimuth angle ψ, specifies the unit surface normal vector $\hat{n}$.

2.2.3 Invariants of Polarization Transformations

The state of polarization for light originating from a scene may be distorted as it travels through the medium that occupies the space between the source and the polarimetric sensor. These distorted signatures usually make the tasks of scene analysis and target detection/recognition more challenging since these tasks typically rely on knowledge of the undistorted objects in a scene. Undoing these polarization transformations is a difficult, sometimes impossible task because the specifics of the transformations are generally unknown. In the following, we present an approach for extracting scene-related attributes that will remain unchanged under polarization transformations, thus negating the need for specific knowledge of the transformations themselves. Such invariant attributes are therefore desirable for scene analysis and target classification.

Let S be the Stokes vector for the emitted IR waves and S' be the Stokes *vector for* the waves that are the output of the polarimetric IR distortion. Denote M^{ir} as the Mueller matrix [1] for the polarimetric distortion. Then, the following holds according to the Mueller calculus:

$$S' = M^{ir} S \tag{2.35}$$

In many real-world scenarios, one can assume that the first Stokes parameter (signifying the total intensity of electromagnetic wave) attenuates negligibly during transmission through a medium [3]. Moreover, in a number of scenarios the transmission medium does not affect the circular polarization of the electromagnetic waves or only rotates the inclination angles of the polarization ellipse, leaving the ellipticity angle unchanged. An optical rotator is a good examples of such as medium. Hence, one can assume that in these cases only the second and third Stokes parameters (S_1 and S_2) will be affected when they under go a linear transformation. The four unknown transformation parameters are a, b, c, and d:

$$\begin{bmatrix} S'_0 \\ S'_1 \\ S'_2 \\ S'_3 \end{bmatrix} = \begin{bmatrix} 1 & 0 & 0 & 0 \\ 0 & a & b & 0 \\ 0 & c & d & 0 \\ 0 & 0 & 0 & 1 \end{bmatrix} \begin{bmatrix} S_0 \\ S_1 \\ S_2 \\ S_3 \end{bmatrix} \tag{2.36}$$

2.2.3.1 Probabilistic Representation

Any two-dimensional pattern can be viewed as a probability density function by proper normalization [19]. Thus, any pattern $Q(S_1,S_2)$ is a cluster of relevant points in a general S_1 and S_2 coordinate system where a pixel intensity is related to the probability of occurrence of that pixel. If the joint characteristic function of random variables S_1 and S_2 has a Taylor series expansion valid in some region about the origin, it is uniquely determined in this region by the moments of the random variables. The two-dimensional moments of order $p+q$ of a density $Q(S_1,S_2)$ are defined in terms of the Riemann integral as

$$m_{pq} = \int_{-\infty}^{\infty} \int_{-\infty}^{\infty} S_1{}^p S_2{}^q Q(S_1,S_2) dS_1 dS_2 \tag{2.37}$$

It is assumed that $Q(S_1,S_2)$ is bounded and is zero in R^2 space except at a finite part. Based on this assumption, it can be proved that the sequence m_{pq} determines $Q(S_1,S_2)$ uniquely.

The moment-generating function for the two-dimensional moments may be expressed after expansion into a power series as

$$M(u_1,u_2) = \int_{-\infty}^{\infty} \int_{-\infty}^{\infty} \sum_{p=0}^{\infty} \frac{1}{p!} (u_1 S_1 + u_2 S_2)^p \times Q(S_1,S_2) dS_1 dS_2 \tag{2.38}$$

From this, it is seen that moments are related to a set of polynomials in terms of S_1 and S_2. Hence knowing $M(u_1,u_2)$ would imply knowing $Q(S_1,S_2)$ the joint probability density function of the S_1 and S_2 random variables. Now a polarimetric signature of an object is represented at each pixel by an S_0, S_1, S_2 set. However, only S_1 and S_2 are variables when this signature undergoes a polarimetric transformation (Eq. 2.36). The changes in the polarization will imply a transformation of this probability density function $[Q(S_1,S_2)]$.

2.2.3.2 Invariant Algebra

What remains unchanged under these transformations falls under the domain of theory of invariant algebra [4, 19, 22]. The objectives pursued under this branch of algebra, developed in the 19th century by Caylely and Silvester, is the study of the algebraic expressions that remain unchanged under linear transformation of the coordinate systems.

If a binary polynomial of order p has an invariant, then moments of order p also have an algebraic invariant [19]. Using the tools of invariant algebra, seven expressions that are invariant under scale, rotation, and translation can be obtained [19].

The central moments of order $p+q$ are defined as

$$\rho_{pq}(Q) = \sum_{S_1}\sum_{S_2}(S_1 - \overline{S_1})^p (S_2 - \overline{S_2})^q Q(S_1,S_2) \tag{2.39}$$

where the bar signs indicate mean values.

The normalized central moments for $p+q \geq 2$ are defined as

$$v_{pq}(Q) = \frac{\rho_{pq}}{\rho_{00}^{\frac{p+q}{2}+1}} \tag{2.40}$$

The invariants up to second order are shown to be [19]:

$$\Xi_1(Q) = v_{20} + v_{02} \tag{2.41a}$$

$$\Xi_2(Q) = (v_{20} - v_{02})^2 + 4v_{11}^2 \tag{2.41b}$$

$$\Xi_3(Q) = (v_{30} - 3v_{12})^2 + (3v_{21} - v_{03})^2 \tag{2.41c}$$

$$\Xi_4(Q) = (v_{30} + v_{12})^2 + (v_{21} + v_{03})^2 \tag{2.41d}$$

$$\Xi_5(Q) = (v_{30} - 3v_{12})(v_{30} + v_{12})[(v_{30} + v_{12})^2 - 3(v_{21} + v_{03})^2]$$
$$+ (3v_{21} - v_{03})(v_{21} + v_{03})[3(v_{30} + v_{12}) - (v_{21} + v_{03})^2] \tag{2.41e}$$

$$\Xi_6(Q) = (v_{20} - v_{02})[(v_{30} + v_{12})^2 - (v_{21} + v_{03})^2]$$
$$+ 4v_{11}(v_{30} + v_{12})(v_{21} + v_{03}) \tag{2.41f}$$

$$\Xi_7(Q) = (3v_{21} - v_{03})(v_{30} + v_{12})[(v_{30} + v_{12})^2 - 3(v_{21} + v_{03})^2]$$
$$+ (3v_{12} - v_{30}) \times (v_{21} + v_{03})[3(v_{30} + v_{12})^2 - (v_{21} + v_{03})^2] \tag{2.41g}$$

These invariants are global invariants (valid for the entire image) and are not limited only to the target areas.

2.3 Simulation and Experimental Results

For simulation, we used a physics-based IR modeling tool to generate polarimetric images of a tactical scene. In this scene, several aircraft hangars are connected by runways, and the target is parked on the grass. In some images, the target was a M35 truck; in other images, a T72 tank was used. These two vehicles were used as typical examples of different ground-based targets. Any other set of targets can be similarly simulated and used. The range from sensor to target varied from 5 to 12 km. At 5 km, the target and a hangar are well resolved (tens of pixels). At 12 km, the target subtends only 4 pixels.

For the polarimetric sensors, we assumed the use of off-the-shelf components. We included the characteristics of the Santa Barbara Focal plane SBF-119 640×512 pixels, InSb focal plane array. We also included the characteristics of a 400-mm lens.

Among the simplifying assumptions made in this scenario are: (1) All surfaces have the same temperature 24°C. (2) Only two surface materials exist: grass, which emits unpolarized light, and glossy paint, which emits polarized light with characteristics described by Fresnel equations. (We use a complex index of refraction of 1.5 + $i0.15$, which is representative of paint in the wavelength range, 3–5 μ.) (3) No sun is present.

2.3.1 Surface Properties and Geometry

The information we are attempting to capture for both single-sensor and multisensor cases is mainly the orientation of surface patches in the scene. This can be determined for each surface patch by obtaining two angles: azimuth angle of the surface normal and the depression angle of this vector. The azimuth angles can be directly obtained via use of the angle of linear polarization at each pixel location. For the case of a single sensor, one in general cannot determine the depression angle directly since it depends on the value of the index of refraction of the surface patch at that pixel location, which usually cannot be assumed to be known. However, as will be shown, exact knowledge of this value may not be necessary if one is interested in a rough estimate of the surface normals.

On the other hand, as shown next, having access to two or more polarimetric imaging sensors can provide not only an accurate estimate of the surface patch orientation at each pixel location in the scene, but also an estimate of the index of refraction of the surface patch at that pixel location.

2.3.1.1 Single-Sensor Example

For case, a single IR polarimetric imaging sensor views the scene described. For each target scenario, $P_{linear}, S_1, S_2, \psi$, and the resulting ϕ angles [see Eq. (2.26)], images were computed. The depression angles for surface normals ϕ are computed for a range of values for indices of refraction from 1.001 to 10. In each case, these angles are computed for pixels whose P_{linear} values exceed values from 0.3 to 0.6. Figure 2.8 shows $S_0, S_1, S_2, \psi, P_{linear}$, and the 1st of the four resulting ϕ angles for the scene described earlier for a T72 Tank at 5 km ranges.

Table 2.1 shows the variation of estimated surface normal depression angles for 6 different surface patches on the M35 truck and T72 tank. For the surface patches from the truck object, the ϕ angles show that all 3 patches have similar values. However, since their corresponding ψ angles are not identical, it could be an indication that the 3 patches come from different portions of a curved surface. This table also shows that the estimated ϕ values have a standard deviation of around 6.5° for tank patches. However, for truck patches the standard deviation is 6.8°. The variations becomes much smaller at larger values of n. Note that n varies by equal increments from $n_1 = 1.001$ to $n_9 = 10$.

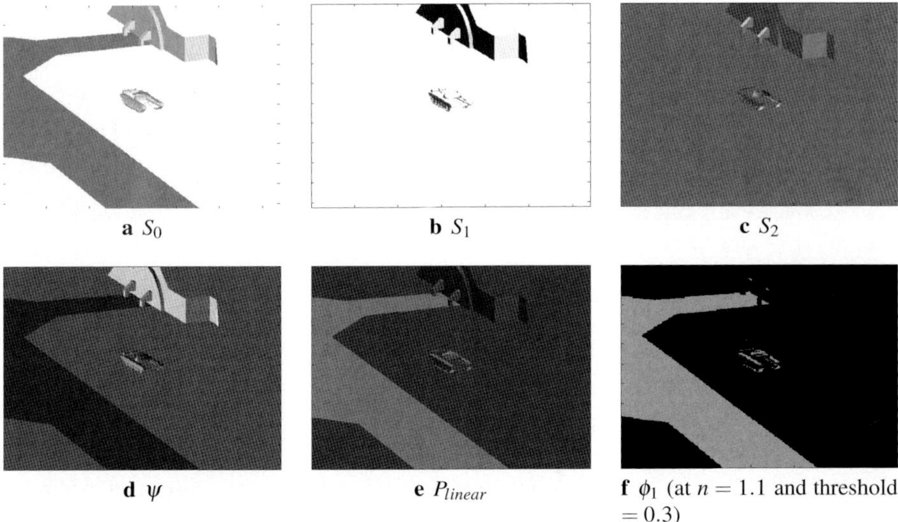

Fig. 2.8 Polarimetric imagery of a scene showing a T72 tank

Table 2.1 Sensitivity of surface normal depression angles (in degrees) with respect to the index of refraction

Emitting surface patches	ϕ at n_2	ϕ at n_4	ϕ at n_6	ϕ at n_8
Surface A : M35 truck	34.63	41.58	44.43	45.97
Surface B : M35 truck	34.63	41.58	44.43	45.97
Surface C : M35 truck	34.63	41.58	44.43	45.97
Surface D : T72 tank	32.84	39.43	42.14	43.61
Surface E : T72 tank	33.75	40.53	43.31	44.82
Surface F : T72 tank	33.93	40.74	43.53	45.05

Even though the variations of ϕ are not negligible, the approach could be useful in providing rough estimates of the actual surface normal depression angles.

2.3.1.2 Dual-Sensor Example

By making use of two sensors, one can obtain more direct estimates of the indices of refraction and depression angles of objects' surface normals. In this case, the images of a scene are captured by two sensors placed at two different geographical positions. First the Stokes vectors for the two different views of the scene were computed at each pixel. Then the two images were registered to establish correspondence among pixels associated with the same points in the scene. This is a critical

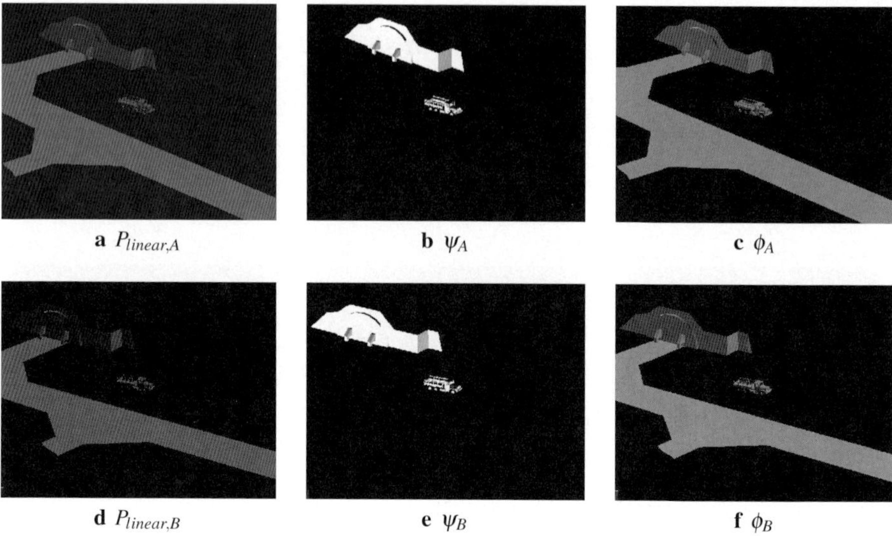

Fig. 2.9 Degree of linear polarization, azimuth, and depression angles of the surface normals at sensor position A (**a–c**) and sensor position B (**d–f**) for a scene containing an M35 truck

stage in the process. Any error in this step will contribute directly to the errors in the following computations. Figure 2.9 shows the degree of linear polarization and azimuth and depression angles of the surface normals for an M35 truck in the scene as sensed by the two sensors. Figure 2.10 similarly shows the degree of linear polarization and azimuth and depression angles of the surface normals for a T72 tank in the scene as sensed by the two sensors. It can be seen that horizontal surfaces such as the top of the vehicles and runways correspond to large depression values, whereas side-looking surfaces such as side of the vehicles and hangars correspond to large azimuth angle values.

2.3.1.3 Laboratory Experiments

The objectives of our laboratory experiments were first to capture real polarimetric signatures (Stokes vector parameters) of various targets placed at different distances from the sensor and then to use the Stokes vector parameters to obtain surface normals and indices of refraction of the selected targets in the laboratory. Our experimental setup consisted of an IR sensor, linear polarimetric wire-grid filters placed in a rotating frame, and a set of targets kept at various temperatures and placed at various distances from the sensor.

One approach used to obtain the first three Stokes parameters was by capturing the outputs of four distinct polarimetric filters (0, 45, 90, and 135°) positioned in front of an infrared (IR) camera. For this we employed a Thorlabs rotating polarizer

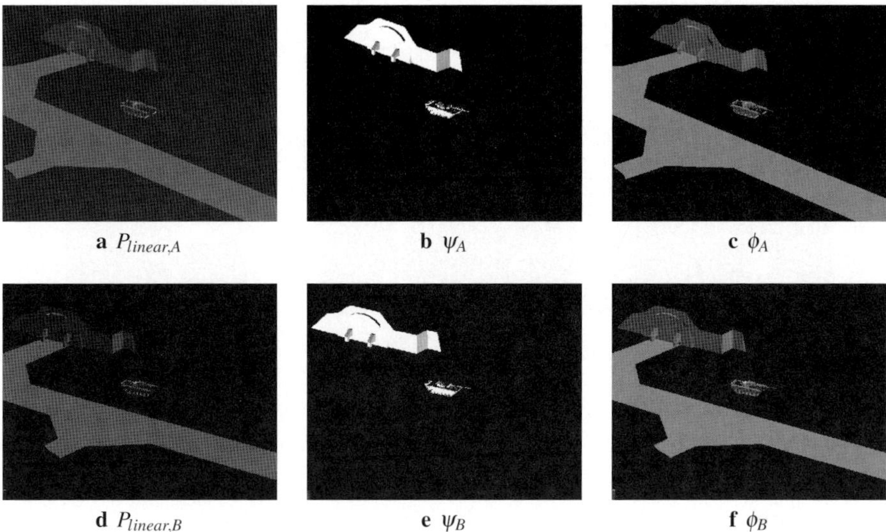

Fig. 2.10 Degree of linear polarization, azimuth, and depression angles of the surface normals at sensor position A (**a–c**) and sensor position B (**d–f**) for a scene containing a T72 tank

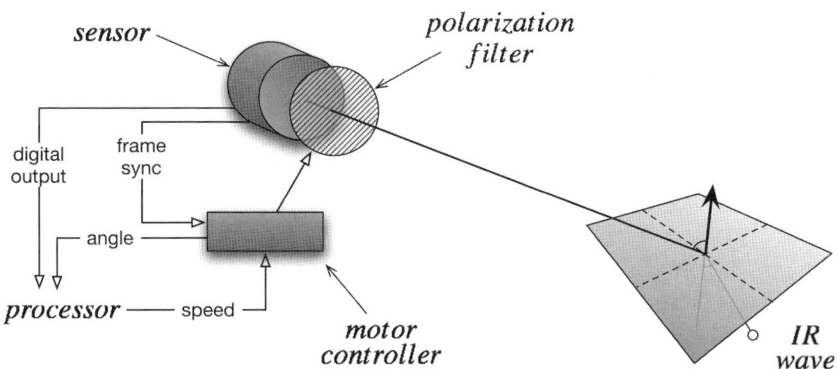

Fig. 2.11 Sensor with rotating polarizer and associated computer-control system

(Fig. 2.11). This system uses both manual and a computer-controlled stepper motor to change the orientation of the polarizer by a fraction of a degree. A computer-controlled rotating polarizer such as the one we used is a reliable tool for capturing Stokes parameters so long as the viewed scene is stationary. In the experiment, we used a polarimetric long-wave (8–14μm) mercury cadmium telluride IR imaging sensor to obtain angle imagery of a rectangular plate (Fig. 2.12) in a laboratory environment.

Fig. 2.12 Hot plate used in the experiment

Using Mueller calculus, one can obtain the outputs of 0, 45, 90 and 135° filters. For the 0° polarizer,

$$
\begin{bmatrix} i_0 \\ S_1^{(1)} \\ S_2^{(1)} \\ S_3^{(1)} \end{bmatrix} = \frac{1}{2} \begin{bmatrix} S_0 + S_1 \\ S_0 + S_1 \\ 0 \\ 0 \end{bmatrix}
\tag{2.42}
$$

For the 45° polarizer,

$$
\begin{bmatrix} i_{45} \\ S_1^{(2)} \\ S_2^{(2)} \\ S_3^{(2)} \end{bmatrix} = \frac{1}{2} \begin{bmatrix} S_0 + S_1 \\ 0 \\ S_0 + S_1 \\ 0 \end{bmatrix}
\tag{2.43}
$$

For the 90° polarizer,

$$
\begin{bmatrix} i_{90} \\ S_1^{(3)} \\ S_2^{(3)} \\ S_3^{(3)} \end{bmatrix} = \frac{1}{2} \begin{bmatrix} S_0 - S_1 \\ -S_0 + S_1 \\ 0 \\ 0 \end{bmatrix}
\tag{2.44}
$$

And, for the 135° polarizer,

$$
\begin{bmatrix} i_{135} \\ S_1^{(4)} \\ S_2^{(4)} \\ S_3^{(4)} \end{bmatrix} = \frac{1}{2} \begin{bmatrix} S_0 - S_1 \\ 0 \\ -S_0 + S_1 \\ 0 \end{bmatrix}
\tag{2.45}
$$

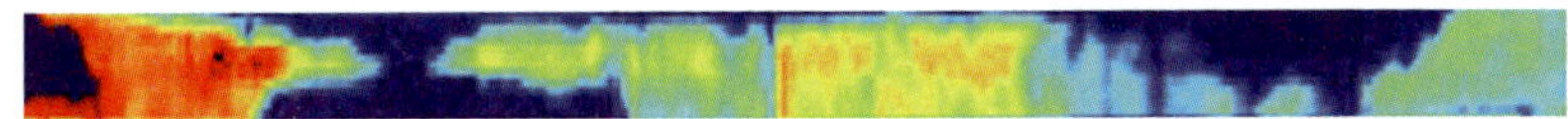

Fig. 2.13 Index of refraction image (gray levels indicate various indices of refraction associated with different residues on the front edges of the hot plate)

By algebraic manipulation, one can use the light intensities (i) at the output of these four filters to obtain the S_0, S_1, and S_2 values:

$$S_0 = \frac{1}{2}(i_0 + i_{45} + i_{90} + i_{135}) \tag{2.46}$$

$$S_1 = i_0 - i_{90} \tag{2.47}$$

$$S_2 = i_{45} - i_{135} \tag{2.48}$$

Figure 2.13 shows the depression angle image of the surface normals. Differently shaded patches on the surface of the plate indicate the presence of different indices of refraction (different paints). The actual values of these indices of refraction can be computed directly from these depression angles and their corresponding percent of polarization values. By direct measurements of the depression angles of the surface normals, estimates of the indices of refraction of various paints on the surface of the plate, useful for their classification, were obtained.

2.3.2 Polarimetric Invariants

Using the same physics-based IR modeling tool described, we generated polarimetric images of a tactical scene. As before for each scenario P_{LP}, S_1, S_2, and ψ images as well as S_1 vs. S_2 (obtained by relating the variation of S_1 as a function of S_2) were computed as shown in Figs. 2.14, 2.15, and 2.16 for an M35 truck. Figure 2.17 shows the variations of two of the invariant attributes (1st and 7th) as functions of distortion parameter ι (defined next) for the case of the long-range(12-km) scene containing the M35 truck. Figures 2.18, 2.19, and 2.20 show the calculations for a T72 tank. Similarly, Fig. 2.2 shows the variations of two of the invariant attributes, 1st and 7th, as functions of the distortion parameter ι for the case of the short-range (5-km) scene containing a T72 tank. The plots for other invariant attributes (not shown) were similar in their indication of the invariance with respect to distortion parameter ι.

In our experiments, we considered that polarimetric distortions were only due to a general rotation of $S_2 S_3$ coordinates with rotation angle denoted as ι:

$$M^{ir} = \begin{bmatrix} 1 & 0 & 0 & 0 \\ 0 & \cos(2\iota) & \sin(2\iota) & 0 \\ 0 & -\sin(2\iota) & \cos(2\iota) & 0 \\ 0 & 0 & 0 & 1 \end{bmatrix} \tag{2.49}$$

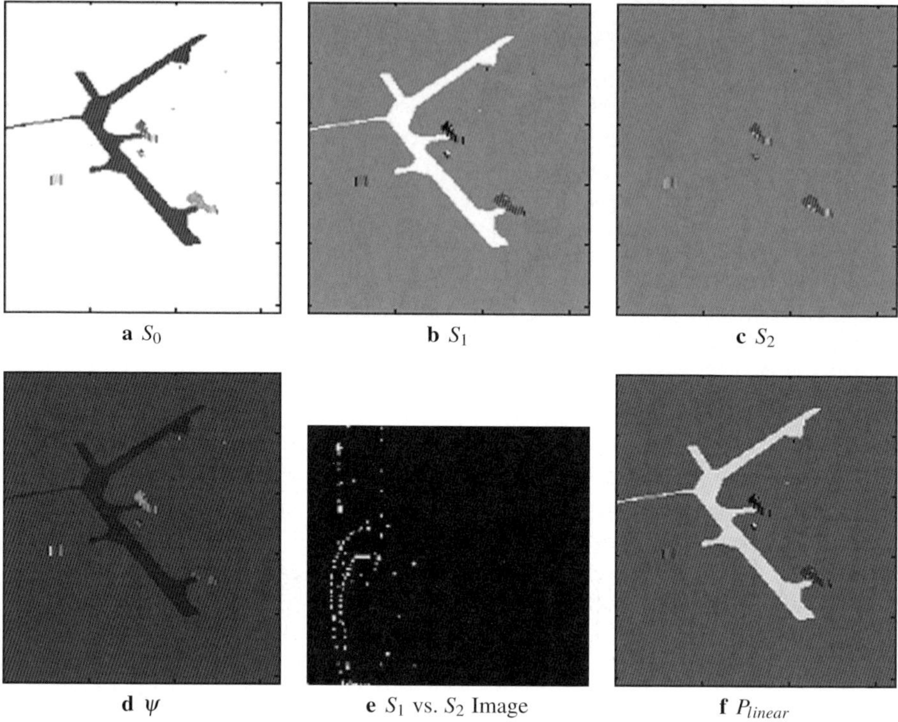

Fig. 2.14 Polarimetric imagery of a scene showing an M35 truck at $\iota = 0°$

The results indicate that (1) the algebraic invariants remain mostly unchanged as the distortion parameter ι varies from 0 to 360°. (2) The angle of linear polarization imagery ψ also seems to be invariant with respect to the rotation angle ι (see Figs. 2.14d, 2.15d, 2.16d, 2.18d, 2.19d, and 2.20d). However, this is an artifact of the display system and is not true. Using Eqs. (2.17), (2.36), and (2.49) one can obtain $\psi_{new} = \psi_{old} - \iota$.

2.4 Summary and Conclusions

We discussed a number of new methods for exploiting information that can be derived from passive polarimetric imagery. First, we presented a method for extracting 3D information and indices of refraction from a scene by means of a pair of polarimetric passive-imaging sensors. Each sensor provides the degree and angle of linear polarization for each pixel of the scene. From angle of linear polarization, we get the azimuth angle of the surface normal vector. Two cases for sensor placement were considered. For the special case when the two sensors have a common azimuth

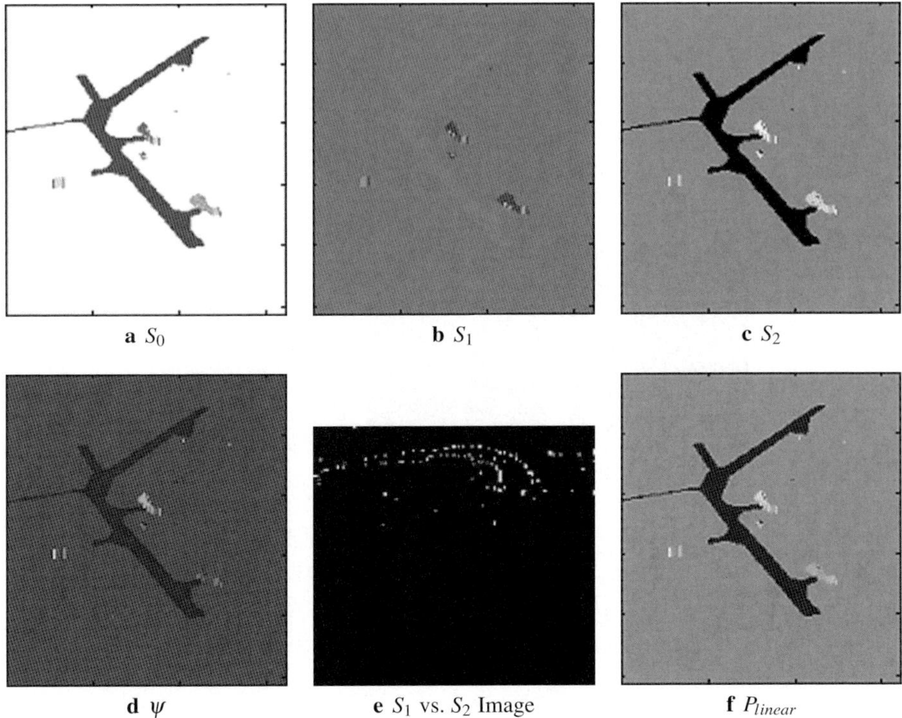

a S_0 **b** S_1 **c** S_2

d ψ **e** S_1 vs. S_2 Image **f** P_{linear}

Fig. 2.15 Polarimetric imagery of a scene showing an M35 truck at $\iota = 45°$

plane, the index of refraction can be found analytically in terms of the degrees of polarization and the angle γ between the lines of sight from the two sensors, from which the depression angles of the surface normal can be computed. For the second and more general case, the surface normal is estimated from the cross product of the azimuth vectors from the two sensors and the depression angle from the dot product of the line-of-sight vectors and surface normal. Once the depression angles are estimated, the index of refraction can be computed. The method is applied on simulated polarimetric IR imagery. The results are significant because by use of two passive polarimetric imaging sensors positioned at different (but with overlapping fields of view) geographical positions, one can obtain both 3D geometrical shape information and indices of the refraction of the objects in the scene.

In this chapter, we also presented an approach to address the problem of extracting attributes that remain unchanged under polarization transformations. It was shown that by starting with Mueller calculus the problem can be reduced to a linear transformation of a subspace of Stokes vector, which then (through the use of invariant algebra) leads to the derivation of a set of polarimetric invariants. To test the approach, we used a phenomenological IR modeling tool to generate physically accurate polarimetric imagery of different scenes containing different objects.

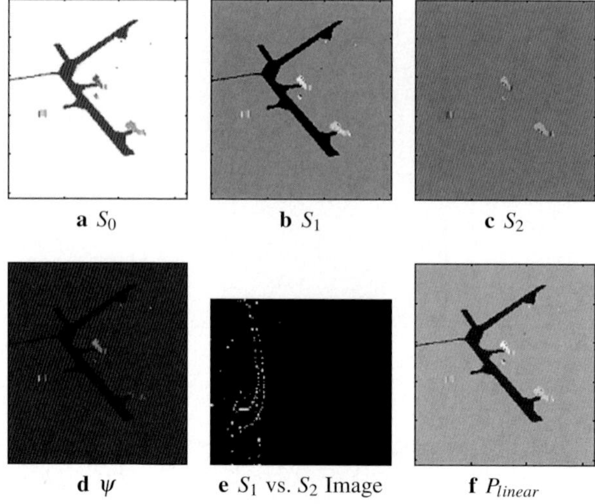

a S_0 **b** S_1 **c** S_2

d ψ **e** S_1 vs. S_2 Image   **f** P_{linear}

Fig. 2.16 Polarimetric imagery of a scene showing an M35 truck at $\iota = 90°$

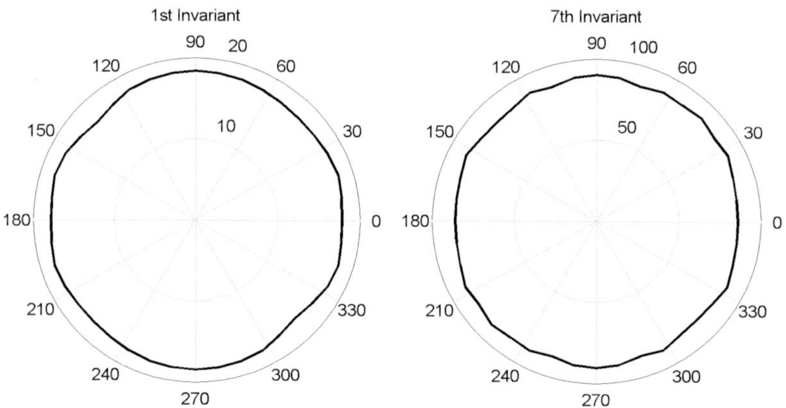

Fig. 2.17 1st and 7th invariants as function of ι for scene containing an M35 Truck shown in Fig. 2.14

The experimental results show that the angle of linear polarization imagery and the derived polarimetric invariant expressions remain unchanged under polarization transformations.

This chapter demonstrated a method for extracting 3D information and indices of refraction from a scene by means of two polarimetric passive IR imaging sensors. Each sensor provides the user with the Stokes parameters for each pixel through the use of a rotating polarizer from which the degree of linear polarization and the angle of linear polarization are computed. The angle of linear polarization gives us

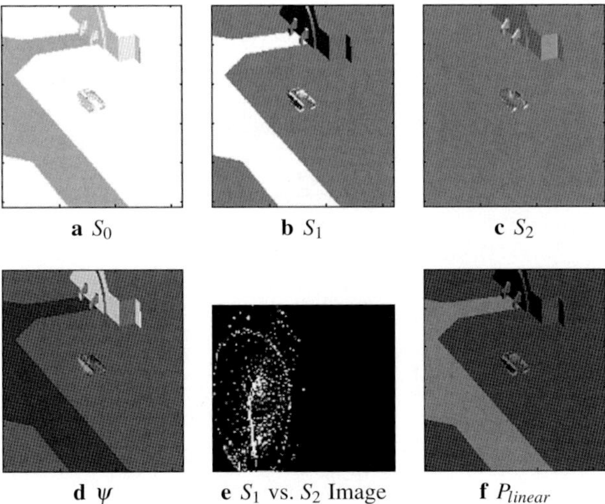

Fig. 2.18 Polarimetric imagery of a scene showing a T72 tank at $\iota = 0°$

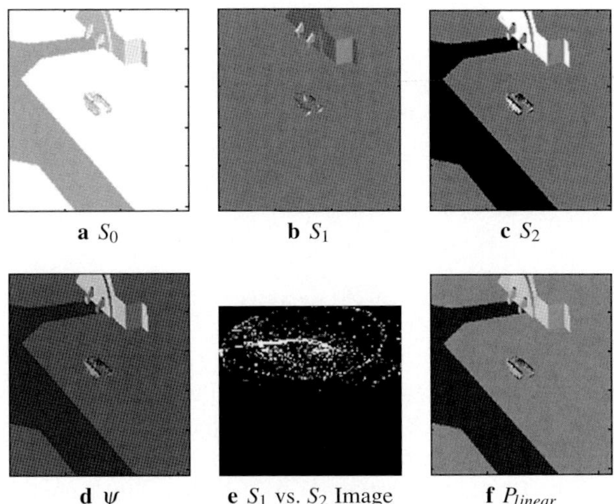

Fig. 2.19 Polarimetric imagery of a scene showing a T72 tank at $\iota = 45°$

the azimuth angle of the surface normal at the associated pixel. Two geometries were examined, one that corresponds to the special case when the surface normal is coplanar with the lines of sight of the two sensors and one that corresponds to a more general sensor placement (where the only real constraint is the number of pixels in the scene that can be registered between the two sensor images). It would also be possible to use two sensors along the same line of sight, provided they observed

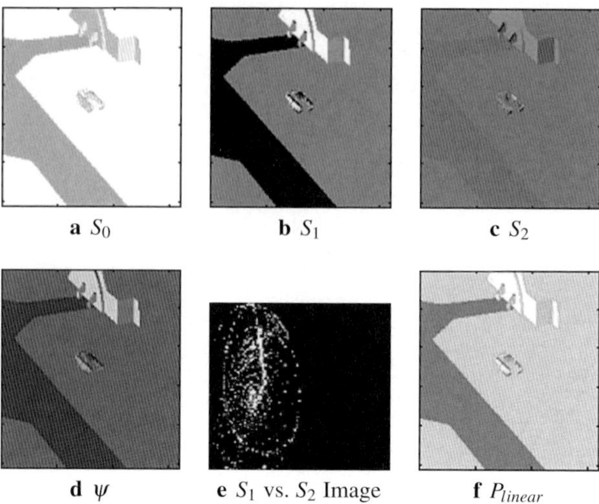

a S_0 **b** S_1 **c** S_2

d ψ **e** S_1 vs. S_2 Image   **f** P_{linear}

Fig. 2.20 Polarimetric imagery of a scene showing a T72 tank at $\iota = 90°$

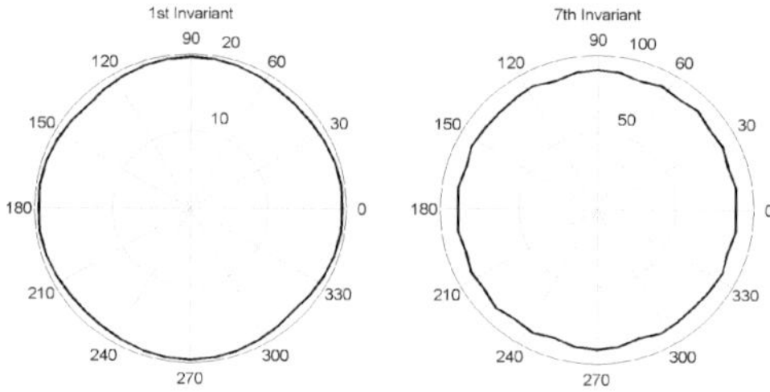

Fig. 2.21 1st and 7th invariants as function of ι for a scene containing T72 tank shown in Fig. 2.18

the scene at different frequencies. In such a case, the frequency dependence of the index of refraction becomes key in extracting further information from the target. In the first (special) case, the degree of polarization from each sensor and the angle between their lines of sight are sufficient to calculate the depression angle (and therefore find the surface normal vector). In the more general case, the observed azimuth angles are used to generate a pair of azimuth vectors, vectors that both lie in the plane of the target surface and that can therefore be used (by means of a cross product) to generate the surface normal vector. In either case, once the surface normal has been found, one can use the angle of incidence to find the index of refraction. Results of the application of this approach on simulated and real-world IR polarimetric data were provided.

Chapter's References

1. M. Born, and E. Wolf, *Principles of Optics*, Cambridge Unviversity Press, Newyork, 1998
2. C.S. Chun, F.A. Sadjadi, Polarimetric laser radar target classification, *Optics Letters*, 30(14):1806–1808, 2005
3. E. Collett, *Polarized Light: Fundamentals and Applications*, Marcel Dekker, New York, 1993
4. E.B. Elliott, *An Introduction to the Algebra of Optics*, Clarendon, Oxford, U.K., 1895
5. S. Huard, *Polarization of Light*, Wiley, New York, 1996
6. A. Ishimaru, *Electromagnetic Wave Propagation, Radiation, and Scattering*, Prentice Hall, Englewood Cliffs, 1991
7. J.D. Jackson, *Classical Electrodynamics,* 3rd ed., Wiley, New York, 1999
8. F.A. Jenkins, and H.E. White, *Fundamentals of Optics,* 2nd ed., McGraw-Hill, New York, 1950
9. O. Matoba, B. Javidi, Three-dimensional polarimetric integral imaging, *Optics Letters* 29:2375–2377, 2004
10. D. Miyazaki, M. Kagesawa, K. Ikeuchi, Transparent surface modeling from a pair of polarization images, *IEEE Transaction on Pattern Analysis and Machine Intelligence*, 26(1):73–82, 2004
11. H. Mott, *Antennas for Radar and Communications: A Polarimetric Approach*, Wiley, New York, 1992
12. F.A. Sadjadi, Improved target classification using optimum polarimetric SAR signatures, *IEEE Transactions on Aerospace and Electronic Systems*, AES-37(1):38–49, 2002
13. F.A. Sadjadi, A. Mahalanobis, Target adaptive polarimetric SAR target discrimination using MACH filters, *Applied Optics*, 45:3063–3070, 2006
14. F.A. Sadjadi, Adaptive polarimetric sensing for optimum radar signature classification using a genetic search algorithm, *Applied Optics*, 45:5677–5685, 2006
15. F.A. Sadjadi, Passive 3D imaging using polarimetric diversity, *Optics Letters*, 32(3):229–231, 2007
16. F.A. Sadjadi, C.S. Chun, Passive polarimetric IR target classification, *IEEE Transactions on Aerospace and Electronic Systems*, AES-37(3):740–751, 2001
17. F.A. Sadjadi and C.S. Chun, Automatic detection of small objects from their infrared stateof-polarization vectors, *Optics Letters*, 28: 531–533, 2003
18. F.A. Sadjadi, C.S. Chun, Remote sensing using passive infrared stokes parameters, *Optical Engineering Journal*, 43(10):2283–2291, 2004
19. F.A. Sadjadi, E.L. Hall, Three-dimensional Moment Invariants, *IEEE Transaction on Pattern Analysis and Machine Intelligence*, 2:127–136, 1980
20. M. Saito, Y. Sato, K. Ikeuchi, H. Kashiwagi, Measuremnet of surface orientations of transparent objects using polarization in highlight, *Journal of Optical Society of America, A*, 16(9):2286–2293, 1999
21. W. Shurcliff, *Polarized Light*, Harvard University Press, Cambridge, MA, 1960
22. B. Sturmfels, *Algorithms in Invariant Theory*, Springer-Verlag, New York, 1993
23. H. Van De Hulst, *Light Scattering by Small Particles*, Wiley, New York, 1957
24. L.B. Wolff, T.E.B. Boult, Constraining object features using a polarization reflectance model, *IEEE Transaction on Pattern Analysis and Machine Intelligence*, 13(7):635–657, 1991

Chapter 3
Vehicle Classification in Infrared Video Using the Sequential Probability Ratio Test

Mark W. Koch and Kevin T. Malone

Abstract This chapter develops a multilook fusion approach for improving the performance of a single-look vehicle classification system for infrared video. Vehicle classification is a challenging problem since vehicles can take on many different appearances and sizes due to their form and function and the viewing conditions. The low resolution of uncooled infrared video and the large variability of naturally occurring environmental conditions can make this an even more difficult problem. Our single-look approach is based on extracting a signature consisting of a histogram of gradient orientations from a set of regions covering the moving object. We use the multinomial pattern-matching algorithm to match the signature to a database of learned signatures. To combine the match scores of multiple signatures from a single tracked object, we use the sequential probability ratio test. Using infrared data, we show excellent classification performance, with low expected error rates, when using at least 25 looks.

Keywords: Infrared video · Vehicle classification · Scale space · Histogram of orientations · Multinomial pattern matching · Multilook fusion · Sequential probability ratio test · Surveillance · Unattended sensor system

3.1 Introduction and Problem

We are interested in developing an unattended sensor system for monitoring high-profile areas that do not warrant continuous human supervision. One of the attractive sensors for the unattended system is an uncooled infrared video imager due to its relatively small size, low power requirement, and day-night imaging capability. One possible application would monitor a choke point such as a roadway for vehicles. Here, we would like the system to detect only vehicles and ignore other moving objects such as animals. This is a challenging problem since vehicles can take on many different appearances and sizes based on their function and the viewing conditions.

R.I. Hammoud (ed.), *Augmented Vision Perception in Infrared: Algorithms and Applied Systems*, Advances in Pattern Recognition, DOI 10.1007/978-1-84800-277-7_3,
© Springer-Verlag London Limited 2009

Vehicles can also vary in details such as size and number of wheels and windows. We want to develop a model that captures the commonality between objects in the vehicle class but can also accommodate the endless variability of the objects.

We would like our vehicle detector to improve its performance as it gets more looks at the moving object and to handle a certain percentage of looks that might be contaminated from background such as temporary occlusions from trees or bushes. We further desire a set of algorithms that work over a large range of camera distances or object scales and naturally occurring environmental conditions, such as different types of background and ambient temperatures.

A video surveillance system viewing a road will detect and track other moving objects besides vehicles. Humans and animals could be detected or possibly flying birds. Windblown objects such as clouds, boxes, and tumbleweeds could also be detected. We want a classifier that can distinguish vehicles from all the other objects without having to specifically train against all possible moving objects. We call this requirement *one-class classification*, and we discuss it in the next section.

3.2 One-Class Classification

In making any decision, we want to control two types of errors: missed detection and false alarm errors. *Missed detection errors* can result from missing a target signature by calling it a nontarget, and *false alarm errors* can result from alarming on a nontarget signature by calling it a target. We choose to structure our classification problem as a *one-class* classifier [16]. For one-class classifiers, we are just interested in one specific target θ_1 represented by the alternative hypothesis. The nontarget class $\bar{\theta}_1$ is represented by the null hypothesis. If we have other targets of interest $\theta_2, ..., \theta_m$, then we would design a one-class classifier for each of them. For the θ_1 one-class classifier, we can further divide the nontargets into two groups: the other targets of interest $\theta_2, ..., \theta_m$ and the unknown class θ_0. For example, in our infrared classification problem, any moving object not belonging to the vehicle class (target) set is an unknown. This allows us to further distinguish between two types of false alarm errors: between-class and out-of-class errors. *Between-class errors* occur when alarming on another target $\theta_2, ..., \theta_m$ by calling it the target θ_1. *Out-of-class errors* occur when alarming on an unknown signature θ_0 by calling it the target θ_1.

For example, suppose we want to detect not only vehicles but also humans. Here, θ_1 would represent the vehicle class and θ_2 the human class. The unknown class θ_0 would represent all the moving objects not in θ_1 or θ_2. This would be any windblown object (tumbleweed, boxes, trash cans, etc.) or any animals. These unknown moving objects, a possible source of the out-of-class errors, are a significant problem in real-world pattern recognition problems in unconstrained environments. A Bayesian classifier approach designed for the vehicle-vs-human problem, while minimizing the between-class errors, would require models of all the possible objects that could be imaged by the sensor to control the out-of-class errors. Otherwise, it would classify an animal as vehicle or human. Modeling "the whole world" of

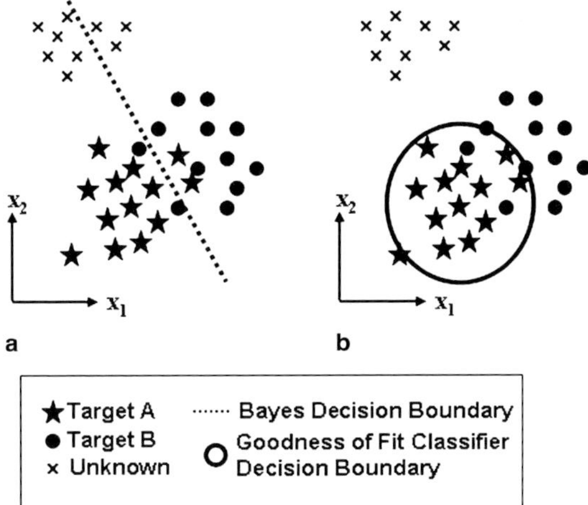

Fig. 3.1 Comparison of Bayes and goodness-of-fit (GOF) classifiers. **a** Bayes classifier, **b** GOF classifier

possible objects is untenable for most realistic systems deployed in unconstrained environments. In this chapter, we use a goodness-of-fit (GOF) classifier to control the out-of-class errors and power analysis [17] to model the unknown class.

Whereas Bayesian classifiers minimize the between-class error, they do nothing to control the out-of-class errors. Figure 3.1 illustrates this potential problem. The figure shows a two-dimensional (2D) feature space with samples from two targets: target A, represented by stars, and target B, represented by filled circles. Assuming normal distributions and equal covariance matrices for the targets, the Bayes decision boundary has a linear form (Fig. 3.1a). Whereas the Bayes classifier minimizes the between-class errors of the A and the B targets, it does not control the out-of-class errors caused by unknown objects represented by x symbols. Depending on which side of the boundary the nontarget falls, the classifier will assign the unknown to one of the known classes and make an out-of-class error. Figure 3.1b shows a GOF classifier that tries to surround the target class. Here, the unknown objects, which have widely differing features from the target class (x symbols), will be classified correctly. In general, the GOF classifier has improved out-of-class errors, but the between-class errors will increase since it is not an optimal Bayes classifier.

3.3 Object Classification

Object classification in video imagery can be considered a topic in pattern recognition. One can divide the large number of possible approaches into two categories: shape-based and motion-based classification.

3.3.1 Shape-Based Classification

Two categories of shape-based classification are global and local features. Global features are based on the entire object, whereas local features represent a parts-based approach and use features of small parts from objects.

3.3.1.1 Global Features

Depending on the distance metric (e.g., Euclidean distance), templates of the entire object, including eigenspace approaches [28], would be considered global features. Creating templates of all the possible vehicles would require a large data collection effort to capture all the possible subclasses of vehicles.

Blob features [3, 12] such as area, perimeter, compactness, length and orientation of the major and minor axis, and aspect ratio are also examples of global features. Lipton et al. [12] use *dispersedness* (ratio of blob perimeter squared to blob area) to classify vehicles and humans in optical imagery. Humans tend to have a higher dispersedness than vehicles since their limbs increase the perimeter at a higher rate then the corresponding area. While Lipton reported good results, he did not consider unknown moving objects such as animals and windblown objects. Presumably, the dispersedness of animals is also larger than vehicles and will cause a larger number of out-of-class errors.

Brown [3] investigated a large number of blob-type and motion-based features (see Section 3.3.2) for classifying humans and vehicles. She used a nearest neighbor classifier with the closest distance as the confidence. While the confidence could be used as a GOF metric, she did not report any out-of-class errors, so one can only presume that she was only concerned with the within-class distinctions. Thus, we suspect she selected the class with closest distance as in Fig. 3.1a. Her excellent within-class results would probably degrade if she used a threshold on her confidence metric. Requiring a confidence above a specific threshold, while reducing out-of-class errors, tends to decrease within-class performance results.

While computation of global features tends to be fast, global features are highly sensitive to image segmentation errors and occlusions. This sensitivity translates to higher missed detection or false alarm errors. Instead, global features might be useful as a prescreener. Here, we could design a weak classifier and accept a higher number of false alarms to achieve low missed detection errors. Thus, the global features could weed out possible detections that are definitely not vehicles and pass on the other detections to more computationally intensive classification algorithms.

3.3.1.2 Local Features

Local features are based on parts of an object and can be divided into two interrelated categories. The categories are coverage and descriptive power.

For coverage, the local features can just describe a single part of the object, describe the entire object, or lie somewhere between these two extremes. Ridder [21] developed a shared weight neural network to detect wheels in infrared imagery. While wheels are present on many vehicles and the number of detected wheels can help in identifying subclasses of vehicles, wheel detection is highly sensitive to occlusion and resolution. Wheels can easily be obscured by tall grass and bushes along the side of the road. Thus, a set of features that cover only a small part of an object tend to be unique or have high descriptive power.

Viola and Jones [29] used an integral image representation to efficiently compute local features and detect faces in optical imagery. To select critical and discriminatory visual features, a learning algorithm based on AdaBoost [24] was used. The small numbers of features are combined with a cascade of weak, but efficient, classifiers. These weak classifiers have a very high probability of detection (PD) and a moderate false alarm rate. By cascading these classifiers, one can produce a very low false alarm rate system that sequentially eliminates background and nontargets while including potential target-like regions. This approach needs a large set of training images to learn the local features that distinguish the target class from the unknown class.

Olson [18] used a template of oriented edges that covers the entire vehicle in Forward Looking Infra-Red (FLIR) imagery. Since he used a generalized Hausdorff metric [7] that requires only a fixed fraction of matching edges, we consider this a local feature approach where the local features are individual oriented edges. Since the geometry of these local features are quite specific, Olson needed multiple templates to describe the vehicles of interest and their orientation. Thus, this approach is more appropriate for a vehicle identification problem and not a classification problem.

The second category of local features is descriptive power. For the oriented edge of approach used by Olson [18], most of the descriptive power is in the geometric relationships between the edges and not in the individual edges themselves. Mikolajczyk et al. [15] and Schneiderman and Kanade [25, 26] put more descriptive power in the local features and less in the geometric relationships by using a histogram-of-features approach. Both divided the detected object into a small set of regions and computed a histogram of features in each region. Mikolajczyk et al., inspired by Lowe's local scale-invariant features for object recognition [13, 14], used features derived from different scale spaces [11], such as edges and blobs, and applied his method to detecting humans in single images. Schneiderman and Kanade were interested in detecting vehicles and faces and based their features on a wavelet transform, which has many similarities to Mikolajczyk et al.'s scale space approach.

By increasing the number and decreasing the size of the regions in the histogram-of-features approach, one can shift the descriptive power from features to geometry and vice versa. This allows one to provide specific details of a target class or become less specific with the goal of increasing generalization.

The local feature methods tend to be more robust to noise, image segmentation errors, and occlusion, but become more computationally intensive, especially if one needs multiple templates to account for different vehicle orientations.

3.3.2 Motion-Based Classification

The articulated motion of humans and animals can have a periodic property. By detecting a periodic component, one could distinguish humans and animals from other moving objects such as vehicles. Cutler and Davis [4] have developed a system to measure periodicity using similarity between multiple frames of a detected object. They reported excellent results on imagery of low resolution and poor quality. Other approaches require the segmentation of arms and legs [2]. Unfortunately, if the segmentation fails due to noisy or low-resolution data, then the entire algorithm also fails. Reducing the segmentation failures often comes at the expense of algorithm processing speed.

Javed and Shah [8] developed a recurrent motion image to distinguish vehicles from people in surveillance imagery. The recurrent motion image measures the amount of repeated motion from a tracked object. For example, people walking tend to exhibit significant repeated motion due to the periodic component of this movement, whereas moving vehicles have little or no repeated motion.

While motion can clearly separate vehicles from articulated objects such as humans and animals, it is not clear how well it performs in distinguishing vehicles from windblown objects.

3.3.3 Shape-and-Motion-Based Classification

Of course, one can combine the best of the shape- and motion-based classifiers to effectively classify moving objects. Viola et al. extended their face detection approach [29] to detect pedestrians in video imagery [30]. The new approach learns not only appearance (shape) features, but also motion features based on the difference between two consecutive frames. The AdaBoost algorithm is again used to learn the critical features from a large training set, and a cascade classifier is used to detect pedestrians and eliminate possible false detections. Brown [3] (discussed in Section 3.1) used the recurrent motion image [8] along with global blob-type features to develop a system that tries to distinguish vehicles from humans.

3.4 Overall Approach

For this chapter, we assume we have a detector tracker that provides a sequence of detected thermal image chips to the classifier. The detector models the background while slowly adapting to temperature changes and detects objects by looking for differences between the current frame and the background model [3]. The detector should account for swaying branches and grass and allow for sensor noise and uncertainties at background edges. The detector produces an image *chip* or a smaller subimage containing the object. The tracker associates the chips to a *track*

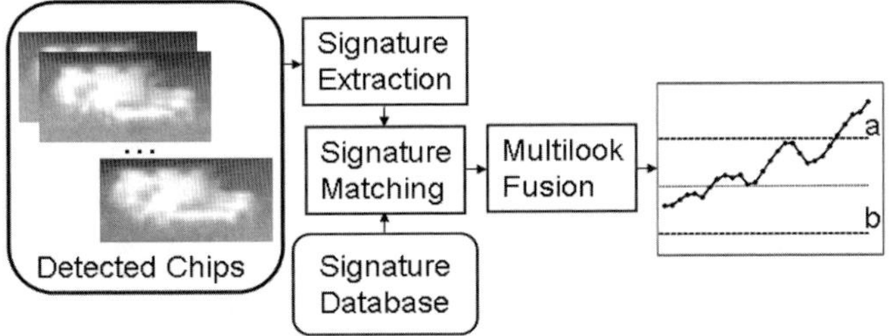

Fig. 3.2 Overall block diagram of vehicle classification approach

by using the predicted position and velocity [5]. A track thus represents a sequence of detected chips belonging to the same object.

Figure 3.2 shows the overall block diagram of our vehicle classification approach based on the histogram-of-features approach [13–15]. For each detected chip, we extract a signature by dividing each chip, into a set of regions and extracting a histogram of orientations from each region. By dividing the chip into the same number of regions and using a scale space approach for gradient estimation, we can handle objects at different distances from the infrared video camera.

After extracting a signature from a detected chip, we compare the signature to a database of learned signatures. The database contains the different modes of vehicle signatures. These modes could arise from vehicles at different aspects, for example, a head-on view, or from very specialized vehicles such as a bus or a tank.

We use multinomial pattern matching (MPM) [27] to compare a signature to a template. The MPM algorithm was originally designed for a target recognition problem and quantizes the signature to produce an instance of a multinomial random variable. This rank quantization gives MPM very good generalization and rejection characteristics. The MPM test also has a known distribution given the target class, which is very important for our multilook fusion approach.

Our contribution puts the matching of signatures into a statistical framework and uses the sequential probability ratio test (SPRT) [31] to perform multilook temporal fusion of the detected chips in each track. By combining the decision from each detected chip, we can, on average, increase our confidence about the vehicle/nonvehicle decision. The SPRT uses two thresholds to make a decision. These thresholds, a and b, are computed based on the desired false alarm and missed detection errors. If the accumulation of evidence from the detected chips goes above the upper (a) threshold, then a vehicle decision is made. If the accumulation of evidence goes below the lower (b) threshold, then a nonvehicle decision is made. If the current accumulation of evidence is between the two thresholds, then there is not enough evidence to make a decision at the required error rates. We use the extensions described by Koch et al. [9] to handle the nonvehicle class, dependencies between looks, and contamination of looks.

Section 3.5 discusses the vehicle classifier for the single-look case, and Section 3.6 describes the SPRT framework for multilook fusion. Finally, Sections 3.7 and 3.8 discuss the experimental results and conclusion and future work, respectively.

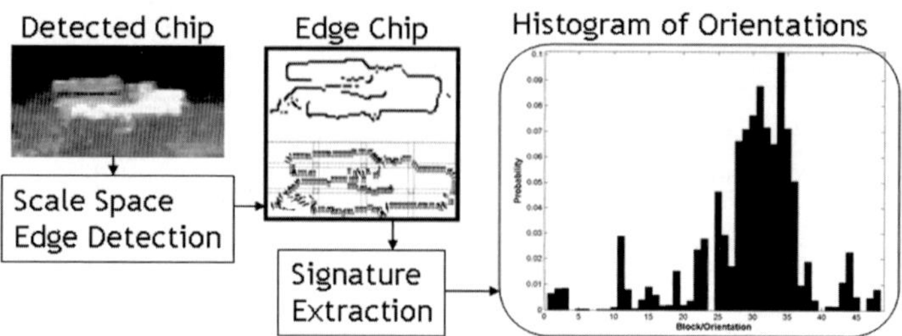

Fig. 3.3 Block diagram for single-look signature extraction

3.5 Single-Look Vehicle Classifier

Figure 3.3 shows a block diagram of our signature extraction approach for a detected chip from a single look. The first step applies scale space feature extraction and computes the gradient at many different scales. This allows the handling of vehicles at different distances from the camera and different sizes of vehicles. For each pixel in the chip, we select the *best* scale to represent the gradient. Thus, we have a chip that gives the gradient magnitude and another chip for the gradient orientation. We divide the resulting chips into $N \times M$ equal-size rectangles and quantize the orientations into K possible angles. The signature is then a histogram of orientations that represent the frequency of each of the K orientations in each of the $N \times M$ regions.

3.5.1 Scale Space Feature Extraction

A scale space representation gives a set of images at different resolution levels. We use Lindeberg's approach [11] for creating a scale space gradient map. Let the function $I : \Re^2 \rightarrow \Re$ represent the image

$$I(\mathbf{r}) = I(x,y) = \lambda, \tag{3.1}$$

where $\mathbf{r} \in \Re^2$, $\lambda \in \Re$. Lindeberg created multiple scales by convolution with a Gaussian kernel $G_\sigma(\mathbf{r})$ with standard deviation σ. Thus,

$$L(\mathbf{r}, \sigma_k) = I(\mathbf{r}) \otimes G_{\sigma_k}(\mathbf{r}) \tag{3.2}$$

represents a scale space image at resolution σ_k. The multiplicative factor ξ takes us from one scale to the next: $\sigma_{k+1} = \xi \sigma_k$.

Gaussian kernels are attractive for a number of theoretical and computational reasons. Theoretically, a Gaussian does not add any new *details* with increasing σ. Thus, as σ increases the information content of the image decreases (image becomes flatter), and local peaks or valleys do not, respectively, get higher or deeper. Also the convolution properties of a Gaussian allow us to easily compute the next resolution level from the previous level. Thus, we have

$$L(\mathbf{r}, \sqrt{2}\sigma) = L(\mathbf{r}, \sigma) \otimes G_\sigma(\mathbf{r}). \tag{3.3}$$

Computationally, the 2D Gaussian is a separable function and we can implement it by convolving a one-dimensional (1D) Gaussian along the rows followed by a Gaussian convolved along the columns.

The following equation gives the gradient magnitude for the scale space image:

$$\nabla L(\mathbf{r}, \sigma_k) = \sigma_k^{1/2}(L_x^2(\mathbf{r}, \sigma_k) + L_y^2(\mathbf{r}, \sigma_k))^{1/2} \tag{3.4}$$

where $L_x(\mathbf{r}, \sigma_k)$ is the partial derivative of $L(\mathbf{r}, \sigma_k)$ with respect to x. The scale factor $\sigma_k^{1/2}$ normalizes the scale space image for the gradient computation [11]. This allows us to compare gradients over multiple scales for each point and pick the maximum gradient for that point. The maximum gradient represents the characteristic scale for a given point [11]. Thus, by selecting the maximum gradient at each point over the different scales, we can automatically select the appropriate scale and avoid having multiple images containing similar gradient structures. This reduces the multi-image representation to a single image.

3.5.2 Signature Extraction

We use histograms of orientations to represent the signature of a vehicle [15,25,26]. The orientation of the gradient is calculated as follows:

$$\Theta(\mathbf{r}, \sigma_k) = \tan^{-1}(L_y(\mathbf{r}, \sigma_k)/L_x(\mathbf{r}, \sigma_k)) \tag{3.5}$$

and we select the orientation with the scale that corresponds to the maximal gradient. We do not distinguish between negative and positive orientations since a vehicle could potentially be hotter or colder than the background. We divide the edge chip into $N \times M$ overlapping regions and quantize the orientations into K levels. A signature is represented by a histogram Y with $B = N \times M \times K$ bins, and thus each bin b contains the estimated probability $y_b = Y(b)$ of the kth orientation being present in the (n,m) region.

3.5.3 Signature Matching

After extracting a signature consisting of a histogram of orientations Y, we want to compare Y to a learned template M_T. A number of dissimilarity measures, from Minkowski-form distance to the Kullback-Leibler divergence [10], have been suggested for comparing two probability distributions [22]. After implementing and comparing a number of dissimilarity measures, we prefer Simonson's MPM test statistic [27, 33] (unfortunately, the references on MPM are generally unavailable). It not only produces the best results, but also we have found it to produce very good generalization and consistently produces a known probability distribution for target data. Knowing the distribution of the test given the target is very important for our multilook fusion approach.

The MPM test statistic uses a multinomial indexing transform $\Gamma : \Re \rightarrow I, I = \{1, 2, ..., Q\}$, that maps the amplitude of a signal into a discrete index representing group membership. Thus, the mapping $\Gamma(y_b)$ takes y_b from bin b of Y and maps it to one of Q quantiles $q = 1, ..., Q$. Here, the smallest of the first $1/Q$ y_b values maps to one, the next smallest $1/Q$ maps to two, and so on to the largest $1/Q$, which maps to Q. Figure 3.4 shows an example for $Q = 4$. The graph at the top of the figure shows the original signature, and the bottom graph shows the transformed signature. Here the bin probability y_b gets mapped to a value of 1, 2, 3, or 4. Thus, a high bin probability y_b would get mapped to 4, a low y_b would get mapped to 1, and the intermediate y_b values would get mapped to 2 or 3. The multinomial transform can be accomplished by rank ordering the $y_b \in Y$ and dividing the resulting ordered set into equal quantiles. This type of quantile mapping tends to give excellent within-class generalization and out-of-class rejection.

Using the training data of the target, we estimate a template of the quantile probabilities $\hat{P}_{bq}$. Here, the probabilities $\hat{P}_{bq}$ represent the observed proportion of vehicle-training signatures for which bin b in the target signature maps to quantile q. Thus,

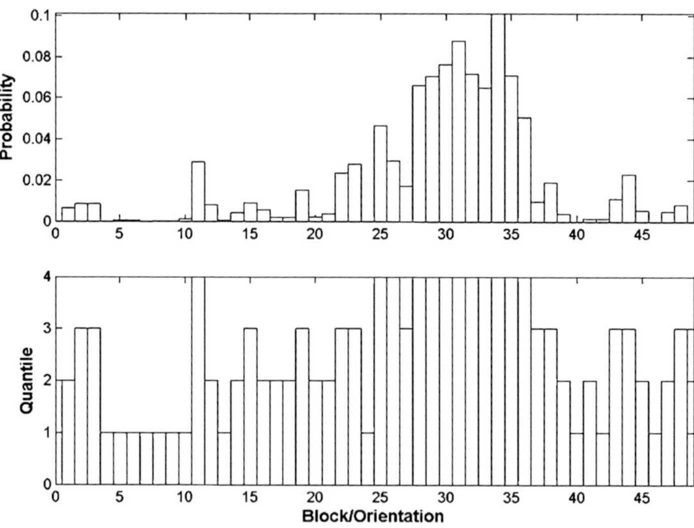

Fig. 3.4 Example of multinomial indexing transform: original signature (*top*), transformed signature (*bottom*)

the template M_T is a $B \times Q$ matrix of quantile probabilities $\hat{P}_{bq}$. Equation (3.6) gives the MPM test statistic for histogram signature $y_b = Y(b)$:

$$Z_{MPM} = \sum_{k=0}^{B-1} \frac{(1 - \hat{P}_{k,\Gamma(y_k)})^2 - \hat{E}_k}{\sqrt{C \times \hat{V}_k}} \tag{3.6}$$

Here, the quantities $\hat{E}_k$ and $\hat{V}_k$ represent, respectively, the estimated expected value and variance of the quadratic penalty $(1 - \hat{P}_{k,\Gamma(y_k)})^2$, and C accounts for the correlations between histogram bins. Low Z_{MPM} scores are consistent with a good match to the target. Using the central limit theorem [19], as the number of elements in a histogram signature B increases, we can show that Z_{MPM} approximates a normal distribution with zero mean and unit variance $[N(0,1)]$ given target data.

The estimates of the mean and variance of the quadratic penalty $\left(1 - \hat{P}_{kq}\right)^2$ are

$$\hat{E}_k = \sum_{q=0}^{Q-1} \tilde{P}_{kq}(1 - \hat{P}_{kq})^2, \tag{3.7}$$

and

$$\hat{V}_k = \sum_{q=0}^{Q-1} \tilde{P}_{kq}(1 - \hat{P}_{kq})^4 - \hat{E}_k^2. \tag{3.8}$$

Here, $\tilde{P}_{bq}$ represents the estimates of the quadratic penalty probabilities for bin b and quantile q. To compute $\tilde{P}_{bq}$, we use a Bayes estimator of the form [23]:

$$\tilde{P}_{bq} = \alpha\, p_{bq}^0 + (1 - \alpha)\hat{P}_{bq}, \tag{3.9}$$

where $0 \le \alpha \le 1$ represents a weight, $\hat{P}_{bq}$ represents the maximum likelihood estimation of the probability p_{bq}, and p_{bq}^0 represents a priori information about p_{bq} and satisfies the properties of a probability. The Bayes estimator prevents the case of estimating a zero probability for a specific bin b and quantile q. For the multinomial distribution, the Dirichlet distribution is the conjugate prior [32].

Using a symmetric Dirichlet [23] prior gives $p_{bq}^0 = 1/Q$, and $\alpha = Qv/(n+Qv)$, where n represents the number of training signatures and v is a single user-specified parameter for the Dirichlet distribution. Thus, the Bayes estimation Eq. (3.9) for the quantile probabilities becomes [23, 27]

$$\tilde{P}_{bq} = (v + n\hat{P}_{bq})/(n + Qv). \tag{3.10}$$

As $n \to \infty$, the Bayes estimate $\tilde{P}_{bq}$ approaches the maximum likelihood estimate $\hat{P}_{bq}$. We estimate $\hat{P}_{bq}$ and v using the training data and a leave-one-out (LOO) estimation technique. The v parameter is selected to give the LOO Z_{MPM} scores a zero mean.

3.6 Multilook Sequential Classifier

To perform multilook temporal fusion of the detected chips from each video frame k, we use the SPRT [31]. Here we have a stream of observations (Z_{MPM} test scores) $x_1, x_2, \ldots$ resulting from the best match of the detected signatures $S_1, S_2, \ldots$ (from multiple image frames) to a database of templates $\{M_T(1) \ldots M_T(D)\}$. From the previous section, $x_i \sim N(0, 1)$ given x_i belongs to a target (vehicle class). In previous work [9], we have used power analysis [17] to model the nontarget (nonvehicle class) as $N(\mu_0, \sigma_0)$. In Appendix 1, we outline the derivation to show that $N(\mu_0, 1)$ is the worst-case nontarget since it requires the largest number of observations to make a decision. Here, for a signature with $\mu < \mu_0$ we accept that the target and signature are so close that the errors we make have no practical consequence, and this preference increases with decreasing μ. For a signature with $\mu > \mu_0$ we call this a nontarget, and this preference increases with increasing μ. Thus, we have the two hypotheses

$$
\begin{aligned}
H_0: & \quad \mu = \mu_0 > 0, \ \sigma = 1 & \Rightarrow \text{Nontarget} \\
H_1: & \quad \mu = 0, \quad\quad\ \sigma = 1 & \Rightarrow \text{Target}
\end{aligned}
\tag{3.11}
$$

The SPRT takes the observations and uses the log likelihood ratio to accumulate evidence. Assuming independent observations, the following equation represents the SPRT accumulation of evidence $Z(n)$ and log likelihood ratio z_i for the hypothesis test:

$$
Z(n) = \sum_{i=1}^{n} z_i, \ z_i = \log\left(\frac{f(x_i|T)}{f(x_i|\bar{T})}\right).
\tag{3.12}
$$

Here, the functions $f(x_i|T)$ and $f(x_i|\bar{T})$ represent the probability density function (PDF) of an observation x_i, given the target T and the nontarget $\bar{T}$, respectively. Thus, for our problem we have $f(x_i|T) \sim N(0, 1)$ and $f(x_i|\bar{T}) \sim N(\mu_0, 1)$. The SPRT uses two decision boundaries (a, b) to make a decision

$$
\begin{aligned}
&\text{Reject } H_0 & &\text{If } Z(n) \geq a \\
&\text{Accept } H_0 & &\text{If } Z(n) \leq b \\
&\text{Get more data} & &\text{If } b < Z(n) < a
\end{aligned}
\tag{3.13}
$$

These decision boundaries can be obtained using the desired false alarm rate α and the missed detection rate β:

$$
a = \log\frac{1-\beta}{\alpha} \ \text{and} \ b = \log\frac{\beta}{1-\alpha}.
\tag{3.14}
$$

Since the test almost never ends exactly at the boundaries, the equations in (3.14) are an approximation. It has been shown that the SPRT, on average, uses the smallest number of observations to make a decision [31].

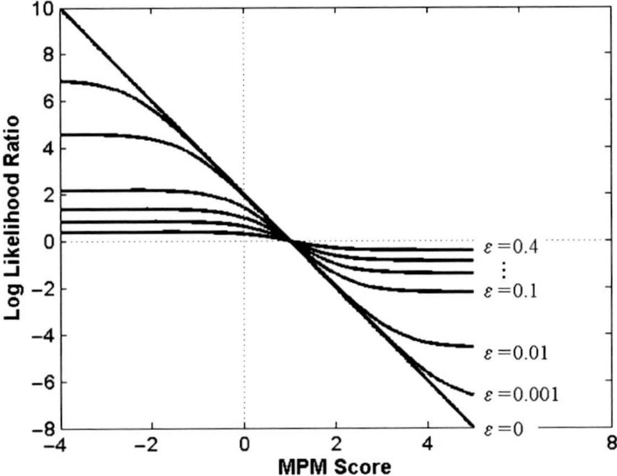

Fig. 3.5 Effect of contamination parameters on the log likelihood ratio. Here $\varepsilon = \varepsilon_0 = \varepsilon_1$

To handle contaminated observations, we use a modified log likelihood ratio [9]:

$$z(i) = \left(\frac{(1 - \varepsilon_1) f(x_i|T) + \varepsilon_1 f(x_i|\bar{T})}{\varepsilon_0 f(x_i|T) + (1 - \varepsilon_0) f(x_i|\bar{T})} \right). \tag{3.15}$$

Here, ε_0 or ε_1 represent the fraction of observations from $\bar{T}$ or T that could be contaminated. As $\varepsilon_0 / \varepsilon_1$ increases from 0 to 0.5, the positive/negative weight of evidence becomes clipped, so that the evidence does not count as much but still has some weight in the final outcome. Even if we do not expect any contamination, small values of ε_0 and ε_1 give us a classifier robust to unexpected observations or outliers. Figure 3.5 shows $\varepsilon = \varepsilon_0 = \varepsilon_1$ and how ε affects the log likelihood ratio.

To handle dependence between observations, we scale the cumulative log likelihood ratio $Z(n)$ by κ [9]:

$$Z(n) = \frac{1}{\kappa} \sum_{i=1}^{n} z(i). \tag{3.16}$$

The quantity κ represents the ratio of the number of observations n to the effective number of independent observations n^*. It can be estimated using the autocovariance of z_i (3.15) and assuming weak stationarity for z_i [1]:

$$n^* \approx \frac{n}{\kappa}, \text{ and } \kappa = 1 + 2 \sum_{\tau=1}^{\Delta} \rho_\tau \text{ for } n \gg \Delta. \tag{3.17}$$

Here, ρ_τ represents the normalized autocovariance at lag τ. Note, when no correlation exists, $\kappa = 1$, giving $n^* = n$. In practice, ρ_τ usually has a value greater than zero, giving $\kappa \geq 1$, but theoretically its estimate could be negative, especially for small ρ_τ. This could potentially give $\kappa < 1$. For the estimate of $\rho_\tau < 0$, a conservative approach is to make $\hat{\rho}_\tau = 0$ [6]. In Appendix 2, we provide an overview of the derivation to show that using κ gives the same SPRT termination probabilities for dependent observations as independent observations.

The multilook fusion time complexity is linear in the number of frames containing the detected object. Thus for each frame i, we have an observation x_i representing a Z_{MPM} test score. Each observation is converted to evidence z_i (3.15) and then accumulated (3.16). As evidence accumulates, the system compares the resulting sum to the a and b thresholds (3.14). The comparison provides a classification decision or a request for more image frames containing the detected object. A clever hardware implementation could use lookup tables to do the evidence conversion.

3.7 Data and Results

In this section, we discuss the data collected and the results of the multilook classification algorithms.

3.7.1 Infrared Video Data

To test our algorithms, we use infrared video data collected over many different experiments and in realistic conditions. Figure 3.6 shows a collage of chips from different target vehicles collected using Thermovision's A10 infrared camera with 160×120 resolution and 8-bit precision. Based on detected motion, we adjusted the images such that the vehicles are always pointing to the right. The data contain different types of civilian and military vehicles traveling at various speeds and at different distances (50 to 150 m) from the camera. We also reduced the image sizes by one half to simulate vehicles farther from the camera. We processed approximately 21,000 frames of data to get 1,792 frames containing an entire vehicle. From these frames, we selected 259 frames for training and used the remaining 1,533 frames for test with another 1,533 frames from the size reduction operation. The training frames were selected to cover a wide range of targets at different scales.

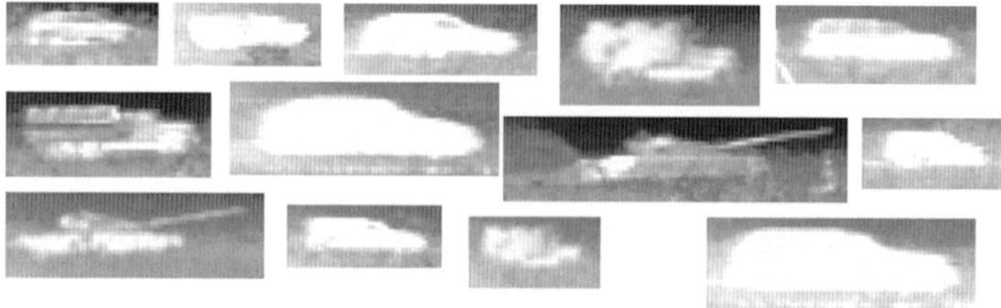

Fig. 3.6 Target examples: military and civilian vehicles

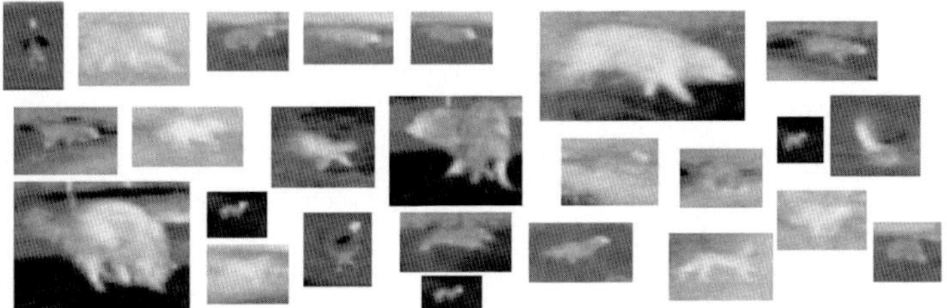

Fig. 3.7 Nontarget examples: various animals

Table 3.1 Algorithm parameters and their values

Parameter	Type	Value
σ_o	Gradient	0.5
ξ	Gradient	1.4
Scales	Gradient	10
(N, M, K)	Gradient	$(3, 4, 4)$
Q	MPM	4
υ	MPM	1.03
C	MPM	5.9
$\varepsilon_0\ \&\ \varepsilon_1$	SPRT	0.05
μ_0	SPRT	4
κ	SPRT	2.03
$\alpha\ \&\ \beta$	SPRT	1×10^{-3}

Figure 3.7 shows a collage of nontarget imagery used to test the vehicle classifier algorithm. The data were acquired using the same camera and under similar conditions. Animal subjects were used as the primary confuser. The distances of the animals from the video camera were 35 to 50 m. We processed 5,100 frames of data with 1,918 frames containing the entire animal.

3.7.2 Algorithm Parameters

Table 3.1 shows the parameters used for the experiments. The parameters υ, C, and κ were learned from the vehicle-training data. The SPRT error rates α and β were selected to give low numbers of missed detections and false alarms. Larger values would give decisions using fewer frames but the possibility of more errors. The SPRT contamination parameters ε_0 and ε_1 were selected to give the SPRT some robustness to outliers but are still relatively small since the data does not contain much contamination. The gradient and signature extraction parameters were selected based on previous experience. The parameter μ_0 controls the sensitivity of the system to targets and nontargets. The value $\mu_0/2$ is the *break point* of the evidence assignment. MPM scores (Z_{MPM}) less than $\mu_0/2$ give positive SPRT evidence for the target, and (Z_{MPM}) greater than $\mu_0/2$ gives negative target evidence. Thus, as μ_0 increases, the system will detect targets using fewer frames at the expense of potentially increased errors on nontargets. As μ_0 decreases, we risk operating the system at a suboptimal level and not having enough frames to classify a target.

3.7.3 Results

Figure 3.8a shows the PDF of the LOO scores using the training data. The PDF was estimated using Parzen density estimation [20]. The dotted curve shows the PDF of a standard $N(0, 1)$ density. Comparison of the two curves shows that the MPM scores are approximately normal with zero mean and unit standard deviation.

Figure 3.8b shows the operating characteristic (OC) curve for making a decision using a single look based on Z_{MPM}. The OC curve plots the probability of false alarm (PFA) vs. the PD for different decision thresholds on Z_{MPM}. The curve shows a 52% PD at zero PFA and 100% PD at 26% PFA.

To improve the performance of the system, we sequentially combine looks from each frame. When the trace (cumulative log likelihood ratio) goes above the a threshold, a target decision is made, and when it falls below the b threshold, a nontarget decision is made. Figure 3.9a shows the SPRT traces from video of targets. The dotted lines represent the a and b thresholds. The figure shows that it takes at most 25 frames to correctly classify all of the test target sequences at the error rates specified by α and β.

Figure 3.9b shows the SPRT traces from video of nontargets. One of the traces has a period of about a second where a dog at a large distance from the camera looks more like a vehicle than an animal. The dog is stationary during this time, but as it starts to move the system quickly recovers. The figure shows that all of the nontarget decisions are made within 10 frames.

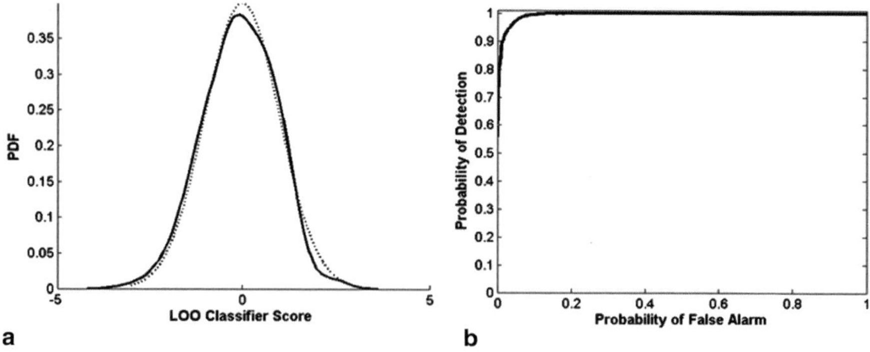

Fig. 3.8 Single-look results. **a** Leave-one-out (Loo) training scores (*solid line*), PDF for a $N(0, 1)$ distribution (dotted line); **b** Single-look operating characteristic

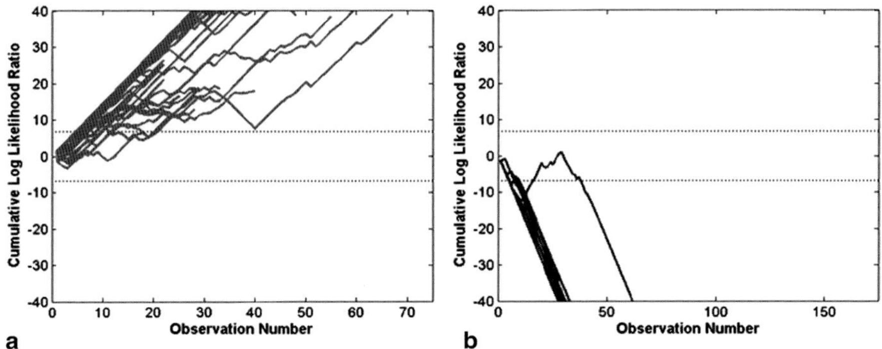

Fig. 3.9 Multilook results. **a** MPM SPRT target traces, **b** MPM SPRT nontarget traces

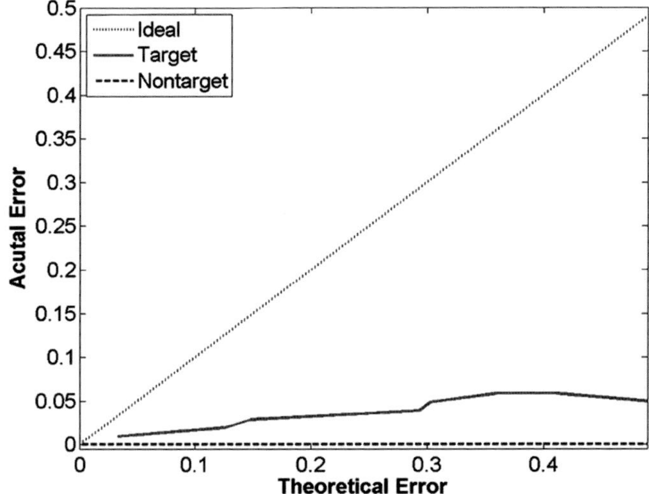

Fig. 3.10 MPM SPRT verification plot: ideal (*diagonal short-dotted line*), actual target errors (*solid line*), actual nontarget errors (*long-dotted line*)

To check the models developed for the SPRT, we use *verification plots* by selecting a theoretical error γ and computing the (A, B) decision boundaries using $\alpha = \beta = \gamma$ and Eq. (3.14). Using the decision boundaries, we estimate the actual error and then plot the theoretical versus the actual error for a range of γ values. While this does not cover all the possible combinations of possible (α, β) errors, it does cover the small errors since $A \approx 1/\alpha$ and $B \approx \beta$ for $\alpha, \beta \ll 1$. Figure 3.10 shows a verification plot for the vehicle classification SPRT (Fig. 3.9a, b). The diagonal short-dotted line shows the theoretical error matching the actual error. The solid and long-dotted curves show the results of comparing the theoretical error to the actual error of the target and nontarget sequences. Here, the theoretical error bounds the actual error. This shows that the SPRT theory is not violated by the models developed for our vehicle classification problem.

3.8 Conclusions and Future Work

We have developed and tested a vehicle classifier for infrared imagery. The classifier takes a sequence of detected chips from a detector tracker and extracts a signature from each chip based on a histogram of gradient orientations selected from a set of regions that cover the detected object. The orientations are computed in a scale space framework based on normalized gradients using a Gaussian kernel to move from one resolution to the next [11]. The scale space framework allows us to deal with vehicles of different sizes and at various distances from the camera.

We match signatures to learned templates using Simonson's MPM algorithm [27]. The MPM approach quantizes the signature to produce a multinomial distribution. The quantization produces good generalization and rejection characteristics. The test statistic is designed to produce a standard normal distribution with zero mean and unit variance and accounts for correlations between histogram bins.

While performance based on a single-look MPM score is very good, we show how Wald's SPRT [31] can be used to fuse the MPM scores from multiple looks producing superior performance. The modified SPRT is designed to handle detections from an unknown class, contamination of observations, and dependence between observations [9]. The result is a system that correctly classifies moving objects at low error rates within 25 looks.

For future work, we would like to investigate the performance of other scale space features such as ridges and blobs [11] and incorporate motion-based features. We would also like to use the SPRT to fuse not only multiple looks but also multiple dissimilarity metrics. In addition, we are interested in extending this work to human classification and eventually use it as a front end to human activities recognition.

Acknowledgments We would like to thank José Salazar, Melissa Koudelka, and the anonymous reviewers for their valuable questions, comments, and suggestions. Finally, we thank Mark Bishop, Art Storer, and Heather Tate for their help in the data collections.

Appendix 1: Worst-Case Nontarget Standard Deviation

Assume we have observations $x_1, x_2, \ldots$ with $x_i \sim N(\mu, \sigma)$ and two hypotheses:

$$
\begin{aligned}
H_0 : \quad & \mu = \mu_0 > 0, \ \sigma = \sigma_0 > 0 \quad && \Rightarrow \text{Nontarget} \\
H_1 : \quad & \mu = 0, \qquad\quad \sigma = 1 \quad && \Rightarrow \text{Target}
\end{aligned}
\tag{3.18}
$$

For these hypothesis test and with observations coming from a nontarget, we outline the derivation to show $\sigma_0 = 1$ requires the largest average number of observations to make a decision, implying $\sigma_0 = 1$ is the worst-case nontarget σ_0.

Using (3.12) and the assumption that the observations have normal distributions we have

$$z = \frac{-x^2}{2} + \frac{(x-\mu_0)^2}{2\sigma_0^2} + \log(\sigma_0) \tag{3.19}$$

and

$$E\{z|H_0\} = \frac{1}{2}\left(1 - \mu_0^2 - \sigma_0^2 + 2\log(\sigma_0)\right). \tag{3.20}$$

From Wald's approximation for the average number of observations given a nontarget hypothesis,

$$E\{n|H_0\} \approx \frac{\alpha \log(A) + (1-\alpha)\log(B)}{E\{z|H_0\}}, \tag{3.21}$$

From Eqs. (3.20) and (3.14) we get:

$$E\{n|H_0\} \approx \frac{2\left[\alpha \log(1-\beta) - \alpha \log(\alpha) + (1-\alpha)\log(\beta) - (1-\alpha)\log(1-\alpha)\right]}{1 - \mu_0^2 - \sigma_0^2 + 2\log(\sigma_0)}. \tag{3.22}$$

Taking the derivative of (3.22) with respect to σ_0, setting $\frac{\partial E\{n|H_0\}}{\partial \sigma_0}$ equal to zero, and solving for σ_0, we get

$$\sigma_0^* = 1. \tag{3.23}$$

The sign of the second derivative is negative for $0 < \alpha, \beta < 1/2$, verifying $\sigma_0 = 1$ gives the largest average observations to make a decision given a nontarget.

Appendix 2: Scaling the Decision Boundaries for Handling Dependence

In this appendix, we outline the derivation of the scaling factor for adjusting the decision boundaries for handling dependence. We design an SPRT assuming independence and then adjust the decision boundaries so that the termination probabilities are the same for the dependence process. We assume the dependence creates log likelihoods $z_1, z_2, \ldots z_n$ (3.12) that are serially correlated and weakly stationary. Let Δ represent the maximum lag with nonzero correlation. Here the normalized autocovariance ρ_τ at lag τ, $|\tau| \leq \Delta$, fully describes the statistics of the dependence process.

Assuming the log likelihoods z have finite first and second moments, then by the central limit theorem, as the number of observations n becomes large the cumulative log likelihood ratio $Z(n)$ in Eq. (3.12) has a normal distribution $N(\mu_{Z(n)}, \sigma_{Z(n)}^2)$. Here,

$$\mu_{Z(n)} = n\mu_z, \quad \text{and} \quad \sigma_{Z(n)}^2 = n\kappa\sigma_z^2 \tag{3.24}$$

where $\mu_z = E\{z\}$, $\sigma_z^2 = E\{(z-\mu_z)^2\}$, and $\kappa \approx \sum_{\tau=-\Delta}^{\Delta} \rho_\tau$ for $n \gg \Delta$ [1]. The quantity κ represents the ratio of the number of observations to the effective number of independent observations [1], and $\kappa = 1$ for independent log likelihoods z.

Using characteristic functions and the normal approximation from the central limit theorem,

$$E\left\{e^{-sZ(n)}\right\} \approx \phi^n(s) \tag{3.25}$$

where

$$\phi(s) = \exp(-\mu_z s + \tfrac{1}{2}\kappa\sigma_z^2 s^2) \tag{3.26}$$

Even though $Z(n)$ is the sum of dependent variables, the serially correlated and weakly stationary assumption allows it to be written as a multiplication of characteristic functions.

The following equation gives Wald's fundamental identity [31]:

$$E\left\{e^{-sZ(N)}\phi^{-N}(s)\right\} = 1 \tag{3.27}$$

where N represents the number of observations required by the test. Wald's fundamental identity [31] was derived assuming independence, but only the property (3.25) is required. Thus we can use (3.27) to determine p_a and p_b, the probability of terminating at decision boundary a and b, respectively. Using Wald's fundamental identity, we have

$$p_a E\left\{e^{-sZ(N)}\phi^{-N}(s)|Z(N) \geq a\right\} + p_b E\left\{e^{-sZ(N)}\phi^{-N}(s)|Z(N) \leq b\right\} = 1. \tag{3.28}$$

Assuming $\mu_z \neq 0$, find $s_0 \neq 0$ such that $\phi(s_0) = 1$. From (3.26), we get

$$s_0 = 2u_z/\kappa\sigma_z^2. \tag{3.29}$$

Using Eq. (3.28) and $s = s_0$, we have

$$p_a E\left\{e^{-s_0 Z(N)}|Z(N) \geq a\right\} + p_b E\left\{e^{-s_0 Z(N)}|Z(N) \leq b\right\} = 1. \tag{3.30}$$

Neglecting excess over the boundaries, $Z(N) \approx a$ when $Z(N) \geq a$ and $Z(N) \approx b$ when $Z(N) \leq b$, we have

$$p_a e^{-s_0 a} + p_b e^{-s_0 b} \approx 1. \tag{3.31}$$

Finally, using $p_a + p_b = 1$ and (3.29), we have the probability of termination at boundary a as

$$p_a \approx \frac{1 - \exp(-2\mu_z b/\kappa\sigma_z^2)}{\exp(-2\mu_z a/\kappa\sigma_z^2) - \exp(-2\mu_z b/\kappa\sigma_z^2)} \tag{3.32}$$

and the probability of termination at boundary b as

$$p_b \approx \frac{\exp(-2\mu_z a/\kappa\sigma_z^2) - 1}{\exp(-2\mu_z a/\kappa\sigma_z^2) - \exp(-2\mu_z b/\kappa\sigma_z^2)}. \tag{3.33}$$

If we scale the decision boundaries a and b by κ, then the termination probabilities are independent of κ. Here we have an SPRT with the same termination probabilities for independence and dependence, where the dependence results from serial correlation and weak stationarity. Thus, the new decision boundaries a' and b' become $a' = \kappa a$ and $b' = \kappa b$, respectively.

Chapter's References

1. V. Bayley and J.M. Hammersley. The effective number of independent observations in an autocorrelated time series. *Supplement to the Journal of the Royal Statistical Society*, 8:184–197, 1946
2. A. Bazin and M.Nixon. Gait verification using probabilistic methods. In *IEEE Applications of Computer Vision*, pages 60–65, Jan 2005
3. L. Brown. View independent vehicle/person classification. In *Proceedings of the ACM Second International Workshop on Video Surveillance and Sensor Networks*, pages 114–123, Oct 2004
4. R. Cutler and L. Davis. Robust real-time periodic motion detection, analysis, and applications. *IEEE Transactions on Pattern Analysis and Machine Intelligence*, 22(8):781–796, 2000
5. L. Forsti. Object detection and tracking in time-varying and badly illuminated outdoor environments. *Optical Engineering*, 37:2550, 1998
6. O. Gilbert. *Statistical Methods for Environmental Pollution Monitoring*. Wiley, New York, 1978, pages 35, 39
7. D. Huttenlocher, R. Lilien, and C. Olson. View–based recognition using an eigenspace approximation to the hausdorff measure. *IEEE Transactions on Pattern Analysis and Machine Intelligence*, 21(9):951–955, Sept 1999
8. O. Javed and M. Shah. Tracking and object classification for automated surveillance. In *Proceedings of the 7th European Conference on Computer Vision*, Vol. 4, pages 343–357, May 2002
9. M.W. Koch, G.B. Haschke, and K.T. Malone. Classifying acoustic signatures using the sequential probability ratio test. *Sequential Analysis Journal*, 23(4):557–583, 2004
10. S. Kullback and R.A. Leibler. On information and sufficiency. *Annals of Mathematical Statistics*, 22(1):79–86, March 1951
11. T. Lindeberg. Scale-space: a framework for handling image structures at multiple scales. In *CERN School of Computing Proceedings*, pages 8–21, Sept 1996
12. A. Lipton, H. Fujiyoshi, and R. Patil. Moving target classification and tracking from real-time video. In *Proceedings of the Fourth IEEE Workshop on Applications of Computer Vision*, pages 8–14, Oct 1998
13. D.G. Lowe. Object recognition from local scale-invariant features. In *Proceedings of the IEEE International Conference on Computer Vision*, Vol. 2, pages 1150–1157, 1999
14. D.G. Lowe. Distinctive image features from scale-invariant keypoints. *International Journal of Computer Vision*, 60(2):91–110, 2004
15. K. Mikolajczyk, C. Schmid, and A. Zisserman. Human detection based on a probabilistic assembly of robust part detectors. *European Conference on Computer Vision*, I:69–81, 2004
16. M. Moya and D. Hush. Network constraints and multi-objective optimization for one-class classification. *Neural Networks*, 9(3):463–474, 1996
17. K. Murphy and B. Myors. *Statistical Power Analysis: A Simple and General Model for Traditional and Modern Hypothesis Tests*. Erlbaum, Mahwah, NJ, 1998
18. C. Olson and D. Huttenlocher. Automatic target recognition by matching oriented edge pixels. *IEEE Transactions on Image Processing*, 6(1):103–113, Jan 1997
19. A. Papoulis. *Probability, Random Variables, and Stochastic Processes*. McGraw-Hill, New York, 1965
20. E. Parzen. On estimation of a probability density function and mode. *Annals of Mathematical Statistics*, 27:1065–1076, 1962

21. D. Ridder, K. Shutte, and P. Schwering. Detection of vehicles in infrared imagery using shared weight neural network feature detectors. In Ivan Kadar, ed., *Proceedings SPIE Signal Processing, Sensor Fusion, and Target Recognition VII*, Vol. 3374, pages 247–258, 1998
22. Y. Rubner, J. Puzicha, C. Tomasi, and J. Buhmann. Empircal evalution of dissimilarity measures for color and texture. *Computer Vision and Image Understanding*, 84:25–43, 2001
23. T.J. Santner and D.E. Duffy. *The Statistical Analysis of Discrete Data*. Springer-Verlag, New York, 1989
24. R. Schapire and Y. Singer. Improved boosting algorithms using confidence-rated predictions. *Machine Learning*, 37(3):297–336, 1999
25. H. Schneiderman and T. Kanade. A statistical method for 3d object detection applied to faces and cars. In *Proceedings IEEE Computer Society Conference on Computer Vision and Pattern Recognition*, Vol. 1, pages 746–751, 2000
26. H. Schneiderman and T. Kanade. Object detection using the statistics of parts. *International Journal of computer Vision*, 56(3):151–177, 2004
27. K.M. Simonson. Multinomial pattern matching: a robust algorithm for target identification. In *Proceedings of Automatic Target Recognizer Working Group*, Huntsville, AL, 1997
28. M. Turk and A. Pentland. Eigenfaces for recognition. *Journal of Cognitive Science*, 3(1):71–86, 1991
29. P. Viola and M. Jones. Robust real-time face detection. *International Journal of Computer Vision*, 57(2):137–154, 2004
30. P. Viola, D. Snow, and M. Jones. Detecting pedestrians using patterns of motion and appearance. *International Journal of Computer Vision*, 63(2):153–161, 2005
31. A. Wald. *Sequential Analysis*. Wiley, New York, 1947
32. S. Wilks. *Mathematical Statistics*. Wiley, New York, 1962
33. M.P. Zumwalt. Robust high range resolution radar for target classification. Master's thesis, Air Force Institute of Technology, Wright-Patterson Air Force, Ohio, 2000

Part II
Thermal Imagery & Vital Sign Detection

Chapter 4
Multiresolution Approach for Noncontact Measurements of Arterial Pulse Using Thermal Imaging

Sergey Y. Chekmenev, Aly A. Farag, William M. Miller, Edward A. Essock, and Aruni Bhatnagar

Abstract   This chapter presents a novel computer vision methodology* for noncontact and nonintrusive measurements of arterial pulse. This is the only investigation that links the knowledge of human physiology and anatomy, advances in thermal infrared (IR) imaging and computer vision to produce noncontact and nonintrusive measurements of the arterial pulse in both time and frequency domains. The proposed approach has a physical and physiological basis and as such is of a fundamental nature. A thermal IR camera was used to capture the heat pattern from superficial arteries, and a blood vessel model was proposed to describe the pulsatile nature of the blood flow. A multiresolution wavelet-based signal analysis approach was applied to extract the arterial pulse waveform, which lends itself to various physiological measurements. We validated our results using a traditional contact vital signs monitor as a ground truth. Eight people of different age, race and gender have been tested in our study consistent with Health Insurance Portability and Accountability Act (HIPAA) regulations and internal review board approval. The resultant arterial pulse waveforms exactly matched the ground truth oximetry readings. The essence of our approach is the automatic detection of region of measurement (ROM) of the arterial pulse, from which the arterial pulse waveform is extracted. To the best of our knowledge, the correspondence between noncontact thermal IR imaging-based measurements of the arterial pulse in the time domain and traditional contact approaches has never been reported in the literature.

4.1 Introduction

Arterial pulse is generated by mechanical contractions of the heart triggered by electrical activation. Doppler ultrasound and arterial tonometry are traditional approaches to track the hemodynamics changes in the arteries. This work attempts to

---

* A patent application has been filed. Provisional patent has been approved.

R.I. Hammoud (ed.), *Augmented Vision Perception in Infrared: Algorithms and Applied Systems,* Advances in Pattern Recognition, DOI 10.1007/978-1-84800-277-7_4,
© Springer-Verlag London Limited 2009

build an alternative noncontact computer vision methodology for accurate measurement of arterial pulse based on the thermal infrared (IR) imaging of arterial pulse propagation along the superficial arteries of the human body.

4.1.1 Related Work

Several noncontact technologies have been recently demonstrated to measure the heart rate in humans. Among active techniques, the use of radar- [2] and laser- [1] based monitors have been proposed. In the radar vital signs monitor, the heart rate is sensed with the use of an active radar detector. In this approach [2], a beam of radio waves is directed toward the body of a subject, and the reflected signal that contains the information on movement of the body is analyzed to extract the heart rate information. However, exposure of the general public to radiation remains a significant concern with this technique. Another active approach utilizes a laser Doppler vibrometer that emits a laser beam perpendicular to the surface of human skin. Motion of the skin is analyzed to measure heart rate. Again, the issues of active sensors, laser targeting, and motion artifacts remain significant limitations of this approach.

Passive sensors have a short but bright history in the monitoring of human vital signs. This area has been greatly advanced by a group of researchers from the University of Houston, Texas [3, 4], who reported the first practical demonstration of the capability of thermal IR imaging for monitoring human vital signs. In their work [4], they applied thermal imagery for measuring heart rate from the superficial blood vessel network. In their approach, they estimate the dominant heart rate frequency by averaging the power spectra of each pixel in a preselected segment of the superficial vessel. One of the limitations of their approach is that the skin segment has to be manually selected for measuring heart rate. Another limitation is that only the dominant heart rate frequency, not the actual heart rate wave form with spatial beat-to-beat information is reported. Their work is the most related to our investigation presented here. The differences between their and our approach, from the physiological, signal-processing, and sensor suit prospectives, are discussed in Section 4.7.

4.1.2 Research Issues

The idea behind the measurements of the arterial pulse using a thermal IR camera resides in the capability of the camera to capture the heat variations on the skin caused by the propagation of the arterial pulse. In spite of the simplicity of this idea, there are many nontrivial research issues that need to be resolved to permit reliable measurements. The time-varying conglomerate of radiated heat patterns are difficult to interpret in terms of a desirable physiological signature. Low-magnitude arterial pulse hides in the numerous heat patterns observed on the skin, and a proper strategy is required for faithful and descriptive measurements. First, the proper thermal IR imaging system needs to be used. Various physics and engineering-related issues of imaging in the thermal IR band with regard to the biometrics measurements need to be addressed. The next crucial step is physiological modeling. The accuracy and robustness of the pulse measurements dramatically depend on how faithfully the pulsatile nature of the blood flow and heat dissipation are modeled. After these strongly depend on how the steps, the signal-processing algorithms come into play to solve detection, tracking, and pulse recovery issues.

The pulsation we measure occurs in superficial arteries, not in the veins, which are typically located closer to the skin and appear brighter in thermal imagery. The arterial pulsating heat contribution produced by a superficial artery is of a very low magnitude in comparison with other heat patterns observed on the skin. This poses a problem for accurate detection of the region of measurement (ROM). This issue can be approached in a manual fashion, such as by palpating with a finger, or by proposing an automatic mechanism based on the pulsatile nature of the arterial pulse. In our approach, the issue of ROM detection is addressed in conjunction with the pulse recovery procedure. Our algorithm performs test measurements of arterial pulse from several locations on the skin within a region of interest (ROI). However, only the one, most reliable location, is output as the arterial pulse waveform, and the corresponding location is referred to as the ROM. Another concomitant problem is the tracking of the ROM in thermal video. Since a superficial artery does not produce sharp thermal imprints, the surrounding skin typically offers poor features for accurate tracking of the small skin patch identified as the ROM. The problem is worsened by continuous heat variations on the skin, which affect the features selected for tracking the ROM. This requires the construction of particular time-invariant tracking features or the development of some learning mechanism to update the tracking features in time.

The rest of this chapter is organized as follows: Section 4.2 discusses some basic concepts of imaging in the thermal IR band. Section 4.4 describes the anatomy of the superficial arteries of the body most suitable for the thermal IR measurements. Section 4.5 covers the multiresolution framework applied as a preprocessing step for the arterial pulse recovery in our methodology, which is presented in Section 4.7. Section 4.6 introduces the wavelet approach for detection of ROM and reconstruction of the arterial pulse from preprocessed thermal IR data. In Section 4.8, we discuss our experiments and results. Finally, the conclusion of our study is presented in Section 4.9.

4.2 Thermal Imaging

It is well known that different energies of electromagnetic (EM) radiation deliver different phenomena. The radiation at energies of thermal IR is remarkable because this is where the human body emits most of its radiation encoding valuable biometric information. This information, if properly processed, may be used in numerous biomedical applications, such as arterial pulse measurements. A solid understanding of the physics of light, image formation principles, choice of imaging IR band, and instrumentation is crucial for successful processing and interpreting the desired biometrics signatures through thermal IR imaging. This section briefly discusses salient features of thermal IR.

4.3 Thermal Radiation

Humans possess the ability to perceive the world in different colors. What we actually see is EM radiation from a particular, very narrow frequency band called the *visible* or the *optical spectrum*. Based on the energy of the light, the entire EM spectrum is typically divided into the following bands: gamma rays, X-rays, ultraviolet (uv), visible, IR, microwaves, millimeter waves and radio waves. The light from the bands other than visible cannot be observed by unaided eye, as illustrated in Fig. 4.1a, and special instrumentation is needed to register and to visualize this part of the spectrum.

Since the physics of light depends on its energy or frequency, it would be helpful to point out the relationship discovered by Einstein [5], which states that radiation with higher frequency carries more energy per photon/wave than that of lower frequency:

$$E = h\omega \tag{4.1}$$

where constant h is Planck's constant (6.63×10^{-34} J $\cdot$ s), E represents the energy of a light harmonic, and ω stands for its frequency. X-rays and gamma rays, the waves with the highest frequency in Fig. 4.1a, are the most energetic representatives of the EM radiation spectrum. The gamma rays are produced in nuclear reactions and are characterized by extremely high energies (i.e., very short wavelengths). This radiation may travel long distances and propagate through the matter, actively interacting

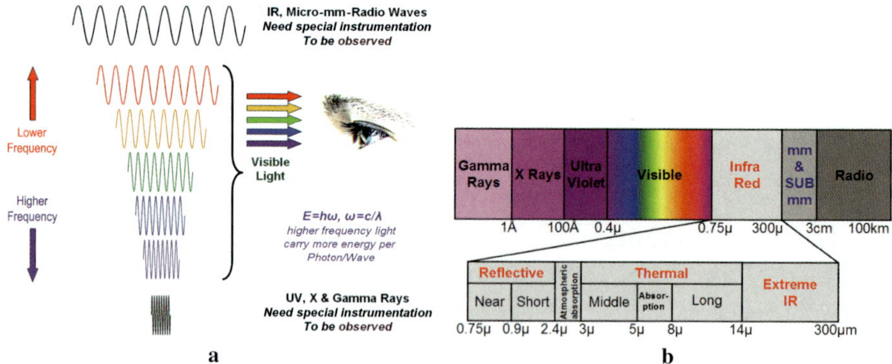

Fig. 4.1 Properties of light: **a** EM spectrum **b** IR bands

with its molecules and atoms. X-rays are produced by very hot objects and are characterized by their excellent penetration ability, which has been extensively exploited in medical imaging to visualize the contrast between different types of living tissues. UV light is less energetic and is widely used for sanitizing purposes in medical facilities to kill undesirable bacteria and microorganisms. Visible light plays a key role in the everyday interaction of numerous species of life on Earth, including that of humans. This is where we actually can see the world with unaided eye. Most current computer vision-based approaches are concentrated in the visible spectrum, mimicking human vision. In the last decade, the IR spectrum has become more popular for a variety of industrial, military, and remote-sensing applications. The fact that the human body mostly emits in the thermal IR band makes the imaging in this band an essential tool for human physiological analysis (medical thermography), surveillance applications, and various biometrics applications. The upper range of the EM spectrum is known as the radio-frequency band. This band is mainly used for communication applications.

The IR band is typically divided into reflective, thermal, and extreme bands, as illustrated in Fig. 4.1b. Both reflective and thermal bands are further subdivided into near wave infrared (NWIR; 0.75–2.4 μm), short-wave infrared (SWIR; 0.9–2.4 μm), middle-wave infrared (MWIR; 3–5 μm) and long-wave infrared (LWIR; 8–14 μm). The extreme infrared (14–300 μm) is sometimes referred to as very long wave infrared (VLWIR) in the literature.

The NWIR and SWIR light is produced by hot objects like bulbs or IR lasers. Normally, the surrounding environment is at a much lower temperature to be able to radiate significantly within these IR bands. To increase the contrast when imaging in NWIR and SWIR, some sort of artificial illumination sources can be applied to make use of reflected energy from illuminated objects on a scene. This is the key difference between imaging in reflective and thermal IR bands. Thermal IR imaging needs no artificial illuminating of any kind and because of this is frequently referred to as *passive*. Most of the objects at room temperature, including the human body, are good sources of thermal IR energy. This means that both MWIR and LWIR are good candidates for biomedical applications. However, there are some important differences between imaging in these bands. The MWIR band is significantly more susceptible to ambient illumination than the LWIR band. This means that measurements in LWIR are inherently less affected by ambient illumination of hot objects like the sun, bulbs, fire, and so on. This makes LWIR more advantageous for measurements of the pulsatile heat patterns of human vital signs. The strength of the MWIR band is in lower optical diffraction and lower background radiation than that in LWIR, which actually results in imaging that is sharper and has more contrast.

4.4 Anatomy

The circulatory system consists mainly of the lungs, heart, blood vessels, and circulating blood. In addition to its role in the transport of materials, the circulatory system is responsible for the distribution of heat throughout the body (Fig. 4.2a).

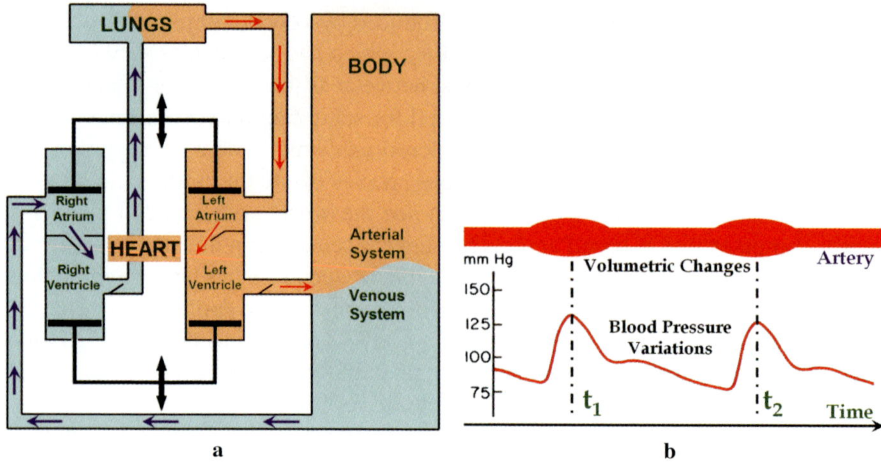

Fig. 4.2 The circulatory system: **a** heart as a two-sided stroke pump, **b** arterial pulse wave propagation along the arterial tree

The heart can be viewed as a two-sided stroke pump. The right heart pumps the blood through the lungs; the left heart pumps the blood through the rest of the body. The heart contractions produce sequential changes in the arterial blood pressure. These changes propagate along the arterial tree, causing deformation of arterial walls. These sequential deformations, known as the *arterial pulse*, can be palpated with a finger from the main superficial arteries of the body (Fig. 4.2b). The shape and the speed of the arterial pulse wave change when it propagates along the arterial tree. This is mainly because of the change in mechanical properties of arterial walls (elastic tapering) and the decrease of the cross-sectional areas of the arteries (geometric tapering) [6]. For each cycle (Fig. 4.2b), there is a systolic upstroke (ascending limb). This upstroke represents the rapid increase in arterial pressure and flow due to the opening of aortic and pulmonary valves during the ventricular systole. The maximum sharp peak at the end of the ascending limb is the *systolic peak pressure*. The dicrotic notch located on the systolic decline is caused by the closure of the aortic valve during early diastole. The pressure then gradually drops until the end-diastolic pressure [7].

There are several locations on the human face and neck where the arterial pulse can be palpated (Fig. 4.3a, b). These ROIs are the carotid artery complex and superficial temporal artery (STA). As can be seen from the thermal imagery of the human face and neck in Fig. 4.3c, the carotid complex and branches of the STA are readily accessible superficial arteries for the thermal IR-based measurements. However, each ROI illustrated in this figure has its own advantages and disadvantages regarding thermal IR measurements. The carotid ROIs are more predisposed toward deformation artifacts than those in the vicinity of the STA complex. The deformations are mainly due to muscle contraction, swallowing, breathing, the heartbeat, and speaking. Therefore, the approach for measurements on the carotid complex should

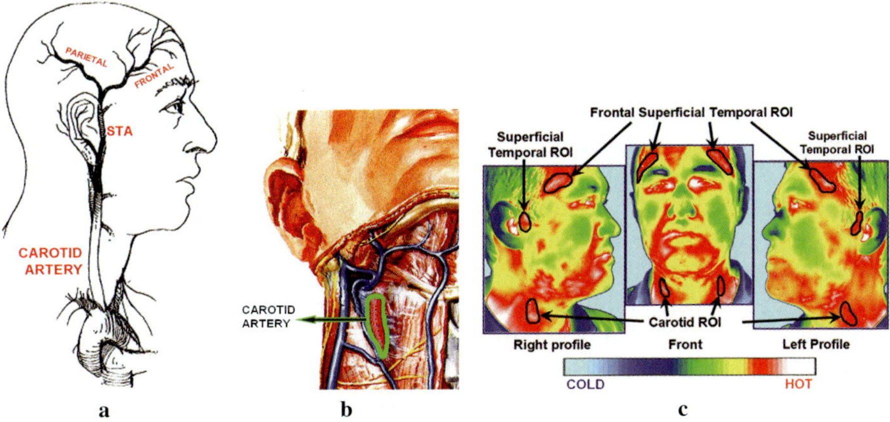

Fig. 4.3 Anatomy and thermal imaging of facial and neck ROIs. (**a**), (**b**) Anatomy of the STA and carotid artery complex (courtesy Exergen Corporation, Watertown, MA, www.exergen.com), **c** thermal imagery of neck and face ROIs suitable for measurement of the arterial pulse

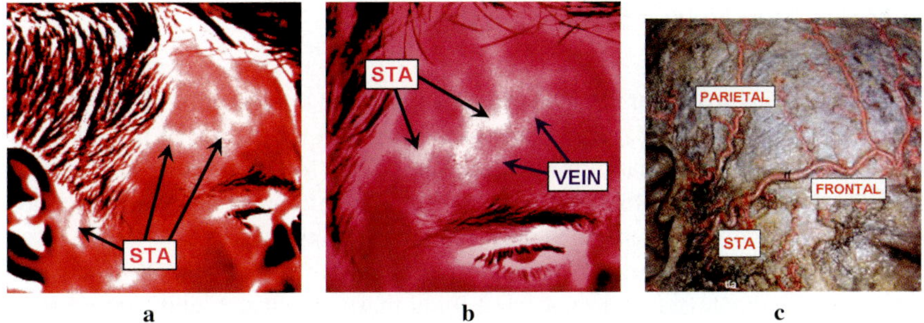

Fig. 4.4 Thermal imaging and anatomy of the STA: (**a**) The locations of the STA where the artery is most superficial and accessible for arterial pulse measurements, (**b**) a closer view of the forehead region in thermal imagery with the frontal branch of the STA and the facial vein clearly observed and seen to be parallel with the vein straighter and less tortuous than the STA, **c** dissection of the STA and its branches (Adapted from [8])

cope not only with rigid movements of the ROI but also with the artifacts deforming the ROI as well. In contrast to the carotid ROIs, the STA ROIs are less susceptible to deformation due to anatomy of the regions. The anatomy of both the carotid artery complex and the STA is illustrated in Fig. 4.3a, b. The arterial pulsation within the carotid complex is significantly stronger than that in the STA branches. This logically may result in a stronger and richer thermal IR signal measured within this ROI. However, factors such as deformation artifacts and obesity may impose a real challenge for practical measurements.

The STA can be thought of as the narrowing continuation of the external carotid artery, which becomes quite superficial after it passes the ear area marked on the Fig. 4.4a as the STA. The STA ROIs are ideally suitable for thermal-based arterial pulse measurements; they are easily accessible, contain no mucous membranes, and have a negligible amount of fat and muscle tissues. In the upper head, the STA splits into parietal and frontal branches as shown in the Fig. 4.4c. For a comprehensive anatomical study of the STA, see [8]. In that study, the STA and its branches were dissected in 27 specimens. High variability in topology of the STA was observed. However, the absence of the frontal branch of the STA was not encountered. In only one sample was the STA's parietal branch missing and the STA continued as a frontal branch. The diameter of the frontal branch was bigger than those of the parietal branch in 15 samples. The diameters of both the frontal and parietal branches were equal in four samples. In the remaining eight samples, the diameter of the parietal branch was bigger than that of the frontal one.

Typically, the STA and facial veins produce prominent thermal imprints in thermal images, as shown in Fig. 4.4a, b. The facial veins (as other veins of the body) are parallel to arteries. In contrast to the arteries, however, the veins typically tend to be straighter and less tortuous. This is illustrated in Fig. 4.4b.

4.5 Multiscale Image Decomposition

One of the central ideas behind the measurements of the arterial pulse in our methodology is the decomposition and analysis of thermal images at a set of fine-to-coarse scales. The goal of multiscale image decomposition (MSD) is to identify the decomposition subbands at which the pulse propagation effect is the most pronounced and the influence of irrelevant noisy heat patterns is minimal.

Figure 4.5 illustrates the raw thermal images of the frontal branch of the STA. The images are displayed as three-dimensional thermal maps. X and Y are the spatial axes, and Z is the axis corresponding to the normalized temperature. From this figure, it can be seen that the heavy noise is comparable to approximately one fourth of the direct current (DC) thermal difference between the artery and the surrounding tissue in our data acquisition system. According to our experiments, the direct measurement of the arterial pulse from the raw thermal data is impossible, at least in

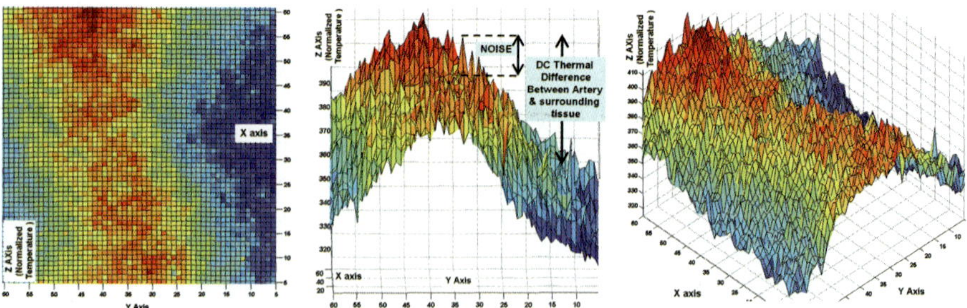

Fig. 4.5 The issue of the noise artifact in thermal IR imaging of superficial arteries; three different views of the raw thermal data captured from the frontal branch of the STA

our laboratory settings. These results suggest that the wavelet model for MSD and analysis would be useful as a preliminary step for the arterial pulse recovery. The rest of this section introduces the basics of multiresolution theory that we apply in our approach for thermal image analysis.

4.5.1 Multiresolution Analysis

In our MSD procedure, we apply the multiresolution analysis (MRA) framework introduced by Mallat [9]. This is an elegant signal analysis tool by which the scaling functions create a series of approximations differing by a factor of 2 from the nearest neighboring approximation. The wavelet functions then encode the difference in information between adjacent approximations. The scaling functions represent a set of expansion functions composed of integer translations and binary scalings of real, square-integrable functions $\varphi_{j,k}(x) = 2^{j/2}\varphi(2^j x - k)$, which meet the fundamental MRA requirements formulated as follows

1. The scaling function is orthogonal to its integer translates.
2. The subspaces spanned by the scaling function at low scales are nested within those spanned at higher scales: $V_{-\infty} \subset \cdots \subset V_{-1} \subset V_0 \subset V_1 \subset \cdots \subset V_{\infty}$, where $V_j = \overline{Span_k(\varphi_{j,k}(x))}$.
3. The only function that is common to all V_j is $f(x) = 0$. This is the function of no information: $V_{j=-\infty} = 0$.
4. Any function can be represented with arbitrary precision.

The wavelet functions $\psi_{j,k}(x) = 2^{j/2}\psi(2^j x - k)$ are then defined so that their integer translations and binary scalings span the difference between any two adjacent scaling subspaces, such as $V_{j+1} = V_j \oplus W_j$, where $W_j = \overline{Span_k(\psi_{j,k}(x))}$. This is graphically illustrated in Fig. 4.6a.

For computation of the discrete wavelet transformation (DWT) for M-sampled one-dimensional (1D) signal $f(m)$:

$$V_\varphi(j_0,k) = \frac{1}{\sqrt{M}}\sum_m f(m)\varphi_{j_0,k}(m), \quad W_\psi(j,k) = \frac{1}{\sqrt{M}}\sum_m f(m)\psi_{j,k}(m), j > j_0 \quad (4.2)$$

is not as computationally efficient as its fast implementation version, the fast wavelet transformation (FWT) proposed by Mallat [9]. For the 1D case, the FWT algorithm is summarized in Fig. 4.6b. As shown in the figure, the subsequent approximations and detail coefficients $V_\varphi(j,k)$ and $W_\psi(j,k)$ at scale j are computed by convolving the approximation coefficients $V_\varphi(j+1,k)$ at scale $j+1$ with wavelet and scaling filters $h_\psi(n)$ and $h_\varphi(n)$:

$$V_\varphi(j,k) = h_\varphi(n) * V_\varphi(j+1,n)\,|_{n=2k}, \quad W_\psi(j,k) = h_\psi(n) * W_\psi(j+1,n)\,|_{n=2k} \quad (4.3)$$

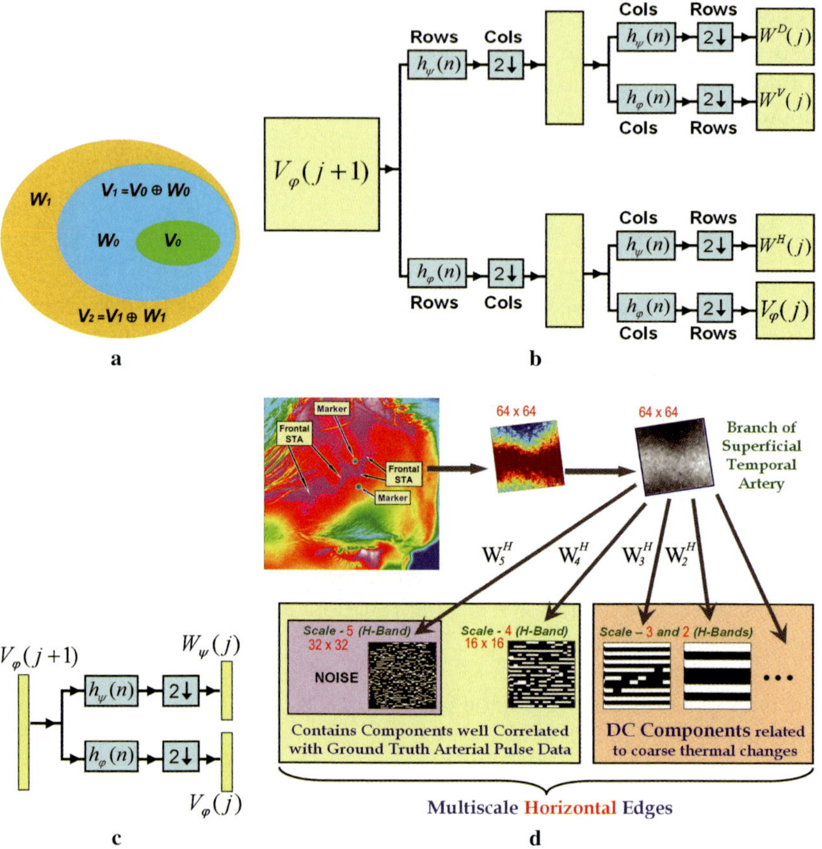

Fig. 4.6 MSD and image analysis: **a** nested scaling and wavelet function spaces, **b** 1D FWT diagram, **c** 2D FWT diagram, **d** MSD for arterial pulse measurements

Splitting the information available at scale $j+1$ into approximation and detailed coefficients at scale j requires downsampling by a factor of 2 to avoid redundancy. This is achieved by evaluation of the convolutions at instants $n = 2k$. The $h_\varphi(n)$ and $h_\psi(n)$ can be considered as the low- and high-pass orthogonal filters splitting the input signal content into coarse and fine wavelet coefficients at multiple scales. The decomposition scales include only the additional details that are not already available at preceding levels.

The 1D case is generalized to the two-dimensional (2D) case for image analysis by constructing a 2D scaling function $\varphi(x,y) = \varphi(x)\varphi(y)$ and three 2D wavelet functions, $\psi^H(x,y) = \psi(x)\varphi(y)$, $\psi^V = \varphi(x)\psi(y)$, and $\psi^D = \psi(x)\psi(y)$ obtained by the product of 1D scaling and wavelet functions. The 2D wavelet functions measure the directional variations of image intensity: $\psi^H(x,y)$ captures the variations along columns (*horizontal edges*), $\psi^V(x,y)$ captures the variations along rows (*vertical edges*), and $\psi^D(x,y)$ corresponds to the diagonal variations. The 2D FWT scheme is illustrated in Fig. 4.6c. The image approximation $V_\varphi(j+1)$ at scale $j+1$ is decomposed into three wavelet bands $W_\psi(j)$ and the image approximation $V_\varphi(j)$ at scale j. This can be viewed as a two-step process. The first step is the filtering of each row of the $V_\varphi(j+1)$ with orthogonal low-pass and high-pass filters followed by downsampling along the columns by the factor of two. During the second step, the filtering is applied to the columns and downsampling performed along the rows. At each scale, the image is decomposed into horizontal, vertical, and diagonal details. Each band of wavelet coefficients contains only details that are not already available in other wavelet bands. This idea has been extensively utilized in our approach, for which the intention is to identify the wavelet subspace where the thermal variations due to the pulse propagation are the most detectable over the background of other heat phenomena irrelevant to the process of arterial pulsation. We hypothesize that thermal variations in the direction perpendicular to the direction of the pulse propagation should be most observable. Thus, according to this hypothesis the candidate wavelet subspace should be the one containing the horizontal heat variations along the artery. Figure 4.6d explains the idea. A 64×64 pixel patch of the STA is illustrated in the top of the figure. The patch is rotated to be parallel with the direction of the horizontal edge-sensitive wavelets, which capture the intensity variation along the columns. The multiscale horizontal wavelet bands of the patch are displayed in the bottom of the Fig. 4.6d. According to the experimental results, the horizontal edge variations at $Scale-4$ contain the components that are well correlated with the ground truth oximetry sensor pulse data. The H data at $Scale-5$ illustrated in the figure in just a few cases yield a good match with oximetry sensor measurements. However, in most of the cases this scale is dramatically affected by the high-frequency noise, which makes the measurements produce inaccurate results. Lower scales such as $Scale-3$ and $Scale-2$ relate to a very coarse representation of the STA branch. The variations at these scales reflect very poorly the arterial pulse propagation events and therefore are not of interest.

4.6 Continuous Wavelet Analysis

The classical Fourier transform represents a useful and powerful tool for signal analysis and synthesis. However, it is more applicable to stationary signals, the signals whose characteristics do not change with time. The time-varying properties of non-stationary signals are more observable via the tools capable of representing a signal jointly in both time and frequency domains. Flexible wavelet functions can be easily applied for analysis of signal structures at different scales and time/space locations. In our method, the continuous wavelet analysis (CWA) [10] framework is applied to solve two tasks: automatically detect local maximums with the use of the ROM detection procedure and filter the final arterial waveforms out from the noise, preserving important signal structures with as minimal distortion as possible.

4.6.1 Continuous Wavelet Transformation

The continuous wavelet transformation (CWT) of the function *f(t)* is defined [10] as

$$CWT(a,b) = \frac{1}{\sqrt{C_\psi}} \frac{1}{\sqrt{a}} \int_{-\infty}^{+\infty} \psi^*(\frac{t-b}{a}) f(t) dt \tag{4.4}$$

where

$$C_\psi = \int_{-\infty}^{+\infty} \frac{|\Psi(\omega)|^2}{|\omega|} d\omega < \infty \tag{4.5}$$

is the constant defined by the Fourier transform of the wavelet function $\psi(t)$:

$$\Psi(\omega) = \int_{-\infty}^{+\infty} \psi(t) \exp(-i\omega t) dt \tag{4.6}$$

Condition (4.5) implies that $\Psi(\omega) = 0$ when $\omega = 0$.

The function $\psi(t)$ is called the *mother wavelet*. By shifting in time and dilating or compressing this function in the frequency domain, one obtains a set of self-similar functions

$$\psi_{a,b}(t) = \psi(\frac{t-b}{a}) \tag{4.7}$$

where $a \sim frequency^{-1}$ is the scale that provides dilating or compressing, b is the time shift, and t is "time," which can also be mass, energy, or the like depending on the problem. In contrast to DWT, big scales in CWT relate to the coarse representation of a signal, whereas small scales constitute its fine details.

The self-similar functions $\psi(t)$ have to confirm the following conditions:

$$\int_{-\infty}^{+\infty} \psi(t) dt = 0 \tag{4.8}$$

which implies that $\psi(t)$ are oscillating functions bounded at the origin, and

$$\int_{-\infty}^{+\infty} \psi_k(t) * \psi_l^*(t) dt = \delta_{k,l} \tag{4.9}$$

meaning that the $\psi(t)$ represent an orthonormal basis. The indexes k, l specify particular functions from the orthonormal basis (set of functions ordered by index). If the integrals (4.5) and (4.6) are bounded, then there exists an inverse transform, which gives us a reconstructed signal:

$$f(t) = <f(t)> + \frac{1}{\sqrt{C_\psi}} \frac{1}{a^2} \int_a \int_b CWT(a,b) \psi(\frac{t-b}{a}) db da \tag{4.10}$$

In Eq. (4.10), the average value of a signal needs to be added to the inverse wavelet transform to obtain a reconstructed signal. This is because an average value of any wavelet is zero. The "Mexican hat" (MH) wavelet is defined as

$$\psi(t) = (1 - t^2)\exp^{-t^2/2} \tag{4.11}$$

was applied in our approach. This wavelet function is characterized by its good localization in the spacial domain and small number of oscillations. This property of MH suits it perfectly for the task of isolation of local signatures of the arterial pulse in multiscale noisy thermal data. A family of MH wavelets at four different scales is plotted at the top of Fig. 4.7a. An example of CWT is presented in Fig. 4.7b. A sample signal consisting of four Gaussian functions of different width at different locations is plotted in the Fig. 4.7a:

$$f(t) = G_1(t) + G_2(t) + G_3(t) + G_4(t) \tag{4.12}$$

with means and variances as $(\mu_1 = 120, \sigma_1 = 50)$, $(\mu_2 = 300, \sigma_2 = 25)$, $(\mu_3 = 405, \sigma_3 = 12)$, $(\mu_4 = 460, \sigma_4 = 5)$. The corresponding CWT of this signal is given in Fig. 4.7b. It can be noticed that the CWT maps a 1D signal into a 2D one. The CWT provides a signal representation in both scale and space domains. The magnitude of the CWT at different scale-space locations relates to the amount of correlation between features stored in a signal and a family of analyzing wavelet functions applied to construct the CWT. In Fig. 4.7b, all four Gaussians can be clearly observed in the CWT plane. They can be located along both the space and scale axes. The MH wavelet was applied to compute the CWT, and as a result of this, the Gaussian structures within the CWT can be located with more certainty/resolution along the space axis rather than scale. The narrower Gaussians occupy finer regions of the CWT, while the wider ones reside in coarse domains of the plane. A logarithmic scale was used to plot the structures in CWT plane. This example gives a clear idea of how the detection of structures in both scale and space can be implemented using the CWT. We just need to apply the wavelet that matches the feature of interest.

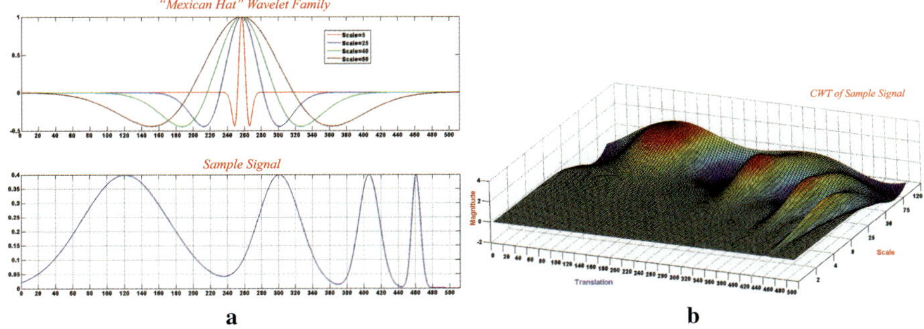

Fig. 4.7 CWT of a sample signal: **a** "Mexican hat wavelet" at different scales on the top and sample signal containing Gaussians of different scale, **b** CWT of sample signal illustrated in **a**

4.6.2 Sample CWA

Sample CWA can be demonstrated on a model signal with known parameters generated similar to the thermal waveforms typically measured from various ROM configurations in the proposed methodology. The top of Fig. 4.8 illustrates the original model signal as a sum of Gaussians, aimed to mimic the arterial pulsation peaks:

$$f(t) = G_1(t) + G_2(t) + G_3(t) + G_4(t) + G_5(t) + G_6(t) + G_7(t) + G_8(t) + G_9(t) + G_{10}(t) \quad (4.13)$$

with variances $\sigma_{i=1...10} = 5$, and means $\mu_1 = 30$, $\mu_2 = 80$, $\mu_3 = 130$, $\mu_4 = 180$, $\mu_5 = 230$, $\mu_6 = 280$, $\mu_7 = 330$, $\mu_8 = 380$, $\mu_9 = 430$, $\mu_{10} = 480$. The Gaussian noise ($\mu_{noise} = 0$ and $\sigma_{noise} = 0.4$) shown in the middle of Fig. 4.8 was added to the original

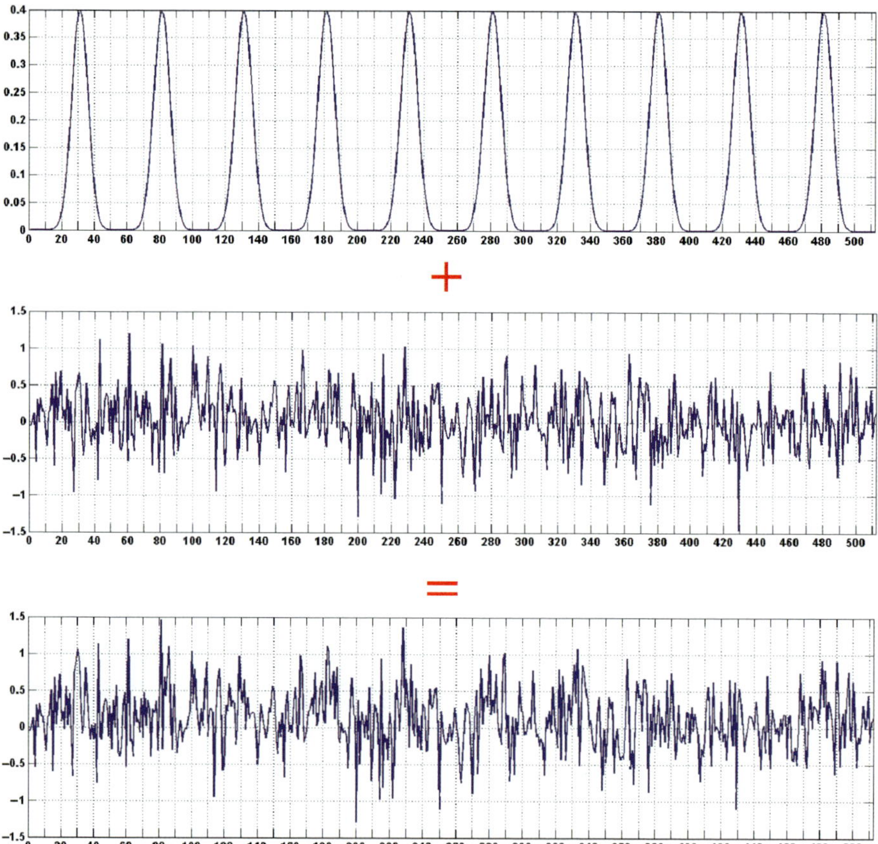

Fig. 4.8 Modeling the arterial pulse waveform: (*Top*) model signal created by a sum of Gaussian functions, (*middle*) the Gaussian noise to be added to the model signal illustrated on the top, (*bottom*) the sum of the noise and model signal waveforms

model signal, and the resultant noisy waveform is displayed in the bottom of the same figure. The Gaussian peaks cannot be observed any more. They are hidden by the severe noise pattern. This resultant noisy waveform is similar to the raw thermal patterns measured in our approach. Recall that when our methodology is applied to real data, the goal of the CWA is to detect the arterial pulse features from noisy data. The periodicity detection (PD) algorithm presented in this section compares the periodicity of CWA features from various waveforms measured at different locations of ROI and based on this select the most faithful data sample/ROM configuration. The CWA filtering procedure is next applied to this final data sample to filter the noise and reconstruct the arterial pulse waveform with as minimum corruption of arterial structures as possible. The purpose of this example is to show the capability of CWA to detect the arterial pulse features (applied in the PD algorithm in the next section) from a severely corrupted waveform and demonstrate the logic behind the reconstruction procedure.

The CWTs of the original model signal and the signal with the noise added are illustrated in the top and middle of Fig. 4.9 respectively. In CWT plane related to the model signal, we can locate all the generated Gaussian peaks in both scale and space domains. Since the MH was applied, the space resolution outperforms the certainty along the scale axis. However, there still is a very prominent band of scales ($a = 10..30$) where the bulk of the energy relevant to the model arterial pulses is concentrated in a mixture with noisy structures. The CWT plane (the middle one in Fig. 4.9) related to the noisy waveform provide a clear representation of the content of this noisy signal. At fine scales of this plane, the very powerful (sharp spikes) noise is clearly observed. Some of the noise patterns continue from fine-scale area toward the coarse region, where the pulses of interest reside. These noise patterns interfere with the features of interest and partially destroy them. However, noticeable periodic structures related to the model signal features still can be observed. The bottom of Fig. 4.9 represents the noisy signal CWT plane after the filtering procedure was applied. The filtering algorithm that we apply removes all the structures that begin at fine scales but degrade at some intermediate scale and have no continuation toward the coarse region. The rationale behind this is to remove noisy patterns while preserving the structures of interest with as little corruption as possible. In contrast to our CWA filtering algorithm, application of a straight thresholding procedure in CWT plane would remove all the structures below a certain thresholding scale related to both the noise and the arterial pulse, leaving some of the noise structures combined with the structures of interest above the selected threshold. The proposed filtering scheme works selectively, removing most of the noise structures and letting features of interest survive. The Fourier-based filtering techniques would be inapplicable in our framework. First, thresholding in the frequency domain would smear the arterial pulse peaks and possibly produce false peaks due to the Gibbs phenomena. This is extremely undesirable in our framework. Another motivation to use the CWA is that the CWT plane does provide a powerful tool for signal analysis. In contrast to the Fourier spectrum, the CWT spacial information can be directly used to detect the periodic structures at different scales and, based on their periodicity

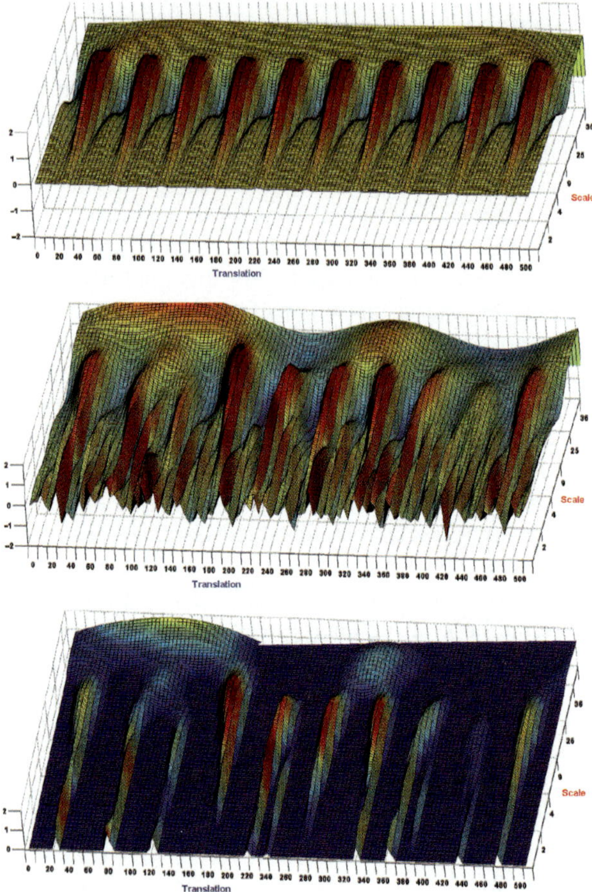

Fig. 4.9 Model CWA: *Top plane* is the CWT of the model signal presented in the Fig. 4.8 (*top*), *middle* is the CWT of the noise signal from Fig. 4.8 (*middle*), *bottom* is the CWT of noisy model signal illustrated in the middle of the figure after the filtering algorithm proposed in this work is applied. Most of the noise structures are removed while a significant part of the model arterial pulse signal is preserved.

measures, to select the ROM. The Fourier analysis cannot provide such a level of flexibility for the signal analysis required for our work.

After the filtering is complete, the inverse CWT is applied to those scales that contain the most periodic structures; as can be seen from the bottom of the Fig. 4.9, this is the region of scales: $a = 10..20$. The resultant reconstructed waveform is illustrated in the middle of Fig. 4.10. The top of this figure illustrates both the original model signal (blue waveform) and the one reconstructed using the CWA framework (red waveform). As we can see, the reconstructed signal matches fairly well with the original model waveform, although it is not an identical match. In the bottom of Fig. 4.10, the time locations of the pulse maximums extracted from the recon-

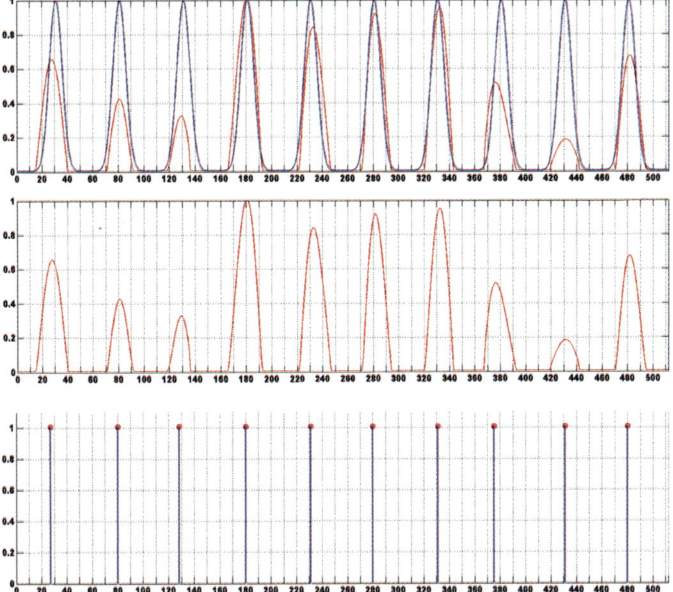

Fig. 4.10 Model signal recovery: *Middle* waveform is the reconstruction of the model signal from the filtered CWT plane illustrated in the bottom of Fig. 4.9. Validation of the recovered waveform with the originally generated model signal is illustrated on the *top*; the blue is the original signal, and the red is the recovered one. Spacial locations of the maximums extracted from the reconstructed waveform are illustrated in the *bottom*. This sequence of peak locations is the input for the PD algorithm

structed waveform are displayed. These time locations are the input for the PD algorithm discussed in the rest of this section.

One of the strong features of the CWA framework presented in this section is that we do not specify the scale region for the reconstruction. The scale band is selected automatically after the CWT filtering procedure is applied. The band of the scales that yield the most periodic locations of the CWT maximums is selected for the reconstruction.

4.6.3 Periodicity Detection (PD Algorithm)

The PD algorithm applied for automatic ROM selection can be summarized as follows: The input for the algorithm is a set of N local maximums (similar to those illustrated in the bottom of Fig. 4.10), $Max[t_1], Max[t_2], \ldots, Max[t_N]$, detected at a certain scale a within the CWT plane. The following steps outline the algorithm:

- Compute the average distance between two maximums: $L = (t_n - t_1)/(N - 1)$
- Set $PM = 0, i = 0$

- Do

$$i = i + 1$$
$$L_i = t_{i+1} - t_i$$
$$PM = PM + |L - L_i|$$

   Until $i = N - 1$
- Output PM

The output of the algorithm is the periodicity measure (PM) aimed to reflect the degree of periodicity of the input data.

The purpose of the PD algorithm is to test different ROM configurations and identify the location for measuring the arterial pulse. This is achieved by comparing the PM computed for waveforms recorded from different ROM configurations. The performance of the algorithm is demonstrated in Fig. 4.11. Three different sample configurations (location, orientation, and size) of ROM are illustrated in this figure. The final most optimal ROM is identified for the ROM configuration presented in Fig. 4.11a. The other two configurations illustrated in Fig. 4.11b, c represent the intermediate ROM configurations, which yield bigger PM and therefore are not selected to be the final location for measurement of the arterial pulse. The ROMs are illustrated as white slits (1D pixel arrays) on the surface of skin covering the STA artery. It should be understood that the actual measurements are performed not on the raw thermal imagery data but on the wavelet subspaces. In Fig. 4.11, the ROMs are illustrated over the raw thermal imagery just for visualization. For each ROM illustrated in Fig. 4.11, the corresponding thermal waveform is computed. The CWT is next applied to these waveforms to detect the most periodic subwaveforms by zooming into different wavelet subspaces of the raw signal. These periodic subwaveforms are processed to identify the locations of maximums that are fed into the PD algorithm. The PD algorithm outputs a PM for each ROM configuration based on the input series of maximums. In particular, the following PMs were computed for each of the ROM specification: PM = 32 for the ROM in Fig. 4.11a, PM = 40 for the ROM in Fig. 4.11b, and PM = 94 for the ROM illustrated in Fig. 4.11c. Based on results presented in Section 4.8, the PM proves to be a powerful measure to automatically identify the optimal location for measurement of the arterial pulse.

4.7 Method

We formulate the problem of measurement of the arterial pulse in the form of registering the so-called thermal delegates of the arterial pulse propagation events along the most accessible branches of the superficial arteries. The goal is to catch those changes in the thermal patterns, the delegates, that are directly linked with a sequence of arterial pulse waves propagating throughout the artery. Our experiments suggest the monitoring of the fine edges along the superficial arteries that capture the radial heat redistributions caused by the arterial pulse propagation. The use of the fine edges as thermal delegates of arterial pulse propagation events and actual recovering of the arterial pulse are discussed in this section.

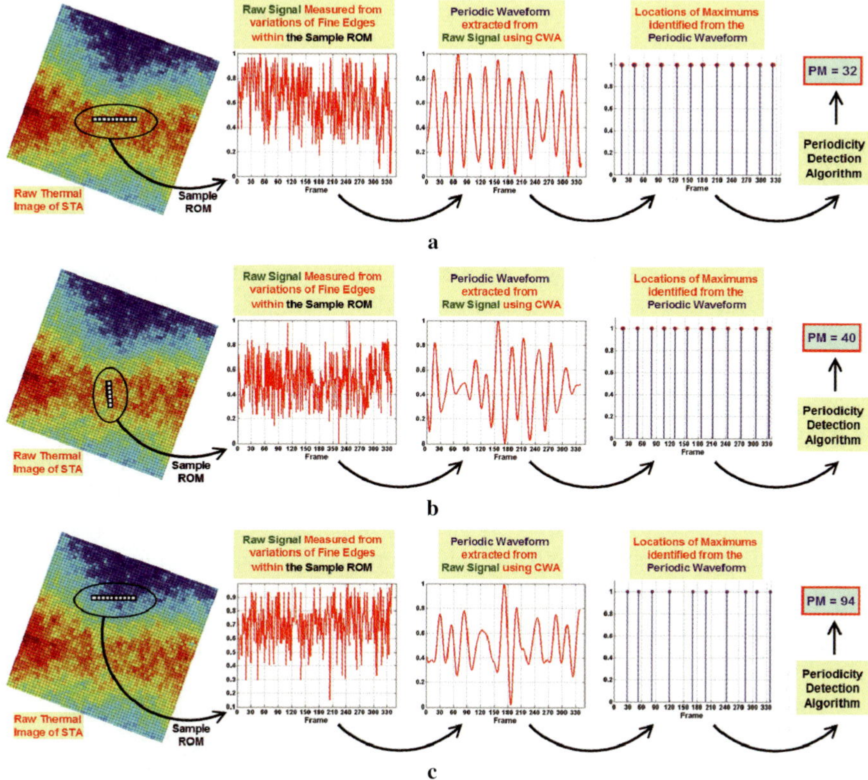

Fig. 4.11 Automatic detection of ROM: **a** sample ROM configuration: case 1, **b** sample ROM configuration: case 2, **c** sample ROM configuration: case 3

4.7.1 Thermal Delegates of the Arterial Pulse

The arterial blood pressure wave is created during ventricular systole and propagates along the arterial tree, pumping the blood across the circulatory system. Propagating along the arterial network, the pressure wave causes the distention of arterial walls, which can be palpated as an arterial pulse from the major superficial arteries of the body. The aortic pulse wave velocity reported in the literature [11] varies around 8.7 m/s. The shape and the speed of the arterial pulse wave changes as it propagates along the arterial tree. This is mainly because of the change in mechanical properties of arterial walls (elastic tapering) and a decrease in the cross-sectional area of the arteries (geometric tapering).

In our model, we simply consider the arterial tree as a network of cylindrical tubes. Our hypothesis is that volumetric response of the arterial wall to a pressure variation causes redistribution of heat along the artery. We further assume that these heat variations are more pronounced in the radial, rather than longitudinal, direction.

This assumption is roughly sketched in Fig. 4.12a. The heat variations are further passed to the skin surface, where they can be registered by a thermal camera. Since the length of the pulse wave is much longer than the length of the ROM, it seems logical to detect the pulse components in some striplike ROM oriented along the artery (Fig. 4.12b). In this case, the radial heat distortions caused by the pulse propagation would be in phase at each of the ROM pixels, and their averaging at each frame would yield the desired result. Unfortunately, this scheme is impractical on raw thermal imagery. Irrelevant heat patterns on the skin destroy or dramatically corrupt thermal patterns produced by arterial pulse propagation. We suggest constructing ROMs in a wavelet representation of the original thermal images. This way it may be feasible to separate different heat phenomena into different wavelet subbands and successfully recover the signal of interest from one of the subbands without overlap by others. The coarse wavelet layers should not be of interest since the periodic arterial pulse waves do not contribute significantly to the overall heat change on the skin. The prominent pulse components with respect to other patterns should reside in the horizontal components of the fine wavelet layers, which can be referenced as fine edges since they encode fine variations of the heat distribution in the direction perpendicular to arterial pulse propagation. This model is quite different from the model proposed previously by Pavlidis et al. [4], who measured the longitudinal temperature variations along the vessel in the raw thermal IR imagery.

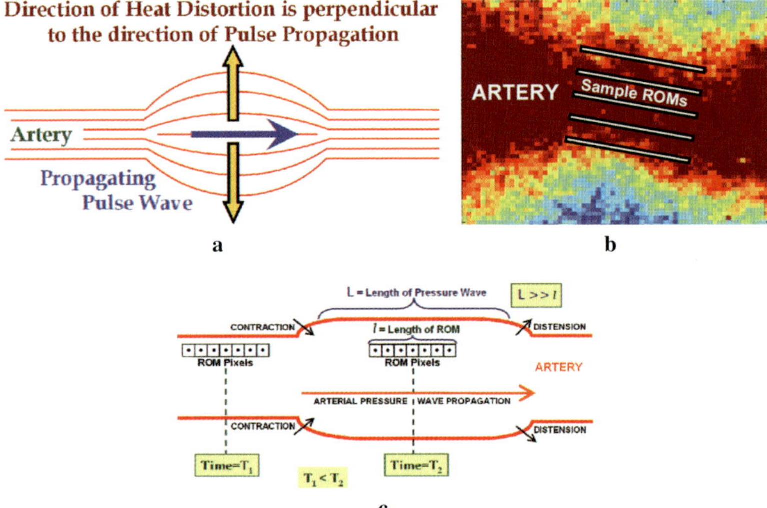

Fig. 4.12 Distortion of heat distribution along the artery caused by the arterial pulse propagation. **a** Rough sketch of difference in thermal map along a linear segment of an arterial tube when it is dilated by the increase in blood pressure and contracted after the pulse wave is moved forward. **b** Thermal image of a segment of the frontal branch of the STA and a sample ROM configuration. **c** Our model for registration of arterial pulse propagation events from the periodic distortions of fine radial components of heat distribution along the artery

4.7.2 Measurement of Arterial Pulse

We use the Mallat [9] wavelet framework and Haar filters to decompose thermal images into disjoint bands encoding horizontal (H), vertical (V), and diagonal (D) edges and residual coarse band. Depending on the image resolution, we choose a certain fine scale at which the data from the H band is used to compute the mean value within the ROM (Fig. 4.12c) for all frames (k) and plot it with respect to time/frame, obtaining 1D arterial pulse waveforms. In particular, the multiscale edges computed at each frame k are as follows:

$$W_j^H(m,n|k) = I(k) * \psi_j^H \tag{4.14}$$

The ψ_j^H is a wavelet function that captures the variations along the columns of a thermal image. $I(k)$ is kth thermal image, and j is a decomposition level. The magnitudes of the horizontal components are averaged within the ROM, producing the *arterial pulse* value $AP(k)$ at each frame k:

$$AP(k) = \frac{1}{ROM} \sum_{(m,n) \in ROM} [W_j^H(m,n|k)^2]^{\frac{1}{2}} \tag{4.15}$$

Figure 4.12c schematically illustrates a sample ROM configuration. By ROM configuration, we mean that all frames are processed with certain $\psi_{j=j_{fixed}}^H$ and certain frame orientation with respect to the $\psi_{j=j_{fixed}}^H$. Other ROM configuration parameters are the ROM's size and its location. In Fig. 4.12c, the ROM pixels are displayed as a 1D array along the longitudinal direction of the artery. Time T_1 corresponds to the relaxed, undilated state of the artery. At time T_2, the artery is distended by the blood pressure increase. This causes the noticeable redistribution of heat in the radial direction. The horizontal edges shift and change the values within the ROM. Longitudinal heat distortion due to the artery distention is not readily observable. However, some other heat processes can vary the longitudinal heat distribution and may corrupt the signal. Properly tuned wavelet scale, frame orientation with respect to wavelet direction, ROM's size, and location yield the waveform with the best PM. During this optimization, the intermediate arterial pulse waveform candidates corresponding to different configuration states are analyzed with the 1D Mexican hat wavelet, known for its excellent spatial resolution. The Mexican hat is used to isolate local fine-to-coarse minima and maxima of the waveforms and to compute their PMs. Modeling the arterial pulse as a quasi-periodic signal within a short period of time makes the automatic detection of the optimal ROM possible when the PD algorithm as discussed in Section 4.6 is applied. The CWA approach as discussed in Section 4.6 is not intended to recover a particular harmonic such as is done in the Fourier transform-based filtering techniques. What is recovered is the structures belonging to a particular scale. Their time behavior is not restricted to a particular spacing rate. This makes our approach quite valuable in dynamic real-life applications for which the momentary changes in the heart rate are of particular importance.

4.8 Experiments and Results

Eight subjects (seven males and one female) with ages ranging from 25 to 45 years old participated in our study consistent with internal review board approval. Four Caucasians, two African Americans, and two Asians of Middle East origin were tested. The experimental setup mainly consisted of a long-wave Phoenix IR camera from FLIR [12], a multiparameter vital signs monitor from Smiths Medical [13], and a personal computer to store and process data. The experimental setup is illustrated in Fig. 4.13. All subjects were seated stationary on a chair 1 to 4 ft away from the camera, zoomed to display their foreheads. The acquisition time for the arterial pulse measurement was 20 to 40 s with a frame rate of 30 fps. The oximetry sensor was attached to record the peripheral pulse waveforms (Fig. 4.13), which served as the ground truth data in this study. The thermal and oximetry data acquisition were synchronized to observe the correlation between the two modalities. The results of the arterial pulse measurements on the frontal branch of the STA for all eight subjects are presented in Fig. 4.14.

The 12-s arterial pulse waveforms measured from the thermal IR data are plotted in red. The corresponding oximetry data are plotted in blue. The frequency content of the waveforms is given underneath each plot. Since the subjects were relaxed and with no apparent heart defects, the heartbeats appeared nearly equally spaced in the time domain with small variations, and the corresponding power spectrum has one dominant frequency component associated with the heart rate. In all eight cases, we obtained an excellent correlation between thermal-based and the oximetry pulse data with a perfect match to heart rate.

In the work presented in this chapter, we do not address the issue of tracking the ROI. The conceptual proof of our model and a solid experimental validation were the primary goal of this work. In the experiments, to cope with the motion artifact, aluminum foil markers were attached to the skin in the field of view of the camera

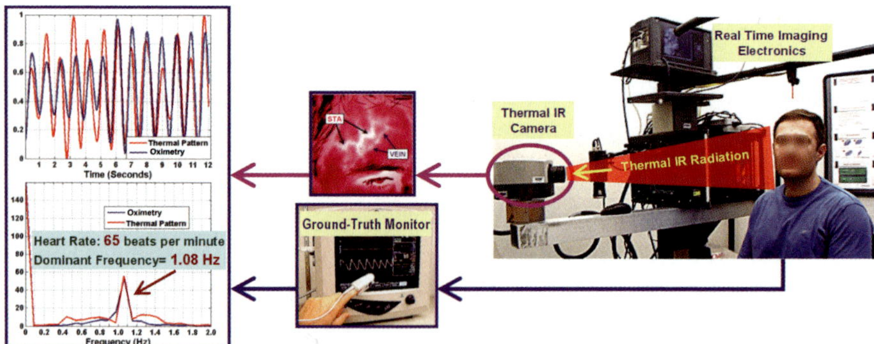

Fig. 4.13 Experimental setup: The thermal sensor used in the experiments for this work was a long-wave Phoenix IR camera from FLIR [13]. The multiparameter contact vital signs monitor Advisor was used for ground truth measurements [14]

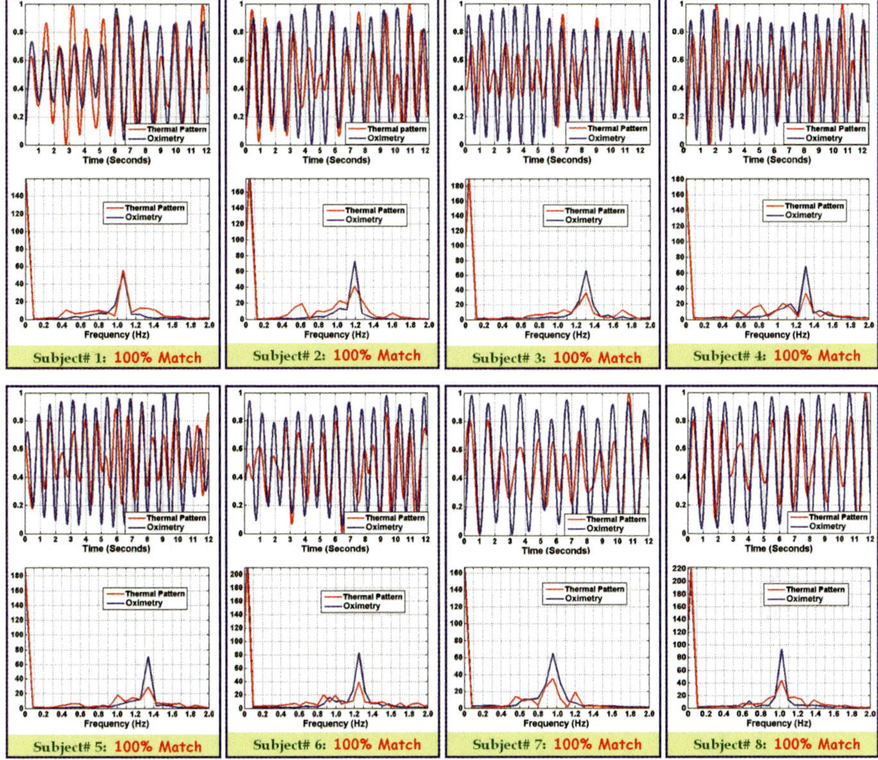

Fig. 4.14 Experimental results: All measurements were performed on the frontal branch of the STA. Eight people of different age, race, and gender have been successfully tested with 100% accuracy in matching the heart rate with that obtained with an oximetry sensor

as shown in Fig. 4.15a. The metal markers look significantly colder in contrast to the surrounding tissue and can be accurately segmented and tracked from frame to frame.

Another important issue is the localization of the STA. In our approach, we refer to this issue as the identification of the optimal ROM configuration as discussed in Section 4.6. The arteries are typically located deeper in the tissue than the veins. In addition, the arteries tend to be more tortuous and may not produce a sharp thermal imprint. Moreover, there can be a case when an artery hides in the venous network, as illustrated in Fig. 4.15c. In this case, it is difficult to manually distinguish between the artery and the vein and to make an appropriate ROM selection. In Fig. 4.15b, there are sample video and thermal images of the forehead region where the venous network produces a sharper and brighter imprint than the STA. A segment of STA goes in parallel with a vein. This segment cannot be seen on video and could be barely seen in the thermal image. In our approach, the region close to the prominent superficial vessel (which could be a vein or an artery) is automatically scanned with

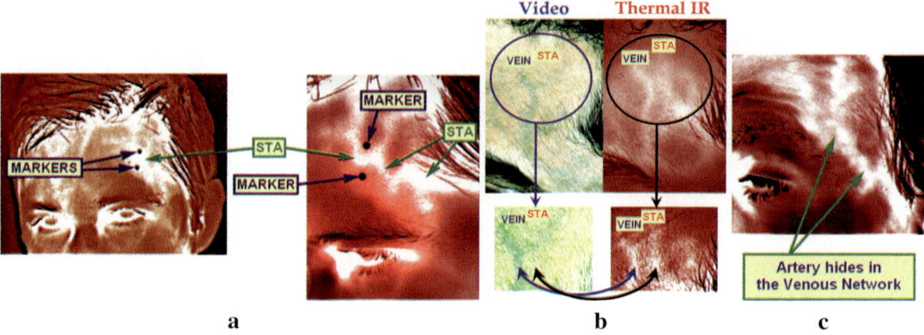

Fig. 4.15 Tracking and STA localization in thermal IR imagery: **a** Aluminum foil markers were used to eliminate most of the motion artifacts. **b** Video and corresponding thermal image where a vein and the STA go in parallel. **c** A thermal image of the forehead where the STA hides in the venous network

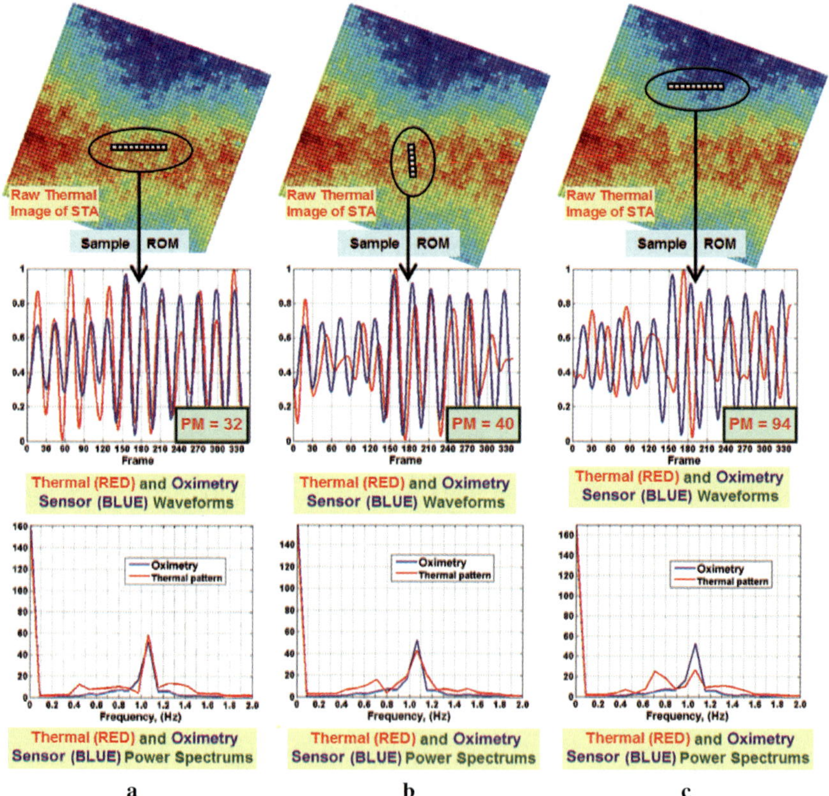

Fig. 4.16 PM as a reliable tool for automatic detection of ROM: The arterial pulse waveform with the smallest PM has the best correlation with oximetry sensor readings. **a** Sample ROM configuration: case 1, **b** sample ROM configuration: case 2, **c** sample ROM configuration: case 3

different configurations of the ROM (such as scale, orientation, size, and location), and the optimal configuration yields the location where the arterial pulse is the most pronounced.

Figure 4.16 demonstrates that the PM is a quite powerful measure to automatically identify the optimal location for measurement of the arterial pulse. Three different configurations (location, orientation, and size) of ROM and corresponding measurements of arterial pulse matched against the oximetry sensor readings are illustrated in these figures. Figure 4.16a shows an excellent match of thermal-based measurement with the oximetry data in both spatial and frequency domains. Measurement from ROM in the Figure 4.16b contains more irrelevant thermal patterns to the arterial pulse and as a result leads to a more corrupted signal of interest. This makes the reconstructed waveform more disturbed and more different from the oximetry signal (several pulses are time shifted). The PD algorithm applied to this waveform yields a bigger PM value than that for the waveform within the ROM in Fig. 4.15a, and therefore this ROM configuration is not selected for the measurement of the arterial pulse. The reconstructed waveform for the ROM in Fig. 4.16c does not match the ground truth oximetry sensor signal in the time domain; however, in the frequency domain we still can observe an excellent match for one of the prominent frequency's peaks related to the heartbeat rate harmonic. The other prominent peak is irrelevant to the arterial pulse. In this case, it is impossible to correctly estimate heart rate from such ROM configuration. This ROM (Fig. 4.16c) is located off the artery, and therefore the heat waves produced by pulse propagation are significantly corrupted and time shifted before they reach this region. The corresponding reconstructed waveform from this region is unacceptably damaged. From a comparison of reconstructed signal with the oximetry sensor waveform in the same Fig. 4.16c, it can be seen that several beats are even missing. Matching in the frequency domain still provided a good match for one of the prominent frequency's peaks with actual heart rate frequency maxima. However, in addition there is a lower-frequency artifact peak that relates to some other irrelevant heat process on the skin. The PM computed for this ROM is much higher than that in the previous two cases: 94 versus 32 and 40.

The results presented in this chapter were obtained from the frontal branch of the STA. As an alternative, the branch of the STA before it splits into parietal and frontal branches, located close to the ear, can also be used for arterial pulse measurements. In fact, according to [8], this segment of the STA is bigger in diameter than both of its branches and is accessible by the thermal IR camera as well.

4.9 Conclusions

A novel model was presented for measuring the arterial pulse from the superficial arteries of the body using passive thermal IR sensors. The measurements are performed in a noncontact and nonintrusive way. The proposed method assumes that the arterial pulse propagation causes measurable periodic distortions in the fine components of heat distribution along a superficial artery in a radial direction perpendicular to the direction of pulse propagation. Our model is quite different from previous models and is supported by strong experimental validation. Eight people of different age, race, and gender were successfully tested with 100% accuracy in matching the heart rate with that obtained with an oximetry sensor.

The strength of our model is the ability to measure interbeat intervals, not just estimates of the heart rate frequency. Accurate monitoring of heart rate and heart rate variability in both the time and frequency domains is essential for measuring cardiovascular and autonomic regulatory functions and mental and emotional loads and for assessment of health in general. Although, there several limitations, such as tracking and accurate peak localization, remain, we believe that our novel approach will form a basis for future advancements in the area of thermal IR-based measurements of human vital signs that will find extensive application in the fields where noncontact and nonintrusive monitoring of heart rate and its variability is needed.

Chapter's References

1. M. Sharpe, J. Seals, A. MacDonald, S. Crowgey, Non-contact vital signs monitor, U.S. Patent 4958638, http://www.freepatentsonline.com/4958638.html
2. A. Lynn, L. John, W. Ohley, Non-contact waveform monitor, U.S. Patent 7128714, http://www.freepatentsonline.com/7128714.html
3. N. Sun, M. Garbey, A. Merla, and I. Pavlidis, Estimation of blood flow speed and vessel location from thermal video, *IEEE Computer Society Conference on Computer Vision and Pattern Recognition*, Vol. 1, Washington, DC, June 27–July 2, 2004, pp. 356–363.
4. N. Sun, M. Garbey, A. Merla, and I. Pavlidis, Imaging the cardiovascular pulse, *IEEE Computer Society Conference on Computer Vision and Pattern Recognition*, Vol. 2, June 2005, 20–25, pp. 416–421
5. R. Serway, *Physics for Scientists and Engineers,* 3rd ed., Saunders, 2003
6. L. John, *The Arterial Circulation. Physical Principles and Clinical Applications,* Humana Press, Totowa, NJ, 2000
7. W. Ganong, *Review of Medical Physiology*, Appleton and Lange, East Norwalk, CT, 1991
8. Y. Atamaz and F. Govsa, Anatomy of the superficial temporal artery and its branches: its importance for surgery, in *Surg. Radiol. Anat.*, 28(3): 248–253, 2006
9. S. Mallat, A theory for multiresolution signal decomposition: the wavelet representation, *IEEE Trans. Pattern Anal. Machine Intell.*, 11(7): 674–693, 1989
10. S. Mallat, *A Wavelet Tour of Signal Processing*, 2nd ed., Ecole Polytechniue, Paris, Courant Institute, New York University, 2001
11. R.N. Mackey, K. Sutton-Tyrrell, P. Vaitkevicius, P. Sakkinen, M. Lyles, H. Spurgeon, E. Lakatta, and L. Kuller, Correlates of aortic stiffness in elderly individuals: a subgroup of the Cardiovascular Health Study, *Am. J. Hypertens.*, 15:16–23, 2002
12. Indigo Systems, http://www.indigosystems.com
13. Smiths Medical, http://www.smiths-medical.com/catalog/multi-parameter-monitors/advisor/advisor-vital-signs-monitor.html
14. L. Geddes, *Cardiovascular Devices and Their Applications,* Wiley-Interscience, 1984
15. T. Togawa, Non-contact skin emissivity: measurement from reflectance using step change in ambient radiation temperature, *Clin. Phys. Physiol. Meas.*, 10: 39–48, 1989
16. R. Gonzalez, R. Woods, *Digital Image processing,* 2nd ed., Pearson, Delhi, 2001

Chapter 5
Coalitional Tracker for Deception Detection in Thermal Imagery

Jonathan Dowdall, Ioannis Pavlidis, and Panagiotis Tsiamyrtzis

Abstract We propose a novel tracking method that uses a network of independent particle filter trackers whose interactions are modeled using coalitional game theory. Our tracking method is general; it maintains pixel-level accuracy, and can negotiate surface deformations and occlusions. We tested our method in a substantial video set featuring nontrivial motion from over 40 objects in both the infrared and visual spectra. The coalitional tracker demonstrated fault-tolerant behavior that far exceeds the performance of single-particle filter trackers. Our method represents a shift from the typical tracking paradigms and may find application in demanding imaging problems across the electromagnetic spectrum.

Keywords: Tracking · Particle filter · Coalitional game theory · Thermal imaging

5.1 Introduction

The extraction of high-level information from video through the use of computer vision algorithms has become increasingly important over the past decade. A diverse array of applications use this technology, including quality control in the manufacturing sector [1, 2], surveillance in the security industry [3, 4], biomedical measurements for health care [5–7], and behavioral analysis [8–10]. Of key importance to all these computer vision applications is the ability to detect and track objects in their respective input video streams. The problem of tracking can be cast as guessing how things change over time. Specifically, tracking involves modeling how the parameters of the object modulate in successive input frames by using prior knowledge. When this is done accurately, it can be useful in a number of applications for which knowing the current state of a given target object is important. An intriguing line of computer vision research focuses on measurements of physiological signals on facial tissue. The measurements are performed on infrared imagery and are used

R.I. Hammoud (ed.), *Augmented Vision Perception in Infrared: Algorithms and Applied Systems,* Advances in Pattern Recognition, DOI 10.1007/978-1-84800-277-7_5,

in biomedical [5–7] and behavioral applications [8–10]. Although a large body of work has been devoted to facial tracking research [11–13], we found the existing methods insufficient to achieve the high degree of accuracy required in imaging measurements of facial tissue. This was our initial motivation for exploring a novel tracking paradigm.

5.1.1 Prior Work

Computer vision tracking has been dominated by sequential Monte Carlo methods (particle filtering) [14] for the last several years. Among the most popular particle filter tracking methods is the CONDENSATION algorithm, which was introduced by Isard et al. circa 1998 [15–17].

An interesting tracking methodology based on deformable templates was also developed in parallel. Typical deformable templates focus on tracking object contours, not surfaces [18]. Therefore, they cannot adequately address out-of-plane tracking, like the case of left-right facial rotation.

Alternative tracking methodologies employ specific models of the target to provide better accuracy [19–21]. Unfortunately, this increased accuracy comes at the expense of speed and generality. A noteworthy modeling approach is known as active appearance modeling, and it takes into account both shape and texture [22, 23]. For example, Dornaika et al. [23] first recovered the three-dimension (3D) head pose using a deformable wire frame and then local motion associated with some facial features using active appearance model search. Such 3D active appearance models can potentially perform quality tracking in demanding facial-imaging applications in the visual spectrum. However, their performance may break in thermal infrared imagery due to thermal diffusion and the resulting fuzzy image edges. In such an environment, appearance models may have hard time maintaining 3D-2D (two-dimensional) correspondences, which are partly based on thermal gradients.

Tracking in the thermal infrared spectrum is of particular interest to us because recent research demonstrated that many vital signs, including blood flow [5], pulse [6], and breathing function [7], can be measured in this modality. The success of these measurements depends strongly on a reliable tracking method to register the motion of facial tissue.

Our method aims to achieve what sophisticated modeling methods reportedly achieve, but it is more general and robust. It does not employ a single explicit 3D model but many generic and cooperating 2D particle filter trackers, which are spatially distributed over the target's surface. Our effort can be seen as a first step toward developing a tracking methodology that is able to accurately track a wide array of targets across imaging modalities.

There has been some other work on multiple trackers that work together to follow *multiple objects* [24–27]. In contrast, we employ multiple trackers to track a *single object*.

5.2 Tracking Methodology

Our goal is to develop a general tracking methodology that can accurately monitor the motion of the target's surface even in the presence of deformation or partial occlusion. Many existing general tracking methods monitor the target's outline (not surface). This is a different and far easier problem.

We arrived at a fault-tolerant surface-tracking method that works on both infrared and visual video without resorting to explicit modeling. It uses a network of particle filter trackers that influence each other (see Fig. 5.1). Each individual tracker is unreliable at times, but the combination of many neighboring trackers produces robust performance. The intertracker influence is modeled as a coalitional game in which each tracker is a player, and the goal of the game is to propagate one's influence in subsequent frames of video. Within this framework, the winning coalition of trackers is used to calculate the state of the tracked object.

5.2.1 Tracking Network

We use a network of trackers to achieve accurate surface tracking and fault tolerance. Tracking is maintained even if all but one of the trackers fail in the tracking network. The trackers are each assigned a different portion of the target's surface to track (see Fig. 5.2). By default, the trackers are configured in a regular grid, although alternative configurations are possible through a feature selection mechanism. Intertracker communication allows trackers that are correctly tracking the target to

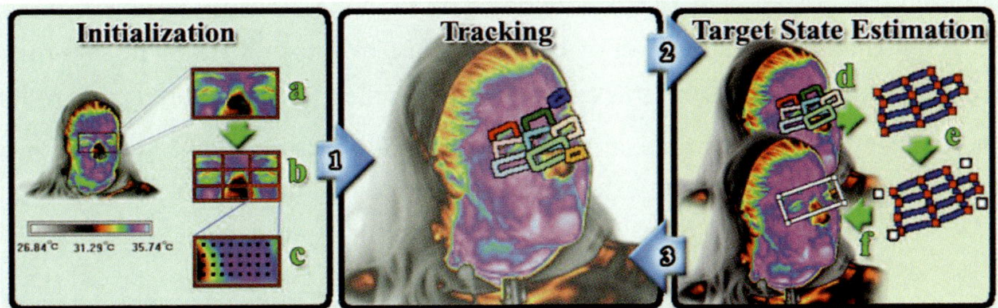

Fig. 5.1 Overview of the tracking method. Initialization consists of the following steps: **a** extraction of the user-selected region of interest from the input video, **b** subdivision of the Region of interest into the tracking network, and **c** individual tracker template creation. Tracking: the individual trackers in the tracking network follow their targets. Target state estimation consists of the following steps: **d** the winning coalition is produced, **e** the deformation mesh is calculated from the winning coalition, and **f** the deformation mesh is used to calculate the target state. The method proceeds from initialization to tracking (*arrow 1*) to target state estimation (*arrow 2*) and back to tracking (*arrow 3*). In the latter transition, the winning coalition is passed back to the tracking stage to distribute the intertracker influence.

Fig. 5.2 Example of a 3×3 tracking network on a visual image. Each tracker in the network is shown in a different color. Each tracker is tracking a separate part of the target

"tip-off" other trackers that have become lost regarding the true location of the target's surface. This intertracker influence is realized within a statistical framework and is managed by the coalitional game model described in Section 5.2.2.

The idea arose naturally in the effort to address the problem of facial tissue tracking in the infrared. As the subject's head moves (e.g., left and right), part of the facial surface is occluded at times. Trackers that correspond to the occluded part of the face are aided by trackers that correspond to the exposed part.

In our implementation, each tracker in the tracking network is an individual particle filter tracker. We denote the state of each individual tracker i at time t by $x_i^{(t)}$ and its associated image observation by $z_i^{(t)}$. The target tracker's prior will be formed using intrasamples $s_{(i,i)}^{(t)}$ from tracker i and intersamples $s_{(i,j)}^{(t)}$ that correspond to the (intertracker) influence of tracker i from tracker j. The intersamples $s_{(i,j)}^{(t)}$ are generated based on the initial relationship between the trackers involved in the exchange:

$$s_{(i,j)}^{(t)} = T_{x_i^{(0)}}^{x_j^{(0)}} x_j^{(t)} \tag{5.1}$$

where $s_{(i,j)}^{(t)}$ is the intersample generated by tracker j for tracker i, and $T_{x_i^{(0)}}^{x_j^{(0)}}$ is the transformation that gives a sample for tracker i given a state for target j at time t. The transformation $T_{x_i^{(0)}}^{x_j^{(0)}}$ is computed during initialization for every possible tracker pair $\left(x_j^{(0)}, x_i^{(0)} \right)$.

5.2.2 The Coalitional Game

The tracking network is a versatile architecture for tracking objects, but it does not have any intrinsic method to generate the final target state or to manage tracker interaction. The simplest solution would be to allow every tracker to influence all of the other trackers. Unfortunately, this is not an optimal solution because trackers that have lost their target would be allowed to influence other trackers in the network that have not gone awry. This also highlights the problem of determining which trackers in the network are correctly tracking their targets and which ones have strayed away. What is needed is a mechanism that can determine the validity of each of the trackers, compute the target's state vector based on the valid trackers, and finally propagate the influence of the valid trackers to keep the network correctly tracking the target surface.

There are many optimization algorithms one can use to manage the network of trackers. We chose to optimize tracker interaction using a game theoretic solution for two main reasons: It naturally fits the problem space, and it is relatively simple. Game theory [28–31] has been successfully used to analyze topics ranging from simple deterministic games, to complex economic models [32, 33], and even to international negotiations [34, 35]. Our adaptation was to view the trackers as players in a cooperative game [36, 37] in which the objective was to increase their influence by forming coalitions with other trackers. The winning coalition would then be used to compute the state vector of the target and subsequently propagate its influence onto the entire tracking network.

Specifically, the members $m_j^{(t)}$ of the winning coalition C^t influence every other tracker i in the tracking network by adding intersamples $s_{(i,j)}^{(t)}$. Trackers that are not members of the winning coalition cannot propagate any influence at all. The intuitive affinity of the problem space to cooperative gaming is apparent in the example of facial tissue tracking. There, the winning coalition is composed mostly of trackers that correspond to the exposed part of the face. These are trackers that feature high-quality information and give a "helping hand" (influence) to the "clueless" trackers that correspond to the occluded or deformed part of the face.

The coalitional form of an $N-$tracker game is given by the pair (Ω, Π), where $\Omega = \{1, 2, \ldots, N\}$ is the set of trackers and Π is a real-valued function, called the *characteristic function* of the game, defined on the set of all coalitions (subsets of Ω), which has cardinality 2^N and satisfying $\Pi(\emptyset) = 0$ [28]. In other words, the empty set has value zero. The size of a coalition C will be denoted from now on by k, where $k \in \{1, 2, \ldots, N\}$, and there are $\binom{N}{k}$ coalitions of size k. The quantity $\Pi(C_K)$ may be considered as the value, or worth, or power, of coalition $C_k \subset \Omega$ when its members act together as a unit.

The definition of a coalitional game is quite general and leaves the specification of the characteristic function to the game designer. We designed a characteristic function for the tracking game that encompasses four scores. These scores are calculated from the trackers participating in the coalition under consideration at time t:

- template match $\alpha^{(t)}$
- geometric alignment $\beta^{(t)}$
- interframe projection agreement $\gamma^{(t)}$
- interframe membership retention $\delta^{(t)}$

The characteristic scores support the fact that quality tracking is characterized by consistency in the content and geometric configuration of the individual trackers. Specifically, the template match score rewards trackers that maintain consistent imaging content. The geometric alignment score favors coalitions whose members have geometric alignment analogous to the original ($t = 0$) configuration. The interframe projection agreement score is a continuity constraint. It improves robustness by penalizing abrupt (and improbable) changes of the projected state of the target between successive frames. The interframe membership retention score is also a continuity constraint. It reflects the tendency of the winning coalition from the previous time step to retain its members.

The template match score $\alpha_{C_k}^{(t)}$ for a coalition C_k of size k at time t is given by

$$\alpha_{C_k}^{(t)} = \frac{1}{k} \sum_{i=1}^{k} \alpha_{m_i}^{(t)} \tag{5.2}$$

where $\alpha_{m_i}^{(t)}$ refers to the template match score (a number in $[0,1]$) of member m_i in the coalition C_k at time t.

For the second and third scores, we first need to define the function that measures the geometric alignment between two target projections (see Fig. 5.3), as are computed from samples s_i and s_j:

$$G\left(\mathcal{F}(s_i), \mathcal{F}(s_j)\right) = G(S_i, S_j) = \omega \times \left[1 - \frac{\sqrt{(S_{ix} - S_{jx})^2 + (S_{iy} - S_{jy})^2}}{M_d} \right]$$

$$+ (1 - \omega) \times \left[1 - \frac{|S_{i\theta} - S_{j\theta}|}{M_\theta} \right] \tag{5.3}$$

where $\mathcal{F}(s)$ is a function that transforms the tracker sample s into its corresponding target projection S, (S_{ix}, S_{iy}) are the (x, y) coordinates of the center of target projection S_i, while $S_{i\theta}$ is the angle of rotation about the center of target projection S_i. M_d is set to the maximum movement allowed by the target in a single frame, while M_θ is the (positive) maximum rotation allowed by the target in a single frame. The weight ω appropriately penalizes the center and angle discrepancies. Ideally, the target projections in Fig. 5.3 c should have coincided (perfect alignment), so that the combined projection of the two tracker samples is reminiscent of the original target shape. Note that the upper bound for $G(.,.)$ is 1 (when the two target projections are identical), but the lower bound is not necessarily 0. This would have been the case if we chose M_d and M_θ to be the maximum observed values at time t, but this would

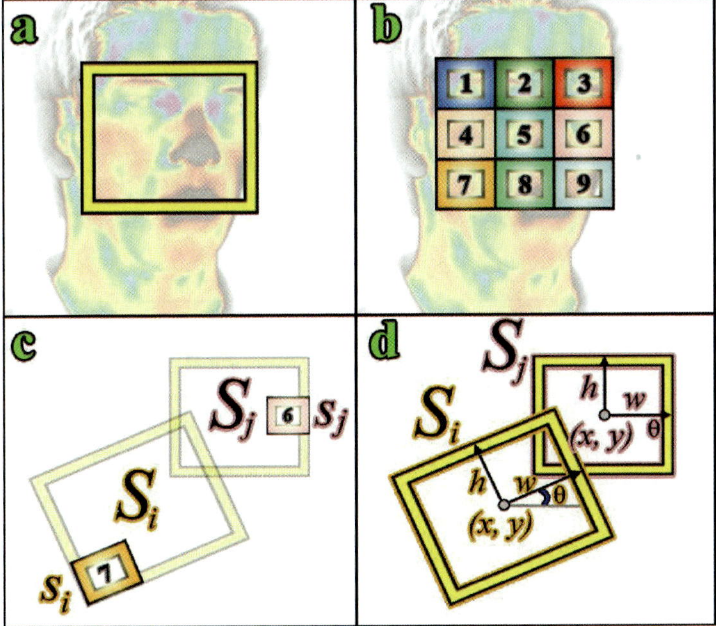

Fig. 5.3 Geometric alignment of tracker's target projections: **a** target projection at $t = 0$, **b** tracker network overlaid on the initial target projection, **c** trackers 6 and 7 at a subsequent time along with their corresponding target projections, **d** parameterization of target projections to facilitate measurement of geometric alignment

have slowed the computation. Besides, we do not mind giving negative scores to some tracker pairs (i.e., penalizing as opposed to rewarding them) whose geometric alignment is very bad.

Having defined the geometric alignment function for a pair of samples [see Eq. (5.3)], we use it to compute the geometric alignment score $\beta_{C_k}^{(t)}$ of a coalition of size k:

$$\beta_{C_k}^{(t)} = \frac{f(k)}{\binom{k}{2}} \sum_{i=1}^{k-1} \sum_{j=i+1}^{k} G\left(S_i^{(t)}, S_j^{(t)}\right) \tag{5.4}$$

where $S_i^{(t)}$ and $S_j^{(t)}$ are target projections corresponding to the samples with the highest template match scores for coalition members m_i, m_j, respectively. Regarding the function $f(k)$, we have $f(1) = 0$, and it is nondecreasing for $k = 2, 3, \ldots, N$. The $\beta_{C_k}^{(t)}$ is analogous to the average of the geometric alignment of all possible tracker pairs in the coalition. In general, as the size of the coalition increases, the average of the geometric alignment function of the members of the coalition decreases. To compensate for that loss, we introduced the linear function $f(k)$, whose role is to reward higher-order coalitions as opposed to lower-order ones.

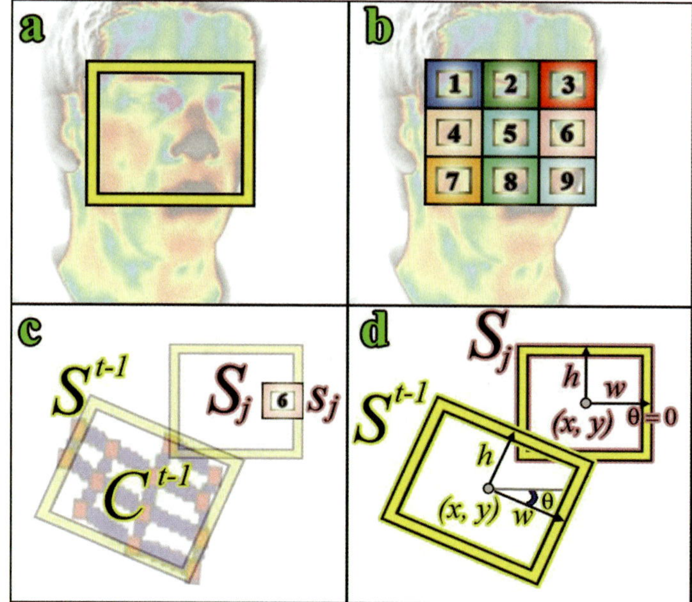

Fig. 5.4 Interframe projection agreement: **a** target projection at $t = 0$, **b** tracker network overlaid on the initial target projection, **c** the target projection at time $t - 1$ and tracker 6 with its corresponding target projection at time t, **d** parameterization of target projections to facilitate measurement of interframe projection agreement

We also use the geometric alignment function for a pair of samples [see Eq. (5.3)] to compute the interframe projection agreement $\gamma_{C_k}^{(t)}$ score (see Fig. 5.4):

$$\gamma_{C_k}^{(t)} = \frac{1}{k} \sum_{i=1}^{k} G\left(S_i^{(t)}, S^{(t-1)}\right) \tag{5.5}$$

where $S_i^{(t)}$ is the target projection corresponding to the sample with the highest template match score for coalition member $m_i^{(t)}$ at time t; $S^{(t-1)}$ is the target projection corresponding to the target state at time $t - 1$ (previous frame). The interframe membership retention score $\delta_{C_k}^{(t)}$ for a coalition C_k of size k at time t is given by

$$\delta_{C_k}^{(t)} = \frac{1}{k} \sum_{i=1}^{k} \Delta\left(m_i^{(t)}, C^{t-1}\right) \tag{5.6}$$

where m_i^t is the ith member of coalition C_k at time t, C^{t-1} is the winning coalition from the previous time step, and Δ is defined as

$$\Delta(m, C) = \begin{cases} -1 & \text{if } m \text{ is not a member of } C \\ +1 & \text{if } m \text{ is a member of } C \end{cases} \tag{5.7}$$

where m is a tracker, and C is a coalition.

Having defined the four scores, we proceed with the definition of the character-istic game function $\Pi^{(t)}(C_k)$:

$$\Pi^{(t)}(C_k) = \omega_\alpha \times \alpha_{C_k}^{(t)} + \omega_\beta \times \beta_{C_k}^{(t)} + \omega_\gamma \times \gamma_{C_k}^{(t)} + \omega_\delta \times \delta_{C_k}^{(t)} \tag{5.8}$$

where $\omega_\alpha, \omega_\beta, \omega_\gamma, \omega_\delta$ are the weights (values range in [0,1] and sum to 1) assigned to the four scores. Note that because of the function $f(k)$ in the geometric alignment score, the characteristic score may exceed the value of 1. This may happen when we have higher-order coalitions and quite good geometric alignment.

For every size of coalition $k \in \{1, 2, \ldots, N\}$, we have $\binom{N}{k}$ different coalitions of size k, out of which we select the one with the highest payoff. To avoid complicating the symbology, C_k will continue to denote the preferred coalition of size k. Thus, we decide the winning coalition C^t at time t to be

$$C^t = \arg\max_{C_k} \Pi^{(t)}(C_k) \tag{5.9}$$

In coalitional game theory, sometimes the characteristic function is a nondecreas-ing function of the size of the coalition (i.e., superadditivity) [28]. In our case, this is not desirable because there may exist trackers that have lost their targets. In other words, the grand coalition (i.e., the coalition where all players/trackers participate) is not always the optimal to use. Thus, we need to give rewards to coalitions in such a way that the winning coalition is the coalition whose members best approximate the target. This is achieved by reducing the characteristic function of the coalition if it acquires poor trackers. Superadditivity is also related to the $f(k)$ function since if $f(k)$ increases, say exponentially, then the geometric alignment score will dominate the other three scores, allowing superadditivity. In our case, having a linear $f(k)$ worked fairly well.

5.2.3 Target State Estimation

We compute the final target state S^t from the winning coalition C^t in two steps. In the first step, we compute the deformation mesh M^t from the winning coalition. The deformation mesh M^t is composed of a set of points $A = (a_1, \ldots, a_m)$, which are distributed over the selected target region during the initialization step. Each point is linked to anywhere between one through four trackers depending on its spatial location; one on the corners, two on the borders, and four on the inside. For each point, a transformation matrix $T_{a_i}^{c_j}$ is computed that, when applied to the center c_j of tracker j, gives the location of the point a_i:

$$a_i = \frac{1}{\sum_{j=1}^{n_i} \omega_j} \sum_{j=1}^{n_i} \begin{bmatrix} c_{jx} \\ c_{jy} \end{bmatrix} T_{a_i}^{c_j} \omega_j \tag{5.10}$$

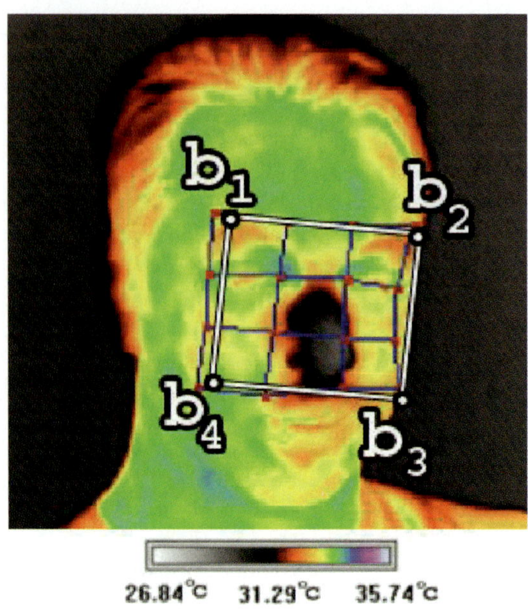

Fig. 5.5 Border points of the target projection. The target projection is shown in white. The deformation mesh is shown in blue, and the deformation mesh points are shown

where a_i is one of the points in the deformation mesh M^t, n_i is the number of trackers linked to the mesh point a_i, and ω_j is the weight associated with tracker j. If the tracker j is a member of the winning coalition, the associated weight is the tracker's template match score $\left(\omega_j = \alpha_{m_j}^{(t)}\right)$; otherwise, it is 0 $\omega_j = 0$. Next, the four border points outlining the target projection in the clockwise direction $B = (b_1, \ldots, b_4)$ are computed from the mesh points A (see Fig. 5.5).

$$b_i = \frac{1}{\omega_{tot}} \sum_{j=1}^{m} \begin{bmatrix} a_{jx} \\ a_{jy} \end{bmatrix} T_{b_i}^{a_j} \omega_{a_j} \tag{5.11}$$

where m is the number of points in the deformation mesh, b_i is one of the border points, a_j is one of the points in the deformation mesh M^t, $T_{b_i}^{a_j}$ is the transformation from point a_j to b_i, ω_{a_j} is the weight associated with mesh point a_j, which is the summation of each of its n_j member tracker weights:

$$\omega_{a_j} = \sum_{k=1}^{n_j} \omega_k \tag{5.12}$$

and ω_{tot} is the total weight of all mesh points a_j:

$$\omega_{tot} = \sum_{k=1}^{m} \omega_{a_k} \tag{5.13}$$

The second step is to compute the final target state S^t from the deformation mesh M^t by using the border points B. The target parameter vector $P = (p_1, \ldots, p_5)$, is defined as follows:

- p_1 is the x coordinate of the target center.
- p_2 is the y coordinate of the target center.
- p_3 is the rotation about the center of the target.
- p_4 is the width of the target.
- p_5 is the height of the target.

The parameter vector P is computed from the border points B of the winning coalition C^t as follows:

$$p_1 = \frac{1}{4} \sum_{i=1}^{4} b_{ix} \tag{5.14}$$

$$p_2 = \frac{1}{4} \sum_{i=1}^{4} b_{iy} \tag{5.15}$$

$$p_3 = \frac{1}{|C|} \sum_{i=1}^{|C|} c_{i\theta} \tag{5.16}$$

$$p_4 = \frac{\sqrt{(b_{1x} - b_{2x})^2 + (b_{1y} - b_{2y})^2}}{2} + \frac{\sqrt{(b_{3x} - b_{4x})^2 + (b_{3y} - b_{4y})^2}}{2} \tag{5.17}$$

$$p_5 = \frac{\sqrt{(b_{1x} - b_{4x})^2 + (b_{1y} - b_{4y})^2}}{2} + \frac{\sqrt{(b_{2x} - b_{3x})^2 + (b_{2y} - b_{3y})^2}}{2} \tag{5.18}$$

where $|C|$ is the cardinality of the winning coalition.

5.2.4 Configuration of Tracking Network

By varying the number of trackers and their relative spatial location, one could produce a large number of possible configurations for the tracking network. We narrow down this to a manageable subset by considering only uniform grids over a rectangular region (see Fig. 5.6).

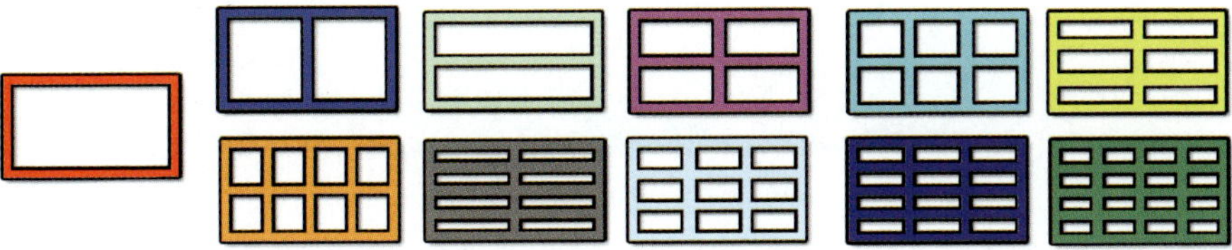

Fig. 5.6 Tracking network configurations selected for evaluation

An important consideration when picking a network configuration is the number of coalitions that must be evaluated to make a final target prediction because the number of coalitions increases with the number of trackers in the network (see Fig. 5.7).

To determine the relative performance of each of the configurations, we used each configuration to track the same target (i.e., a face) in a thermal video sequence. The true target position was annotated in each frame of the thermal video sequence to allow computation of the tracking errors from the various configurations (see Fig. 5.8).

The results show correlation between the number of trackers and tracking accuracy (see Fig. 5.9). Another trend in the data is that extra columns of trackers within the network configuration seem to be more beneficial than extra rows (see Fig. 5.9). This is explained by the facial motion exhibited in the particular experiment. The subject looks side to side and thereby deforms out of plane in the horizontal axis. Therefore, tracking configurations that add more detail along the horizontal dimension (i.e., more columns) perform better in the experiment.

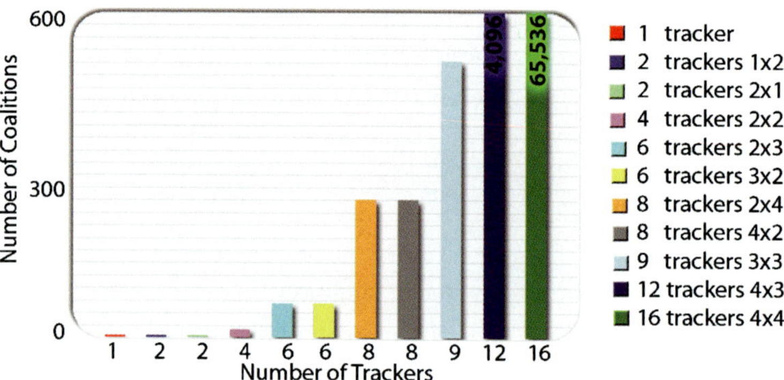

Fig. 5.7 Number of coalitions for each network configuration

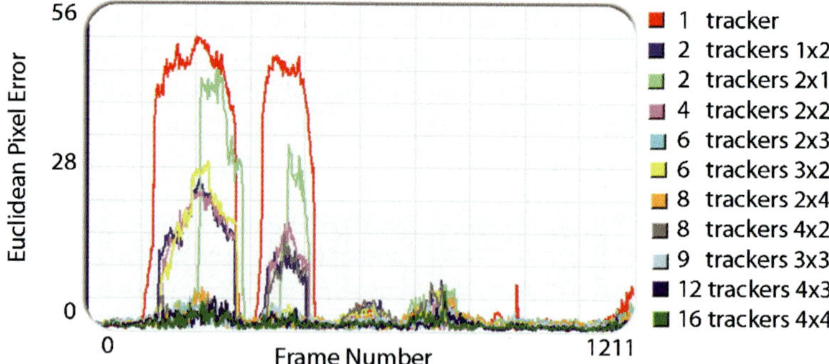

Fig. 5.8 Tracking errors of network configurations along the timeline; two large spikes correspond to out-of-plane movement by the target

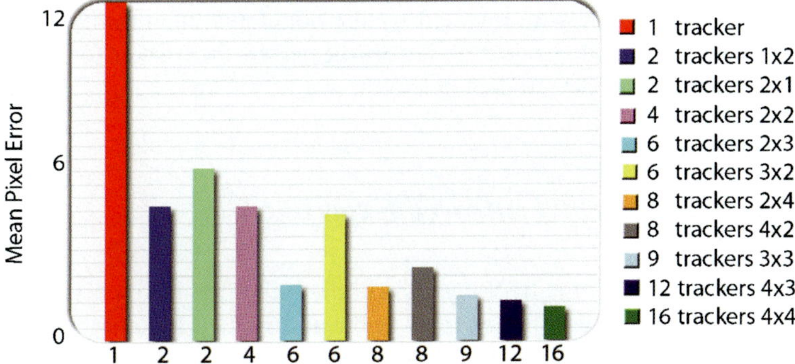

Fig. 5.9 Mean tracking error of network configurations as computed from the data in Fig. 5.8

Not surprisingly, there is not one tracking network configuration that fits every situation. The tracking network configuration must be selected based on the intended application and the type of target motion expected. In our case, we chose as a compromising solution a 3×3 tracking network configuration because we had a diverse assortment of target motions within our data set.

5.3 Experimental Design

An important consideration in our experimental design was exact quantification of the tracker's performance. For this, we needed an environment that would provide automatic ground-truthing. The cornerstone of our experimental design, however, was the provision to test our tracking method on video input from at least two different bands of the electromagnetic spectrum, one reflected and one radiated. The motivation was to demonstrate that the methodology is general enough to handle both. To satisfy this specification, we performed experiments using visual band video (reflected) and midwave infrared video (radiated). The underlying implication was that if the tracker worked on both visual and midwave infrared video, then the tracker would be general enough to be adapted to other radiated bands, such as long-wave infrared, as well as other reflected bands, such as the near infrared.

5.3.1 Design of Simulated Tracking Environment

We used a simulated tracking environment to precisely quantify the tracker's performance. The environment was initialized to a frame of thermal video, and then the tracker was initialized to the target. The target to be tracked was then translated about the image plane while simultaneously undergoing transformations. Because the target transformations were dictated by the simulated environment, we could measure the true target state against its state projected by the tracker for each frame. Every simulated run was 200 frames in length.

5.3.2 Design of Thermal Infrared Experiment

For the purpose of testing the tracking algorithm on thermal infrared video, we selected a data set that was used in previous publications [12]. It consists of 39 video clips, each containing a main human subject undergoing an interview. We chose to track 1,000 frames of video from each of the subject clips, for a total of 39,000 frames of video. The chosen video segments featured a temporary occlusion of the main subject by another subject who was passing through the field of view. More important, the clips featured out-of-plane rotation of facial tissue as subjects were rotating their heads left or right, up or down. We chose a single-particle filter tracker to compare against the coalitional tracker. Both the single-particle filter and coalitional trackers featured identical parameterization. Both the single-particle filter and the coalitional network were tasked to track exactly the same facial tissue of each subject. The ground-truthing of this experiment was the reconciliation of the observations of two independent operators.

5.3.3 Design of Visual Experiment

To demonstrate that the tracking methodology can also be applied to visual band video, we performed experiments on a series of visual videos, each containing a different type of target. These targets included faces and cans.

5.4 Experimental Results

5.4.1 Results of Simulated Tracking Environment

We first measured the accuracy of the coalitional tracker (see Fig. 5.10). The coalitional tracker maintained a mean error of about 1 pixel, which is sufficient for demanding applications, such as physiological measurements [5]. Moreover, the coalitional tracker exhibited consistent performance over 20 identical trials (see Fig. 5.10). This is extremely important because it would be impossible to extract useful physiological measurements if the tracker gave inconsistent results each time it was run. However, due to the stochastic nature of particle filtering, it is very difficult to altogether eliminate minute variability from the tracking result. To determine

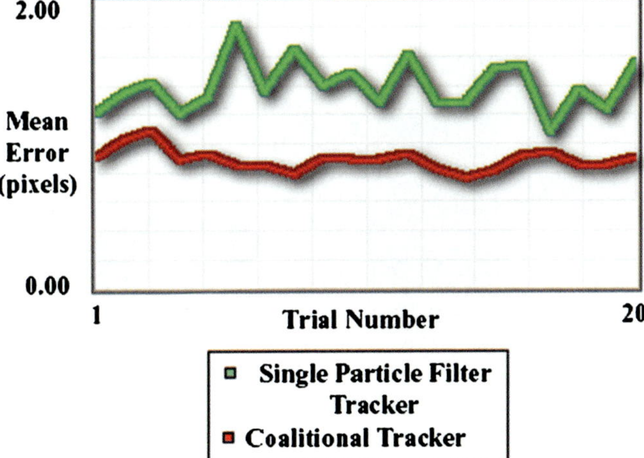

Fig. 5.10 Error and stability analysis of single-particle filter (green) versus coalitional tracking (red). Both trackers were used to track the same target in 20 identical trials using the simulated tracking environment

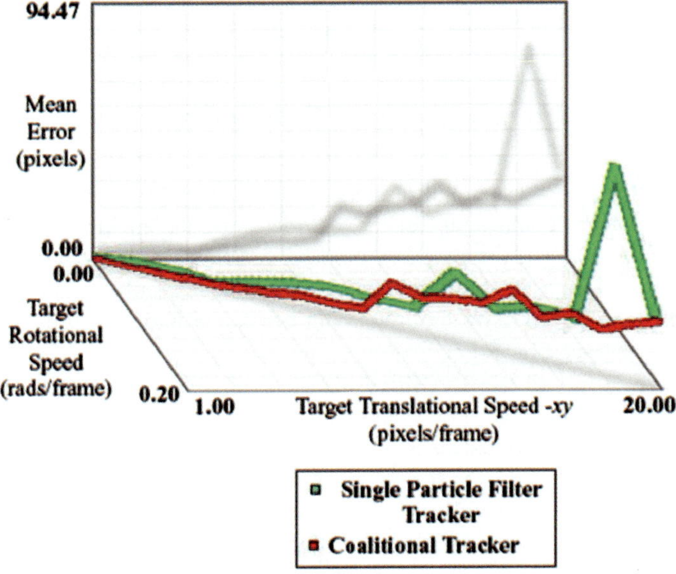

Fig. 5.11 Error analysis of single-particle filter (green) versus coalitional tracking using the tracking network (red). Both trackers were used to track the same target in 20 trials using the simulated tracking environment. Each trial involved increasingly faster translational and rotational target motion

the operational limits of the tracker, we measured its error under increasingly faster target motion in the simulated tracking environment (see Fig. 5.11). The superior performance of the coalitional tracker in complex and fast transformations is evi-

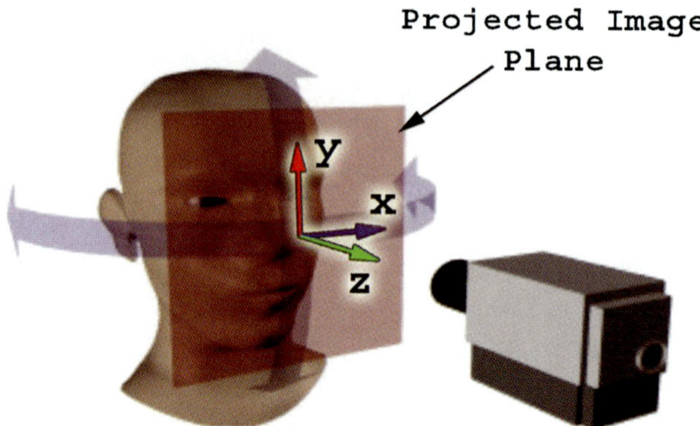

Fig. 5.12 Out-of-plane facial rotation. Any rotation that is not about the z-axis is considered out-of-plane rotation

dent. The coalitional tracker was also capable of negotiating out-of-plane facial rotations (see Fig. 5.12) much more successfully than the single-particle filter tracker (see Fig. 5.13).

5.4.2 Results of Thermal Infrared Experiment

The results from the thermal infrared experiment (see Fig. 5.14 and Table 5.1) clearly show that the coalitional tracker provides superior tracking over the single-particle filter tracker. The proposed method proved robust in typical (see Fig. 5.15) and difficult (see Fig. 5.16) operational scenarios. The few failures of the coalitional tracker were mainly caused by significant out-of-plane rotation or substantial occlusion of the target (see Fig. 5.17).

A rare case of failure is exemplified in Fig. 5.18, when the subject experienced rapid physiological changes on a grand scale. The subject in the figure underwent facial temperature increase in excess of 2°C within 6 min due to a state of high anxiety. This problem is due to the template measurement method, which assumes that the target's projection will not change dramatically over time. One possible solution to this problem is to dynamically update the template as presented in [39].

We extracted a sample physiological measurement from subject S2 (inventory reported in [12]) and compared it against the respective ground truth signal. The measured signal is the mean temperature of the subject's periorbital area through the course of the video clip. It is evident that the coalitional tracker enables the acquisition of a signal nearly identical to the ground truth (see Fig. 5.19), an indication of its fitness for accurate physiological measurements.

Fig. 5.13 Out-of-plane rotation comparison. *Left*, single-particle filter tracker (green); *right*, coalitional tracker (red). **a** Initial frame, **b** and **c** intermediate frames, **d** final frame in a 1-min thermal clip. The poor performance of the single-particle filter tracker is evident

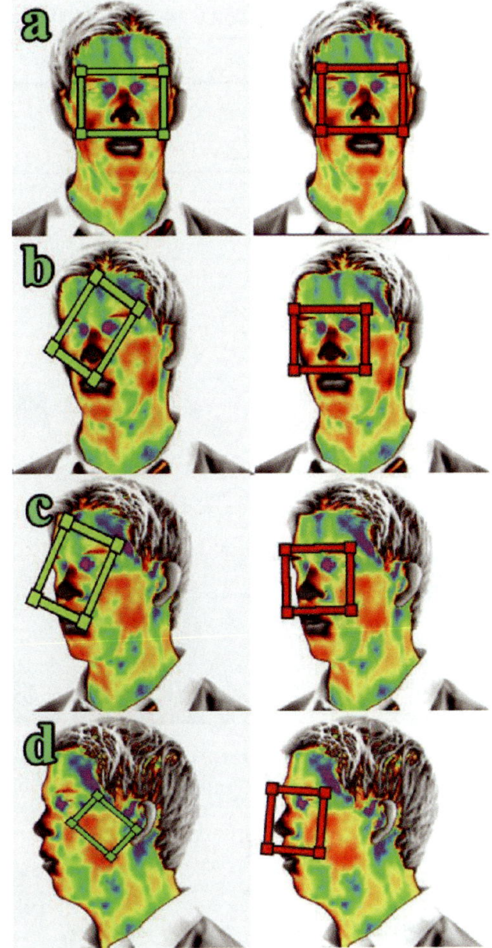

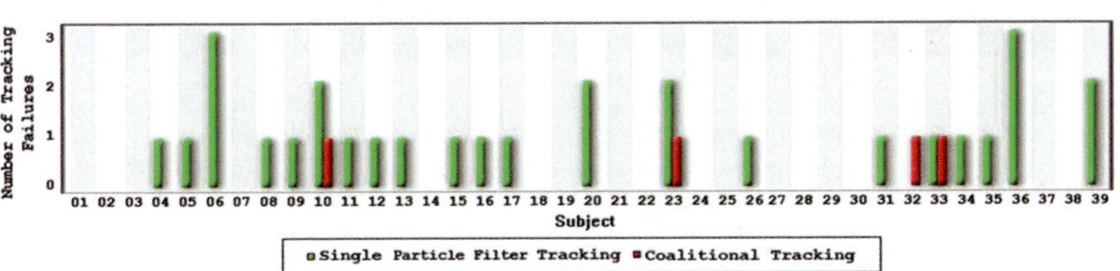

Fig. 5.14 Tracking failure graph for the 39 video clips in the thermal data set. For each clip the number of single-particle filter and coalitional tracking failures is shown in green and red, respectively. The absence of red bars in some video clip entries indicates perfect performance of the coalitional tracker

Table 5.1 Causation of tracking failures in the thermal data set

Reason for failure	Coalitional tracker failures	Single-tracker failures
Target rotation	1	18
Partial occlusion	2	9
No recovery	1	2
Total	4	29

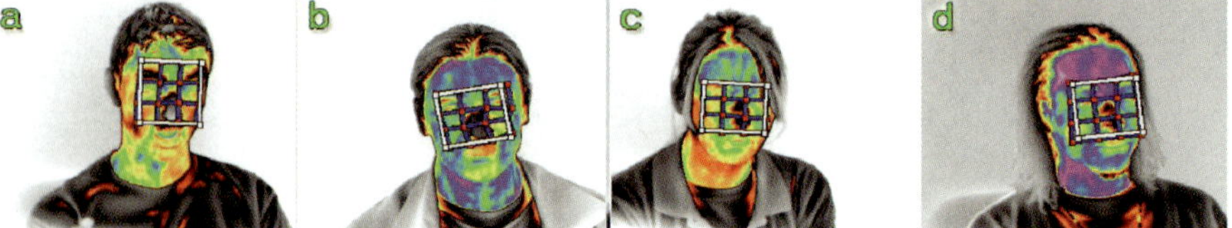

Fig. 5.15 Typical facial-tracking examples from the thermal data set. The selected subjects represent different ethnicities and both genders

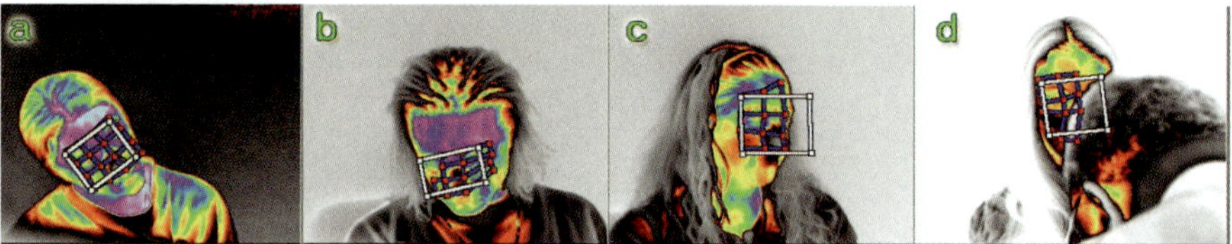

Fig. 5.16 Successful coalitional tracking in the presence of difficult circumstances in the thermal spectrum: **a** target rotating in plane, **b** target rotating out of plane, **c** target rotating out of plane, **d** target partially occluded

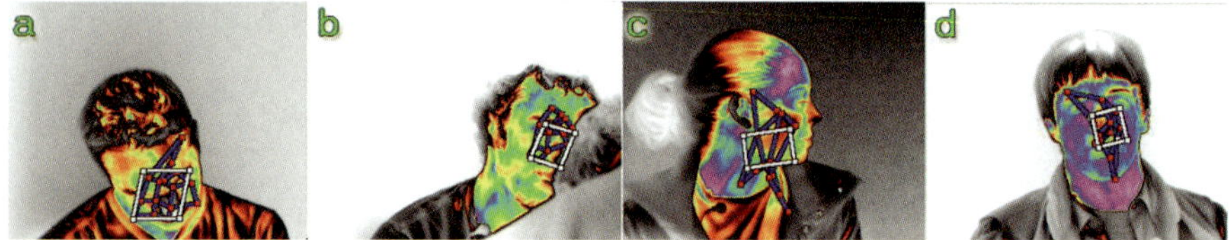

Fig. 5.17 Tracking failures in the thermal spectrum. **a** and **c** The target has rotated out of plane beyond the tracker's ability to compensate. **b** The original target (periorbital area) is largely occluded. **d** The target has undergone extreme physiological changes relative to the initial tracking frame (see Fig. 5.18 for more details)

5.4.3 Results of Visual Experiment

The coalitional tracker performed robustly in several visual band experiments with various objects (faces and cans). The template was composed of 3-tuples (red, green, and blue reflectance values) instead of temperatures. The motion patterns included translation, rotation, and scaling (see Fig. 5.20 and Fig. 5.21).

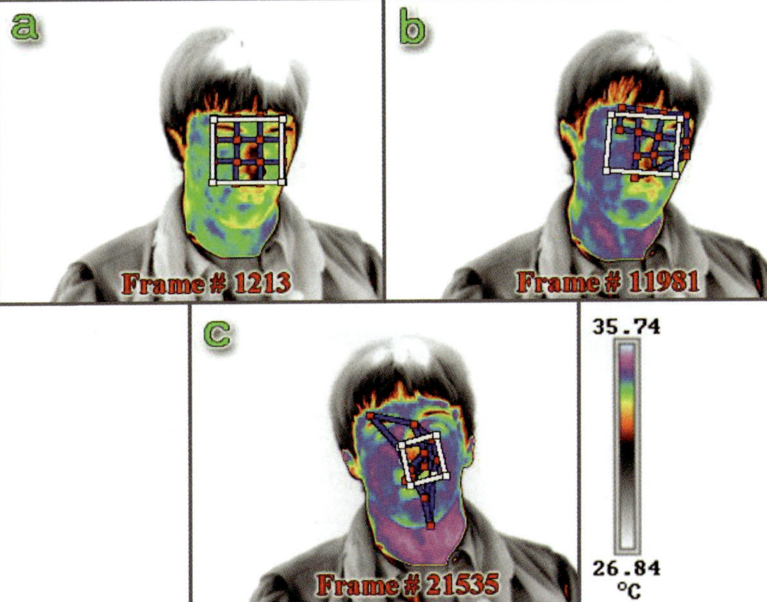

Fig. 5.18 Coalitional tracker performance under substantial physiological changes. **a** Tracker initialization. **b** The subject's face undergoes a substantial thermal change in the middle of the video clip. The tracker is still performing correctly, but the winning coalition is composed of fewer trackers that are able to follow their targets. **c** Toward the end of the clip, the subject's facial thermal profile continues to change dramatically, and the coalitional tracker is off target

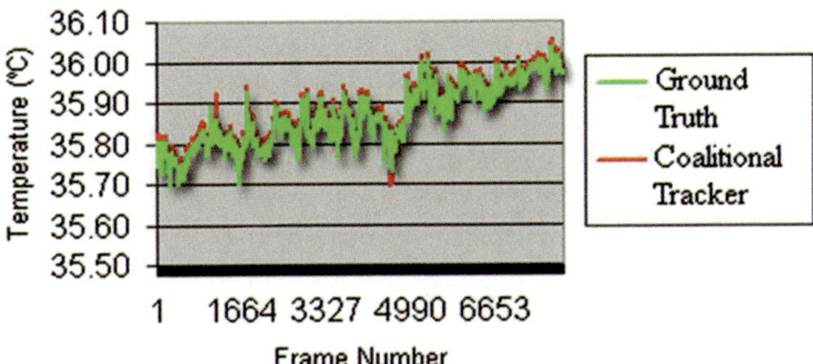

Fig. 5.19 Physiological signal extracted using coalitional tracking (in red) versus the ground truth signal

5.5 Application Perspective

Coalitional tracking is a fairly general framework and can be applied to a variety of problems across imaging modalities. Nevertheless, it was originally developed for a particular modality (i.e., thermal infrared) and for specific applications (i.e., physiological measurements on the face). To measure the success of coalitional tracking, it

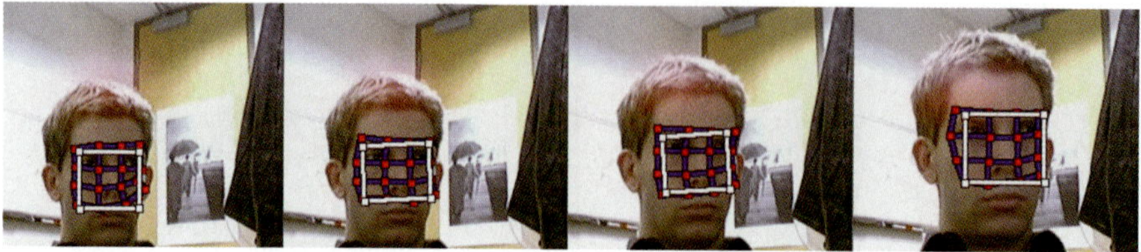

Fig. 5.20 Example of tracking a face experiencing scaling and translation in the visual spectrum. The frames are shown chronologically from left to right. The deformation mesh is shown in blue, and the white rectangle represents the projected target state

Fig. 5.21 Example of tracking an object experiencing scaling and translation in the visual spectrum. The frames are shown chronologically from left to right. The deformation mesh is shown in blue, and the white rectangle represents the projected target state

is important to understand its original application framework and impact. We touch on two major applications for which coalitional tracking is now used routinely: lie detection and sleep studies.

Levine et al. [8] reported a physiological sign of stress manifested as increased blood flow in the orbital muscle. Pavlidis et al. [9] demonstrated the potential of this stress sign as a lie detection indicator in the context of a well-designed interrogation. The importance of this cannot be overestimated. It was the first time that a localized physiological sign of cholinergic origin was identified on the face as "polygraph ready." The facial locale is ideal for casual observation as it is typically exposed. Furthermore, the sympathetic relevance of the periorbital sign ranks very high because the face is heavily innervated with neuronal pathways. In summary, we had a primary stress indicator easily observable, but unfortunately not easily measurable. To begin, there was significant difficulty in sensing blood flow in the orbital muscle because of the delicate nature of the tissue. A popular method for sensing blood flow is ultrasound. Imagine, for example, the examiner rubbing the eyes of the subject with an ultrasound wand to get a blood flow measurement on the orbital muscle. This would clearly be impractical, particularly in the context of psychophysiological experiments. The problem was solved with the introduction of thermal imaging as the modality of choice for such measurements. Superficial blood flow under thin facial tissue emits a heat signature due to convection, which can be captured and analyzed by a thermal imaging sensor package. Thermal imaging did not only solve the sensing problem in a practical way, but also superbly supported psychophysiological experiments because it is totally unobtrusive.

The fact that the measurement was done at a distance might have been a blessing from the psychological point of view, but it posed a challenge from the medical point of view. What was needed was a virtual probe to isolate the area of interest in the image — a nontrivial segmentation problem. Moreover, a tracking method had to be developed to keep this virtual probe in the orbital area irrespective of head motion. Both segmentation and tracking had to maintain pixel-level accuracy for the measurement to remain valid. The tracking problem, which is of interest here, was especially challenging due to the functional nature of thermal infrared imaging and the real-time requirements of the application. Thermal imaging of the face depicts physiological changes. Therefore, it is highly dynamic, unpredictable, and difficult to model. Despite the absence of strong models, tracking still has to be accurate enough to support valid medical measurements. Plus, it has to be highly efficient as the technology created an opportunity for lie detection "on the fly," which the polygraph community wanted to fully exploit.

Coalitional tracking solved these conflicting requirements and facilitated research and development in a big way. The secret of its success is that it efficiently optimized the behavior of many weak model trackers to achieve robustness and accuracy reminiscent of strong model trackers in structural imaging domains (e.g., computed tomography). Coalitional tracking was used to measure stress in three major government experiments involving multiple lie detection interviews of more than 150 subjects. This accounts for hundreds of recording hours and millions of frames. All the thermal videos used in the experiments detailed in Section 5.4.2 are a small subset of the lie detection inventories.

Starting in February 2007, coalitional tracking was also applied with great success in sleep studies at the University of Texas Medical School. The physiological measurement of interest in this case was breathing (see Fig. 5.22). Patients in sleep studies suffer from chronic respiratory diseases that manifest themselves during sleep. A prime example of such a disease is obstructive sleep apnea, for which breathing is suspended for a few seconds, several times every minute. This creates temporary asphyxiation, which triggers the "fight-or-flight" response. As the phenomenon repeats itself every few seconds, it results in an almost permanent load

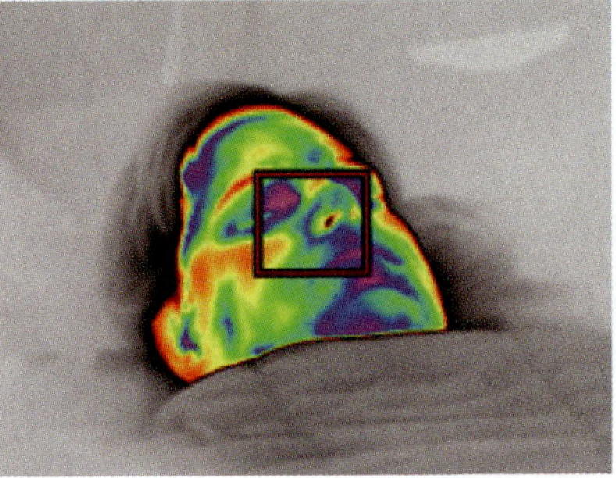

Fig. 5.22 Coalitional tracker monitoring a subject's nasal region during inventory of sleep study experiments at the University of Texas Medical School

on the cardiovascular system and poor-quality sleep. Both have serious long-term repercussions on the health of the patient. Diagnosis of sleep apnea involves monitoring of the patient's sleep for several nights in the lab. During these times, the patient is heavily instrumented (see Fig. 5.23), a highly uncomfortable proposition for anyone, but especially for people who suffer from sleep problems. Therefore, there is strong motivation to unwire the patient to the extent possible.

Coalitional tracking helped to reliably extract the breathing signal through thermal imagery. The monitoring periods exceeded one h for every patient. The accuracy of the imaging computation was ascertained against the clinical gold standard (i.e., thermistor). In contrast to the lie detection application in which the subject's head moves in moderate amounts all the time, during sleep studies the patient's head exhibits minute motor motion (i.e., due to breathing) and occasionally abrupt large-scale motion (i.e., turning). The different motion profiles in the two applications represent a comprehensive testing of the tracker's abilities.

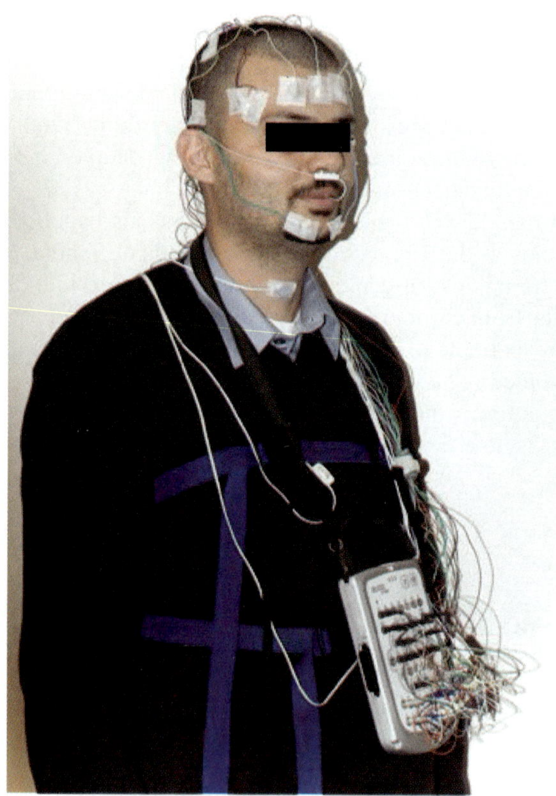

Fig. 5.23 Patient wired for sleep study (Courtesy University of Texas Medical School)

5.6 Conclusion

We have proposed a novel tracking method. Our method uses a spatially distributed network of trackers whose interactions are modeled using coalitional game theory. The output of the method provides pixel-level tracking accuracy, even in the presence of multidimensional target transformation.

We tested our method in thermal and visual video sets featuring faces and objects. We compared the performance of the proposed coalitional tracker with that of a single-particle filter tracker. The coalitional tracker exhibited superior performance in both regular and challenging tasks. The strength of the method comes from the redundancy that is elegantly encoded in its game theoretic structure. Detailed quantification and ground truth verification indicated that the new method provides accuracy appropriate for demanding medical imaging applications. Equally important is the fact that the method appears to be general and flexible enough to use in imaging applications across the electromagnetic spectrum.

5.6.1 Future Work

The particular adaptation of game theory to tracking presented in this chapter is but one of many possible approaches that might be adopted. For example, the problem of tracking could be alternatively viewed as a noncooperative game in which the trackers compete with each other, and the final solution could then be modeled as a Nash (strategic) equilibrium [29]. Plus, active research areas in game theory, such as stochastic and differential games [41], could potentially be adapted for use in tracking.

An important area that is amenable to improvement is the current static template scheme. Although it works well in the thermal infrared band, where emission of most objects does not change dramatically in short observation periods (e.g., a few minutes), it is potentially vulnerable in the visual band, where reflected light may change dramatically over a split second depending on the angle of incidence. A dynamic template mechanism will eliminate this vulnerability.

The current method is based on deterministic management of probabilistic trackers. A future method may be developed that will be based on probabilistic management of probabilistic trackers. This can be realized within a Bayesian framework in which the posterior weight of each tracker in the coalitional game would be computed from its prior and an appropriate likelihood function. Since this will add probabilistic memory into coalition membership, one can eliminate the membership retention factor in the current characteristic function, which in essence crudely plays the same role.

Acknowledgments We would like to thank the National Science Foundation (grant IIS-0414754) and Dr. Ephraim Glinert for supporting this effort in its early phase. We would also like to thank Dr. Andrew Ryan, Dr. Dean Pollina, Dr. Troy Brown, and the Department of Defense (multiple research contracts) for supporting the late phase of the research. The views expressed by the authors in this chapter do not necessarily reflect the views of the funding agencies.

Chapter's References

1. D.A. Gonzalez, F.J. Madruga, M.A. Quintela, and J.M. Lopez-Higuera, Defect assessment on radial heaters using infrared thermography, NDT & E International, 38(6):428-432, September 2005
2. M. Burrell, Computer vision for high-speed, high-volume manufacturing, in Proceedings of the 1993 International Conference on Systems, Man, and Cybernetics, 3:349–354, October 17–20, 1993
3. I. Pavlidis, V. Morellas, P. Tsiamyrtzis, and S. Harp, Urban surveillance systems: from the laboratory to the commercial world, Proceedings of the IEEE, 89(10):1478–1497, October 2001
4. R.T. Collins, A.J. Lipton, H. Fujiyoshi, and T. Kanade, Algorithms for cooperative multi-sensor surveillance, Proceedings of the IEEE, 89(10):1456–1477, October 2001
5. M. Garbey, A. Merla, and I. Pavlidis, Estimation of blood flow speed and vessel location from thermal video, in Proceedings of the 2004 IEEE Computer Society Conference on Computer Vision and Pattern Recognition, 1:356–63, June 27–July 2, 2004
6. N. Sun, M. Garbey, A. Merla, and I. Pavlidis, Imaging the cardiovascular pulse, in Proceedings of the 2005 IEEE Computer Society Conference on Computer Vision and Pattern Recognition, 2:416–21, June 20–25, 2005
7. J. Fei, Z. Zhu, and I. Pavlidis, Imaging breathing rate in the CO_2 absorption band, in Proceedings of the 27th Annual International Conference of the IEEE Engineering in Medicine and Biology Society, September 1–4, 2005
8. J. Levine, I. Pavlidis, and M. Cooper, The face of fear, Lancet, 357(9270), June 2, 2001
9. I. Pavlidis, N.L. Eberhardt, and J. Levine, Human behavior: seeing through the face of deception, Nature, 415(6867):35, January 3, 2002
10. I. Pavlidis and J. Levine, Thermal image analysis for polygraph testing, IEEE Engineering in Medicine and Biology Magazine, 21(6):56–64, November–December 2002
11. C. Eveland, D. Socolinsky, and L. Wolff, Tracking human faces in infrared video, Image and Vision Computing, 21:578–590, July 2003
12. P. Tsiamyrtzis, J. Dowdall, D. Shastri, I. Pavlidis, M.G. Frank, and P. Ekman, Lie detection—recovery of the periorbital signal through tandem tracking and noise suppression in thermal facial video, in Proceedings of SPIE Sensors, and Command, Control, Communications, and Intelligence (C3I) Technologies for Homeland Security and Homeland Defense IV, E.M. Carapezza, editor, p. 5778, March 29–31, 2005
13. S. Krotosky, S. Cheng, and M. Trivedi, Face detection and head tracking using stereo and thermal infrared cameras for "smart" airbags: a comparative analysis, in Proceedings of the 7th International IEEE Conference on Intelligent Transportation Systems, 1:17–22, 2004
14. A. Doucet, N. DeFreitas, and N. Gordon, editors, Sequential Monte Carlo Methods in Practice, Springer-Verlag, 2001
15. M. Isard and A. Blake, Condensation — conditional density propagation for visual tracking, International Journal of Computer Vision, 19(1):5–28, 1998
16. M. Isard and A. Blake, ICONDENSATION: unifying low-level and high-level tracking in a stochastic framework, in Proceedings of the 5th European Conference on Computer Vision, 1:893–908, June 2–6, 1998
17. J. MacCormick and M. Isard, Partitioned sampling, articulated objects, and interface-quality hand tracking, in Proceedings of the 7th European Conference on Computer Vision, 1843:3–19, 2000

18. Y. Zhong, A.K. Jain, and M.P. Dubuisson-Jolly, Object tracking using deformable templates, IEEE Transactions on Pattern Analysis and Machine Intelligence, 22(5):544–549, May 2000
19. Y. Shi and W. Karl, Real-time tracking using level sets, in Proceedings of the 2005 IEEE Computer Society Conference on Computer Vision and Pattern Recognition, 2:34–41, June 20–25, 2005
20. C. Zimmer and J. C. Olivo-Marin, Analyzing and capturing articulated hand motion in image sequences, IEEE Transactions on Pattern Analysis and Machine Intelligence, 27(11):1838–1842, November 2005
21. S. Goldenstein, C. Vogler, J. Stolfi, V. Pavlovic, D. Metaxas, Outlier rejection in deformable model tracking, in Proceedings of the 2004 Conference on Computer Vision and Pattern Recognition, June 19–26, 2004
22. T.F. Cootes, G.J. Edwards, C.J. Taylor, Active appearance models, IEEE Transactions on Pattern Analysis and Machine Intelligence, 23(6):681–685, June 2001
23. F. Dornaika and J. Ahlberg, Efficient active appearance model for real-time head and facial feature tracking, Proceedings of the 2003 IEEE International Workshop on Analysis and Modeling of Faces and Gestures, pp. 173–180, October 13, 2003
24. C. Cheng, R. Ansari, and A. Khokhar, Multiple object tracking with kernel particle filter, in Proceedings of the 2005 IEEE Computer Society Conference on Computer Vision and Pattern Recognition, 1:566–573, June 20–25, 2005
25. Y. Ting and W. Ying, Decentralized multiple target tracking using netted collaborative autonomous trackers, in Proceedings of the 2005 IEEE Computer Society Conference on Computer Vision and Pattern Recognition, 1:939–946, June 20–25, 2005
26. M. Isard and J. MacCormick, BraMBLe: a Bayesian multiple-blob tracker, in Proceedings of the 8th IEEE International Conference on Computer Vision, 2:34–41, July 7–14, 2001
27. J. MacCormick and A. Blake, A probabilistic exclusion principle for tracking multiple objects, International Journal of Computer Vision, 39(1):57–71, 2000
28. T.S. Ferguson, game theory, Chapter 4, http:www.math.ucla.edutomGame_TheoryContents.html
29. K. Ritzberger, Foundations of Non-Cooperative Game Theory, Oxford University Press, New York, 2002
30. A. Rapoport, N-Person Game Theory: Concepts and Applications, University of Michigan, 1978
31. E. Rasmusen, Games and Information: An Introduction to Game Theory, Blackwell, 1989
32. T.G. Fisher et al., Managerial Economics: A Game Theoretic Approach, Routledge, 2002
33. C. Schmidt, editor, Game Theory and Economic Analysis: A Quiet Revolution in Economics, Routledge, 2002
34. P. Ordeshook, Game Theory and Political Theory: An Introduction, Cambridge University Press, Cambridge, U.K., 1986
35. S. Brams, Game Theory and Politics, Free Press, New York, 1975
36. S. Hart, editor, Cooperation: Game-Theoretic Approaches, Springer-Verlag, New York, 1997
37. M. Mareš, Fuzzy Cooperative Games, Physica-Verlag, 2001
38. S. Baker and I. Matthews. Equivalence and efficiency of image alignment algorithms, in Proceedings of the IEEE Conference on Computer Vision and Pattern Recognition, 1:1090–1097, 2001
39. I. Matthews, T. Ishikawa, S. Baker. The template update problem, IEEE Transactions on Pattern Analysis and Machine Intelligence, 26(6):810–815, June 2004
40. Y. Adini, Y. Moses, S. Ullman, Face recognition: the problem of compensating for changes in illumination direction, IEEE Transactions on Pattern Analysis and Machine Intelligence, 19(7)721–732, 1997
41. M. Bardi, T. Raghavan, T. Parthasarathy, editors, Stochastic and Differential Games: Theory and Numerical Methods. Annals of the International Society of Dynamic Games, Birkhauser, 1998

Chapter 6
Thermal Infrared Imaging in Early Breast Cancer Detection

Hairong Qi and Nicholas A. Diakides

Abstract The application of thermal infrared (TIR) imaging in breast cancer study started as early as 1961. However, it has not been widely recognized due to the premature use of the technology, the superficial understanding of the infrared (IR) images, and its poorly controlled introduction into breast cancer detection in the 1970s. Recent advances in image-processing ability and pathophysiological-based understanding of IR images, coupled with the new-generation IR technology, have spurred renewed interest in the use of infrared breast imaging. This chapter provides a survey of recent achievements from these aspects.

6.1 Introduction

Temperature is a long-established indicator of health. The Greek physician Hippocrates wrote in 400 B.C., "In whatever part of the body excess of heat or cold is felt, the disease is there to be discovered" [59]. The ancient Greeks immersed the body in wet mud, and the area that dried more quickly, indicating a warmer region, was considered the diseased tissue. The use of hands and thermometers to measure heat emanating from the body remained well into the sixteenth through the eighteenth centuries. Now, we still rely on thermometers a lot when performing a health examination. Since the British astronomer Sir William Herschel discovered the existence of infrared (IR) radiation in 1800, major advances have taken place with IR imaging that do not need direct contact with the patient.

IR radiation occupies the region between the visible and microwave ranges of the spectrum. All objects in the universe emit radiation in the IR region as a function of their temperature. As an object gets hotter, it gives off more intense infrared radiation, and it radiates at a shorter wavelength [27]. The human eye cannot detect IR rays, but they can be detected by using IR cameras and detectors. In general, IR radiation covers wavelengths that range from $0.75\,\mu\text{m}$ to $1{,}000\,\mu\text{m}$, among which the human body emissions that are traditionally measured for diagnostic purposes

R.I. Hammoud (ed.), *Augmented Vision Perception in Infrared: Algorithms and Applied Systems,* Advances in Pattern Recognition, DOI 10.1007/978-1-84800-277-7_6,
© Springer-Verlag London Limited 2009

only occupy a narrow band at wavelengths of 8 μm to 12 μm [61]. This region is also referred to as the *long-wave IR* (LWIR) or *body infrared rays*. Another terminology that is widely used in medical IR imaging is *thermal infrared* (TIR), which covers wavelengths beyond about 1.4 μm. Within this region, the infrared emission is primarily *heat* or *thermal* radiation, hence the term *thermography*. The image generated by TIR imaging is referred to as the *thermogram*.

The first documented application of IR imaging in medicine was in 1956 [35], when breast cancer patients were examined for asymmetric hot spots and vascularity in IR images of the breasts. Since then, numerous research findings have been published [18, 36, 39], and the 1960s witnessed the first surge of medical application of IR technology [14, 17], with breast cancer detection as the primary practice. However, IR imaging has not been widely recognized in medicine, largely due to the premature use of the technology, the superficial understanding of IR images, and its poorly controlled introduction into breast cancer detection in the 1970s [32].

6.1.1 Breast Cancer and Imaging Modalities

According to American Cancer Society's report on cancer facts and figures [55], breast cancer is the most commonly diagnosed cancer in women, accounting for about 30% of all cancers in women. In 2008, an estimated 182,460 new cases of invasive breast cancer were expected to occur among women in the US; about 1,990 new cases were expected in men. On the other hand, research [41] has shown that if detected earlier (tumor size less than 10 mm), the breast cancer patient has an 85% chance of cure as opposed to 10% if the cancer is detected late. Other research also showed evidence of early detection in saving life [8, 11, 12].

Many imaging modalities can be used for breast screening, including mammography, which uses X-ray; (IR); magnetic resonance imaging (MRI); computed tomography (CT); ultrasound; and positron emission tomography (PET) scans. Figure 6.1 shows breast scan examples using mammography, IR, and ultrasound. Although mammography has been the baseline approach, several problems still exist that affect diagnostic accuracy and popularity. First, mammography, like ultrasound, depends primarily on structural distinction and anatomical variation of the tumor from the surrounding breast tissue [32]. Unless the tumor is beyond a certain

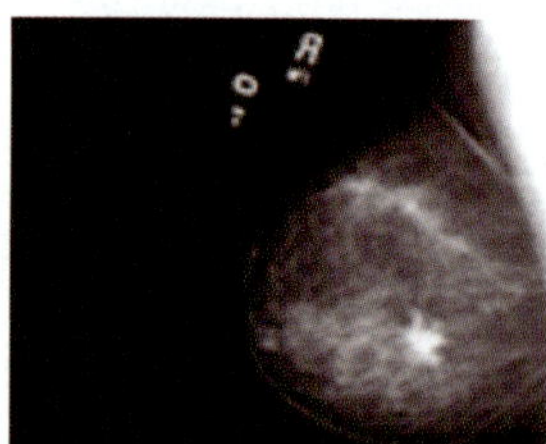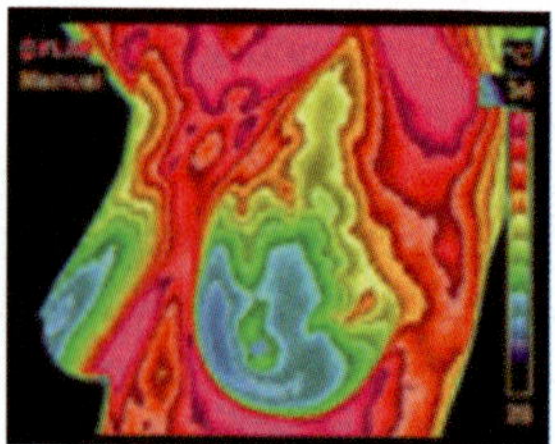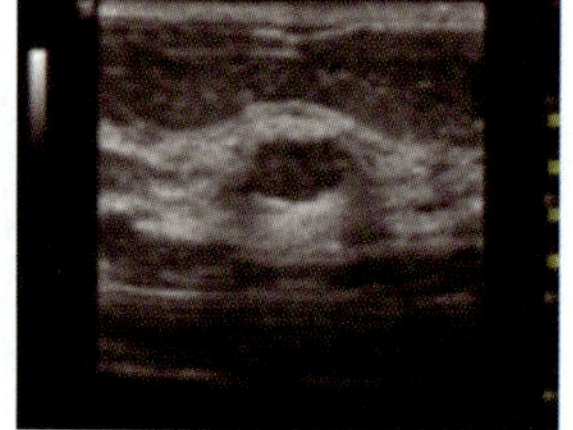

Fig. 6.1 Breast scan using *(left)* mammography, *(middle)* IR, and *(right)* ultrasound [7]

size, it cannot be imaged as X-rays essentially pass through it unaffected. Second, the mammogram sensitivity is higher for older women (age group 60–69 years) at 85% compared with younger women (<50 years) at 64% [41], whose denser breast tissue makes it more difficult for mammography to pick up suspicious lesions. Third, patients who go through mammography screening are exposed to X-ray radiation, which can mutate or destroy the tissue it penetrates. A new study in the British medical journal *The Lancet* [44] showed that screening actually leads to more aggressive treatment, increasing the number of mastectomies by about 20% and the number of mastectomies and tumorectomies by about 30%. The cancer and facts and figures [55] also stated that "after continuously increasing for more than two decades, female breast cancer incidence rates decreased by 3.5% per year from 2001–2004. . . . It may . . . reflect a slight drop in mammography utilization; according to the National Health Interview Survey, mammography rates in the past two years in women 40 and older decreased from 70.1% in 2000 to 66.4% in 2005." Finally, mammography is relatively expensive now and is less convenient (e.g., long duration and uncomfortable contact).

Even though other modalities like MRI and PET scan could provide valuable information for diagnosis, they are not popularly adopted for various reasons, including high cost, complexity, and accessibility issues [31]. Compared to mammography, MRI, CT, ultrasound, and PET scans, which are also called the *after-the-fact* (a cancerous tumor is already there) *detection technologies,* IR imaging is able to detect breast cancers 8–10 years earlier than mammography [13, 42]. Keyserlinkg reported [32] that the average tumor size undetected by IR imaging is 1.28 cm vs. 1.66 cm by mammography. Breastthermography.com further stated on their Web site [8] that, "Breast thermography has the ability to warn women up to 10 years before any other procedure that a cancer may be forming." In addition, IR imaging is noninvasive, nonionizing, risk free, patient friendly, and the cost is considerably low. These features, together with its early detection capability, have enabled IR imaging to be a strong candidate as a complementary diagnostic tool to traditional mammography.

In his recent live interview with ABC news [8], Dr. Amalu from San Francisco stated that, "Sensitivity* of thermogram alone is close to 90% compared to mammography with a sensitivity of 80%." He also indicated that this statistic is supported by "approximately 800 studies in the medical literature . . . where over 300,000 women participated." Further details are provided from breastthermography.com [7], and some of the most important statistics (i.e., sensitivity and specificity)† are listed in Table 6.1.

Recently, advances in a couple of related areas have pushed forward a series of activities to reappraise the role of IR imaging in medicine [2, 21, 23, 27, 31, 32]. These advances, including the development of smart image-processing algorithms and the pathophysiological-based understanding of IR images, coupled with the

---

* *Sensitivity* is defined as the probability of making a correct cancer diagnosis of a patient with cancer, which equals $1 - Miss\ rate.$

† *Specificity* is defined as the probability of identifying normal patients without cancer, which equals $1 - False\ alarm\ rate.$

Table 6.1 Sensitivity and specificity comparison among mammography, IR, and ultrasound [7]

	Mammography	IR	Ultrasound
Sensitivity	Average 80% in women over age 50, drops to 60% in women under age 50	Average 90% in all age groups	Average 83% in all age groups
Specificity	Average 75%	Average 90%	Average 66%

new-generation IR technology, will provide a cost-effective, noninvasive, nondestructive, and patient-friendly approach to health monitoring and examination, as well as assisting diagnosis. In this chapter, we discuss these new developments with a focus on the understanding of IR images (Section 6.2) and smart algorithm designs (Section 6.3).

6.2 Pathophysiological-Based Understanding of IR Imaging

Infrared imaging is a physiological test that measures the subtle physiological changes that might be caused by many conditions, such as contusions, fractures, burns, carcinomas, lymphomas, melanomas, prostate cancer, dermatological diseases, rheumatoid arthritis, diabetes mellitus and associated pathology, deep venous thrombosis (DVT), liver disease, bacterial infections, and more. These conditions are commonly associated with regional vasodilation, hyperthermia, hyperperfusion, hypermetabolism, and hypervascularization [3,5,19,28,53,57,61], which generate a higher-temperature heat source. Unlike imaging techniques such as X-ray radiology and CT that primarily provide information on the anatomical structures, IR imaging provides functional information not easily measured by other methods. Thus, correct use of IR images requires in-depth physiological knowledge for its effective interpretation.

Cancer cells result from permanent genetic change in a normal cell triggered by some external physical agents, such as chemical agents, X-rays, ultraviolet (UV) rays, and so on. All types of cancer cells have an imbalanced metabolic activity that leads to the utilization of a large amount of blood glucose and the release of large amounts of lactate into the blood. In addition, the high metabolic rate of cancer cells causes an increase in local temperature as compared to normal cells. These factors have enabled IR imaging as a viable technique to visualize the abnormality. The IR image provides more dynamic information of the tumor since the tumor can be small in size but can be fast growing, making it appear as a high-temperature spot in the IR image [20,60].

The heat emanating on the surface from the heat source and the surrounding blood flow can be quantified using the Pennes' bioheat equation [46]. This equation includes the heat transfer due to conduction through the tissue, the volumetric metabolic heat generation of the tissue, and the volumetric blood perfusion rate, whose strength is considered to be the arteriovenous temperature difference [41]. The equation is given as

$$k\Delta^2 T - c_b w_b (T - T_a) + q_m = 0 \tag{6.1}$$

where k is conductivity, q_m is the volumetric metabolic rate of the tissue, $c_b w_b$ is the product of the specific heat capacity and the mass flow rate of blood per unit volume of tissue, T is the unknown tissue temperature, and T_a is the arterial temperature. In theory, given the heat emanating from the surface of the body measured by TIR imaging, by solving the inverse heat transfer problem, we can obtain the heat pattern of various internal elements of the body. Different methods of solving the bio-heat transfer equation have been presented in the literature [9, 24]. Although it is possible to calculate the thermal radiation from a thermal body by thermodynamics, the complexity of the boundary conditions associated with the biological body makes this approach impractical.

Besides the usage of imaging techniques, molecular marker detection of the existence of malignant cells is a recent beneficial development from the advances in molecular biology. Tremendous biological variability exists in lesions most likely to progress to invasive breast cancer and the ways in which they grow and spread. According to [34], pathologists have long understood that breast lesions exist on a histological continuum. Cells in the stem cell compartment, the normal terminal duct lobular units, begin to undergo a hyperplastic proliferative response, which is the first phase in the continuum. Understanding the biology of breast cancer and performing diagnosis based on the molecular biology instead of the morphology would likely lead to even earlier breast cancer detection before it becomes malignant.

In principle, tumor gene expression profiles can be used as molecular fingerprints that accurately classify the tumor. However, according to a recent survey [56], the usage of genomic data has not had a great impact, although it is expected to do so in the future. The major challenges in gene expression data analysis come from the high dimension of the data set versus a much smaller amount of samples and the noise in the gene expression. Several attempts to overcome this problem have been reported [48, 52].

As a conclusion to this section, we would like to mention an interesting work conducted recently, although its influence on diagnosis is yet to be investigated. Alexjander and Deamer [1] proposed to study the sound (rhythms and frequencies) made within the human body by accessing the infrared frequencies of DNA bases. Imagine if we can "hear" the body, would a pleasing pattern to the ear indicate a healthy subject? Or would different patterns present a sign of a certain disease?

6.3 Smart Image-Processing Approaches to IR Images

Computer-aided diagnosis (CAD) has been playing an important role in the analysis of IR images, as human examination of images is often influenced by various factors like fatigue, carelessness, and more. The detection accuracy is also confined by the limitations of the human visual system. On top of all these factors, a shortage of qualified radiologists also put an urgent demand on the development of CAD technologies.

Studies reported in [34] showed that a significant number of cancers (30–65%) can be visualized on prior mammograms in retrospective review. Double reading of mammograms by two radiologists can improve the detection rate of cancer (which yields a 4–15% increase in cancers detected) but is expensive and time consuming. The goal of CAD is to improve detection rates in a more efficient and cost-effective manner.

Currently, research on smart image-processing algorithms on IR images tends to improve the detection accuracy from three perspectives: smart image enhancement and restoration algorithms; asymmetry analysis of the thermogram, including automatic segmentation approaches; and feature extraction and classification.

6.3.1 Smart Image Enhancement and Restoration Algorithms

One of the problems with thermograms that has put IR imaging in a somewhat disadvantage situation is its lack of resolution due to blur compounded by rather high levels of noise. Snyder et al. [54] developed an algorithm to increase the effective resolution of thermograms by a 2:1 ratio while removing the noise and preserving edges in the image. This algorithm is based on a minimization strategy known as *mean field annealing*, which takes into account processes of blur, noise, and image correlations to make an optimal estimate of the missing pixels.

Researchers at the Massachusetts Institute of Technology (MIT) attempted to enhance the resolution of IR images through another route. The minimally invasive optical biopsy system developed at MIT [6] uses infrared light in conjunction with an intravenously injected dye and special computer software to create a clear, high-contrast image that could easily allow physicians to detect breast masses and determine if they are benign or malignant.

To eliminate the effect of various thermal environmental conditions, Kakuta et al. [30] developed a "human thermal model" such that IR images taken under different conditions can be compared through normalization of skin surface temperature. The model is based on a numerical calculation of the bioheat transfer equations, and a 16-cylinder-segment model is used as the geometry of the human body.

Kaczmarek and Nowakowski [29] proposed the use of active dynamic thermography (ADT), commonly adopted in nondestructive testing of materials, to further enhance the image quality. ADT analyzes thermal transients after the application of external thermal excitation. Some preliminary results have shown the promise of this approach.

Dynamic thermography is another technique recently proposed to better understand IR images. Because of the interference from complex vascular patterns and the existence of cold tumors, in breast cancer detection researchers have proposed to monitor the thermal recovery process after exposure of cold stress by *sequential* thermography or by digital subtraction thermography. Studies [43] have shown an increase in sensitivity of breast imaging.

Anbar et al. [4] proposed the dynamic area telethermometry (DAT) technique. It has been demonstrated as applicable to any quantitative pathophysiological assessment. The authors demonstrated that using classical fast Fourier transform (FFT) and elementary statistics, the large amount of sequential observations can be reduced to a single quantitative diagnostic parameter without the participation of human experts. Other related work was also reported in [10, 16].

6.3.2 Asymmetry Analysis

Comparisons between contralateral images are routinely done by radiologists. When the images are relatively symmetrical, small asymmetries may indicate a suspicious region. This is the underlying philosophy in the use of asymmetry analysis for mass detection in a breast cancer study [15] as well as in anomaly detection of other parts of the human body. Unfortunately, these small asymmetries might not be easy to detect, and it is important to design an automatic approach to eliminate human factors. There have been a few articles addressing techniques for asymmetry analysis of mammograms [15, 63, 64].

Head et al. [22, 37] analyzed the asymmetric abnormalities in IR images. In their approach, the image is segmented first by operator. Then, breast quadrants are derived automatically based on a unique point of reference (i.e., the chin and the lowest, rightmost, and leftmost points of the breast). Qi and Head [50] developed an automatic approach to asymmetry analysis in IR images. It includes automatic segmentation and pattern classification. Hough transform is used to extract the four feature curves that can uniquely segment the left and right breasts. The feature curves include the left and the right body boundary curves and the two parabolic curves indicating the lower boundaries of the breasts. Mabuchi et al. [40] designed a computerized thermographic system, which would produce images of the distribution of temperature differences between the affected side and the contralateral healthy side. Because there is no standard skin surface temperature, the system measures the body surface temperature of each pixel in the affected area and subtracts from it the body surface temperature of the corresponding pixel in the symmetrically located contralateral healthy area to generate the difference image. Figure 6.2 is a side-by-side comparison of thermograms of a normal patient and a patient with ductal carcinoma in the left breast.

6.3.3 Feature Extraction and Classification

On segmentation, different features can be extracted from the segments. Asymmetric abnormalities can then be identified based on mature pattern classification techniques. In this process, feature extraction is crucial to the success of computer-aided

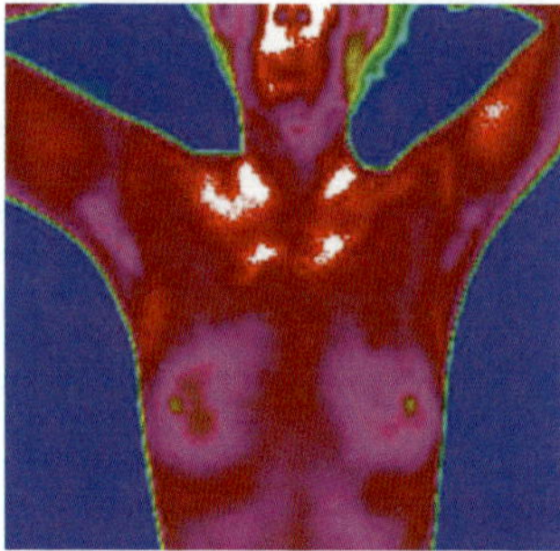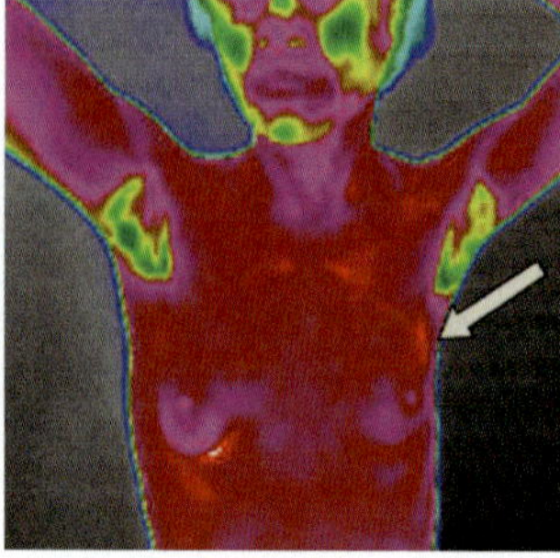

Fig. 6.2 Thermal signatures of a normal patient (*left*) with nearly symmetric temperature readings and of a patient with ductal carcinoma in the left breast (*right*) with apparent asymmetric thermal distribution

diagnosis. The high-order statistics (e.g., variance, skewness, and kurtosis) and joint entropy were shown in [33] to be the most effective features to measure the asymmetry, while low-order statistics (e.g., mean) and entropy do not assist asymmetry detection. Jakubowska et al. [26] also addressed the importance of using statistical parameters (1st and 2nd order) in extracting thermal signatures for asymmetry analysis.

Szu et al. [58] proposed a new paradigm shift that uses at least two dual-band (mid and long) IR imaging cameras operating simultaneously on the patient. This system enables a smart brainlike neural network algorithm, the Lagrange constraint neural network (LCNN), to achieve submillimeter scaling of the close-up breast imaging for vascular and angiogenesis effects as well as stage-zero detection of ductal carcinoma in situ.

The techniques mentioned are just samples of activities reported in recent conferences, workshops, and symposia. Another trend worth mentioning is the transition of automatic target recognition (ATR) algorithms developed for military application to medicine. "From tanks to tumors" [25, 45] has been the theme of this transition, and the rich collection of ATR algorithms that the military has sponsored will greatly improve the state-of-the-art of CAD development.

6.3.4 The Thermal-Electric Analog

Liu et al. [38, 51] presented a new method for analyzing a thermal system based on an analogy to electrical circuit theory, referred to as *thermal-electric analog*. Figure 6.3 illustrates this analogy, where the heat source S inside the human body can be simulated as a battery with voltage U_S; the heat loss inside the heat source can be simulated as the heat loss on a resistor R_S. The *temperature* of the heat source can then correspond to the *voltage* of the battery, and the *heat current* to the *circuit current*. Similarly, we can map the heat source in the air (outside the human body) as

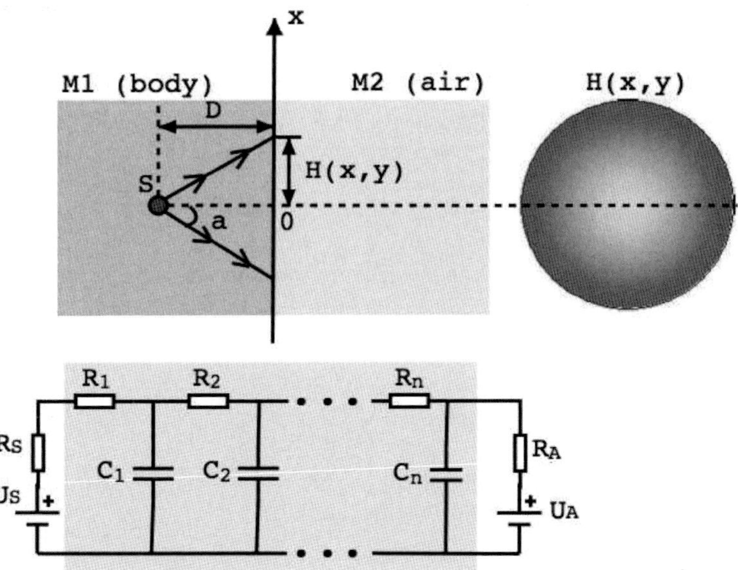

Fig. 6.3 The thermal-electric analog

U_A and the heat loss as R_A. The set of R_i's and C_i's corresponds to the unit heat resistance and heat capacity along each radiation line. The circuit in Fig. 6.3 only shows the analogy for one radiation line. In the study of breast cancer, it is reasonable to assume that the medium between the heat source (S) and the surface is homogeneous. Therefore, the radiation pattern sensed by the IR camera at the surface should have a Gaussian-like distribution as shown in Fig. 6.3. The surface temperature $H(x,y)$ corresponds to the output voltage. $U(x,y)$ can then be calculated by Eq. (6.2):

$$U(x,y) = U_S - \frac{\sum_{i=1}^{n} R_i}{R_S + R_A + \sum_{i=1}^{n} R_i} \times (U_S - U_A) \tag{6.2}$$

and

$$n = \left\lfloor \frac{D}{R_0 \cos a} \right\rfloor, a = \arctan \frac{D}{\sqrt{x^2 + y^2}}$$

where $\lfloor m \rfloor$ represents the largest integer less than m, n is the number of resistors used in the circuit, D is the depth of the heat source, and R_0 is the unit heat loss (or the *heat resistance rate*) in a certain medium. Different parts of the human body have different heat resistance rates, as shown in Table 6.2.

The thermal-electric analog provides a convenient way to estimate the depth of the heat source. This method does not require a direct solution to the inverse heat transfer problem. It can be used to estimate the depth of the heat source and to help understand the metabolic activities undergoing within the human body. The method

Table 6.2 Heat resistance rate of different body parts

Body parts	Heat resistance rate
Fatty tissue	0.1 to 0.15°C/cm
Muscle	0.2°C/cm
Bone	0.3 to 0.6°C/cm

has been used in early breast cancer detection and has achieved high sensitivity.
A diagnosis protocol adopted includes the following six steps:

1. Growth pattern of lymph nodes in the armpits
2. Size of the abnormal area
3. Appearance of the abnormal area
4. Vascular pattern
5. Nipples and areola pattern
6. Dynamic diagnosis with outside agents (antibiotic, etc.)

6.4 New-Generation Infrared Technologies

Infrared technology owes its origin to military research and development in the
Vietnam era for airborne applications. We focus our discussion on the advances
in the detector technologies, especially uncooled camera development.

TIR imaging generally requires the use of a cooling system in the form of a
nitrogen or compressed air cooling bottle, which contains crystals like germanium,
whose electrical resistance is very sensitive to heat. Due to their size, weight, and
complexity, these systems were limited to fixed deployment like tripod mounting.

The advance in solid-state models has made a new class of sensors possible,
the uncooled detector design. The Defense Advanced Research Projects Agency
(DARPA) issued a Broad Agency Announcement (BAA) in 1999 [47] that solicits
proposals for increasing the performance of the uncooled IR sensors to their theoretical limit. The objective for the thermal sensitivity is set at less than 10 millikelvin
with a pixel size less than or equal to $25\,\mu$m. As far as the array size, high-
performance arrays for long-range systems can be as large as $960 \times 1,280$ elements,
while arrays for microsensors may be as small as 240×320 elements [47].

Two technologies were developed at about the same time to make uncooled sensors a reality, the barium strontium titanate (BST) technology by Raytheon Corporation and the microbolometer technology by Honeywell Corporation. BST cameras
use a ferroelectric detector that converts the infrared energy to a change in *capacitance*. The BST detectors incorporate a mechanical chopper wheel/motor assembly
that rotates at 30 times per second during operation to enable the sensor to refresh itself 30 times a second. The images produced can be rather choppy, with dark ghosts
produced around hot images and multiple images of the same object smeared across
the screen during movement of the camera.

The microbolometer technology, which is thermal-electric in nature, converts infrared energy to a change in *resistance* instead of capacitance as in the BST technology. The microbolometer-based cameras do not require continuously moving parts and thus can provide high-quality images without the choppiness and ghosting associated with the chopper wheel used in BST cameras. The pictures are also smoother and clearer since automatic brightness control instead of mechanical controls is achieved using advanced digital signal-processing techniques.

Because the uncooled cameras do not require a cooling system, they are much lighter, smaller, more reliable, and less expensive compared to the cooled cameras. Currently, the uncooled cameras are approaching the thermal sensitivity of the cooled ones ($0.05°C$ or $0.02°C$ of uncool vs. $0.01°C$ of cool) and are very popular in breast imaging. Wiecek conducted a brief comparison [62] between uncooled thermal and deeply cooled QWIP (quantum well IR photodetector) detectors and discussed the limits in both technologies.

6.5 Summary

This chapter discussed recent research achievements in early breast cancer detection. The objective was to show that due to the advances in image-processing techniques, the pathophysiological-based understanding of theromograms, and the new-generation IR sensing technology, IR imaging is mature for use as a first-line supplement to both health monitoring and clinical diagnosis. We have established a Web site [49] to facilitate researchers working in the field of medical thermography to exchange research findings. We welcome contributions to enrich this list of collections.

Chapter's References

1. S. Alexjander and D. Deamer. The infrared frequencies of DNA bases: science and art. *IEEE Engineering in Medicine and Biology Magazine*, 18(2):74–79, Mar–Apr 1999
2. M. Anbar. Quantitative and dynamic telethermometry—a fresh look at clinical thermology. *IEEE Engineering in Medicine and Biology Magazine*, 14(1):15–16, Jan–Feb 1995
3. M. Anbar. Clinical thermal imaging today. *IEEE Engineering in Medicine and Biology Magazine*, 17(4):25–33, July–Aug 1998
4. M. Anbar, L. Milescu, A. Naumov, C. Brown, T. Button, C. Carly, and K. AlDulaimi. Detection of cancerous breasts by dynamic area telethermometry. *IEEE Engineering in Medicine and Biology Magazine*, 20(5):80–91, Sept–Oct 2001
5. M. Bale. High-resolution infrared technology for soft-tissue injury detection. *IEEE Engineering in Medicine and Biology Magazine*, 17(4):56–59, July–Aug 1998
6. M. Braunstein, R.W. Chan, and R.Y. Levine. Simulation of dye-enhanced near-IR transillumination imaging of tumors. In *Proceedings of the 19th EMBS Annual International Conference*, Vol. 2, pages 735–739, Chicago, 1997
7. BreastThermography. Mammography/thermography/ultrasound—what's the difference? http://www.breastthermography.com/mammography_thermography.htm

8. BreastThermography. Medical infrared imaging for breast cancer and early detection. `http://www.breastthermography.com`

9. C.L. Chan. Boundary element method analysis for the bioheat transfer equation. *ASME Journal of Heat Transfer*, 114:358–365, 1992

10. I. Fujimasa, T. Chinzei, and I. Saito. Converting far infrared image information to other physiological data. *IEEE Engineering in Medicine and Biology Magazine*, 19(3):71–76, May–June 2000

11. P. Gamigami. *Atlas of Mammography: New Early Signs in Breast Cancer*. Blackwell Science, Boston, 1996

12. M. Gautherie. Thermobiological assessment of benign and malignant breast diseases. *American Journal of Obstetrics and Gynecology*, 147(8):861–869, 1983

13. M. Gautherie. *Atlas of Breast Thermography with Specific Guidelines for Examination and Interpretation*. PAPUSA, Milan, 1989

14. J. Gershen-Cohen, J. Haberman, and E.E. Brueschke. Medical thermography: a summary of current status. *Radiology Clinics of North America*, 3:403–431, 1965

15. W.F. Good, B. Zheng, Y. Chang, X.H. Wang, and G.S. Maitz. Generalized procrustean image deformation for subtraction of mammograms. In *Proceeding of SPIE Medical Imaging–Image Processing*, Vol. 3661, pages 1562–1573, SPIE, San Diego, 1999

16. Yu.V. Gulyaev, A.G. Markov, L.G. Koreneva, and P.V. Zakharov. Dynamical infrared thermography in humans. *IEEE Engineering in Medicine and Biology Magazine*, 14(6):766–771, Nov–Dec 1995

17. J. Haberman. The present status of mammary thermography. *Ca—A Cancer Journal for Clinicians*, 18:314–321, 1968

18. R.S. Handley. The temperature of breast tumors as a possible guide to prognosis. *Acta Unio Int Contra Cancrum*, 18:822, 1962

19. J.R. Harding. Investigating deep venous thrombosis with infrared imaging. *IEEE Engineering in Medicine and Biology Magazine*, 17(4):43–46, July–Aug 1998

20. G.A. Hay. *Medical Image: Formation, Perception and Measurement*. Institute of Physics and Wiley New York, 1976

21. J.F. Head and R.L. Elliott. Infrared imaging: making progress in fulfilling its medical promise. *IEEE Engineering in Medicine and Biology Magazine*, 21(6):80–85, Nov–Dec 2002

22. J.F. Head, C.A. Lipari, and R.L. Elliott. Computerized image analysis of digitized infrared images of the breasts from a scanning infrared imaging system. In *Proceedings of the 1998 Conference on Infrared Technology and Applications XXIV, Part I*, Vol. 3436, pages 290–294, SPIE, San Diego, 1998

23. J.F. Head, F. Wang, C.A. Lipari, and R.L. Elliott. The important role of infrared imaging in breast cancer. *IEEE Engineering in Medicine and Biology*, pages 52–57, May–June 2000

24. T.R. Hsu, N.S. Sun, and G.G. Chen. Finite element formulation for two dimensional inverse heat conduction analysis. *ASME Journal of Heat Transfer*, 114:553–557, 1992

25. J.M. Irvine. Targeting breast cancer detection with military technology. *IEEE Engineering in Medicine and Biology Magazine*, 21(6):36–40, Nov–Dec 2002

26. T. Jakubowska, B. Wiecek, M. Wysocki, and C. Drews-Peszynski. Thermal signatures for breast cancer screening comparative study. In *Proceedings of the 25th Annual International Conference of the IEEE EMBS Conference*, Vol. 2, pages 1117–1120, 2003

27. B.F. Jones. A reappraisal of the use of infrared thermal image analysis in medicine. *IEEE Transactions on Medical Imaging*, 17(6):1019–1027, Dec. 1998

28. B.F. Jones and P. Plassmann. Digital infrared thermal imaging of human skin. *IEEE Engineering in Medicine and Biology Magazine*, 21(6):41–48, Nov–Dec 2002

29. M. Kaczmarek and A. Nowakowski. Analysis of transient thermal processes for improved visualization of breast cancer using IR imaging. In *Proceedings of the 25th Annual International Conference of the IEEE EMBS Conference*, Vol. 2, pages 1113–1116, 2003

30. N. Kakuta, S. Yokoyama, and K. Mabuchi. Human thermal models for evaluating infrared images. *IEEE Engineering in Medicine and Biology Magazine*, 21(6):65–72, Nov–Dec 2002

31. J. Keyserlingk. Time to reassess the value of infrared breast imaging? *Oncology News Internatinal*, 6(9), 1997

32. J.R. Keyserlingk, P.D. Ahlgren, E. Yu, N. Belliveau, and M. Yassa. Functional infrared imaging of the breast. *IEEE Engineering in Medicine and Biology*, pages 30–41, May–June 2000

33. P. T. Kuruganti and H. Qi. Asymmetry analysis in breast cancer detection using thermal infrared images. In *Proceedings of the 2nd Joint EMBS-BMES Conference*, Vol. 2, pages 1129–1130, Oct 2002

34. Institute of Medicine L. Newman for the Committee on the Early Detection of Breast Cancer, National Cancer Policy Board and Commission on Life Sciences. *Developing Technologies for Early Detection of Breast Cancer: A Public Workshop Summary*. National Academies Press, Washington, DC, 2000

35. R.N. Lawson. Implications of surface temperature in the diagnosis of breast cancer. *Canadian Medical Association Journal*, 75:309–310, 1956

36. R.N. Lawson and M.S. Chughtai. Breast cancer and body temperatures. *Canadian Medical Assocition Journal*, 88:68–70, 1963

37. C.A. Lipari and J.F. Head. Advanced infrared image processing for breast cancer risk assessment. In *Proceedings for 19th International Conference of IEEE/EMBS*, pages 673–676, IEEE Chicago, Oct 30–Nov 2 1997

38. Z. Liu and C. Wang. Method and apparatus for thermal radiation imaging. US Patent No. 6,023,637, 2000

39. K. Lloyd-Williams and R.S. Handley. Infrared thermometry in the diagnosis of breast disease. *Lancet*, (2):1378–1381, 1961

40. K. Mabuchi, T. Chinzei, I. Fujimasa, S. Haeno, K. Motomura, Y. Abe, and T. Yonezama. Evaluating asymmetrical thermal distributions through image processing. *IEEE Engineering in Medicine and Biology Magazine*, 17(2):47–55, Mar-Apr 1998

41. E.Y.K. Ng and N.M. Sudarshan. Numerical computation as a tool to aid thermographic interpretation. *Journal of Medical Engineering and Technology*, 25(2):53–60, Mar–Apr 2001

42. E.Y.K. Ng, L.N. Ung, F.C. Ng, and L.S.J. Sim. Statistical analysis of healthy and malignant breast thermography. *Journal of Medical Engineering and Technology*, 25(6):253–263, Nov–Dec 2001

43. Y. Ohashi and I. Uchida. Applying dynamic thermography in the diagnosis of breast cancer. *IEEE Engineering in Medicine and Biology*, pages 42–51, May–June 2000

44. O. Olsen and P.C. Gotzsche. Cochrane review on screening for breast cancer with mammography. *Lancet*, 9290:1340–1342, Oct 2001

45. J.L. Paul and J.C. Lupo. From tanks to tumors. *IEEE Engineering in Medicine and Biology Magazine*, 21(6):34–35, Nov–Dec 2002

46. H.H. Pennes. Analysis of tissue and arterial blood temperature in resting human forearm. *Journal of Applied Physiology*, 2:93–122, 1948

47. DARPA MTO Program. Baa99-30: Uncooled thermal imaging sensors. Commerce Business Daily, July 12, 1999

48. H. Qi. Feature selection and knn fusion in molecular classification of multiple tumor types. In *International Conference on Mathematics and Engineering Techniques in Medicine and Biological Sciences (METMBS)*.

49. H. Qi. Thermal infrared imaging in early detection of breast cancer—a survey of recent research. http://aicip.ece.utk.edu/research/irsurvey.htm, 2003

50. H. Qi and J. Head. Asymmetry analysis using automatic segmentation and classification for breast cancer detection in thermograms. In *Proceedings of the 23rd Annual International Conference of the IEEE EMBS*, Vol. 3, pages 2866–2869, Turkey, Oct 2001. IEEE.

51. H. Qi, P.T. Kuruganti, and Z. Liu. Early detection of breast cancer using thermal texture maps. In *IEEE International Symposium on Biomedical Imaging: Macro to Nano*, pages 309–312, Washington, DC, 2002.

52. S. Ramaswamy, P. Tamayo, R. Rifkin, et al. Multiclass cancer diagnosis using tumor gene expression signatures. *Proceedings of the National Academy of Sciences of the United States of America*, 98(26):15149–15154, Dec 2001

53. E.F.J. Ring. Progress in the measurement of human body temperature. *IEEE Engineering in Medicine and Biology Magazine*, 17(4):19–24, July–Aug 1998

54. W.E. Snyder, H. Qi, L. Elliott, J. Head, and C.X. Wang. Increasing the effective resolution of thermal infrared images. *IEEE Engineering in Medicine and Biology Magazine*, 19(3):63–70, May–June 2000
55. American Cancer Society. Cancer facts and figures. `http://www.cancer.org/docroot/STT/stt_0.asp`
56. R. Stevens, C. Goble, P. Baker, and A. Brass. A classification of tasks in bioinformatics. *Bioinformatics*, 17(2):180–188, 2001
57. T. Szabo, L. Fazekas, L. Geller, F. Horkay, B. Merkely, T. Gyongy, and A. Juhasz-Nagy. Cardiothermographic assessment of arterial and venous revascularization. *IEEE Engineering in Medicine and Biology Magazine*, 19(3):77–82, May–June 2000
58. H. Szu, I. Koprival, P. Hoekstra, N. Diakides, J. Buss, and J. Lupo. Lagrange constraint neural net de-mixing enabled multispectral breast imaging. In *IEEE EMBS*, 2002
59. Thermology. `http://www.thermology.com/history.htm`.
60. D.J. Watmough. The role of thermographic imaging in breast screening, discussion by C.R. Hill. In *Medical Images: Formation, Perception and Measurement 7th L.H. Gray Conference: Medical Images*, pages 142–158, 1976
61. J. Whale. An introduction to dynamic radiometric thermal diagnostics and dielectric resonance management procedures. *Positive Health*.
62. B. Wiecek. Advances in infrared technology—quantum well versus thermal detectors. In *Proceedings of the Second Joint EMBS/BMES Conference*, Vol. 2, pages 1135–1136, 2002.
63. F.F. Yin, M.L. Giger, K. Doi, C.J. Vyborny, and R.A. Schmidt. Computerized detection of masses in digital mammograms: automated alignment of breast images and its effect on bilateral-subtraction technique. *Medical Physics*, 21:445–452, 1994
64. B. Zheng, Y.H. Chang, and D. Gur. Computerized detection of masses from digitized mammograms: comparison of single-image segmentation and bilateral image subtraction. *Academic Radiology*, 2(12):1056–1061, Dec 1995.

Part III
Hyperspectral Imagery

Chapter 7
Hyperspectral Image Analysis for Skin Tumor Detection

Seong G. Kong and Lae-Jeong Park

Abstract This chapter presents hyperspectral imaging of fluorescence for noninvasive detection of tumorous tissue on mouse skin. Hyperspectral imaging sensors collect two-dimensional (2D) image data of an object in a number of narrow, adjacent spectral bands. This high-resolution measurement of spectral information reveals a continuous emission spectrum for each image pixel useful for skin tumor detection. The hyperspectral image data used in this study are fluorescence intensities of a mouse sample consisting of 21 spectral bands in the visible spectrum of wavelengths ranging from 440 to 640 nm. Fluorescence signals are measured using a laser excitation source with the center wavelength of 337 nm. An acousto-optic tunable filter is used to capture individual spectral band images at a 10-nm resolution. All spectral band images are spatially registered with the reference band image at 490 nm to obtain exact pixel correspondences by compensating the offsets caused during the image capture procedure. The support vector machines with polynomial kernel functions provide decision boundaries with a maximum separation margin to classify malignant tumor and normal tissue from the observed fluorescence spectral signatures for skin tumor detection.

Keywords: Hyperspectral imaging · Skin tumor detection · Fluorescence spectroscopy · Support vector machines

7.1 Introduction

When cells lose the ability to grow and divide normally, they may grow out of control to form a tumor. While tumors can be benign, malignant tumors refer to cancers. Skin cancer is one of the most common cancers in the United States, and its incidence is increasing dramatically. Cancers that develop from melanocytes, the pigment-producing cells of the skin, are called *melanomas* [1]. Melanomas can spread quickly to other parts of the body through the lymph system or through

R.I. Hammoud (ed.), *Augmented Vision Perception in Infrared: Algorithms and Applied Systems,* Advances in Pattern Recognition, DOI 10.1007/978-1-84800-277-7_7,

the blood. According to the statistics of the American Cancer Society, more than 1 million people in the United States are diagnosed with nonmelanoma skin cancer every year, and 59,580 persons were diagnosed with melanoma in 2005 [2]. These estimates do not include approximately 1.3 million cases of basal and squamous cell skin cancers that exist in the same time period. For most skin cancer patients, including those with melanoma and nonmelanoma skin cancers, early diagnosis and thorough treatment, such as complete resection, are the keys to gaining a favorable prognosis.

Current diagnostic methods for skin cancers rely on physical examination of the lesion in conjunction with skin biopsy as a definitive test. A biopsy involves the removal of tissue samples from the body for examination under a microscope to assist in diagnosis. For large lesions, the biopsy may require substantial tissue removal. Although this protocol for skin lesion diagnosis has been accepted as the gold standard, the biopsy is an invasive, and often subjective, diagnostic technique that requires both trained professionals and significant waiting time. Laboratory results for the determination of histopathology of a suspected tumor may generally take several days. Since suspicious areas are identified by visual inspection alone, there are a significant number of false positives that undergo biopsy. Conversely, many malignant lesions can also be overlooked. There is an urgent need for objective criteria that would aid the clinician in evaluating whether biopsy is required.

Hyperspectral imaging [3] combines conventional imaging and spectroscopy to acquire spatial information from an object measured at a sequence of individual wavelengths across a sufficiently broad spectral range, which can be used to uniquely characterize and identify given material over a sufficiently broad spectral band. Hyperspectral imaging sensors provide continuous spectral signatures for each image cell obtained from dozens or hundreds of narrow, adjacent spectral channels at a high spectral resolution. Such a high-resolution measurement of spectral signatures enables hyperspectral imaging to reveal critical information that many conventional imaging techniques may not be able to deliver. Hyperspectral imaging has found a wide range of scientific and industrial applications that include remote sensing, environmental monitoring, geological search or mineral mapping, atmospheric composition analysis, military target recognition, medical diagnostics [4,5], and food safety inspection [6,7].

This chapter presents a noninvasive skin tumor detection technique based on the analysis of spectral signatures measured from hyperspectral fluorescence imaging. With high-resolution spectral information at multiple wavelengths, hyperspectral imaging enables reliable detection of subtle differences in the reflectance or fluorescence signature characteristics of biological tissues. In this study, the fluorescence signals emitted by the skin tissue are collected and analyzed to determine whether the skin region of interest (ROI) is tumorous. Many compounds emit fluorescence in the visible spectral range when fluorophores within a tissue medium are excited with external ultraviolet (UV) radiation. Different tissues exhibit unique spectral characteristics of absorption and reflection at particular bandwidths. The altered biochemical and morphological state of the neoplastic tissue is reflected in the spectral characteristics of the measured fluorescence. A pattern classifier, trained with a set of labeled sample data, classifies fluorescence signatures of the object to determine if tumorous cells exist. Among a number of spectral signature classification methods, the support vector machines (SVMs) [8,9] provide an optimal classifier through structural risk minimization in terms of generalization capability. The SVM classifier finds a decision boundary with maximum separating margin between the classes of spectral signatures.

7.2 Hyperspectral Fluorescence Imaging

7.2.1 Principles of Hyperspectral Imaging

Hyperspectral imaging describes the spatial and spectral information of an object in a form of "data cube." Spatially coregistered pixels are combined into a vector representing the spectral signature of the object. A hyperspectral image can be represented by a three-dimensional (3D) volume of data $I(m,n,\lambda)$, where (m, n) denotes the spatial coordinates of a pixel location in the image, and λ indicates the spectral coordinate. The value stored in $I(m,n,\lambda_i)$ is the spectral intensity, such as reflectance or fluorescence of the pixel (m,n) at the wavelength λ_i corresponding to the ith spectral band ($i = 1, 2, \ldots , L$), where L is the number of spectral bands. Spectral resolution is a measure of the narrowest spectral feature that can be resolved by sensor. Many hyperspectral imaging sensors generate hundreds of spectral bands in the visible to near-infrared region of the spectrum at a spectral resolution of 10 nm. Figure 7.1 shows the 3D nature of hyperspectral fluorescence image data. The fluorescence signals are measured in the range of 425 to 711 nm from a chicken sample excited with 330-nm UV radiation.

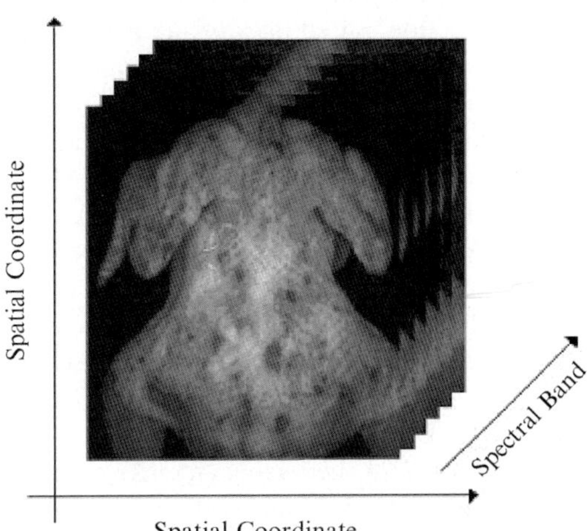

Fig. 7.1 Representation of 3D nature of hyperspectral fluorescence image data

Fluorescence is a phenomenon that, when light is absorbed by a molecule at a given wavelength, the molecule emits light at a longer wavelength. Fluorescence is the result of a three-stage process that occurs in certain molecules called *fluorophores* or *fluorescent dyes*. A *fluorescent probe* is a fluorophore designed to localize within a specific region of a biological specimen or to respond to a specific stimulus. In the excitation stage, a photon is supplied by an external source such as an incandescent lamp or a laser and absorbed by the fluorophore, creating an excited state that exists for a short duration, typically 1–10 ns. During this time, the energy is partially dissipated, yielding a relaxed singlet excited state from which fluorescence emission originates. However, not all the molecules initially excited by absorption return to the ground state by fluorescence emission. The *fluorescence quantum yield*, which is the ratio of the number of fluorescence photons emitted to the number of photons absorbed, is a measure of the relative extent to which these processes occur. A photon is emitted, returning the fluorophore to its ground state. Due to energy dissipation during the excited-state lifetime, the energy of this photon is lower, and therefore of longer wavelength, than the excitation photon. The difference in energy or wavelength, called the *Stokes shift*, is fundamental to the sensitivity of fluorescence techniques because it allows emission photons to be detected against a low background, isolated from excitation photons.

7.2.2 Hyperspectral Imaging Experiment

The Advanced Biomedical Science and Technology Group at Oak Ridge National Laboratory (ORNL) in Tennessee has developed a hyperspectral imaging system capable of reflectance and fluorescence imaging for environmental sensing and medical diagnostics [10, 11]. Figure 7.2 shows a schematic diagram of hardware components of the ORNL hyperspectral imaging system setup capable of fluorescence imaging for skin cancer detection. This system consists of fiber probes for image signal collection, an endoscope, an acousto-optic tunable filter (AOTF) for wavelength selection, a laser excitation source, and endoscopic illuminator (Olympus CLV-10) equipped with a 300-W CW xenon arc lamp source, a charge-coupled device (CCD) color camera (Sony CCD-Iris) for reflection detection, an intensified charge-coupled device (ICCD) camera (IMAX-512-T-18 Gen. II) for fluorescence imaging, and a spectrometer for single-point spectroscopic measurements. That hyperspectral imaging system captures hyperspectral images of 21 spectral bands of wavelength ranging from 440 to 640 nm.

Both fluorescence and reflected lights can be collected through the endoscope into the AOTF device via collimating lenses. A mirror placed in front of the AOTF projects the acquired images onto the ICCD camera for fluorescence imaging and onto the CCD camera for reflection measurement. Excitation for the reflectance images is accomplished using an endoscopic illuminator. The reflection source was coupled to a gastrointestinal endoscope (Olympus T120) equipped with an imaging

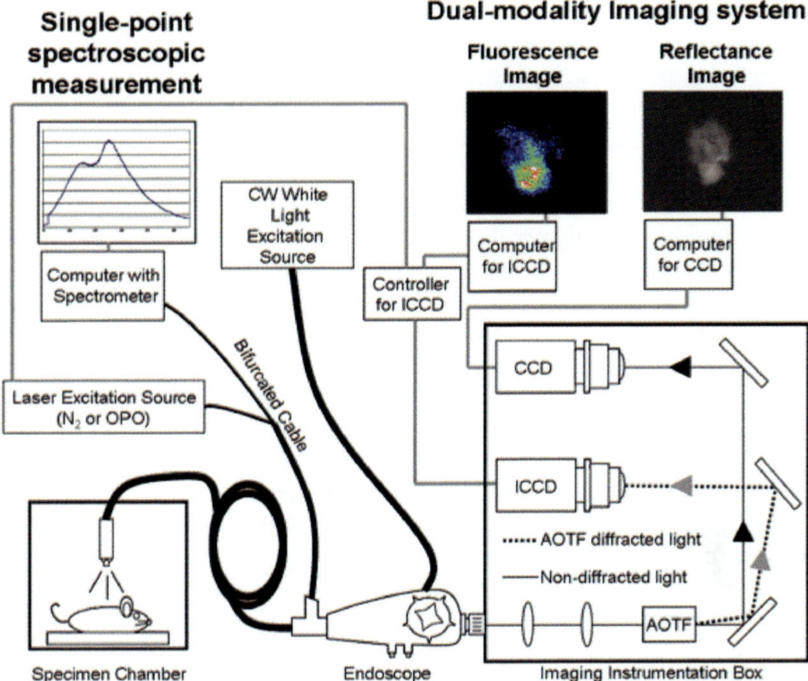

Fig. 7.2 Hardware components of the ORNL hyperspectral imaging system

bundle. Excitation for the fluorescence spectra and images is accomplished using an LSI pulsed nitrogen laser (VSL-337) with a maximum repetition rate of 20 Hz.

The fluorescent light emitted by the tissues is diffracted by the AOTF (Brimrose TEAF10-0.4-0.65-S) at a 6° angle from the undiffracted (zero-order) beam, thus separating the reflected image from the fluorescent image. Individual wavelengths diffracted by the AOTF are thus sent to the ICCD. A Brimrose AOTF controller (VFI-160-80-DDS-A-C2) controls the AOTF. The controller sends a radio-frequency (RF) signal to the AOTF based on input provided by the Brimrose software. Wavelength selection takes place in microseconds, enabling ultrafast modulation of wavelength output to the ICCD. Wavelength-specific images were taken between 440 and 640 nm every 10 nm.

Fluorescence images were acquired by gating the ICCD camera. A timing generator incorporated into the ICCD camera's controller (ST-133) allowed the ICCD to operate in the pulsed mode with a wide range of programmable functions. A 500-ns delay between the laser trigger and the detector activation was programmed to synchronize the laser and the detector. The intensifier was gated for 500 ns, during which a 5-ns laser pulse was delivered to the tissues. An image was captured 20 times per second, integrated by internal software, and output to a screen once per second. This allowed real-time fluorescence detection. Single-point spectroscopic measurements were taken to validate the experimental results. These measurements were taken at locations corresponding to the fluorescence imaging data. An Ocean Optics (USB2000) spectrometer was used as the wavelength-selection device rather than the AOTF.

The hyperspectral image data used in this experiment were laser-induced fluorescence images taken from a mouse sample. An adult mouse was injected subcutaneously with $100\,\mu L$ of Fischer rat 344 rat tracheal carcinoma cells (IC-12) to induce tumor formation. A nude mouse was used to allow the implanted cancer cells to be able to grow due to the lack of an immune system in the animal. The lack of hair on the mouse enabled imaging directly without the need for depilation. The subdermal injection was carried out as close to the skin surface as possible to allow tumor formation close to the skin surface. This also allowed experiments to be performed in vivo rather than after tissue extraction. Experiments performed on whole animals are a vast improvement over those performed on tissue samples due to closer simulation of in vivo human cancerous conditions. After injection, the nude mouse was incubated for a period of 4 days to allow tumor formation to occur. Once a tumor of approximately 1 cm was observed, the mouse was anesthetized for approximately 30 min to permit data collection.

The measured hyperspectral fluorescence images consist of 165×172 pixels with 21 spectral bands from the wavelength λ_1 (440 nm) to λ_{21} (640 nm) with 10-nm spectral resolutions in the spectral region. Such a fine spectral resolution provides sufficient information for the study of tumor detection. Figure 7.3a shows a reflectance image of a skin tumor region taken from a mouse sample. Figure 7.3b is a fluorescence image of the sixth band (490 nm) of the same spot. The lower part of the fluorescence image (a bright U-shaped area) corresponds to normal tissue, and the dark part above the normal tissue is a tumor region. Original spectral band images captured by the hyperspectral imaging system are not spatially aligned since the AOTF diffracts the light at different wavelengths [12, 13].

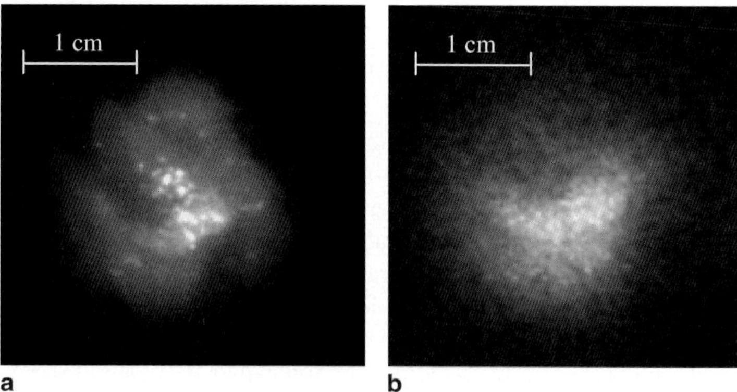

Fig. 7.3 Skin tumor region of a mouse skin sample: **a** reflectance image, **b** fluorescence image at 490-nm wavelength

The AOTF is an optical bandpass filter whose passband is electronically tunable using the acousto-optic interaction inside an optical medium. The AOTF uses birefringent crystal whose refractive index is changed on interaction with an acoustic wave. The most common types that operate from the near UV through the shortwave infrared region use tellurium dioxide (TeO_2) crystals. A piezoelectric oscillator bonded to the crystal converts a RF electrical signal into a high-frequency acoustic wave that propagates through the crystal. Changing the frequency of the transducer signal applied to the crystal alters the period of the refractive index variation of the birefringent crystal; therefore, the AOTF selects and transmits a single wavelength from the incoming light. This quick wavelength tuning, limited only by the acoustic transit time across the crystal, is important to record many spectral bands of hyperspectral image. The RF amplitude level applied to the transducer controls the filtered light intensity level. AOTFs show a fast response time (in microseconds), have spectral resolution of 1–2 nm, and exhibit a high extinction ratio.

In AOTFs, the acoustic and optical waves propagate at quite different angles through the crystal; therefore, there exists an offset between neighboring spectral band images. Figure 7.4a shows the amount of offsets at different wavelengths that the AOTF generates during the image-capturing procedure. The offsets are estimated by maximizing the mutual information between the given band images and the reference band image. The spectral band of the 490-nm wavelength was selected as a reference band with zero offset. A positive (negative) offset value indicates that the band image is shifted to the left (right) by the amount of pixels with respect to the reference band. Figure 7.4b illustrates that a spectral band image at 540 nm is spatially coregistered with the reference image at 490 nm after translating the 540-nm image to the left by the amount of offset (14 pixels).

All the spectral band images are spatially coregistered with the reference band image at 490 nm to obtain exact pixel-to-pixel correspondence by eliminating the offsets before extracting spectral signatures. Figure 7.5 shows eight sample spectral band images from wavelengths 450 to 590 nm after the registration discussed. The

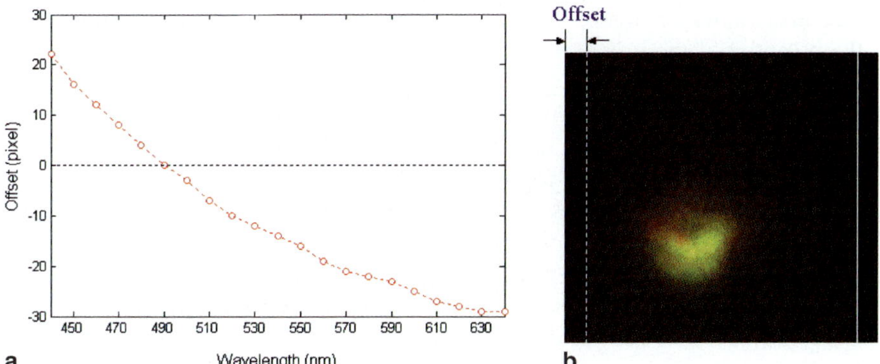

Fig. 7.4 a Amount of offsets of spectral band images caused by the AOTF with respect to the reference band at 490 nm, **b** registered band images of 490 and 540 nm

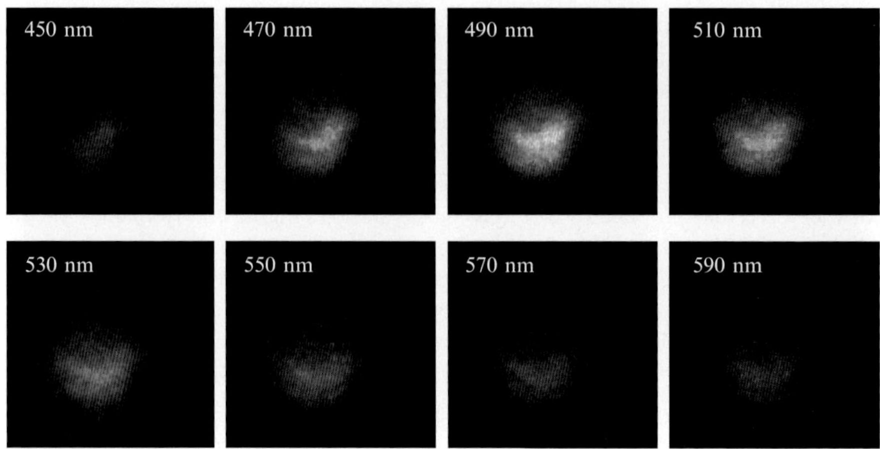

Fig. 7.5 Spectral band images from 450- to 590-nm wavelength

reference spectral band at 490 nm gives the strongest response in terms of fluorescence intensity. Each of these images represents the reflected (absorbed, transmitted) energy of the materials within the pixel at different wavelengths and bandwidths. For any given material, the amount of energy that is reflected will vary with wavelength. This important property allows us to separate distinct material types based on their response values for a given wavelength. By comparing the spectral signature of different materials, we may be able to distinguish between them, whereas we might not be able to if we only compared them at one wavelength.

7.3 Spectral Signatures of Normal and Malignant Skin Tissues

The experiment with a mouse sample having a tumor region provides an illustration of the potential of the proposed hyperspectral fluorescence imaging technique for in vivo skin cancer diagnosis. Optical imaging of reflectance demonstrates nearly uniform responses to both normal tissue and skin tumor areas, which makes visual classification of malignant tumors difficult based on reflectance images. In this study, N_2-pumped laser at 337-nm wavelength was used as an excitation source of the fluorescence signal. A nitrogen laser provides the lowest excitation wavelength, 337 nm, which corresponds to the highest energy. Longer wavelengths can be selected for excitation by using the tunable dye laser system. However, the use of shorter wavelengths tends to excite more components in tissues. This wavelength produces fluorescence spectra of spectral signatures useful to discriminate normal and malignant tissues.

Figure 7.6 shows fluorescence spectra taken from a mouse sample in terms of two categories of interest: normal tissue and malignant tumor. The spectral signatures

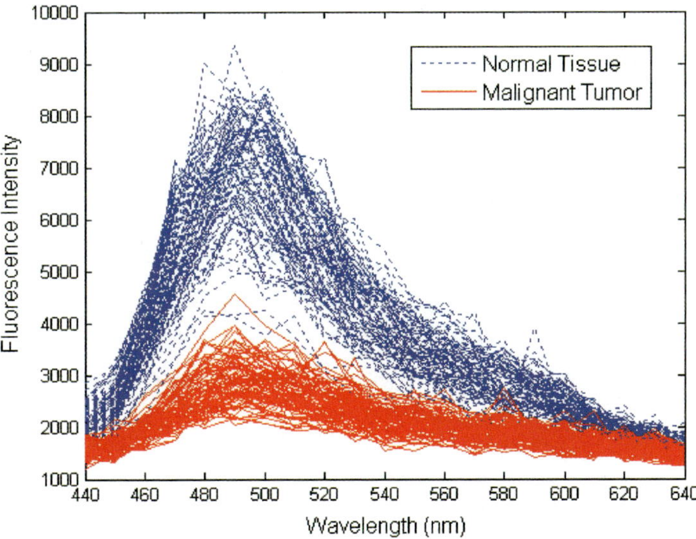

Fig. 7.6 Spectral signature of normal tissue and malignant tumor with fluorescence intensity measured with 337-nm wavelength laser excitation

show a peak response near the 490-nm wavelength, while the fluorescence intensity of malignant tumor is weaker than that of the normal tissue [14]. The higher fluorescence intensities in the normal tissue can be attributed to several factors, depending on what collection wavelength region is described. In the wavelength region of 450 to 510 nm, the higher fluorescence intensity seen in the normal tissues is due primarily to a decrease in collagen and elastin from normal to malignant tissues and a decrease in nicotinamide adenine dinucleotide (NADH) levels in the malignant tissues. The decreases in fluorescence seen in the images in the wavelength region between 500 and 600 nm can be attributed to whole blood absorption in this region. An increase in blood volume in the malignant tissues causes more of the fluorescence from fluorophores such as lipopigments and flavins to be absorbed.

The intensity of the fluorescence signal may not be a consistent parameter according to several factors, such as blood flow, hemoglobin absorption, and surface morphology. The distance between the tissue surface and probe can also affect the intensity of recorded fluorescence signal. The spectral profile reveals more consistent tissue characteristics. In Fig. 7.7, the spectral signature curves of normal tissue (Fig. 7.7a) and malignant tumor (Fig. 7.7b) normalized with respect to the area under the curve are shown. The normalization procedure significantly reduced the within-class variances. The spectral signatures after normalization demonstrate a good separation to differentiate malignant tumors from normal tissues.

Laser-induced fluorescence spectroscopy has been suggested as an instant, noninvasive diagnostic tool to detect skin tumors based on the spectral properties of tissue [15]. Spectroscopy measures the intensity of radiation at a certain wavelength

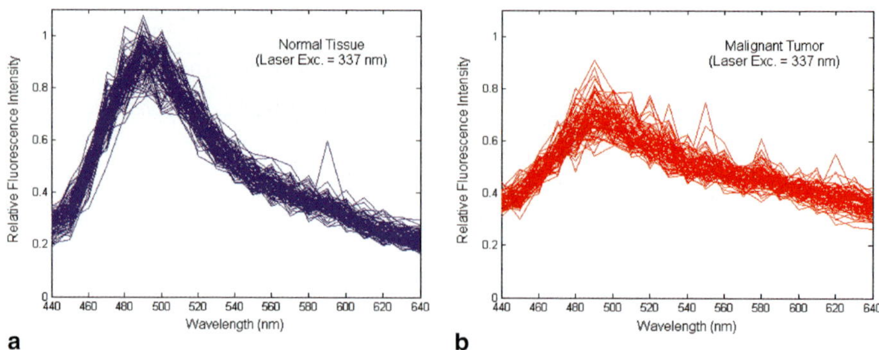

Fig. 7.7 Fluorescence spectra normalized with respect to the area under the spectral curve: **a** normal tissue, **b** malignant tumor

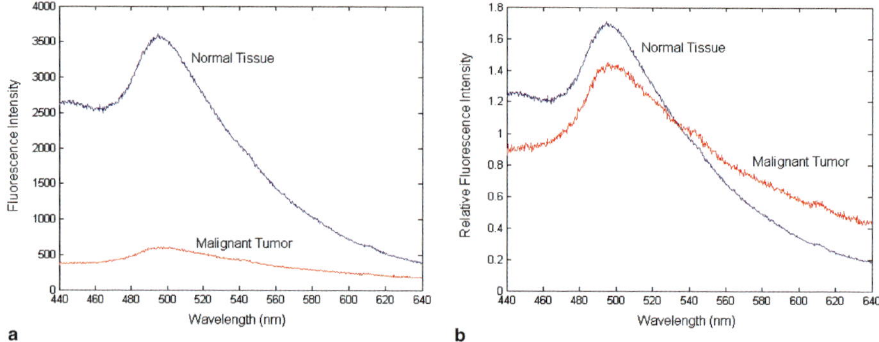

Fig. 7.8 Point spectroscopic measurements of the mouse tissue sample: **a** raw data, **b** normalized spectral signatures with respect to the area under the curve

at a single spot to reveal the spectral signature of the object in a particular wavelength. The disadvantage of such a method is that we can only record the spectrum at a specific spot. To compare the spectral signature curves of Fig. 7.7 to a traditional point spectroscopic analysis, point spectroscopic measurements were taken from the same mouse and are shown in Fig. 7.8. We made 559 measurements of fluorescence intensity for each curve from 440 to 640 nm at a 0.37-nm interval. Notice the correlation between the normal and malignant spectral signature curves of Fig. 7.7 and the point spectroscopic curves of Fig. 7.8. A peak in the normal and malignant spectral signature data is observed around 490 nm in Fig. 7.7 and is confirmed by the point spectroscopic data of Fig. 7.8. This agreement in spectral information lends credence to our imaging system's capability of spectral discrimination.

7.4 Spectral Signature Classification

7.4.1 Classification Models for Hyperspectral Image Analysis

In hyperspectral image classification, high-dimensional spectral information and a relatively small number of labeled training samples often cause difficulties in building classifiers with satisfactory performance. A number of spectral signature classification techniques have been developed to exploit extensive information contained in hyperspectral images. Many popular spectral classification methods compare the pixel spectra with the reference spectra by measuring their spectral similarities. Reference spectra can be derived from various sources, such as a spectral library or ROIs within the hyperspectral image. Such techniques include standard supervised classifiers such as minimum distance and maximum likelihood classifiers as well as the methods specifically developed for hyperspectral images such as the spectral angle mapper (SAM) [16] and spectral feature fitting [17]. The SAM classification method treats spectral signatures as vectors and computes the spectral angle between the questioned and known spectra to classify the underlying pixel into the class of best match. A smaller spectral angle implies greater similarity between the pixel and target spectra. The spectral angle is relatively insensitive to variations in illumination since changes in pixel intensities do not change the direction of the vector but the vector length. Spectral feature fitting matches the pixel and the reference spectra by examining specific absorption features in the spectra. In spectral feature fitting, the user specifies a range of wavelengths within which unique absorption features exist for the chosen target. The pixel spectra are compared to the target spectrum based on the measurements of the feature depth and the feature shape, often using a least-squares technique. The SVMs, based on statistical learning theory [8], offer better generalization ability through structural risk minimization, which often gives higher classification accuracies for the data not used in the training phase [18,19]. The main idea of the SVM classifier is to obtain a decision boundary that maximizes a separation margin. SVMs work well in the presence of heterogeneous classes with only a few training samples available and when class distribution is not balanced [20].

7.4.2 Support Vector Machines

Consider a set of training data $x_i, i = 1, \ldots, M$, in L-dimensional space that belong to either class Z_1 or class Z_2. Let a linear hyperplane $H(x) = w^T x + b = 0$ be a decision boundary that separates the two-class training data. Here, x denotes L spectral intensity values of a hyperspectral image $I(m, n, \lambda_j)$, $j = 1, \ldots, L$, measured at each pixel in the spatial domain. If the patterns are linearly separable, a hyperplane can separate the patterns without any error. Then, a linear decision boundary with a slope parameter w and an offset b satisfies the inequalities in a canonical form:

$$\begin{aligned} \boldsymbol{w}^T \boldsymbol{x}_i + b \geq +1 \quad \text{for } \boldsymbol{x}_i \in Z_1 \\ \boldsymbol{w}^T \boldsymbol{x}_i + b < -1 \quad \text{for } \boldsymbol{x}_i \in Z_2 \end{aligned}$$
(7.1)

The two inequalities can be grouped in one with the class labels $y_i = 1$ for Z_1 and $y_i = -1$ for Z_2.

$$y_i(\boldsymbol{w}^T \boldsymbol{x}_i + b) \geq 1 \text{ for } i = 1, \ldots, M.$$
(7.2)

The distance from the separating hyperplane to the patterns nearest to the hyperplane is called a *margin*, which is given by $1/ \| \boldsymbol{w} \|$ [8]. The hyperplane $H(\boldsymbol{x})$ is optimal in terms of generalization if the corresponding margin is maximized. The problem of finding an optimal decision boundary is a constrained optimization problem that minimizes $\| \boldsymbol{w} \|^2$ with respect to w and b with the constraints given in (7.2). Using the Lagrange multipliers α_i, it becomes to an unconstrained optimization problem:

$$J(\boldsymbol{w}, b, \alpha) = \frac{1}{2} \| \boldsymbol{w} \|^2 + \sum_{i=1}^{M} \alpha_i \left\{ 1 - y_i(\boldsymbol{w}^T \boldsymbol{x}_i + b) \right\}, \ \alpha_i \geq 0.$$
(7.3)

By using the Karush-Kuhn-Tucker (KKT) conditions, the optimization of $J(\boldsymbol{w}, b, \alpha)$ is converted to a dual-quadratic problem that maximizes

$$W(\alpha) = \sum_{i=1}^{M} \alpha_i - \frac{1}{2} \sum_{i=1}^{M} \sum_{j=1}^{M} \alpha_i \alpha_j y_i y_j \boldsymbol{x}^T \boldsymbol{x}_i$$
(7.4)

subject to the constraints

$$\sum_{i=1}^{M} y_i \alpha_i = 0, \quad \alpha_i \geq 0 \quad \text{for } i = 1, \ldots, M.$$
(7.5)

Given α, a solution to $W(\alpha)$, the optimal hyperplane $H^*(\boldsymbol{x})$ is given by

$$H^*(\boldsymbol{x}) = \sum_{i \in S} \alpha_i y_i (\boldsymbol{x}_i^T \boldsymbol{x}) + b$$
(7.6)

where S is a set of support vector indices and $b = y_i - \boldsymbol{w}^T \boldsymbol{x}_i$. The patterns $\boldsymbol{x}_i$ with nonzero α_i that contribute to the determination of the optimal hyperplane are called *support vectors*. Figure 7.9 illustrates two separating hyperplanes in a linearly separable set, where the decision boundary $H_1(\boldsymbol{x})$ offers a larger margin between the two classes than that of $H_2(\boldsymbol{x})$.

For a linearly nonseparable pattern set, the nonnegative slack variables ζ_i are introduced into the derivation of the linearly separable SVM:

$$y_i(\boldsymbol{w}^T \boldsymbol{x}_i + b) \geq 1 - \zeta_i \quad \text{for } i = 1, \ldots, M.$$
(7.7)

Patterns with $\zeta_i \geq 1$ are misclassified patterns. A new optimization problem is introduced to deal with misclassification errors.

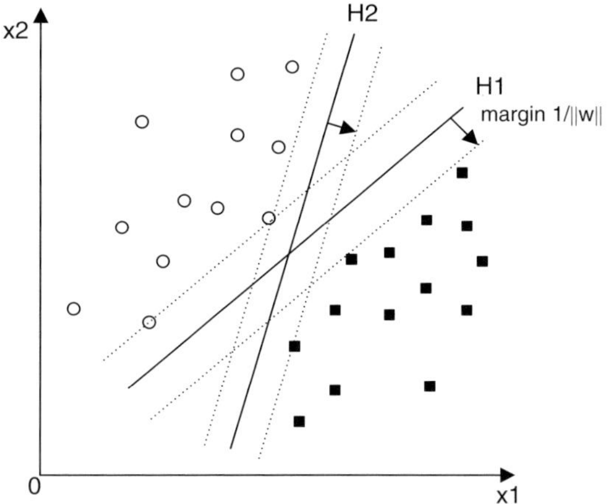

Fig. 7.9 Linear decision boundary of the SVM with a maximum separating margin

$$J(\mathbf{w},b,\zeta) = \frac{1}{2} \| \mathbf{w} \|^2 + C\sum_{i=1}^{M} \zeta_i \qquad (7.8)$$

where C is a regularization factor that determines the balance between the margin maximization and the misclassification minimization. The dual-quadratic optimization problem is obtained in a procedure identical to that of the linearly separable set except that α_i is bounded by the regularization factor like $C \geq \alpha_i \geq 0$ $(i = 1,\ldots,M)$. Large C values can produce the classifiers capable of classifying training patterns correctly at the cost of small margin, while small C leads to simple classifiers against misclassification errors.

Another approach is to map the patterns in the original space into a higher-dimensional space using a nonlinear function $g(x) = [g_1(x),\ldots,g_l(x)]^T$. An optimal hyperplane is determined in the feature space in a way that the patterns that could not be classified correctly in the original input space become separable in the feature space. The SVM training and classification do not need to deal with patterns in the high-dimensional feature space explicitly because the nonlinear mapping occurs in the form of inner products. That is, a kernel function $K(x_i,x_j) = g^T(x_i)g(x_j)$ is used instead of direct computations of $g(x)$. After the dual problem is solved, the resulting decision function is given by

$$H^*(x) = \sum_{i \in S} \alpha_i y_i K(x_i,x) + b \qquad (7.9)$$

$$b = y_i - \sum_{i \in S} \alpha_i y_i K(x_i,x_j) \qquad (7.10)$$

The selection of kernel type and free parameters such as the standard deviation of a radial basis kernel should be determined through cross validation or a series of sufficient preexperiments to achieve satisfactory generalization performance.

7.4.3 Experiment Results

Experiments are carried out to examine whether the SVM classifier is capable of detecting tumor skins correctly and, if so, how accurately it can identify the location, size, and shape of tumors. Each pixel, a subset of spectral bands is selected out of 21 spectral bands to avoid the high dimensionality and is then used as the features for classifiers. The subset of bands is chosen in terms of a divergence criterion, which is defined as the total average information for discriminating two classes Z_1 and Z_2 given by

$$\int_{-\infty}^{+\infty} (p_1(\boldsymbol{x}) - p_2(\boldsymbol{x})) \, \ln \frac{p_1(\boldsymbol{x})}{p_2(\boldsymbol{x})} \, d\boldsymbol{x}, \tag{7.11}$$

where $p_i(\boldsymbol{x})$ is the probability density function of $\boldsymbol{x}$ of class Z_i [21, 22]. The more bands there are, the larger the divergence is, but the most significant six bands are selected for each pixel in the experiment. To train and evaluate the classifier, 1,600 samples, 600 pixels from tumors and 1,000 pixels from normal tissues, were selected from a ROI in the hyperspectral data of a given specimen. A circular region was chosen on the sample on which the light of the endoscopic camera of the hyperspectral imaging system is projected. The selected samples were partitioned into five equal-size subsets for cross validation. Each subset was evaluated against a classifier trained on the samples in the other subsets. Overall error rate was obtained by summing errors from the five evaluation sets. All the pixels in the ROI were used as test samples. Figure 7.10 shows the SVM decision boundaries obtained using the linear and polynomial kernel functions. In each case, 160 samples of the features of fluorescence intensities of band 5 (480 nm) and band 11 (540 nm). The decision boundary seems to be nearly optimal in terms of discrimination of the two-class data. The projected decision boundary depends on the regularization factor and the order of the polynomial. The standard deviation of the radial basis kernel and the regularization factor C were chosen empirically in terms of the average evaluation errors.

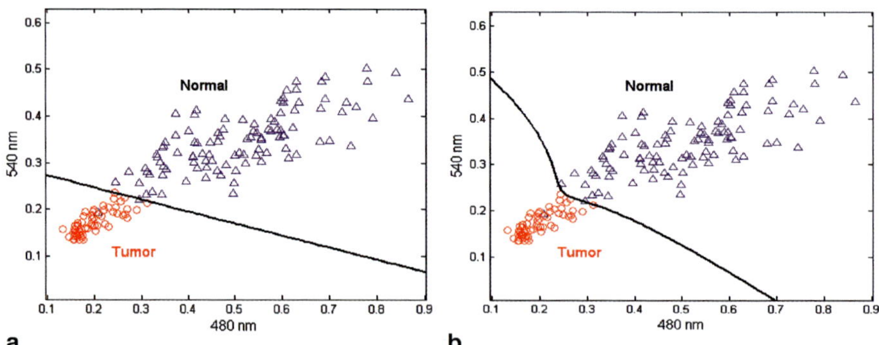

Fig. 7.10 Decision boundaries of SVM classifiers projected onto a 2D space of 480 and 540 nm: **a** linear decision boundary, **b** nonlinear decision boundary with a third-order polynomial kernel function

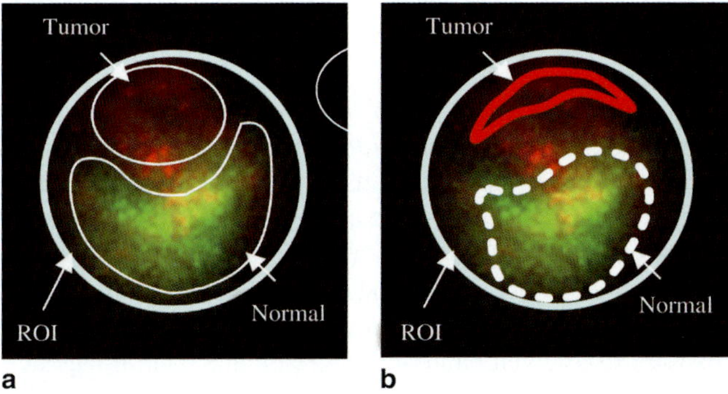

Fig. 7.11 A mosaic of fluorescence and reflectance for an image pair. Fluorescence intensity of the tumor region shows weaker response than that of normal tissue. **a** Ground truth malignant tumor and normal tissues identified by a tumor expert, **b** tumor and normal tissue regions identified by the SVM classifier

A mouse tumor expert identified the tumor location on mouse skin. Fluorescence shows better discrimination for the skin tumor region. A circular ROI was uniformly illuminated with an endoscopic light. Normal tissue showed a relatively stronger response compared to a tumor region in terms of fluorescence intensity. Figure 7.11 visualizes a mosaic image of reflectance and fluorescence at the 490-nm wavelength. The relatively darker area in fluorescence intensity marked by a circle above normal tissue indicates a malignant tumor region. Figure 7.11b shows tumor and normal skin areas that the classifier identified from the test samples. For classification with high confidence, the pixels whose output was larger than 0.9 were assigned to the tumor area (solid line) and pixels whose output was smaller than 0.1 were regarded as the normal area (dotted line). The pixels that have intermediate outputs between the two thresholds were excluded. The classifier identifies and discriminates the tumor skin (upper circular area in Fig. 7.11a) and normal skin (a bright lower U-shaped area in Fig. 7.11a) regions in the ROI correctly. However, the tumor area identified has slightly different shape and size from the original one, while the normal area is identified correctly in size and shape. The lower part of the actual tumor area is missed under classification with high confidence. The reasons seem to be that the light of the endoscopic camera was not projected onto the ROI uniformly, and some tumor skins have similar spectral characteristics to normal ones and vice versa. To reduce the misclassification errors, it would be helpful to utilize spatial features of tumor skins such as sizes and shapes in addition to the spectral features. The fusion of spectral and spatial information increases tumor detection accuracies with lower false-positive rates.

The SVM classifiers can support an approximate of posterior probabilities by feeding the outputs into a sigmoid function, ranging from 0 to 1. If the output of classifier is greater than a threshold, the sample is classified as a tumor skin; otherwise, it is classified as a normal one. By moving a threshold on the continuous output of SVM classifier, we can determine the degree to which a sample is classified as tumor skin gradually.

7.5 Conclusions

This chapter presented a hyperspectral fluorescence imaging technique for noninvasive detection of malignant tumors on mouse skin. Hyperspectral imaging reveals high-resolution spectral information useful for the classification of malignant skin tumors. Fluorescence data captured by the hyperspectral imaging system involves a higher level of contrast and thus provides a greater possibility of detecting tumors. Hyperspectral images generated by acousto-optic tunable filters typically demonstrate spatial offsets between the band images. The offsets are estimated using the mutual information of the given band and the reference band image at 490 nm. All the band images are spatially registered to obtain exact pixel correspondences by compensating the offsets. Raw spectral signatures measured from a mouse sample are normalized with respect to the area under the curve. Normalized fluorescence intensities significantly reduce the within-class variances and demonstrate a good separation to discriminate malignant tumors from normal tissues. The SVMs with polynomial kernel functions provide nonlinear decision boundaries with a maximum separating margin to classify fluorescence spectral signatures of tumor and normal tissue. The hyperspectral fluorescence imaging technique promises the development of noninvasive in vivo detection of malignant skin tumors without the need of tissue biopsy.

Chapter's References

1. Geller, D. Miller, G. Annas, M. Demierre, B. Gilchrest, and H. Koh, Melanoma incidence and mortality among U.S. whites, 1969–1999, *Journal of the American Medical Association*, 288(14):1719–1720, 2002.
2. *Cancer Facts and Figures 2005*. American Cancer Society, Atlanta, 2005.
3. D. Landgrebe, Hyperspectral image data analysis as a high dimensional signal processing problem, *IEEE Signal Processing Magazine* 19(1):17–28, 2002.
4. K. Rajpoot, N. Rajpoot, and M. Turner, *Hyperspectral Colon Tissue Cell Classification*. SPIE Medical Imaging, San Diego, 2004.
5. D. Ferris, R. Lawhead, E. Dickman, N. Hotzapple, J. Miller, S. Grogan, S. Bambot, A. Agrawal, and M. Faupel, Multimodal hyperspectral imaging for the noninvasive diagnosis of cervical neoplasia, *Journal of Lower Genital Tract Disease* 5(2):65–72, 2001.
6. S.G. Kong, Y.R. Chen, I. Kim, and M.S. Kim, Analysis of hyperspectral fluorescence images for poultry skin tumor inspection, *Applied Optics* 43(4):824–833, 2004.
7. I. Kim, M.S. Kim, Y.R. Chen, and S.G. Kong, Detection of skin tumors on chicken carcasses using hyperspectral fluorescence imaging, *Transactions of the American Society of Agricultural Engineers* 47(5):1785–1792, 2004.
8. V. Vapnik, *Statistical Learning Theory*. Wiley, New York, 1998.
9. N. Cristianini and J. Shawe-Taylor, *An Introduction to Support Vector Machines and Other Kernel-based Learning Methods*. Cambridge University Press, New York, 2000.
10. T. Vo-Dinh, D.L. Stokes, M. Wabuyele, M.E. Martin, J.M. Song, R. Jagannathan, E. Michaud, R.J. Lee, and X. Pan, Hyperspectral imaging system for in vivo optical diagnostics, *IEEE Engineering in Medicine and Biology* 23:40–59, 2004.
11. S.G. Kong, Z. Du, M. Martin, and T. Vo-Dinh, Hyperspectral fluorescence image analysis for use in medical analysis, *Proceedings of the SPIE Conference on Biomedical Optics*, San Jose, January 2005.

12. F. Moreau, S.M. Moreau, D.M. Hueber, and T. Vo-Dinh, Fiber-optic remote multisensor system based on an acousto-optic tunable filter (AOTF), *Applied Spectroscopy* 50:1295, 1996.
13. I.-C. Chang, Electronically tuned imaging spectrometer using acousto-optic tunable filter, *Proceedings of the SPIE International Society for Optical Engineering* 1703:24, 1992.
14. S.G. Kong, M. Martin, and T. Vo-Dinh, Hyperspectral fluorescence imaging for mouse skin tumor detection *ETRI Journal* 28(6):770–776, 2006.
15. B. Albers, J. DiBenedetto, S. Lutz, and C. Purdy, More efficient environmental monitoring with laser-induced fluorescence imaging, *Biophotonics International Magazine* 2(6):42–54, 1995.
16. R.H. Yuhas, A.F.H. Goetz, and J.W. Boardman, Discrimination among semiarid landscape endmembers using the spectral angle mapper (SAM) algorithm, *Third Annual JPL Airborne Geoscience Workshop*, JPL Publication 92-14 1:147–149, 1992.
17. R.N. Clark, G.A. Swayze, A. Gallagher, N. Gorelick, and F.A. Kruse, Mapping with imaging spectrometer data using the complete band shape least-squares algorithm simultaneously fit to multiple spectral features from multiple materials, *Proceedings of the Third Airborne Visible/Infrared Imaging Spectrometer (AVIRIS) Workshop*, JPL Publication 91-28, pp. 2–3, 1991.
18. F. Roli and G. Fumera, Support vector machines for remote-sensing image classification, *Proceedings of SPIE* 4170: 160–166, 2001.
19. F. Melangi, and L. Bruzzone, Classification of hyperspectral remote sensing images with support vector machines, *IEEE Transactions on Geoscience and Remote Sensing* 42(8):1778–1790, 2004.
20. S. Abe, *Support Vector Machines for Pattern Classification*. Springer-Verlag, New York, 2005.
21. Z. Du, M.K. Jeong, and S.G. Kong, Band selection of hyperspectral images for automatic detection of poultry skin tumors, *IEEE Transactions on Automation Science and Engineering* 4(3):332–339, 2007.
22. C. Chang, Q. Du, T. Sun, and M. Althouse, A joint band prioritization and band-decorrelation approach to band selection for hyperspectral image classification, *IEEE Transactions on Geoscience and Remote Sensing* 37(6):2631–2641, 1999.

Chapter 8
Spectral Screened Orthogonal Subspace Projection for Target Detection in Hyperspectral Imagery

Stefan A. Robila

Abstract We introduce several new approaches in increasing spectral screening accuracy and reliability and study their applicability for orthogonal subspace projection-based target detection. Spectral screening is a technique used for selecting a representative subset of spectra from hyperspectral data. The subset is formed such that any two spectra in it are dissimilar, and for any spectrum in the original image cube, there is a similar spectrum in the subset. Spectral screening is performed in a sequential manner; at each step, the subset is increased with a spectrum dissimilar from all the spectra already selected. The procedure, using the spectral angle as similarity measure, is employed in a variety of algorithms for linear unmixing and data compression. We modified this algorithm such that at the selection step the spectrum with the largest distance from the set is selected. While not introducing additional computational complexity, the maximum spectral screening (Max SS) algorithm ensures that the overlap among the representatives is minimized, and that the solution is deterministically obtained, eliminating the need for multiple experiments. We also investigated the alternative approach in which at each step the spectrum with the smallest distance (but larger than a threshold value) is selected. In a distinct direction, we also studied the efficiency of the two algorithms by designing detection filters obtained as the classification projector matrices based on the spectral subset using regular or kernel-based orthogonal subspace projections (KOSP and OSP, respectively). The developed algorithms were tested on Hyperspectral Digital Collection Experiment (HYDICE) hyperspectral data using the spectral angle and the spectral information divergence. The results indicated that Max SS outperforms regular spectral screening detection and minimum spectral screening (Min SS).

Keywords: Hyperspectral data · Target detection · Spectral screening · Orthogonal subspace projection · Kernel orthogonal subspace projection · Spectral angle · Spectral information divergence

R.I. Hammoud (ed.), *Augmented Vision Perception in Infrared: Algorithms
and Applied Systems,* Advances in Pattern Recognition, DOI 10.1007/978-1-84800-277-7_8,
© Springer-Verlag London Limited 2009

8.1 Introduction

Remote sensing is generally described as the measurement, from a distance, of spectral features of Earth's surface and atmosphere [1]. In general, the sensed data are collected as images (*spectral images* or spectral bands), with each image corresponding to intervals of wavelengths. Each element from the image (pixel) is associated with a certain area of the scene surveyed and with its spectral response [2]. A collection of spectral images over several wavelength intervals for the same scene is called a *multispectral* (in case of few wide spectral bands) or *hyperspectral* (in case of many narrow spectral bands) *image.* In processing multispectral data, it is a common practice to define *pixel vectors* or *spectra* as the vectors formed of pixel intensities of the same location, across the bands [3] as represented in Fig. 8.1. Since each pixel corresponds to a certain region of the scene surveyed, a pixel vector will represent the spectral information for that region.

Advances in sensor technology development have resulted in increased spectral and spatial resolution. This in turn allows for significant improvements in the accuracy of the results provided through hyperspectral image processing [4]. Due to its richness in information, hyperspectral imagery allows for detection of targets covering areas smaller than a pixel or separation of objects and shapes otherwise undistinguishable in regular images. At the basis of multispectral/hyperspectral data processing is the idea of spectra separability. If a material spectrum is easily separable from the background, then the material will be easily detected in the image. If two adjoining materials have significantly dissimilar spectra, then the edge between them will also be easily detected.

Unfortunately, clear separability cannot be easily achieved. Each spectrum is formed of tens to hundreds of values collected within narrow adjacent wavelength intervals and often exposes strong local correlation [2]. For many human-made and natural materials, within-class spectral variance is often larger than between-class variance, limiting the performance of traditional pattern classification approaches. New approaches that take in consideration both the nature of the data as well as the complexity of working with large three-dimensional (3D) images are needed.

We present a class of methods for target detection based on spectra separability. The unsupervised algorithms sequentially generate a set of spectra that are dissimilar from each other in terms of a spectral distance measure. The spectral subset is

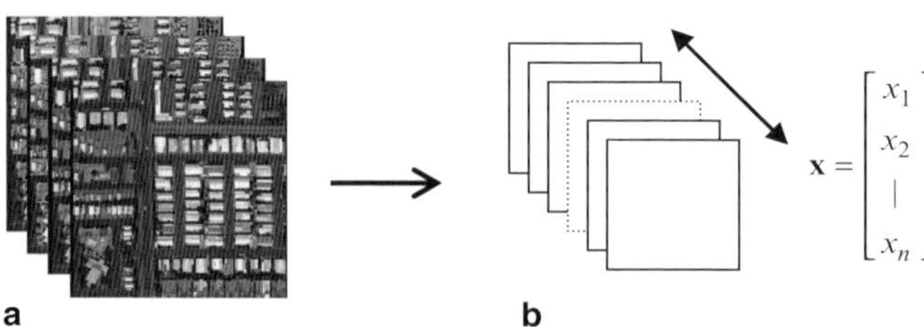

Fig. 8.1 **a** multispectral image, **b** pixel vectors (spectra) are formed of the pixel values for the same coordinates

obtained through a modified selection algorithm known as *spectral screening* [5] or *exemplar selection* [6]. In spectral screening, the subset is built gradually, with each newly added spectra dissimilar from the ones already selected. If we eliminate from the subset the desired target spectra, we end up with a collection of signatures that characterize the background (with respect to the spectra). The spectral subset is thus used to build a target detection filter that will identify pixels that closely match the selected spectra. Two distinct approaches could be employed. While a basic orthogonal projection is the first choice, we were also interested to see how the use of kernel maps influences the detection accuracy. In principle, kernel-based orthogonal subspace projection (OSP; as introduced in [7]) should improve the accuracy given that the target data and the undesired signatures are not orthogonal.

Unfortunately, even when the desired target spectrum is included as the first element in the subset, spectral screening produces results that vary according to the order the data pixels are processed [8]. Thus, the method is perceived as unstable and requires multiple runs. Our approach investigates what happens when the extreme (the "farthest"/"closest" from the already-selected spectra) spectrum is selected every time. Compared with the original technique, the "extreme" screening has the particular advantage of being deterministic in nature and producing unique solutions.

An important component in spectral screening is the distance measure used. In this chapter, we compare the spectral angle (SA) with another recently introduced distance: the spectral information divergence (SID) [9].

Section 8.2 presents the spectral screening algorithm and the various distance measures used. Section 8.3 introduces the maximum and minimum spectral screening (Max SS and Min SS, respectively) algorithms, followed by the creation of detector filters using the spectral subsets and based on regular and kernel-mapped OSP (Section 8.4). Section 8.5 presents a sequence of experiments that allowed us to assess the efficiency of both the detectors and various distance measures.

8.2 Spectral Screening

8.2.1 Algorithm

Given a spectral distance measure $d(.,.)$ and a threshold or error value β, we say that two spectra $\mathbf{x}$ and $\mathbf{y}$ are *similar* with respect to d and β if $d(\mathbf{x},\mathbf{y}) \leq \beta$ and are *dissimilar* otherwise [10].

Based on this, we provide an algorithm that creates a subset of dissimilar spectra (see Fig. 8.2). At the end of the process, the subset S_1 will contain spectra with the following two properties [11]:

$$\forall \mathbf{x}, \mathbf{y} \in S_1, d(\mathbf{x},\mathbf{y}) > \beta. \tag{8.1}$$

$$\forall \mathbf{x} \in S, \exists \mathbf{y} \in S_1, d(\mathbf{x},\mathbf{y}) \leq \beta. \tag{8.2}$$

<table>
<tr><td>1.</td><td>Let S = set of all spectra</td></tr>
<tr><td>2</td><td>Let $S_1 = \varnothing$</td></tr>
<tr><td colspan="2">Initial Step:</td></tr>
<tr><td>3.</td><td>Take $\mathbf{x} \in S$</td></tr>
<tr><td>4.</td><td>$S = S - \{\mathbf{x}\}$</td></tr>
<tr><td>5.</td><td>$S_1 = S_1 \cup \{\mathbf{x}\}$</td></tr>
<tr><td colspan="2">Iterative Step:</td></tr>
<tr><td>6</td><td>While S not empty</td></tr>
<tr><td>7</td><td>Take $\mathbf{x} \in S$</td></tr>
<tr><td>8.</td><td>$S = S - \{\mathbf{x}\}$</td></tr>
<tr><td>9.</td><td>If for all $\mathbf{y} \in S_1$ $d(\mathbf{x},\mathbf{y}) > \beta$ then</td></tr>
<tr><td>10.</td><td>$S_1 = S_1 \cup \{\mathbf{x}\}$</td></tr>
<tr><td></td><td>End (while)</td></tr>
</table>

Equation (8.1) suggests that any two spectra in the subset are dissimilar (with respect to d and β). The second equation indicates that every spectrum in the data S has a "representative" in the subset, that is, a spectrum that is similar to it.

The name *spectral screening* refers to the screening of data for extracting the subset [5]. Alternative definitions have identified this process as *exemplar selection* [6]. The number of selected spectra depends on the distance used and on the threshold β. As the value of β decreases, the number of spectra that are found to be similar also decreases, increasing in turn the subset size. Moreover, the number of selected spectra depends on the order they are processed. Studies on the variation of the subset size for various data types (including Hyperspectral Digital Collection Experiment [HYDICE], Aviris, and Hyperion) have reported experiments in which the order of the spectra was randomly permuted prior to screening. On repetition of screening for several times, it was noticed that the subset size variation ranged from 2 to 10% from the average size. In addition, as the threshold values where increased, the variation also increased [11].

Spectral screening was first introduced as the exemplar selection step within the ORASIS system [12]. Since the system was designed to work with sensors that collected the data one spectrum or one line at a time, the algorithm had the extra advantage of being able to start running before the entire image cube was collected. The screened subset was then employed for end member extraction for the linear mixing model [12]. The abundances for the spectra from the original data are obtained by computing and applying filter vectors to it [6]. The algorithm can be seen as a compression tool since the data are reduced in terms of the number of spectra. The correspondence between the original spectra and the subset spectra is maintained through a codebook entry. For a particular spectrum, the entry refers to the subset spectrum that is similar to the spectrum and triggered its "elimination" in the screening algorithm [12].

An interesting modification of the spectral screening is described in [6]. In this case, once a similar spectrum is found in the subset, the process continues the search until the "most similar" one is found. The new algorithm is described as "best fit," as compared to the original "first fit." While this approach leads to a better representation of the data, it also results in significant increases in the computational time of spectral screening. In best fit, each spectrum needs to be compared to all the spectra already in the subset. Based on this reasoning, the best-fit algorithm should be used for problems that do not require fast computation. In the case of time constraints, the original algorithm is preferable.

8.2.2 Spectral Distance Measures

Many multispectral/hyperspectral applications use measures to assess the similarity (or distance) between spectra or between a spectrum and a group of spectra. Distance measures are used in classification as well as in target detection. In the following, we discuss several of them and analyze their properties and their previous use with hyperspectral imagery.

8.2.2.1 Spectral Angle

Given two vectors of the same dimension $\mathbf{x}$ and $\mathbf{y}$, the *spectral angle* is defined as the arccosine of their dot product [5]:

$$SA(\mathbf{x}, \mathbf{y}) = \arccos \left(\frac{\langle \mathbf{x}, \mathbf{y} \rangle}{\|\mathbf{x}\|_2 \|\mathbf{y}\|_2} \right). \tag{8.3}$$

where $<.,.>$ represents the dot product of the vectors and $\|.\|$ the Euclidean norm.

The SA has most of the properties that define a distance metric. The SA is always greater than or equal to zero; it is symmetric and follows the triangle inequality [13]. Unlike a regular metrics distance, it is possible to have a zero SA even when the two vectors are not identical because the measure is invariant to scalar multiplication.

When the pixel vectors represent reflectance values, two spectra with zero angle would correspond to the same material under different illumination conditions (poor illumination leads to shorter segments, strong illumination will yield longer segments) [13]. The SA was used as a method for mapping the spectral similarity of image spectra to the reference spectra [13] and has been implemented in most of the widely used remote-sensing software packages, such as ENVI.

An important attribute for the SA is that its values fall within a well-defined interval. In the case of reflectance values, since the pixels are all positively defined, the interval for the SA is between 0 and $\pi/2$ irrespective of the amplitude of the various bands. This is important for algorithms that need to rely on threshold values for distance among spectra since this threshold can be set without prior knowledge of the particular scene.

8.2.2.2 Spectral Information Divergence

Given two n-dimensional vectors $\mathbf{x}$ and $\mathbf{y}$, the *spectral information divergence* is defined as [14]

$$SID(\mathbf{x},\mathbf{y}) = \left\langle \frac{\mathbf{x}}{sum(\mathbf{x})} - \frac{\mathbf{y}}{sum(\mathbf{y})}, \log\left(\frac{\mathbf{x}}{sum(\mathbf{x})}\right) - \log\left(\frac{\mathbf{y}}{sum(\mathbf{y})}\right) \right\rangle. \tag{8.4}$$

where the *sum(.)* function refers to the sum of the values composing the vectors $\mathbf{x}$.
SID is derived from the Kullback-Leibler information measure:

$$SID(\mathbf{x},\mathbf{y}) = D(\mathbf{x}||\mathbf{y}) + D(\mathbf{y}||\mathbf{x}). \tag{8.5}$$

where $D(\mathbf{x}||\mathbf{y})$ is defined as

$$D(\mathbf{x}||\mathbf{y}) = \sum_{i=1,n} p_i \log \frac{p_i}{q_i}. \tag{8.6}$$

and

$$(p_1,p_2,..,p_n) = \frac{\mathbf{x}}{sum(\mathbf{x})} \quad (q_1,q_2,..,q_n) = \frac{\mathbf{y}}{sum(\mathbf{y})}. \tag{8.7}$$

are probability mass functions associated with the two vectors.

The SID has also properties similar to a metric. It is always greater than or equal to zero; it is symmetric and follows the triangle inequality. In addition, the normalization by the sum also leads to invariance to scalar multiplication. This means that, in the context of hyperspectral imagery, SID is invariant to illumination conditions.

The SID has been shown to provide a relatively better quantification of similarity than the SA. Unlike the SA, which is based on vector theory, SID is derived from information theory. Combinations of the SA and SID (by using tan and sin functions) have been proposed and shown to increase the accuracy over both measures [14].

8.3 Spectral Screening Using Extremes

A problem encountered when using spectral screening is related to the overlap of the similarity sets. Each spectrum from the original data is eliminated based on its similarity with one of the subset spectra. However, it is likely that the original spectrum is similar to more than one of the selected spectra. The association of one individual with several clusters could affect the results of any subsequent processing [15]. The "best-fit" approach was designed as an answer to this problem by continuing the comparison between the spectrum and the subset elements until a best match was identified. However, it is also possible that future subset elements will be the best match for the eliminated spectrum. In this case, the best fit no longer works properly. A simple improvement would be to postpone the computation of the codebook until after the creation of the full spectral screened subset. This approach is computationally expensive, at least doubling the overall execution time.

8.3.1 Maximum Spectral Screening

Instead of focusing on the best match between a data spectrum and the subset spectrum, we suggest analyzing the problem from the point of view of subset generation. In the original algorithm, at every iterative step (line 6 in Fig. 8.2), the next subset element is chosen as the first one found to be dissimilar to all the spectra already included in the subset.

Consider the situation when such a spectrum $\mathbf{x_1}$ was identified. Then, for all the spectra $\mathbf{x}$ already in S_1 we have

$$d(\mathbf{x}, \mathbf{x_1}) = \beta + \varepsilon. \tag{8.8}$$

where ε is positive. The value of ε can be used as a measure of overlap: the smaller the ε, the larger the overlap. In this sense, and assuming that the triangle inequality holds for the distance measure $d(,)$, the two spectra $\mathbf{x}$ and $\mathbf{x_1}$ will not have any overlap for all $\varepsilon > \beta$, that is, when the distance between the two spectra is at least twice the value of the threshold.

To reduce the possibility of such overlaps, we suggest a modification in the choice of the next candidate for the screened subset. Here, instead of randomly picking the next dissimilar sample, we will choose the one that is the farthest away from all the already selected spectra. In other words, the algorithm chooses at each step the most dissimilar spectrum left.

In case the subset contains only one spectrum $\mathbf{x}_1$, the farthest away will be easily identified as

$$\bar{\mathbf{x}} = \arg\left(\max_{\mathbf{x}\mid d(\mathbf{x},\mathbf{x_1})>\beta} (d(\mathbf{x},\mathbf{x_1}))\right). \tag{8.9}$$

In case several spectra $(\mathbf{x}_1, \mathbf{x}_2, \ldots, \mathbf{x_k})$ are already selected in the subset, we find the most distant spectra as the one with the largest value for the product of the distances from it to $\mathbf{x}_1, \mathbf{x}_2, \ldots, \mathbf{x_k}$, respectively:

$$\bar{\mathbf{x}} = \arg\left(\max_{\mathbf{x}\mid \forall i, d(\mathbf{x},\mathbf{x_i})>\beta} \left(\prod_{i=1}^{k} d(\mathbf{x},\mathbf{x_i})\right)\right). \tag{8.10}$$

Note that Eq. (8.10) would be applied only to the spectra from the original data that are not similar with $\mathbf{x}_1, \mathbf{x}_2, \ldots, \mathbf{x_k}$. In addition, by the use of the product, we ensure that the next selected spectrum is "equally distant" from all the spectra. This was done to avoid possible skewing of the selection toward one direction.

The choice for the next spectrum stays at the basis of the maximum spectral screening (Max SS) algorithm. Max SS is obtained by modifying regular spectral screening through the introduction of the cumulative distance associated with each pixel. The attribute will be used to compute the product of the distances between the spectrum and the samples selected in S_1. At each iterative step, all the spectra left in S that are similar (according to d and β) to the last spectrum chosen for S_1 are eliminated. For each dissimilar spectrum remaining, the cumulative product distance is updated by multiplying it with the distance between them and the last-selected spectra. The sample with the maximum cumulative distance is selected to be included in S_1 and eliminated from S. The algorithm terminates when all the spectra in S are eliminated.

Conceptually, the Max SS algorithm does not change the meaning of spectral screening. The spectrum chosen at each step could coincide with the one provided by the regular spectral screening algorithm if the spectral data are permuted in a certain order. Given the sequence of choices, however, the event would be extremely unlikely to occur. Because of this, Max SS can be seen as producing an extreme case of spectral screening. We note that while the algorithm cannot totally eliminate the overlap, it tries to minimize this occurrence by consistently choosing extremes. Because of this, it is expected that the size of the spectral screened subset for Max SS will be larger than the average size for the subset obtained through spectral screening. It cannot be postulated, however, that the size would be the largest possible.

Max SS also has the advantage of being deterministic up to the starting spectrum. Once the initial spectrum is selected, the choice for the next one is deterministic, and the full spectral screened set is unique. This is significantly different from regular spectral screening, in which each run yields different subsets.

8.3.2 Minimum Spectral Screening

Max SS always selects the spectrum farthest from the ones already in the subset. An alternative approach is to select the one that is closest to the ones in the subset and still dissimilar to each.

In case the subset contains only one spectrum x_1, the closest will be easily identified as

$$\bar{x} = \arg \left(\min_{x|d(x,x_1)>\beta} \left(d(x,x_1) \right) \right). \tag{8.11}$$

In case several spectra $(x_1, x_2, \ldots, x_k)$ are already selected in the subset, we find the most distant spectrum as the one with the smallest value for the sum of the distances from it to $x_1, x_2, \ldots, x_k$, respectively:

$$\bar{x} = \arg \left(\min_{x|\forall i, d(x,x_i)>\beta} \left(\sum_{i=1}^{k} d(x,x_i) \right) \right). \tag{8.12}$$

The choice for the next spectrum stays at the basis of the minimum spectral screening (Min SS) algorithm. Min SS can be obtained from regular spectral screening in the same manner as Max SS. At each step, after the elimination of similar spectra, for each dissimilar spectrum remaining, the cumulative sum distance is updated by adding to it the distance between them and the selected spectrum. The sample with the minimum cumulative sum distance is selected to be included in S_1 and eliminated from S. The algorithm terminates when all the spectra in S are eliminated. Min SS will try to identify spectra that will be as close as possible to the selected ones and yet remain dissimilar. The resulting subset will most likely be larger than the one produced by Max SS. Min SS is also deterministic, up to the choice of the starting spectra.

8.4 Target Detection Using Spectral Screening

Spectral distances have been extensively used in target detection [16, 17]. A popular approach is the spectral angle mapper (SAM), which classifies pixels as targets by measuring the SA between them and chosen spectra and then applying a threshold [18]. Thus, the method is efficient when the spectral variability within the target is very low and is smaller than the chosen threshold. In other situations, projection-based algorithms have proven to be more appropriate. In the following, we discuss two such methods: orthogonal subspace projection (OSP) and kernel orthogonal subspace projection (KOSP).

8.4.1 Orthogonal Subspace Projection

Given s_0, the target spectrum, and $s_1, s_2, \ldots, s_m$, a set of background (or undesired) spectra, the OSP detection filter for s_0 is defined as [19]

$$P_{s_0} = s_0 P_U^{\perp}. \tag{8.13}$$

where $U = [s_1 s_2 \ldots s_m]$, the matrix formed by having the spectra $s_1, s_2, \ldots, s_m$ as columns, and $P_U^{\perp}$ is the undesired target signature annihilator [20, 21]:

$$P_U^{\perp} = I - UU^{\#}. \tag{8.14}$$

and $U^{\#}$ is the pseudoinverse of U:

$$U^{\#} = (U^T U)^{-1} U^T. \tag{8.15}$$

The OSP filter would then be applied to each of the image spectra x:

$$P_{s_0} x = s_0 P_U^{\perp} x. \tag{8.16}$$

The process can be seen as one that eliminates the undesired target spectra. We note that these equations hold for a spectral subset of relatively small size. When the number of background spectra surpasses the number of bands, then the matrix $(U^T U)^{-1}$ can no longer be computed. However, it can be safely assumed that the number of undesired spectra is small compared to the number of bands.

OSP was extensively used in experiments that allow differentiation of both large and subpixel targets. Nevertheless, it still relies on the need for the targets to have signatures as "orthogonal" as possible to the background signatures. Supervised and partially supervised versions exist [20, 21]. In the supervised version, both the target and the background signatures are known. In the partially supervised version, only the target signature is known. Each background signature is obtained as the spectrum that minimizes the orthogonal projection produced by all the previously selected spectra.

8.4.2 Kernel Orthogonal Subspace Projection

Given a nonlinear mapping function Φ:

$$\begin{aligned} \Phi : \mathbb{R}^n &\to F \\ \mathbf{x} &\mapsto \Phi(\mathbf{x}). \end{aligned} \tag{8.17}$$

that maps from the signature space of the original hyperspectral data to a feature space $\mathcal{F}$ given $\mathbf{s_0}$ as the target spectrum and $\mathbf{s_1}, \mathbf{s_2}, \ldots, \mathbf{s_m}$ as a set of background (or undesired) spectra, the KOSP detection filter for $\mathbf{s_0}$ is defined as [7]

$$\mathbf{P}_{\Phi, \mathbf{s_0}} = \Phi(\mathbf{s_0})^T \mathbf{P}_{\mathbf{U}_\Phi}^\perp. \tag{8.18}$$

where

$$\mathbf{U}_\Phi = [\Phi(\mathbf{s_1}), \ldots, \Phi(\mathbf{s_m})]. \tag{8.19}$$

$$\mathbf{P}_{\mathbf{U}_\Phi}^\perp = \mathbf{I}_\Phi - \mathbf{U}_\Phi \mathbf{U}_\Phi^\# = \mathbf{I}_\Phi - \mathbf{B}_\Phi \mathbf{B}_\Phi^T \tag{8.20}$$

In these equations, $\mathbf{U}_\Phi^\#$ is the pseudoinverse of $\mathbf{U}_\Phi$, and $\mathbf{B}_\Phi$ is the matrix formed of the eigenvectors of the covariance matrix for the undesired signatures mapped through the function Φ. The KOSP filter would then be applied to each of the image spectra $\mathbf{x}$:

$$\mathbf{P}_{\Phi, \mathbf{s_0}} \Phi(\mathbf{x}) = \Phi(\mathbf{s_0})^T \mathbf{P}_{\mathbf{U}_\Phi}^\perp \Phi(\mathbf{x}). \tag{8.21}$$

The advantage of KOSP is that it nonlinearly maps the data. As such, if the hyperspectral image contains spectra obtained as a nonlinear mixture of original end members, it is expected that KOSP will perform better at unmixing [7, 22]. A significant difficulty in KOSP is the choice of the nonlinear mapping and the fact that the feature space can have a large number of dimensions. This is nicely avoided by finding that a kernel covers any dot product of data in the new feature space:

$$K : \mathcal{F} \times \mathcal{F} \to \mathbb{R}. \tag{8.22}$$

$$K(\mathbf{x}, \mathbf{y}) = \Phi(\mathbf{x}) \cdot \Phi(\mathbf{y}). \tag{8.23}$$

In this case, following a sequence of steps, the KOSP filter can be written in terms of the kernel. For a full description, see [7]. In our case, we followed the suggested choice for kernel function, the Gaussian radial basis function:

$$K(\mathbf{x}, \mathbf{y}) = e^{-\|\mathbf{x}-\mathbf{y}\|^2/c}. \tag{8.24}$$

where c is a constant value. The kernel has the advantage of being translation invariant and associated with a nonlinear map that is smooth [7]. The choice for c did not constitute the target of our study and was set as 5. Future research needs to investigate this further.

8.4.3 Spectral Screening and Orthogonal Projection Target Detection

Applying the classifier from Eqs. (8.16) and (8.21) to the spectra in the original data, we get a detection image $\mathbf{I}$ for the target $\mathbf{s_0}$. The result can be further normalized to the [0,1] interval by subtracting its minimum and dividing by the difference between the maximum and minimum in the image [23]:

$$\mathbf{I'} = \frac{\mathbf{I} - \min_{\mathbf{I}}}{\max_{\mathbf{I}} - \min_{\mathbf{I}}}. \tag{8.25}$$

The normalized abundance fractions are used with a threshold fraction α. If the pixel value is larger than α, the pixel is labeled as target; otherwise, it is labeled as background.

For each of the target detection approaches, spectral screening is used to generate the collection of undesired signatures. Each of the algorithms is characterized by a three-part name:

$$ST - SD - PR. \tag{8.26}$$

where ST refers to the type of spectral screening (regular SS, Max SS, and Min SS, respectively); SD refers to the type of spectral distance used (SA and SID); and PR refers to the type of projection used for detection (OSP or KOSP). For example, *Min SS-SA-KOSP* refers to target detection using spectra generated by the Min SS algorithm and SA, and the KOSP.

An important component in the spectral screening algorithms is the value of the threshold β used to characterize if two spectra are similar. This value depends on the distance, type of data, and the nature of the scene and the desired target. Previous work focused on spectral screening and data compression has established "ideal" threshold values empirically through repeated experiments on various data types and feature extraction algorithms [15,24]. In the case of target detection, we suggest the use of within-target spectral distance (WTSD) as distance threshold indicator. This can be done by taking the maximum distance obtained between the target reference spectra and all the spectra identified as targets (either from lab measurements or from the ground data).

8.5 Experiments

8.5.1 HYDICE Image

The experiments use a Hyperspectral Digital Collection Experiment (HYDICE) image provided by the Spectral Information Technology Application Center. The image, with a size of 175 × 75 pixels was extracted from the Radiance I set [25]. Following the elimination of the water absorption and artifact-corrupted bands, 166 bands were selected for processing.

The scene contains 24 panels centrally located in the image (see Fig. 8.3). The panels are grouped in eight rows of three panels. Each row (labeled r1 through r8) contains panels of the same material and of sizes 3m × 3m, 2m × 2m, and 1m × 1m (in decreasing order from left to right). Given the reported spatial resolution of 0.75m, it is probable that most of the pixels are mixed.

We compare our results with the ones described in [23], in which a desired target detection and classification algorithm (DTDCA) was introduced. DTDCA follows a similar approach as the MSS algorithm. Starting from a desired target spectrum, it iteratively generates a set of undesired spectra in an unsupervised manner. At each step, an orthogonal subspace projector derived from the already selected spectra is generated. The projector is applied to the data, and the spectrum with the maximum projection length is selected as next undesirable spectrum. The set thus obtained is used with Eqs. (8.16) and (8.21) to identify possible target pixels. Compared to Max and Min SS, DTDCA identifies as the next most distinct spectrum the one obtained from the OSP.

Figure 8.3b shows the location of each of the panels. The rows are labeled r1 through r8 in top-to-bottom order. For each row, we selected four pixels that were averaged to compute the target spectrum. The dark pixels in the image correspond to pixel locations of each of the spectra extracted. Figure 8.4 presents the plot of the average spectrum for each of the eight targets. Note that the spectra plots indicate

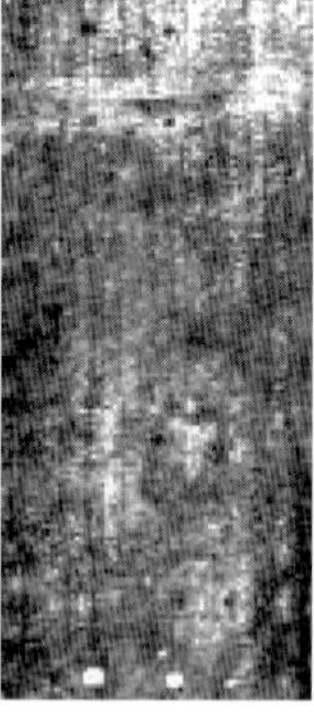
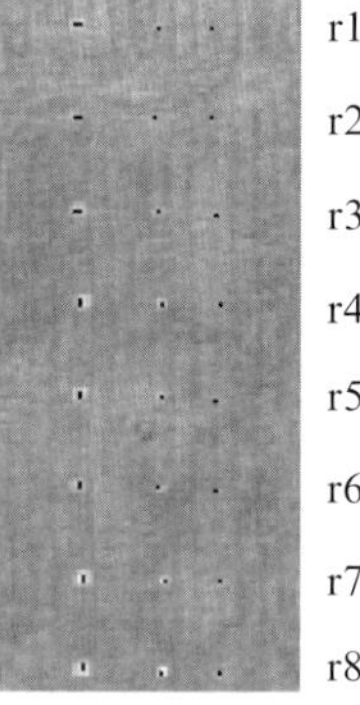

Fig. 8.3 HYDICE data used in the experiments **a** band 29, **b** band 59 where the panels are more visible; black pixels correspond to ground data used to compute target spectra

a　　　　　　　　b

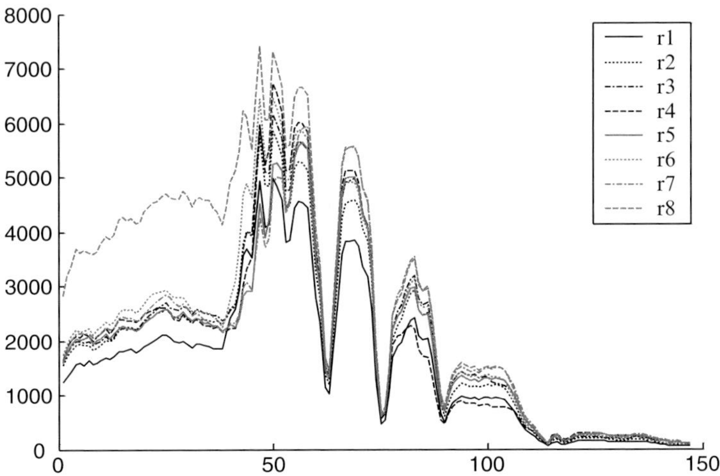

Fig. 8.4 Average spectra for each of the eight targets

Table 8.1 WTSD for each target and distance measure used

	r1	r2	r3	r4	r5	r6	r7	r8	Average
SA	0.13349	0.03487	0.06306	0.09360	0.15928	0.07103	0.17099	0.10672	0.0964
SID	0.01984	0.00215	0.01562	0.01598	0.02668	0.00558	0.03421	0.01433	0.0141

a high level of similarity among the data. The target spectra are labeled according to the row of panels (r1 through r8) from which they were extracted. Table 8.1 lists the WTSD for each of the targets when using SA and SID. The last column lists the average WTSD of the eight within-target distance values. Note that SID provides significantly lower values compared with SA. We also note that the WTSD values are the largest for the first, fifth, and seventh rows, indicating a high possible target spectra variability. This would lead to an increased error in classification.

We applied the spectral screening, Max SS, and Min SS algorithms to the data, with each of the average target spectra as the starting element in the spectral subset, SA and SID as distance measures, and OSP and KOSP as detection filters. Since regular spectral screening does not result in a unique set, to ensure better uniformity we permuted the original data spectra prior to screening. Figure 8.5, Fig. 8.6, and Fig. 8.7 show the classification images for the eight targets when we employed Max SS-SA-OSP, SS-SA-OSP, and Max SS-SA-KOSP, respectively. For chapter succinctness, the visual results for the other detection filters are not included. The images are labeled a through h corresponding to the targets r1 through r8, respectively. In all cases, the targets are identified. In the case of the panels in the first row, we note that the classification also suggests row 6 panels as possible targets, although with slightly lower intensity (Fig. 8.5a). Similar situations occur for the fifth and the seventh targets (Fig. 8.5e, g). These three targets also correspond to the largest WTSD values and required the use of large threshold values in the spectral

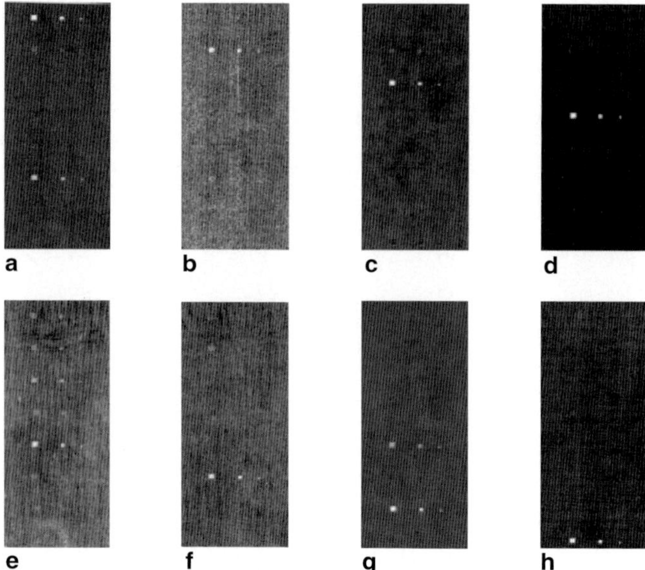

Fig. 8.5 Classification images using Max SS-SA-OSP target detection

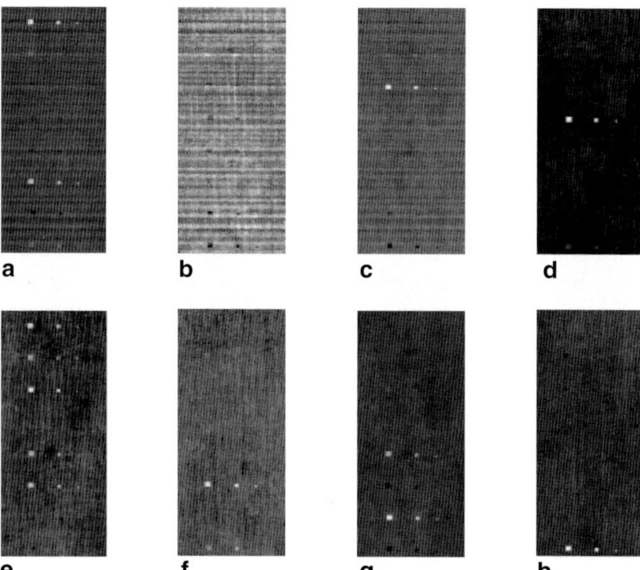

Fig. 8.6 Classification images for SS-SA-OSP target detection

screening. Figure 8.6 shows the result of regular spectral screening starting with the same targets. Comparatively, the classification images have lower quality. In particular, the images for the targets expose either significant error (Fig. 8.6b) or the

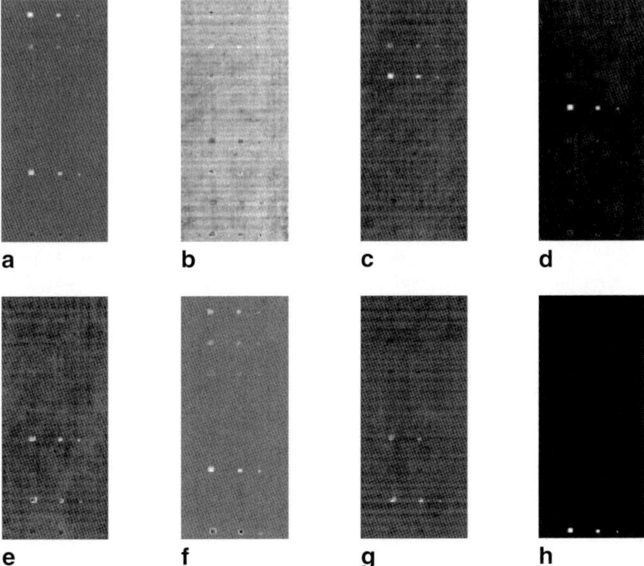

Fig. 8.7 Classification images for Max-SS-SA-KOSP target detection

presence of additional possible targets (Fig. 8.6c, g). Compared to OSP, KOSP displays slightly more accurate images, with a lower number of false detection pixels.

Tables 8.2 and 8.3 show the classification results for the detectors, and Table 8.4 shows the results when using DTDCA on the image. For consistency, we used the α threshold values 0.1, 0.25, and 0.5 as suggested in [23]. As a reminder, the image has a total of 13,125 pixels. The tables list for each algorithm and each α value the number of ground truth pixels correctly identified ($\mathbf{P_T}$) and the total number of pixels identified as targets ($\mathbf{P}$).

In general, the algorithms correctly detected the targets, yielding a small number of false positives. Given the fact that the targets are covering areas larger than the 4, a number of false positives of approximately 20 most likely corresponds only to the panels in the row associated to the target. DTDCA, however, has difficulties in separating the second, fourth, and eight rows and produces a large number of false positives. The algorithm correctly identifies all the ground truth spectra except for one associated to r7.

All spectral screen-based detection filters correctly identified the ground truth targets. For Max SS-SA-OSP/KOSP and Max SS-SID-OSP/KOSP, the number of potential target pixels is quite low for $\alpha = 0.5$. Max SS-SA-OSP best separates the targets r3, r4, r6, and r8 and reasonably separates r1, r2, r7. For r5, a large number of pixels are identified as targets. Max SS-SID-OSP yields relatively similar values with better performance for r2 and r5 but fails to correctly separate r6. It is possible that higher thresholds would reduce the number of false positives. However, for consistency and to match an unsupervised environment, we decided to maintain the same threshold level. In the case of the KOSP-based filters, we note that the behavior

Table 8.2 Results for spectral screening and orthogonal subspace projection (OSP) methods

| | Max SS-SA-OSP | | | | | | Max SS-SID-OSP | | | | | |
| | 0.1 | | 0.25 | | 0.5 | | 0.1 | | 0.25 | | 0.5 | |
	P_T	P	P_T	P	P_T	P	P_T	P	P_T	P	P_T	P
r1	4	12812	4	319	4	39	4	13113	4	521	4	39
r2	4	13112	4	12302	4	40	4	13092	4	11230	4	30
r3	4	12282	4	358	4	21	4	11948	4	221	4	18
r4	4	148	4	31	4	20	4	12879	4	4458	4	21
r5	4	12564	4	2957	4	85	4	13094	4	10126	4	38
r6	4	13071	4	9564	4	24	4	13098	4	11404	4	85
r7	4	13069	4	120	4	35	4	13078	4	118	4	35
r8	4	10247	4	55	4	14	4	11053	4	51	4	16

| | Min SS-SA-OSP | | | | | | Min SS-SID-OSP | | | | | |
| | 0.1 | | 0.25 | | 0.5 | | 0.1 | | 0.25 | | 0.5 | |
	P_T	P	P_T	P	P_T	P	P_T	P	P_T	P	P_T	P
r1	4	10923	4	388	4	42	4	13113	4	11424	4	81
r2	4	13117	4	12869	4	332	4	13098	4	11696	4	172
r3	4	13113	4	13111	4	162	4	13082	4	5223	4	144
r4	4	12915	4	6274	4	26	4	13839	4	3247	4	214
r5	4	13088	4	12092	3	273	4	1122	4	12918	4	73
r6	4	13017	4	10200	4	50	4	13024	4	8523	4	24
r7	4	13088	4	11197	3	133	4	13119	4	13045	3	71
r8	4	12773	4	380	4	23	4	8231	4	40	4	14

| | SS-SA-OSP | | | | | | SS-SID-OSP | | | | | |
| | 0.1 | | 0.25 | | 0.5 | | 0.1 | | 0.25 | | 0.5 | |
	P_T	P	P_T	P	P_T	P	P_T	P	P_T	P	P_T	P
r1	4	13101	4	3207	4	42	4	13120	4	6366	4	54
r2	4	13112	4	10941	4	322	4	13110	4	12786	4	5056
r3	4	13114	4	12240	4	24	4	13103	4	11903	4	124
r4	4	2783	4	31	4	21	4	13063	4	10306	4	414
r5	4	13083	4	11286	4	112	4	12982	4	9319	4	248
r6	4	13114	4	11361	4	44	4	13105	4	11690	4	48
r7	4	13123	4	504	3	35	4	13124	4	6234	3	39
r8	4	12809	4	329	4	15	4	12120	4	212	4	17

of the Max SS is different than DTDCA in terms of the best-identified targets. While DTDCA shows a high positive rate for r2, r4, and r8, MSS is very efficient with r4 and r8, resulting in the lowest number of positives. In addition, r2 is identified better by Max SS-SA/SID. This can be explained by the differences in spectra selection in each case and suggests that DTDCA and Max SS should be used complimentarily. When the targets are easily separable through orthogonal projections, then DTDCA will perform well. When that fails, Max SS provides an attractive alternative. Compared with Max SS, regular spectral screening and Min SS perform poorly in target detection. In addition, for most of the targets, Min SS-OSP is outperformed by SS-OSP. We attribute this to the detection filter used, which is based on orthogonal

Table 8.3 Results for spectral screening and kernel orthogonal subspace projection (KOSP) methods

	Max SS-SA-KOSP						Max SS-SID-KOSP					
	0.1		0.25		0.5		0.1		0.25		0.5	
	P_T	P	P_T	P	P_T	P	P_T	P	P_T	P	P_T	P
r1	4	13124	4	13094	**4**	**55**	4	13095	4	106	**4**	**45**
r2	4	13123	4	13122	**4**	**12888**	4	13124	4	13123	**4**	**11710**
r3	4	13041	4	3947	**4**	**30**	4	13021	4	2260	**4**	**23**
r4	4	1486	4	31	**4**	**19**	4	2167	4	20	**4**	**13**
r5	4	13068	4	6286	**4**	**44**	4	13122	4	1227	**3**	**56**
r6	4	13122	4	13120	**4**	**130**	4	13108	4	1749	**4**	**41**
r7	4	12958	3	3027	**3**	**23**	4	12219	3	2018	**2**	**40**
r8	4	28	4	16	**4**	**14**	4	21	4	14	**4**	**14**

	Min SS-SA-KOSP						Min SS-SID-KOSP					
	0.1		0.25		0.5		0.1		0.25		0.5	
	P_T	P	P_T	P	P_T	P	P_T	P	P_T	P	P_T	P
r1	4	10573	4	87	**4**	**40**	4	13076	4	590	**4**	**43**
r2	4	13124	4	13121	**4**	**12833**	4	13121	4	13115	**4**	**13105**
r3	4	13100	4	6792	**4**	**75**	4	12868	4	1676	**4**	**56**
r4	4	13111	4	3066	**3**	**16**	4	11928	4	88	**3**	**15**
r5	4	13120	4	13103	**3**	**161**	4	13124	4	13090	**4**	**1100**
r6	4	13124	4	13106	**4**	**67**	4	13122	4	13110	**4**	**527**
r7	4	18	3	13123	**2**	**479**	4	13121	3	13103	**1**	**965**
r8	4	13119	4	14	**4**	**14**	4	25	4	16	**4**	**14**

	SS-SA-KOSP						SS-SID-KOSP					
	0.1		0.25		0.5		0.1		0.25		0.5	
	P_T	P	P_T	P	P_T	P	P_T	P	P_T	P	P_T	P
r1	4	13122	4	13094	**4**	**76**	4	13121	4	3298	**4**	**51**
r2	4	13124	4	13107	**4**	**12057**	4	13124	4	13116	**4**	**6458**
r3	4	13111	4	13026	**4**	**495**	4	13116	4	13064	**4**	**391**
r4	4	5467	4	52	**3**	**22**	4	9306	4	534	**4**	**18**
r5	4	13093	4	12184	**3**	**2055**	4	13124	4	13121	**3**	**12691**
r6	4	13121	4	9505	**4**	**54**	4	13114	4	7840	**4**	**51**
r7	4	13099	4	12580	**2**	**593**	4	13122	4	13118	**0**	**160**
r8	4	108	4	17	**4**	**14**	4	29	4	16	**4**	**14**

projections. Since Min SS selects spectra that are as close as possible to the previous selections, most probably these spectra will not be orthogonal. This effect is no longer visible when KOSP is employed.

When comparing OSP with KOSP, we note that the former yields results with a reduced number of false positives. In addition, KOSP fails to correctly identify some of the ground truth spectra. There are several reasons for this behavior. First, in our experiments, we have not investigated other choices for the kernel-mapping function K or try to vary the constant c in Eq. (8.24). Both could lead to significantly different results.

Table 8.4 DTDCA results: number of pixels identified as targets for each classified image α

	0.1		0.25		0.5	
	P_T	P	P_T	P	P_T	P
r1	4	12537	4	490	**4**	**20**
r2	4	12995	4	9742	**4**	**193**
r3	4	13107	4	9692	**4**	**22**
r4	4	13035	4	9823	**4**	**171**
r5	4	13058	4	5592	**4**	**25**
r6	4	13078	4	9394	**4**	**31**
r7	4	12997	4	4641	**3**	**23**
r8	4	13025	4	9351	**4**	**70**

The columns correspond to α values of 0.1, 0.25 and 0.5.

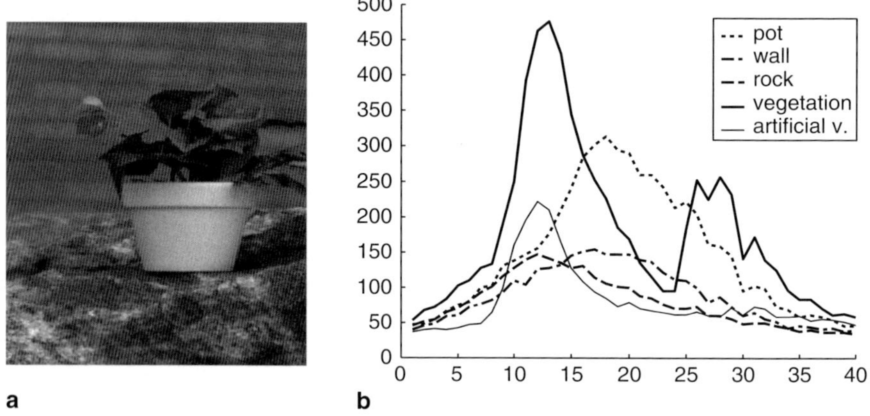

a   b

Fig. 8.8 **a** SOC data, **b** representative spectra from the image

8.5.2 SOC Data

The second experiment uses data produced using an SOC 700 hyperspectral sensor currently available in our lab. The camera is able to produce 640×640 pixel images on 120 bands equally spaced within the range of 400 and 900 nm (i.e., visible to near-infrared range). Forty bands uniformly extracted from the image cube were used (Fig. 8.8a).

The setup was an artificial plant arranged in a light brown ceramic pot. Several real leaves were placed in the plant arrangement (left front side and lower right side). To benefit from full-spectrum illumination, the arrangement was placed outside on a large rock formation. The background was a brown brick wall. The average spectral signatures for the five main elements in the image are shown in the graph in Fig. 8.8b. These averages were obtained by hand selecting several spectra for each feature.

The data were processed using DTDCA, Max SS-SA-OSP, and Max SS-SA-KOSP with the five average spectra as starting spectra. The resulting detection images are presented in Fig. 8.9 with parts a.*, b.*, and c.* corresponding to the three

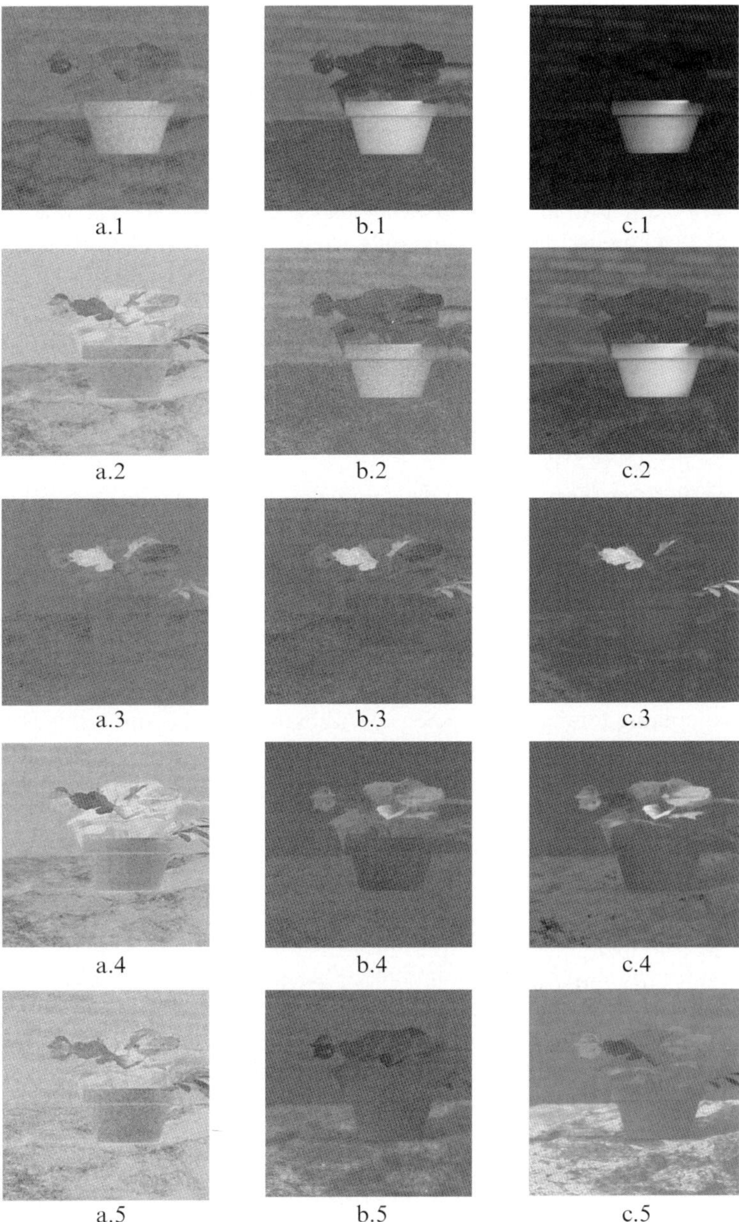

Fig. 8.9 Classification images for Max-SS-SA-KOSP target detection

algorithms in the order given and *.1, *.2, *.3, *.4, and *.5 corresponding to the ceramic pot, background wall, real vegetation, artificial plant, and rock formation, respectively. Note that, unlike in the previous experiment, KOSP outperformed OSP in separating the features. Both the DTDCA and Max SS-SA-OSP failed to correctly detect the five classes and resulted in noisy classification images. Since the two methods differ in how the spectra were collected but are identical in how the projection was done, we conjecture that the issue resides with the relationship among the spectra. Given that all five targets had large variability among the composing spectra, OSP detection was unable to correctly separate them. This factor was significantly reduced when KOSP was used. Since the data were mapped into a new feature space, the within-target variability was probably reduced.

We also note that in all three cases a clear separation of the ceramic pot from the ceramic background was possible only when the pot was chosen as target. An inspection of the two signatures shows that the pot has higher reflectance values but also exposes a similar slope.

8.6 Conclusions

We presented a new class of target detection algorithms for hyperspectral imagery derived from spectral screening and orthogonal projections. Spectral screening is defined as reducing the hyperspectral data to a representative subset of spectra. The subset is formed such that any two spectra in it are dissimilar, and for any spectrum in the original image cube, there is a similar spectrum in the subset. To improve the efficiency of the screening, we modified the spectra selection step such that it chooses either the most distant spectra or the closest spectra still available. This was done by computing for each unselected spectrum the product (sum) of the distances between itself and all the spectra already present in the subset. The most distant spectrum was identified as the one with the largest product value. The algorithm, maximum spectral screening (Max SS), has the advantage of reducing the overlap among the similarity sets for the spectra in the subset. The closest spectrum was identified as the one with the smallest sum value. The algorithm is the minimum spectral screening (Min SS).

Spectral screening, Max SS, and Min SS were adapted for target detection by choosing as the initial spectrum in the subset the target signature. Following screening, a classification procedure was applied that resulted in classification images for the targets. The procedure adopted the classic OSP approach or the more recent KOSP. In addition, in a departure from its traditional use with SA, we investigated the use of spectral screening with SID.

The new methods were tested on hyperspectral data often presented in target detection experiments as well as on in-house data. Qualitative and quantitative results for HYDICE data suggest that Max SS outperforms Min SS and regular spectral screening in identifying correct targets. In addition, Max SS exhibits a performance similar to DTDCA, a target detection algorithm based on OSP. An interesting result was obtained when the within-target variation was considerable, as was the case with the in-house SOC data. In this context, KOSP-based algorithms clearly outperformed OSP ones. The "extreme" spectral-screened subset was a good candidate for use with both OSP and KOSP.

Acknowledgments This work was supported by a 2005–2006 Sun Microsystems Academic Excellence Grant and through Montclair State University's Faculty Scholarship Program.

Chapter's References

1. P.M. Mather, *Computer Processing of Remotely-Sensed Images*, Wiley, New York, 1987
2. J.A. Richards, and X. Jia, *Remote Sensing Digital Image Analysis*, Springer, Berlin, 1999
3. P.J. Ready and P.A. Wintz, Information extraction, SNR improvement, and data compression in multispectral imagery, *IEEE Trans. Commun.*, Com-21, 1123–1130, 1973
4. D. Bannon and D. Milner, Information Across the Spectrum, *Oemagazine*, 3, 18–20, 2004
5. T. Achalakul and S. Taylor, A distributed spectral-screening PCT algorithm, *J. Parallel Distributed Comput.*, 63(3), 373–384, 2003
6. A. Plaza, P. Martínez, R. Pérez, and J. Plaza, A quantitative and comparative analysis of endmember extraction algorithms from hyperspectral data, *IEEE Trans. Geosci. Remote Sensing*, 42(3), 650–663, 2004
7. H. Kwon and N.M. Nasrabadi, Kernel orthogonal subspace projection for hyperspectral signal classification, *IEEE Trans. Geosci. Remote Sens.*, 43, 2952–2962, 2005
8. H. Du, C.-I. Chang, H. Ren, C.-C. Chang, J.O. Jensen, and F.M. D'Amico, New hyperspectral discrimination measure for spectral characterization, *Optical Engineering*, 43(8), 1777–1786, 2004
9. S.A. Robila, Investigation of spectral screening techniques for independent component analysis based hyperspectral image processing, in *SPIE Algorithms and Technologies for Multispectral, Hyperspectral, and Ultraspectral Imagery IX*, Vol. 5093, pp. 241–252, 2003
10. F.A. Kruse, A.B. Lekoff, J.W. Boardman, K.B. Heidebrecht, A.T. Shapiro, P.J. Barloon, and A.F.H. Goetz, The Spectral Image Processing System (SIPS)—interactive visualization and analysis of imaging spectrometer data, *Remote Sens. Environ.*, 44, 145–163, 1993
11. S.A. Robila and A. Gershman, Spectral matching accuracy in processing hyperspectral data, in *IEEE ISSCS*, pp. 163–166, 2005
12. J. Bowles, D. Clamons, D. Gillis, P. Palmadesso, J. Antoniades, M. Baumback, M. Daniel, J. Grossman, D. Haas, and J. Skibo, New results from the ORASIS/NEMO compression algorithm, in *SPIE Imaging Spectrometry V*, Vol. 3753, pp. 226–234, 1999
13. N. Keshava, Distance metrics and band selection in hyperspectral processing with applications to material identification and spectral libraries, *IEEE Trans. Geosci. Remote Sens.*, 42(7), 1552–1565, 2004
14. C.-I. Chang, An information theoretic-based approach to spectral variability, similarity and discriminability for hyperspectral image analysis, *IEEE Trans. Inform. Theory*, 46(5), 1927–1932, 2000
15. S.A. Robila, Using spectral distances for speedup in hyperspectral image processing, *Int. J. Remote Sens.*, 26, 5629–5650, 2005
16. R.H. Yuhas, A.F.H. Goetz, and J.W. Boardman, Discrimination among semiarid landscape endmembers using the spectral angle mapper (SAM) algorithm, *Summaries 3rd Annual JPL Airborne Geosciences Workshop*, Vol. 1, pp. 147–149, 1992
17. P.E. Dennison, K.Q. Halligan, and D.A. Roberts, A comparison of error metrics and constraints for multiple endmember spectral mixture analysis and spectral angle mapper, *Remote Sens. Environ.*, 93, 359–367, 2004

18. M.S. Stefanou, A signal processing perspective of hyperspectral imagery analysis techniques, master's thesis, Naval Postgraduate School, Monterey, CA, 1997
19. J.C. Harsanyi and C.-I. Chang, Hyperspectral image classification and dimensionality reduction: An orthogonal subspace projection, *IEEE Trans. Geosci. Remote Sens.*, 32, 779–785, 1994
20. H. Ren and C.-I. Chang, Automatic spectral target recognition in hyperspectral imagery, *IEEE Trans. Aerospace Electron. Syst.*, 39(4), 1232–1249, 2003
21. H. Ren and C.-I. Chang, A generalized orthogonal subspace projection approach to unsupervised multispectral image classification, *IEEE Trans. Geosci. Remote Sens.*, 38(6), 2515–2528, 2000
22. H. Kwon and N. M. Nasrabadi, Kernel matched subspace detectors for hyperspectral target detection, *IEEE Trans. Pattern Analysis Machine Learning*, 28, 178–194, 2006
23. H. Ren and C.-I. Chang, An experiment-based quantitative and comparative analysis of hyperspectral target detection and image classification algorithms, *IEEE Trans. Geosci. Remote Sens.*, 38, 1044–1063, 2000
24. T. Achalakul and S. Taylor, Real-time multi-spectral image fusion, *Concurrency Comput. Pract. Exper.*, 13(12), 1063–1081, 2001
25. R.C. Olsen, S. Bergman, and R.G. Resmini, Target detection in a forest environment using spectral imagery, in *SPIE Imaging Spectrometry III*, Vol. 3118, pp. 46–56, 1997

Part IV
Face and Facial Expression Recognition in Intensified and Thermal Imagery

Chapter 9
Face Recognition in Low-Light Environments Using Fusion of Thermal Infrared and Intensified Imagery

Diego A. Socolinsky and Lawrence B. Wolff

Abstract This chapter presents a study of face recognition performance as a function of light level using intensified near infrared imagery in conjunction with thermal infrared imagery. Intensification technology is the most prevalent in both civilian and military night vision equipment and provides enough enhancement for human operators to perform standard tasks under extremely low light conditions. We describe a comprehensive data collection effort undertaken to image subjects under carefully controlled illumination and quantify the performance of standard face recognition algorithms on visible, intensified, and thermal imagery as a function of light level. Performance comparisons for automatic face recognition are reported using the standardized implementations from the Colorado State University Face Identification Evaluation System, as well as Equinox's algorithms. The results contained in this chapter should constitute the initial step for analysis and deployment of face recognition systems designed to work in low-light conditions.

Keywords: Face recognition· Biometrics · Night vision

9.1 Introduction

Face recognition by computer is a subject with a rich history and an active research community. While most effort has centered on the use of standard visible still and video imagery, some research has focused on the use of thermal imaging for facial recognition [1–6]. Thermal imagery extends the operational range of a system to complete darkness conditions, but not without trade-offs. One such trade-off is that thermal imagery is difficult to use for anything but cooperative access control applications since enrollment imagery is by and large available only for cooperative subjects. That is, we do not have opportunistically acquired thermal imagery of uncooperative subjects suitable for enrollment into an identification system, whereas such visible imagery is indeed available due to the pervasive presence of visible

R.I. Hammoud (ed.), *Augmented Vision Perception in Infrared: Algorithms*　　　197
and Applied Systems, Advances in Pattern Recognition,　DOI 10.1007/978-1-84800-277-7_9,
© Springer-Verlag London Limited 2009

camera systems throughout urban areas. Even for cooperative subjects, we have a great deal of legacy visible face imagery coming from drivers' licenses, passports, employee IDs, and other forms of identification, but for the most part we lack any corresponding thermal facial imagery. Any imaginable facial identification system utilizing a thermal component would require the collection of thermal face imagery of all subjects to be recognized to be functional. This task would range from cumbersome to impossible, depending on the operational scenario under which the system is to be deployed.

Image intensification provides an alternative technology to thermal imaging for face recognition in low-light levels. This chapter explores the application of such technology and provides initial performance comparisons with visible imagery. Intensified near-infrared imagery (I2) is the most prevalent technology in night vision systems, both civilian and military*. It is a relatively low-cost technology compared to thermal imaging and can function with very little ambient light, with the best systems performing acceptably under overcast moonless night conditions. Since I2 imagery is reflective in nature, it shares many properties with standard visible imagery, and indeed a comparison of visible and I2 images of the same scene under high-light conditions shows them to be very similar (see Fig. 9.1). This suggests the possibility that an automated system could recognize faces acquired with an I2 system based on visible enrollment imagery. Additional questions arise regarding the best way to train and deploy such a system to maximize performance.

The present chapter considers the use of I2 imagery for face recognition under various lighting conditions, ranging from good illumination to near-complete darkness. By carefully controlling the light levels during data collection, we are able to isolate the effect of incident luminance and reason on the relative merits of visible and I2 imagery. We consider the effect of noise and cross-modality matching as these are essential to the problem of deploying a realistic face recognition system in a low-light environment. Even with the caveats stated, face recognition with thermal imagery remains a viable alternative when thermal enrollment data are available. In those cases, the addition of thermal imagery can greatly increase performance and

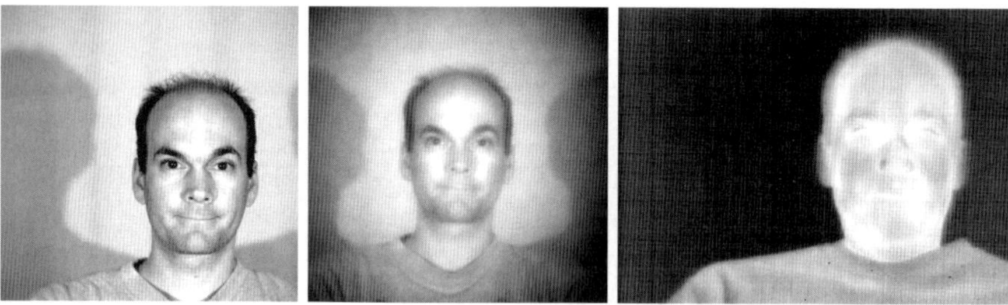

Fig. 9.1 Visible, intensified near-infrared and thermal infrared images of the same subject under bright illumination conditions

---

* Technically, this imagery spans the range from 600 to 900 nm, which overlaps the visible spectrum. However, nighttime luminance favors the near-infrared portion of this range, and thus we refer to it as intensified near infrared.

further extend the operational range of the system. For that reason, this chapter also includes results and analysis on the use of thermal infrared imagery in conjunction with I2 imagery.

9.2 Image Intensification and Thermal Imaging

The most common technology for image intensification is through the use of a microchannel plate (MCP). MCPs are made of several million tightly packed channels about $10\,\mu$m in diameter. Each channel functions as an independent photomultiplier, and since the channels are arranged in a spatially coherent fashion, any light pattern impinging on the input end of the MCP results in the same (intensified) pattern being emitted out the output end. When a photon enters the input end of an individual channel, it releases an electron within the tube. A strong potential difference of several thousand volts is applied between input and output ends of the microchannel, thus accelerating the electrons. As the accelerated electrons travel down the channel, they release more electrons from the tube material as they collide with the inside walls. This effect is called an *electron cascade* and is the crux of the intensification process, as one original electron, through the application of a strong potential difference, generates a much larger number of electrons. Gain factors of 10^6 or more are achievable through this process. Accelerated electrons exit the microchannel and collide against a phosphor screen. On collision, electron release photons from the phosphor, and these photons constitute the final intensified image. Optics in front and behind the MCP allow light to be focused on the front end and the intensified output to be viewed on the back.

Modern image intensifiers are lightweight and operate on very low power. Typical military-grade devices will operate for dozens of hours on standard AA batteries and provide enough photomultiplication to allow for navigation and basic tasks under moonless overcast night conditions.

Thermal imaging technology is quite mature at present. Modern sensors have better sensitivity and higher resolution than previous generations, at a lower cost and with smaller form factor. The sensor used for our data collection is a long-wave infrared (LWIR) microbolometer, sensitive in the 8- to 12-μm range. Each bolometer element in the imaging array consists of a small metal plate whose resistance changes according to the temperature variation induced by incoming infrared radiation. By focusing LWIR light onto the microbolometer array, one is able to create images of the scene, such as those used in this chapter.

9.3 Data Collection and Processing

We performed data collection and a series of experiments aimed at elucidating the role of I2 and thermal imagery in low-light face recognition. Data were collected from a set of 96 student volunteers over two sessions separated by one week to avoid overestimating performance due to same-session artifacts. Imagery was collected with a scientific-grade Dalsa 1M15 visible sensor with 1024×1024 pixel resolution binned into 512×512, an Indigo Merlin LWIR sensor with 320×240 pixels, and another Dalsa 1M15 visible sensor outfitted with an ITT PVS-14 image intensifier. Peak sensitivity for the Dalsa sensor occurs at 820 nm which is comparable to the peak sensitivity of the PVS-14. The Indigo Merlin camera is sensitive in the 8- to 12-µm range. Figure 9.2 shows the arrangement of cameras. The intensified and LWIR sensors were coregistered through the use of a dichroic beam splitter, and the visible sensors were placed above the previous two in a bore-sighted configuration.

Imagery was collected in an interior room below grade, with no windows and two sets of consecutive doors, ensuring no light penetration from the adjacent hallway. Figure 9.3 shows the setup of cameras and lights. During collection, all room lighting, both in the inner and outer rooms, was kept off, and the only source of illumination was provided by sources controlled as part of the experiment. Lighting was controlled with the use of two custom-made fixtures. These fixtures consisted of rectangular boxes closed on five sides and sealed against light leaks in all joints. Each box contains a 20 watt low-power compact fluorescent bulb, selected for its low heat output, with a color temperature of 2700 K. The front side of each fixture featured a slotted channel, also sealed against light leaks, which fit a series of perforated panels with different levels of light transmission. Each panel was made of

Fig. 9.2 Close-up of cameras used for image collection; note PVS-14 device attached in front of the camera on the right-hand side

Fig. 9.3 Camera and lighting setup for image collection

an opaque material and had a hole centered over the bulb, allowing an amount of light proportional to the area of the hole to exit the light fixture. Five different light levels were selected by varying the size of the exit aperture in each panel. This lighting system had two key advantages over the obvious alternative of a rheostat-based system. First, it allowed for repeatable light levels, which is hard to achieve with a variable resistor. Second, it ensured that the color temperature of the light was constant throughout the different light levels, whereas a dimmer induces a red shift as the light level decreases. By using two light sources symmetrically located in front of the subject, we ensured even illumination across the face.

At the brightest light level, the illuminance at the subject's face was 9 lux (Table 9.1), as measured by a Spectra P-2000EL-A light meter. For reference, full moonlight is about 1 lux, and a standard office environment is illuminated to an average of about 300 lux. Second, third, and fourth light levels each decreased by a factor of sixteen, while the fifth light level was about half of the fourth. At the lowest light level, illuminance at the subject's face was about 0.001 lux, which is consistent with starlit conditions. As these low illuminance values cannot be measured with standard light meters, they were estimated from measurements of the brightest light level along with the relative area of the aperture for each light setting. These approximations degraded as the light level decreased as it is harder to control the size of the aperture and possible self-occlusion from the light panel material. As seen in the results, the estimates at the lowest light levels may be optimistic, with the actual light level lower than the estimate. Note that the Dalsa 1M15 camera is very sensitive, so whereas other cameras may have trouble imaging noiselessly at 9 lux illuminance, this is plenty of light for this sensor.

As seen in Fig. 9.4, the subject was seated with the face approximately five ft away from the cameras and flanked by symmetrically positioned light fixtures placed on a line with them. Data were collected with custom software designed to simultaneously collect all three synchronized video streams, visualize them as a color composite, and save them to disk for future analysis (see Fig. 9.5). Five sequences at corresponding light levels were collected for each subject during each collection session. For each sequence, the subject was asked to count out loud for four s while ten video frames were collected. Immediately following, three still shots

Table 9.1 Approximate illuminance levels for different illumination conditions and reference illuminations

Light level	Illuminance (lux)
2	9
3	0.562500
4	0.035156
5	0.002197
6	0.001098
Moonlight	1
Sunlight	75K
Office	300

Fig. 9.4 Subject being imaged under brightest lighting condition

Fig. 9.5 Screenshot of image collection and visualization software

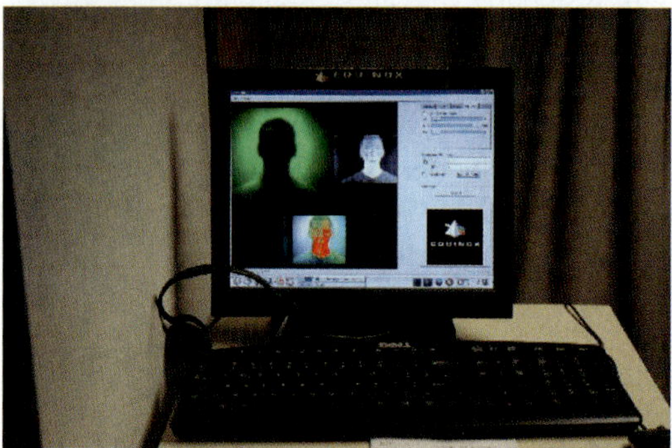

were collected as the subjects were asked to "smile", "frown", and "act surprised". This collection protocol ensured that a degree of expression variability was present in the data.

9.4 Experimental Results and Discussion

We performed a series of recognition experiments aimed at establishing baseline performance for the use of I2 and visible imagery in automatic face recognition, as well as the additional gains afforded by the combined use of thermal imagery. For the sake of standardization, we performed most visible and I2 experiments using the Colorado State University (CSU) Face Identification Evaluation System, version 5.0 [7]. Corresponding results reported were obtained using the Principal Components Analysis (PCA) algorithm with Euclidean distance, as implemented in the CSU system. A simple modification to the standard CSU software was necessary to cope with the fact that all of our imagery was collected with 12-bit cameras. Since the CSU software operates internally with a floating point image format, we only had to modify the loading routine to properly accept 12-bit data. All other aspects of the CSU software were left unchanged, and all experiments used the default settings in the CSU distribution. In addition to the standardized CSU results, we also include the results of using Equinox's own facial recognition system. These results are particularly interesting as they pertain to the fusion of I2 and thermal imagery for enhanced recognition performance in low-light levels.

Training for the PCA algorithm was performed by segregating the first three images from each subject from the first collection date and using that set to estimate the principal component space. We estimated three different PCA spaces, one using only visible images, using only I2 images, and a third using an equal combination of both modalities. Results for the experiments were independent of which of these subspaces was used. That is, experimental results obtained with any one of these training subspaces were statistically equivalent to those obtained with any other. Therefore, we only report results with visible training. This is an interesting result in its own right, indicating that either the exact nature of the PCA subspace is not critical to recognition performance or that visible and I2 imagery at good illumination levels is similar enough to result in indistinguishable PCA subspaces.

For each experiment, gallery images were selected for each subject from the first collection date and disjointly from the training set. Probe images were chosen for each subject from the second collection date to avoid overestimating the recognition rate. Both gallery and probe sets for the reported experiments contain a single image of each subject. We concentrated on the most likely recognition scenario, with well-illuminated visible images as the gallery and probe images coming from different light levels of visible and I2 modalities. This reflects the most natural use of an intensified low-light face recognition system, where high-quality enrollment images are available, for example, from drivers' license pictures, and probe images are acquired under uncontrolled illumination conditions. Other scenarios are interesting as well, but space considerations preclude us from addressing them here.

For the thermal infrared experiments, probe and gallery images were selected in the same manner as discussed. That is, the corresponding thermal image to each visible or I2 image selected was used. All recognition in the thermal infrared was performed using Equinox's algorithm, which was trained using a completely disjoint set of images collected over the last few years and not sharing subjects with the data collected for this study. We also include results of applying Equinox's visible recognition algorithm, which was similarly trained.

Eye coordinates for all images, visible and intensified, were manually located. For the lowest-light level sequences in the visible modality, the face and eyes were not distinguishable in the images. Rather than assign arbitrary eye coordinates and therefore ignore any possible signal, however hard to distinguish, we used predicted coordinates for those frames. The prediction was based on a least-squares approximation of an affine transformation from the corresponding I2 frame. The approximation was computed using eye correspondences from the two highest illumination settings. Since the least-squares fit for all sequences had an average error and standard deviation under 1 pixel, we are confident that the predicted eye coordinates are correct to within the accuracy of manual location.

Figure 9.6 and Fig. 9.7 show the basic results corresponding to well-illuminated visible gallery images with visible and I2 probes at decreasing light levels. Each curve is labeled with its modality (v or i, respectively) and light level (from 2

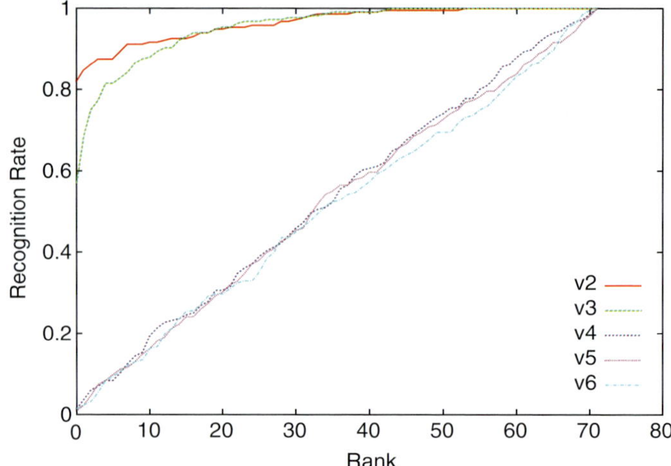

Fig. 9.6 CMC curve for visible probes of varying light levels

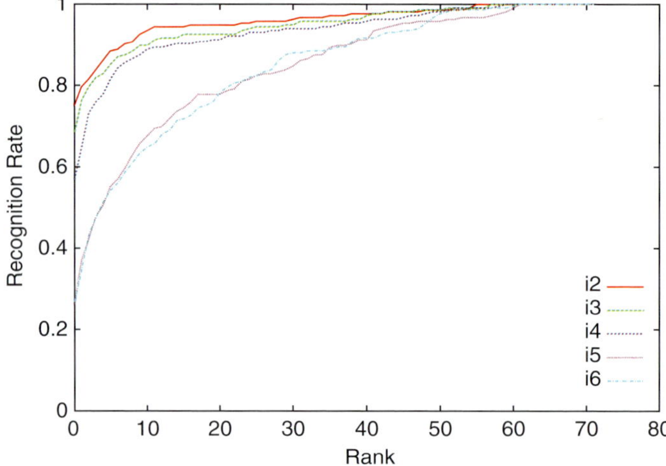

Fig. 9.7 CMC curve for I2 probes of varying light levels

through 6). The most obvious difference between visible and I2 results is the sharp degradation of visible results as the illumination decreases. By contrast, I2 performance degrades much more gracefully, remaining close to the top performance for the first three illumination levels (a factor of approximately 256). The fairly sharp drop seen in the fourth and fifth light levels indicates that our estimate of the illuminance at the subject's face is high for those levels, as discussed. While behavior at the lowest light levels is easily understood, that at the higher light levels shows some interesting features. Figure 9.8 compares the Cumulative Match Characteristic (CMC) curves for visible and I2 probes for the two brightest light levels, those for

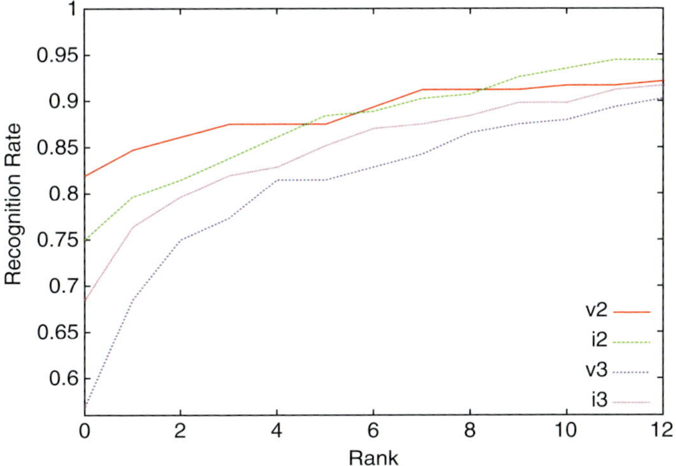

Fig. 9.8 Comparison of visible and I2 CMC curves for first and second light levels. Note that only the first 12 ranks are shown

Table 9.2 Top rank performance summary for different gallery and probe combinations

	V				VB	
	V	I2	VA	I2A	V	I2
2	81.9	75.0	88.9	84.7	87.5	87.5
3	57.9	68.5	81.9	76.4	70.8	76.3
4	1.3	56.9	4.1	69.4	5.5	65.3
5	0.9	26.3	2.8	37.5	2.7	41.7
6	1.4	26.8	1.3	36.1	2.7	30.6

V visible, *I2* intensified NIR, *VB* visible blurred, *VA* time-averaged visible, *I2A* time-averaged intensified NIR

which the visible images are minimally affected by noise. We see that at the highest light level, recognition rates with visible imagery are higher than those with I2 imagery. This is not surprising since while the two modalities are similar, they are not exactly alike. We pay a penalty for the use of I2 probes in combination with visible gallery images. Second, a glance at Fig. 9.1 shows that the I2 image is considerably "softer" than its visible counterpart. This is to be expected from the *Modulation Transfer Function (MTF)* degradation introduced by the MCP in the optical path. As seen in Table 9.2, the penalty at top rank for the brightest light level is about 7 percentage points, or roughly 40% increase in the error rate. By the second light level, performance with I2 imagery is higher than visible imagery by more than ten percentage points.

The lower performance observed for well-illuminated I2 probes versus well-illuminated visible probes motivated the following experiment. If performance degradation is due to MTF effects softening the I2 image, then artificially blurring the visible image to match the sharpness of their I2 counterparts should result in comparable performance. This was actually not the case, but led to the interesting

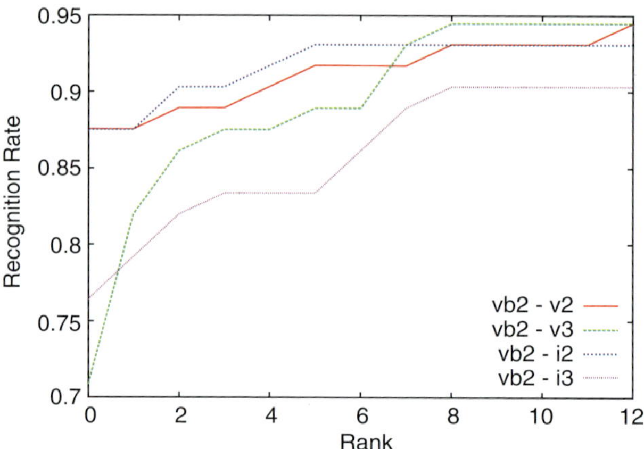

Fig. 9.9 Comparison of visible and I2 CMC curves for first and second light levels, using raw and blurred gallery images; note only the first 12 ranks are shown

results shown in Fig. 9.9. We estimated the necessary artificial blur to be consistent with that obtained by applying a convolution with a Gaussian filter with standard deviation of 1.5 pixels. We applied this blurring to the well-illuminated visible gallery images and reran the previous experiments with visible and I2 probes at all light levels. The most interesting results occurred at the highest light levels, as shown in Fig. 9.9. Applying a blur to the visible gallery increased performance for both modalities and all light levels (see Table 9.2). Interestingly, the performance with visible and I2 probes against a blurred visible gallery is not statistically different. The improvement for I2 probes can be explained by the fact that the blurred visible images simply look more like the raw I2 images. In the case of the visible performance increase (from 81.9 to 87.5%), we can probably ascribe the difference to a mitigation of high-frequency misregistration errors inevitably present due to small errors in eye localization. Regardless, this seems to indicate that, given appropriate preprocessing, I2 and visible images are roughly equivalent at high-light levels, with I2 images providing an unmistakable advantage as the illumination levels decrease.

At the opposite end of the illumination spectrum, we are concerned with relative performance at low-light levels. As Fig. 9.6 and Fig. 9.7 show, performance is clearly vastly superior with I2 imagery as that achieved with visible imagery is essentially equivalent to chance for the lowest three light levels. However, we see that I2 performance also suffers considerably at very low levels of illumination. In this case, this is due to a combination of two noise sources. First and foremost is the noise introduced by the MCP itself, but we also see noise introduced by the CCD (charge-coupled device) imaging the output of the PVS-14 device. Figure 9.10 shows the decrease in noise obtained by the time-averaging process. Note that at standard video rates, this requires only a third of a second of video. For reference, Fig. 9.11 shows a series of I2 images of the same subject at decreasing light levels. Attempts to improve performance by simple image denoising based on Gaussian smoothing and median filter did not yield any significant difference. An alternative

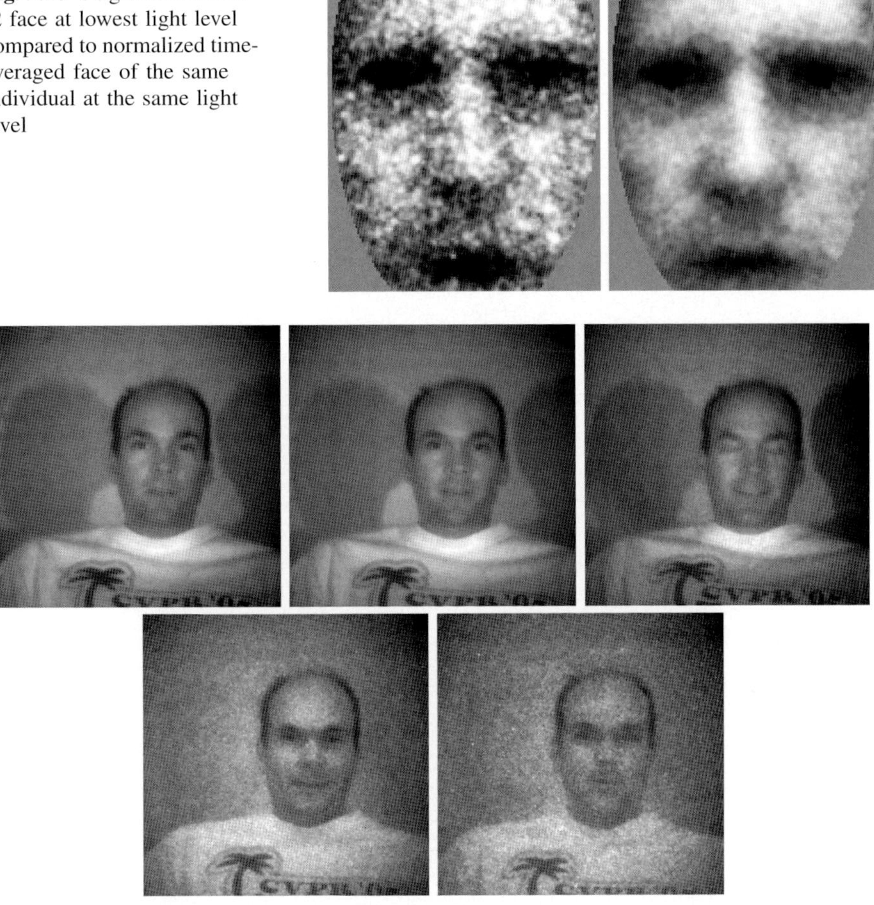

Fig. 9.10 Single normalized I2 face at lowest light level compared to normalized time-averaged face of the same individual at the same light level

Fig. 9.11 I2 images of the same subject under decreasing light levels

means of denoising for video sequences is time averaging of registered frames to obtain a single high-quality image. We applied this process to both visible and I2 images at all light levels by averaging the ten consecutive registered frames from the second collection for each subject.

Figure 9.12 shows performance curves for time-averaged I2 imagery at all light levels. In combination with Table 9.2, we see that time averaging indeed increases performance at most light levels, although at the highest light levels the performance gains afforded by blurring the visible gallery appear a bit larger (but not conclusively significantly so). As Table 9.2 shows, visible probes also benefit from time averaging, although only for the two brightest light levels. After that point, time averaging has negligible effect and underperforms simple blurring. At the highest light level, time-averaged visible imagery is the best performer by a very small margin, which is not likely significant.

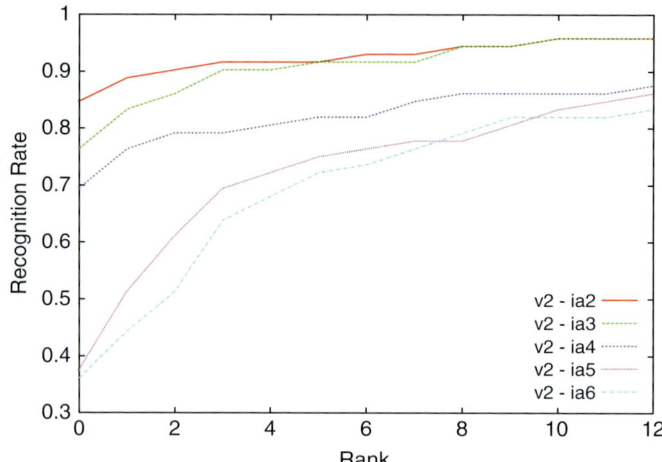

Fig. 9.12 CMC curves for time-averaged I2 imagery at all light levels; note only the first 12 ranks are shown

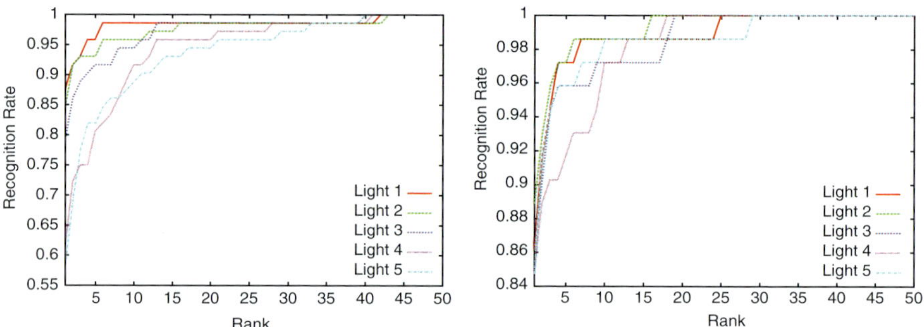

Fig. 9.13 CMC curves for I2 (left) and LWIR (right) only recognition as a function of light level using the Equinox algorithm

The final set of experiments concerns the use of thermal infrared imagery in conjunction with visible and I2 imagery. We should keep in mind that these experiments assume the existence of enrollment imagery in both visible and thermal infrared modalities, which restricts their domain of applicability. Results were obtained using Equinox's face recognition algorithm, which is specially designed to take advantage of the complementary nature of reflective and emissive imagery. Gallery sets consisted of pairs of visible and LWIR images, while probe sets were made of pairs of I2 and LWIR images. Figure 9.13 shows results of applying the Equinox algorithm to recognition of I2 probes with visible gallery (left) and LWIR probes with LWIR gallery (right) as a function of light level. Figure 9.7 should be compared for reference with Fig. 9.13 (left) since they refer to the same data, with different recognition algorithms. Considerably better performance is achieved with the more sophisticated algorithm. Note that, as expected, LWIR performance is

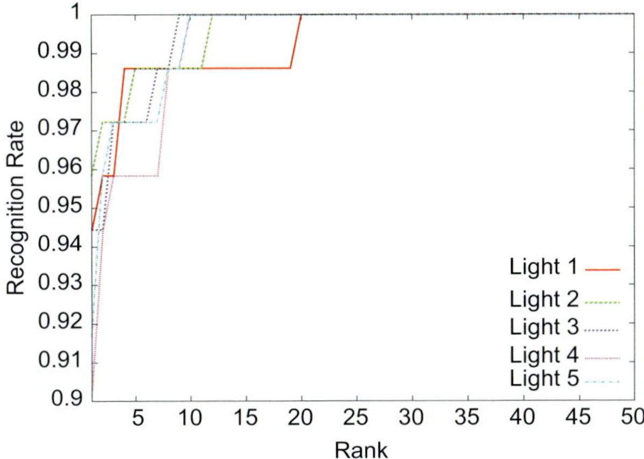

Fig. 9.14 CMC curves for fused I2/LWIR recognition as a function of light level using the Equinox algorithm

essentially insensitive to ambient illumination. We see that recognition performance with LWIR images is better than that achieved with I2 probes and never worse than the best performance achieved through blurring of the visible gallery. This is so despite the fact that illumination was uniform across the face for all light levels.

Figure 9.14 shows recognition performance when using a combination of I2 and LWIR images as probes with a gallery composed of visible and LWIR image pairs. As we can see, results are much better than for single-modality recognition and show less degradation as a function of light level than when using only the reflective modalities. This is comparable to the results reported in the past for fused visible/LWIR recognition and indicates a clear advantage for the fused system whenever thermal enrollment data are available.

Overall, the few experiments performed so far on this rich data set provide some initial understanding of the role of I2 imagery in automatic face recognition as well as its potential for use in conjunction with thermal imagery when available. It is evident from our analysis that the limits of performance for visible imagery are reached quickly as illumination decreases. Intensified imagery expands the range of illuminations under which face recognition is possible. Unfortunately, as illumination levels reach near darkness, recognition with I2 imagery also suffers performance degradation. We should note, however, that this occurs at levels for which an unaided human eye can barely discern the subject's face at a distance of six ft, that is, for a fairly low light level. It is not clear that the levels of performance exhibited by a simple PCA face recognizer on this imagery are sufficient for real-world applications, but we see that a more sophisticated algorithm handily outperforms PCA. Such an algorithm would also improve visible results at high-light levels, but at the lowest illuminations the near or complete absence of visible signal guarantees that visible performance will lag behind I2 performance. Thermal imagery shines under these conditions as it does not depend on light reflection from the subject and rather functions almost exclusively in the emissive domain. For such cases, when thermal enrollment imagery is available, using a combination of imaging modalities is clearly the best alternative.

9.5 Conclusions

We presented an initial set of experiments on a rich, newly collected set of visible, I2, and LWIR imagery of faces at multiple controlled levels of illumination. Data for almost one hundred subjects were collected over two sessions separated by one week. Lighting was carefully controlled by performing the imaging sessions in a lightproofed room and using custom light fixtures capable of producing precise and repeatable light levels. This study constitutes the first of its kind and should provide an experimental baseline for future work in this area.

All experimentation was done using a slightly modified version of the CSU Face Identification Evaluation System, version 5.0, which provides an open standard platform for comparative evaluation of face recognition systems. For simplicity of exposition, all results reported were obtained using the PCA algorithm with Euclidean metric. In addition, results using a more sophisticated algorithm were included for comparison. As could be expected, decreasing levels of illumination had a severe effect on recognition performance with visible imagery, which became equivalent to chance for the middle level in our experimental sequence. Observed performance for I2 imagery was much better than the visible counterpart at the lowest light levels. At the brightest light levels, we discovered that by applying some preprocessing to gallery or probe images, the performance difference between visible and I2 imagery became quite small and likely not significant given the parameters of the experiment. A larger experiment must be conducted to further elucidate this difference. As expected from prior research, the combined use of reflective and emissive modalities greatly improves overall performance.

Despite the considerable advantage afforded by I2 imagery under darkness conditions, performance obtained is lackluster and may not be sufficient for many applications, at least not without the use of a supplementary modality such as thermal infrared. Image enhancement and intensifier noise mitigation might be required to reach acceptable performance levels when using I2 imagery alone. This subject will be taken up in future research and will be reported in upcoming publications.

Chapter's References

1. D.A. Socolinsky and A. Selinger, A comparative analysis of face recognition performance with visible and thermal infrared imagery, in *Proceedings ICPR*, Quebec, Canada, August 2002.
2. D.A. Socolinsky and A. Selinger, Face recognition with visible and thermal infrared imagery, *Computer Vision and Image Understanding*, July–August 2003

3. J. Wilder, P.J. Phillips, C. Jiang, and S. Wiener, Comparison of Visible and Infra-Red Imagery for Face Recognition, in *Proceedings of 2nd International Conference on Automatic Face & Gesture Recognition*, Killington, VT, 1996, pp. 182–187
4. X. Chen, P. Flynn, and K. Bowyer, Visible-light and infrared face recognition, in *Proceedings of the Workshop on Multimodal User Authentication*, Santa Barbara, CA, December 2003
5. B. Abidi, Performance comparison of visual and thermal signatures for face recognition, in *The Biometrics Consortium Conference*, Arlington, VA, September 2003
6. F.J. Prokoski, History, current status, and future of infrared identification, in *Proceedings IEEE Workshop on Computer Vision Beyond the Visible Spectrum: Methods and Applications*, Hilton Head, NC, 2000
7. D. Bolme, R. Beveridge, M. Teixeira, and B. Draper, The CSU face identification evaluation system: its purpose, features and structure, in *Proceedings of the International Conference on Vision Systems*, Graz, Austria, April 2003, pp. 304–311, Springer-Verlag, New York.

Chapter 10
Facial Expression Recognition in Nonvisual Imagery

Gustavo Olague, Riad Hammoud, Leonardo Trujillo, Benjamín Hernández and Eva Romero

Abstract This chapter presents two novel approaches that allow computer vision applications to perform human facial expression recognition (FER). From a problem standpoint, we focus on FER beyond the human visual spectrum, in long-wave infrared imagery, thus allowing us to offer illumination-independent solutions to this important human-computer interaction problem. From a methodological standpoint, we introduce two different feature extraction techniques: a principal component analysis-based approach with automatic feature selection and one based on texture information selected by an evolutionary algorithm. In the former, facial features are selected based on interest point clusters, and classification is carried out using eigenfeature information; in the latter, an evolutionary-based learning algorithm searches for optimal regions of interest and texture features based on classification accuracy. Both of these approaches use a support vector machine-committee for classification. Results show effective performance for both techniques, from which we can conclude that thermal imagery contains worthwhile information for the FER problem beyond the human visual spectrum.

10.1 Introduction

Current trends in modern computer systems reveal an underlying need to enhance human-computer interaction. From voice and handwriting recognition software to bioidentification and vision-based security, these types of systems require novel and imaginative approaches that enhance the communication interface between users and computer applications. Hence, the computer science community is presented with a large set of unresolved and particularly hard problems. One such problem deals with the automatic recognition, by a computer system, of a human's emotional state based on different types of visual information, that is, facial gestures.

It is important to remember that determining what humans think and feel by simply analyzing their facial gestures is in no way a trivial task. Human beings

R.I. Hammoud (ed.), *Augmented Vision Perception in Infrared: Algorithms and Applied Systems,* Advances in Pattern Recognition, DOI 10.1007/978-1-84800-277-7_10,

require years of constant interaction with many people to build efficient discriminative mental models of how we express our emotions. We are able to build rich mental representations of how human emotions are manifested relying on a wide set of contextually relevant information. Once these mental representations are built up, we are able to compare our learned models with what we perceive in a given situation, allowing us to infer a particular person's emotional state. However, generalizing our mental representations can be extremely difficult, leading us to fine-tune our models to people we interact with the most. This is obviously related with the high level of complexity inherent in a human being's psyche and emotional state. For us, simply expressing our emotional state with words can lead us to describe a complex blend of different, and seemingly conflicting, emotions, a direct consequence of the fact that human behavior and psyche cannot be expressed as a contradiction-free axiomatic system.

Extrapolating this process, of human emotion model building, to human-computer interactive systems, where a man-made system will need to determine a person's emotional state, is still beyond current state-of-the-art systems. However, it is widely accepted that a comprehensive solution to this problem would have to consider multi-modal signals extracted from different information channels. Possible useful sources of such information include voice, physiological signal patterns, hand and body movements, and facial expressions. Based on a priori and intuitive assumptions we can expect that visual cues in general and facial gestures in particular, provide important information that could allow us to infer a persons emotional state. This implies that an artificial system will greatly benefit from machine vision techniques that extract relevant visual information from images of a person's face. This problem is known as the Facial Expression Recognition (FER) problem in computer vision literature.

The remainder of this chapter is organized as follows. First, in Section 10.2 the general FER problem is presented, along with the particular instance of FER beyond the human visual spectrum. Then, Section 10.3 gives a brief overview of related work with the problem at hand. Later, Sections 10.4 and 10.5 present our approaches to FER in thermal imagery; In Section 10.4 we describe our automatic facial feature localization and PCA-based classification approach using thermal imagery. In Section 10.5 we present our Evolutionary paradigm based visual learning approach for Facial Expression Recognition. Finally, Section 10.6 is devoted to a brief discussion and exposure of possibly productive future work.

10.2 Facial Expression Recognition

The FER problem has recently received much research attention, as can be noted in surveys by Pantic et al. [1] and more recently by Fasel et al. [2]. FER is centered around extracting emotional content from visual patterns of a person's face using machine vision and pattern recognition algorithms. Techniques developed to solve this problem have used both still images and image video sequences. FER systems that rely on video as their main input offer great insight into the way in which the human face generates different kinds of gestures. Recently, important work in this area has been published, including [3–5]. However, video sequences are not always available in every real-world situation, in which case still images can offer a reliable alternative. Different techniques for FER systems that use still images have been proposed, such as those in [1, 6–9].

A quick review of computer vision literature devoted to FER revealed that most state-of-the-art techniques follow the same basic information-processing approach used by many facial analysis systems [2] (see Fig. 10.1). This approach can be divided into three main parts: Region of interest (ROI) selection, feature extraction, and image classification.

First, appropriate ROIs are selected; these are image areas where feature extraction will take place. State-of-the-art techniques have used holistic methods where the ROI is the entire face [9], and modular or facial feature-based approaches where information is extracted from specific facial regions [6–8]. Second, dimensionality reduction of the selected ROIs is done by some sort of feature extraction procedure. Common feature extraction techniques include principle component analysis (PCA) [8,10], Gabor filter analysis [8], and what are known as hybrid methods [7,9]. Finally, a trained classifier takes the extracted feature vectors of each ROI as input and assigns, as output, each image to one of the predefined training classes. Popular classifiers include neural networks, hidden Markov models and support vector machines (SVMs), to name but a few.

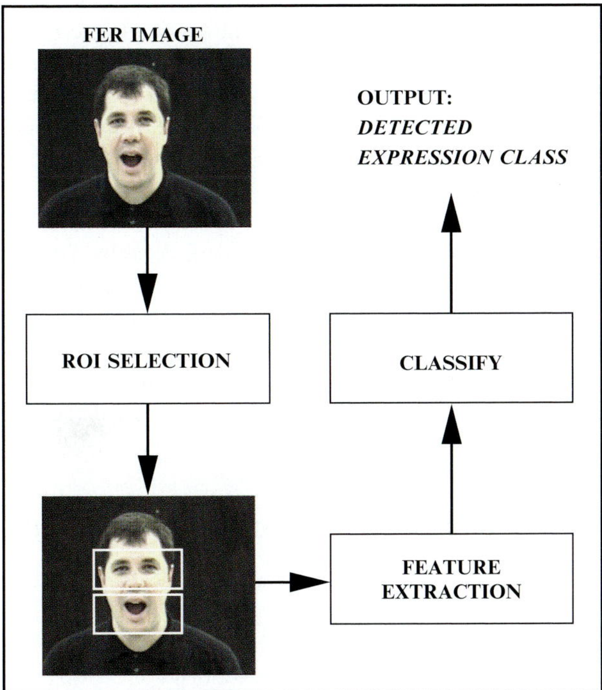

Fig. 10.1 Flow diagram of the FER approach commonly used with still images

10.2.1 FER in Thermal Images

Current research published on FER systems focused on the use of cameras with a single information channel. However, as noted by some researchers [11, 12], facial analysis systems relying on visual spectrum information are highly illumination and pose dependent. Recent works [13–16] have shown the possible usefulness of studying FER beyond the visual spectrum. These references suggested that, given the difficulty of the FER problem, new research in this area should look for new and imaginative ways of applying FER systems based on previously unused sources of information, such as infrared or thermal signals.

10.3 Related Work

Due to the success that thermal or infrared images have had in other facial analysis problems [11, 12, 17], their applicability to the FER problem is now gaining interest. Sugimoto et al. [15, 16], for instance, used predefined ROIs corresponding to areas surrounding the nose, mouth, cheek, and eye regions on the face and applied simulated annealing and template matching for appropriate localization. Features were computed by generating differential images between the average "neutral" face and a given test image followed by a discrete cosine transform. They performed classification using a backpropagation-trained neural network. As another example, Pavlidis et al. [14] used an interesting variant of the classic FER problem. The authors used high-definition thermal imagery to estimate the changes in regional facial blood flow. Their system was designed to be used as an anxiety or lie detection mechanism, similar to a polygraph machine without the subject being aware of the test because of the type of sensors used. Facial regions were handpicked, as in [15, 16], by the system designers.

The following sections present our proposed approaches to FER beyond the human visual spectrum. Section 10.4 presents the work done by Trujillo et al. in [13] based on an automatic feature localization procedure and classification using PCA or eigenimage techniques. Section 10.5 is devoted to a novel visual learning approach that uses texture descriptors extracted from the gray-level coocurrence matrix (GLCM). An evolutionary algorithm is used as a learning engine to search for an optimal set of (1) ROIs on the image and (2) descriptors extracted with the GLCM.

10.4 Automatic Feature Localization and PCA-Based Classification

This section presents the fundamental contribution done by Trujillo et al. in [13]. The aim in [13] was to design a robust FER system (see Fig. 10.2) capable of correct

Fig. 10.2 Functional diagram
of our proposed FER system

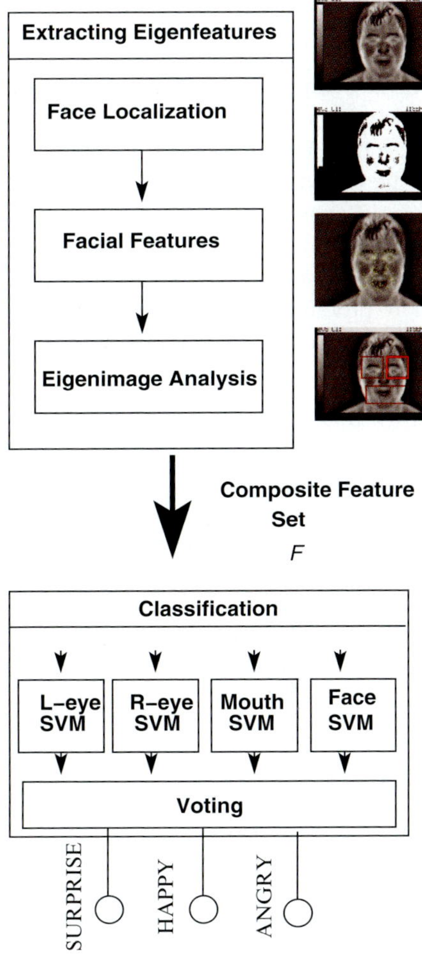

emotional classification in thermal images. The system is invariant to illumination conditions, the presence of partial facial feature occlusion, and pose variation within a certain field of view (see Fig. 10.3). It will distinguish between three different expression classes: surprise, happy, and angry. The proposed FER system can be seen in the block diagram of Fig. 10.2, represented as a four-step process: (1) face localization, (2) facial feature estimation, (3) eigenimage analysis, and (4) classification. First, the position of the subject's face must be estimated within the image. Second, specific facial features are found; these are the mouth and both eyes, making this a combined holistic and facial feature method. Third, dimensionality reduction of each image region is done with PCA. Finally, each image is classified with a SVM committee, one SVM per image region. Each of these steps is presented next.

10.4.1 Face Localization

The first step will identify the approximate position of each subject's face within the image. Figure 10.3 shows how most subjects' faces are not centered within the image frame. This makes facial feature localization a more difficult task. However, due to the property that the higher image intensities correspond to regions with higher thermal content and the fact that in our data set these regions correspond to the face, the process of finding the facial region can be simplified. First, we apply a thresholding operation over the entire image. In this way, facial pixel intensities become more prominent. Considering the n pixels over the threshold value as position vectors $\alpha = (\alpha_x, \alpha_y)$ and computing the geometric centroid $\mu = (\mu_x, \mu_y)$, defined by

$$\mu_j = \frac{\sum_{i=1}^{n} \mu_{j,i}}{n},$$

we approximate the face center with μ. In the case of a nonfrontal view image, μ_x most likely will not be correctly estimated, positioning the point around a cheek area. To correct this problem, we look for the facial pixels along the μ_y direction with the lowest thermal content. These pixels will normally be centered near a person's nose [12]. Now, we can compute a correction factor $\Delta\mu_x$ using the same geometric centroid approach. Figure 10.4 shows how this technique gives an effective approximation of the desired facial center.

10.4.2 Facial Feature Estimation

Several researchers have considered the idea that emotional information is centered around the eye and mouth areas on the face, reducing image analysis only to local areas around these features. To locate them, some researchers followed a manual

Fig. 10.3 Sample images of three different people, showing the three poses used for each facial expression

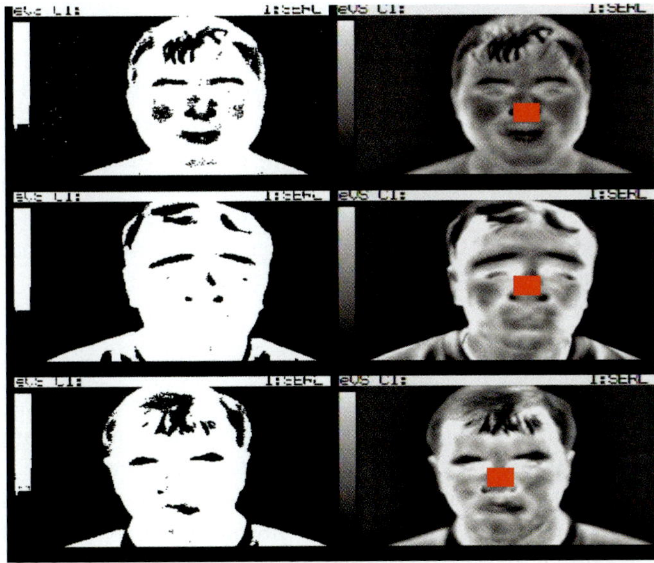

Fig. 10.4 A person's face is made prominent and easily located by applying thresholding on the thermal image

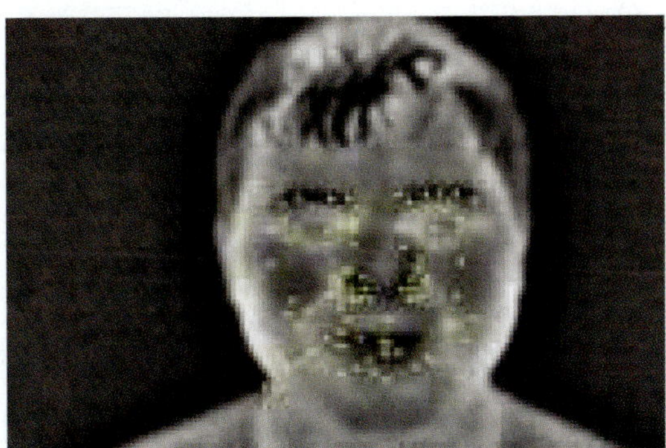

Fig. 10.5 Interest points detected on a person's face; most are grouped around the eyes, mouth, and nose

approach [7, 8], while others have automated the process using special operators [18]. On the other hand, since holistic methods yield effective results, a combined method that uses both local and global information could be a more appropriate approach.

The problem of facial feature localization is to find the eye and mouth areas on the face. Facial features are more difficult to find in thermal images due to the loss of specific feature properties (Fig. 10.3). We propose a method based on k-means clustering of interest points, extracted using the Evolutionary Learning algorithm for interest point detection proposed in [19], to find facial features on the image (Fig. 10.5).

10.4.2.1 Interest Point Clustering

Interest point detection was done over a window W centered around μ. The size of the window was set manually to 120×130. Experimental results showed the window size to be appropriate for our data set. In this way, we obtained a set X of 75 interest points. The set X was not uniformly distributed across the face but grouped around prominent facial features. We defined a set of K clusters and used the k-means clustering algorithm, with a Euclidean distance measure, that minimizes the following criteria:

$$J = \sum_{j=1}^{K} \sum_{n \varepsilon S_j} |x_n - \gamma_j|^2, \tag{10.1}$$

where $x_n \in X$ are each of the interest points, S_j are each of the clusters, and γ_j is the geometrical centroid of cluster S_j. There are several regions of low thermal content that can deceive our clustering approach. Some of those regions are the nose, cold patches on the cheeks, and texture of nonfacial features such as clothes. Our experimental results showed that when setting $K = 9$, two cluster centroids (top left and top right) are positioned over both eye regions, while the rest are grouped around the mouth and jaw. Of the 7 remaining centroids, we discarded the one located higher up on the face and computed an approximate mouth center by performing a second k-means clustering of the remaining 6 cluster centroids. We discarded the highest remaining cluster because it usually is an outlier compared to the other six due to the low temperatures of the nose. We now have a set $M = \{\mu_l, \mu_r, \mu_m, \mu\}$ of the left eye, right eye, mouth, and face approximated centers. Figure 10.6 shows the extracted local/global facial features to be used by our combined eigenfeature classification approach. This is explained next.

10.4.3 Computing Representative Eigenfeatures

We apply an eigenimage representation for each of the 4 facial regions that we extract. Using the eigenimage approach over the entire face and prominent local facial

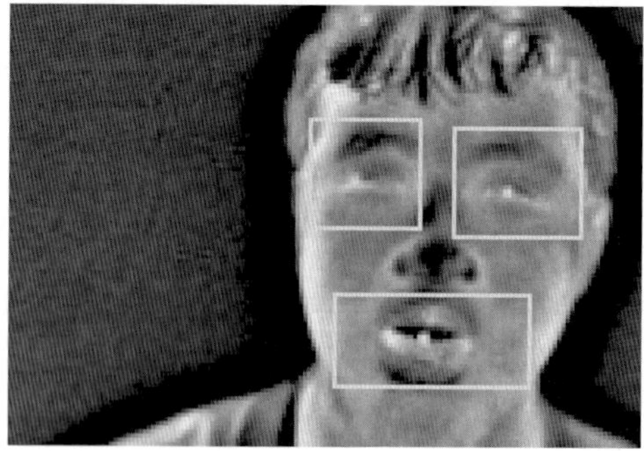

Fig. 10.6 Example output of our interest point clustering approach for facial feature localization

regions that were located with our facial feature localization gives what we call a local/global eigenfeature representation. Eigenimage representation is a popular technique in facial analysis in both visual [10] and thermal spectrum images [20].

10.4.3.1 Principal Component Analysis

Eigenimages are based on PCA, by which images are projected into a lower dimensional space that spans the significant variations (eigenimages) among known facial images. These variations are represented as the eigenvectors v_i of the covariance matrix C. The eigenvalue λ_i associated with v_i is the variance of the data distribution along the v_i direction. Taking Φ_i as the difference between image I_i and an average image Ψ, we have,

$$C = \frac{1}{R} \sum_{j=1}^{R} \Phi_j \Phi_j^T = \frac{1}{R} AA^T, \tag{10.2}$$

$$A = [\Phi_1, \Phi_2, \ldots, \Phi_R], \tag{10.3}$$

where R is the number of images in the training set. Computing the eigenvectors of AA^T turns out to be computationally prohibitive due to its size even for a moderate set of images. Instead, the R eigenvectors and eigenvalues v_i and κ_i of $A^T A$ are computed because the size of the matrix R $\times$ R is smaller [10]. Considering that,

$$A^T A \, v_i = \kappa_i v_i, \tag{10.4}$$

we can obtain the first R eigenvectors of C. If we multiply Eq. (10.4) by $\frac{1}{R}A$ on both sides, and remembering Eq. (10.2) to establish

$$CAv_i = \frac{1}{R} \kappa_i A v_i. \tag{10.5}$$

Now the first R eigenvectors and eigenvalues of C are given by Av_i and $\frac{1}{R}\kappa_i$ respectively.

An image I can now be approximated with the P most significant eigenimages v_i (those with the largest associated eigenvalues), with $P < R$, by

$$I = \Psi + \sum_{i=1}^{P} w_i v_i, \tag{10.6}$$

where $w_i = (I \cdot \mu_i)$ are the projection coefficients of I. The output of the combined eigenfeature approach for image I is a composite feature set $F_i = \{\Theta_l, \Theta_r, \Theta_m, \Theta_f\}$ where each Θ_x is the vector of projection coefficients for the global and the local facial image sections. This establishes our appearance model per expression and per facial feature.

10.4.4 Facial Expression Classification

The idea of computing an eigenimage representation for each ROI gives us a lo-cal/global eigenfeature representation of facial images. With this approach, we can describe each image by a composite feature set F. Because each feature vector $\Theta_x \in F$ corresponds to a different facial region, we propose a SVM committee for classification. In this section, we explain how each SVM learns a separation bound-ary between the eigenfeatures of our three competitive facial expression classes.

10.4.4.1 Support Vector Machines

A machine learning algorithm [21] for classification is faced with the task to learn the mapping $x_i \rightarrow y_i$ of data vectors x_i to classes y_i. The machine is actually defined by a set of possible mappings $x_i \rightarrow f(\mathbf{x}, \alpha)$, where a particular choice of α generates a particular trained machine. To introduce SVM, we can explain the simplest case of a two-class classifier; in this case, an SVM finds the hyperplane that best sep-arates elements from both classes while maximizing the distance from each class to the hyperplane. There are both linear and nonlinear approaches to SVM classi-fication. Thus, suppose you have a set of labeled training data $\{\mathbf{x_i}, y_i\}$, $i = 1, \ldots, l$, $y_i \in \{-1, +1\}$, $x_i \in \mathbf{R}^d$, then a nonlinear SVM defines the discriminative hyper-plane by

$$f(x) = \sum_{i=1}^{l} \alpha_i y_i K(\mathbf{x}_i, \mathbf{x}) + b, \tag{10.7}$$

where $\mathbf{x_i}$ are the support vectors, y_i is its corresponding class membership, and $K(\mathbf{x}_i, \mathbf{x})$ is the kernel function. The sign of the output of $f(\mathbf{x})$ indicates the class membership of $\mathbf{x}$. Finding this optimal hyperplane implies solving a constrained op-timization problem using quadratic programming, where the optimization criterion is the width of the margin between the classes. SVMs are easily combined to handle a multiclass case. A simple, effective combination trains N one-versus-rest classi-fiers (say, *one* positive, *rest* negative) for the N-class case and takes the class for a test point to be that corresponding to the largest positive distance [21].

10.4.5 SVM Committee Classification Approach

Since our problem amounts to classifying 4 different image regions, we implement an SVM committee loosely based on the work done in [22], using a weighted voting decision scheme. In this way, a different SVM is trained for each image region {left eye, right eye, mouth, face}. The composite feature vector F is fed to the SVM committee, where each Θ_x is the input to a corresponding SVM (see Fig. 10.1). The SVM committee uses a weighted voting decision scheme to classify an image, defined by

$$c = argmax^3_{j=1} \sum_{\Theta_x \in F} (K_{\Theta_x,j} \cdot w_{\Theta_x}), \qquad (10.8)$$

where $w_{\Theta_x} \in \{w_{\Theta_l}, w_{\Theta_r}, w_{\Theta_m}, w_{\Theta_h}\}$ are the weights associated with each corresponding committee member, $K_{\Theta_x,j}$ is the output of each committee member with regard to class j (1 if Θ_x is of class j, 0 otherwise), and j is each of the three possible expressions presented in the image database. Higher weights are assigned for the votes cast by the SVMs classifying the mouth and face ($w_{\Theta_m} = 3$ and $w_{\Theta_h} = 2$, respectively) feature vectors, while each of the eye weights are set to 1. This is done to better cope with occluded eyes and to improve the vote in favor of mouth classification due to its high information content and better experimental localization. In the case of a tie, the SVM classifying the mouth region casts the tie-breaking vote.

10.4.6 Experimental Setup

In this section, we describe the image database, the training of our FER system, and obtained results.

10.4.6.1 Thermal Image Database Description

The set of training and probe images was obtained from the IRIS data set in the IEEE Workshop Series on Object Tracking and Classification Beyond the Visible Spectrum (OCTBVS Public Infrared Database) database. Each image is 320 × 240 in bitmap RGB format. For the purposes of our work, we used a gallery set composed of 30 individuals/3 expressions each ("surprise," "happy," and "angry")/3 poses each. The images taken from the data set do not facilitate an automatic FER system because of pose and orientation inconsistencies during image acquisition. Examples of the images are shown in Fig. 10.3. Each row corresponds to the same expression for each person. It's obvious how each person's face has a different orientation, and most faces are not located in the same place within the image frame. Not shown in Fig. 10.3 are the images of people wearing glasses and the large amount of outliers within the data set. Using such an unstructured database is not currently the norm in FER literature. However, we feel that such a database is appropriate to show the benefits of the described approach.

10.4.6.2 System Training

The SVM committee was trained to be able to classify any of the images in the data set. The training phase used each person in the data set without glasses, with manually selected regions, making each SVM biased toward correctly positioned features. We use the 50 leading eigenimages for our PCA analysis. At the end of this learning process, we obtain a separation boundary per facial feature into three regions of membership: surprise, happy, and angry. Consequently, the testing setup is primarily concerned with showing system FER performance under two different criteria: (1) ability to correctly locate and label image features and (2) ability to cope with partial facial occlusions.

10.4.6.3 Testing

In the testing face, the approach described in Fig. 10.1 is applied to each input image. The testing set contains 30 randomly selected images, and testing is concerned with the automatic localization of image features. Table 10.1 shows the confusion matrix of our experimental results. Performance of our FER system clearly degrades when classifying the happy expression of the data set due to three main reasons. First, the training set is small and with many outliers, and given that PCA is a statistical method that assumes Gaussian distribution, it relies on a large amount of data. Second, an SVM builds a separating boundary between classes; if the amount of data used is small, then it is probable that the separating boundary will not be representative. Third, within the training set there are clearly overlapping expressions that make learning the boundary a very difficult problem (Fig. 10.7). Now, to test the system's ability to classify images with people wearing glasses, only images of people with surprise and happy expressions are used because of the difficulties mentioned. A total of 10 different images is used. Example images are shown in Fig. 10.8. Table 10.2 is the corresponding confusion matrix. The system is able to locate mouth and eyes correctly even in these images because it is not specifically

Table 10.1 Confusion matrix of our test results

	Surprise	Happy	Angry
Surprise	10	3	1
Happy	0	4	0
Angry	0	3	9

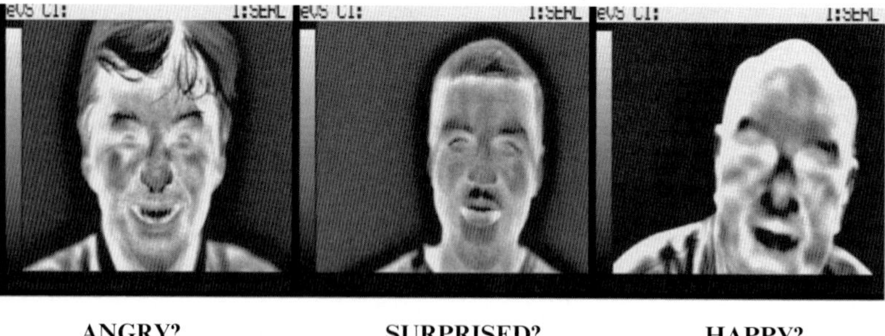

Fig. 10.7 All images in the figure appear to be "happy" but are not labeled as such in the data set

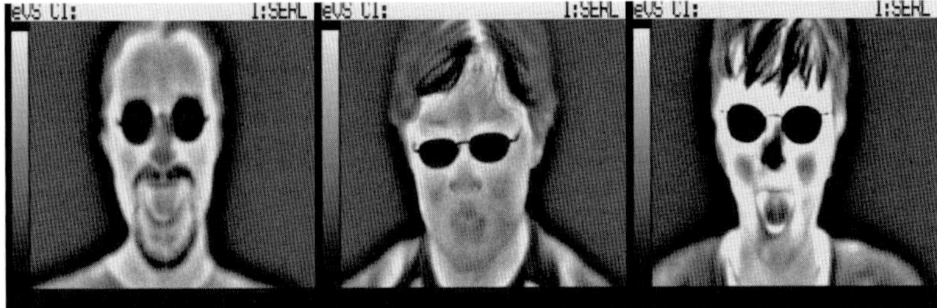

Fig. 10.8 Example images of people with glasses

Table 10.2 Confusion matrix for tests with people wearing glasses

	Surprise	Angry
Surprise	5	1
Angry	0	4

looking for "mouth" and "eyes" on the face, but for interest point locations, which in this cases are related to borders and corners caused by the glasses.

10.4.7 Conclusions

This section presented a local and global eigenimage approach to FER recognition in thermal images. PCA is done both holistically and on specific facial features, which are found using interest point clustering. Experimental results showed that the approach gives robust localization and acceptable classification to varying poses up to a certain point and to partial occlusions caused by people wearing glasses. However, it is important to note how the data set used does not give enough information to appropriately learn a separating boundary between competing classes, specifically due to the fact of overlapping facial expressions and the particularly small size of the set. Hence, to appropriately estimate the value of the approach a more complete data set is required to test for robust classification in the FER problem.

10.5 Evolutionary Learning for Facial Expression Recognition

It is simple to identify two basic questions that a system designer must answer when adhering to the basic process followed by most FER approaches, as seen in Fig. 10.1, (1) Where will image analysis be performed? (2) What type of discriminant features will be extracted? It can be intuitive, while working with visual spectrum images, to give reasonable answers to these questions and highly efficient solutions have been proposed [2]. On the other hand, infrared signals could be unnatural to humans, thus increasing the difficulty of attempting to make the correlation between this type of information and our own mental process. Hence, answering the questions might not be straightforward. Consequently, the appropriate way in which the infrared information could be applied may not be as evident. When confronted with such a task, an appropriate and coherent approach is to use modern meta-heuristics, such as evolutionary computation (EC). EC techniques help guide the emergence of novel and interesting ways of solving nonlinear optimization and search problems [23, 24]. Using EC as a learning engine, it is possible to search a wider space of candidate solutions in order to appropriately answer these, and possibly other, questions. Furthermore, EC can help provide a deeper understanding of how this information should be exploited.

This section presents an evolutionary learning approach that generates an illumination independent FER system. Performing what can be understood as *Visual Learning*, an evolutionary algorithm searches for optimal solutions to the first two steps of the basic FER process (see Fig. 10.1). The first task consists of selecting a set of suitable regions where feature extraction is to be performed. The second task consists of tuning the parameters that define the GLCM used to compute region descriptors, as well as the selection of the best subsets of these descriptors. The output of these two steps is taken as input by a SVM committee similar to the manner discussed. Performance is evaluated experimentally using the same data set shown in Fig. 10.3. Experimental results compare favorably with what a human observer can accomplish, as well as with the previous PCA-SVM approach.

10.5.1 Outline of the Approach

The basic outline of this EC-based approach is depicted in Fig. 10.9 and proceeds as follows:

1. Use a generic algorithm (GA) to perform the search for ROI selection and feature extraction optimization. The GA will select a set Ω of n facial ROIs.
2. Simultaneously, the GLCM parameter set $\pi_{\omega i}$ is optimized $\forall \omega_i \in \Omega$, where $i = [1, n]$. In this step, the GA performs feature construction by tuning the GLCM parameters.
3. Feature selection creates a vector $\gamma_{\omega i} = (\beta_1, \ldots, \beta_m)$ of m different descriptors $\forall \omega_i \in \Omega$, where $\{\beta_1, \ldots, \beta_m\} \subseteq \Psi$. Ψ is the set of all available GLCM-based descriptors. Each $\beta_j \in \gamma_{\omega i}$ represents the mean value of the jth descriptor within region ω_i. The compound set $\Gamma = \{\gamma_{\omega i}\}$ is then used for classification. By evolving both ROI selection and feature extraction within the same GA, we formulate a coupled solution to the first two levels of the basic FER approach.
4. An SVM, ϕ_i is trained for each ω_i, and k-fold cross validation is performed to estimate the classifiers' accuracy. The set $\Phi\{\phi_i\}$ of all trained SVMs represents a committee that performs FER using a voting scheme. The total accuracy of the SVM committee is used as the fitness function for the GA.

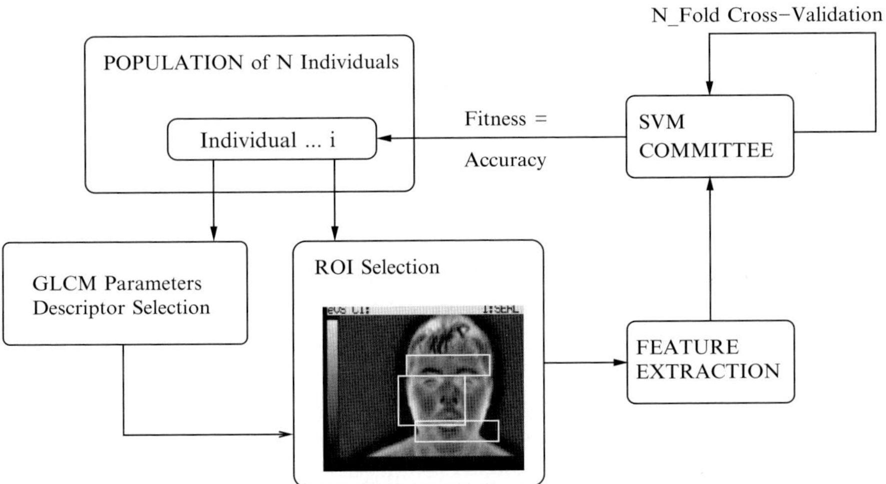

Fig. 10.9 Flowchart diagram of our approach to visual learning

10.5.1.1 Main Research Contributions

The main contributions of the algorithm are:

1. It joins a small but growing list of works that study FER beyond the visual spectrum using images in the thermal spectrum.
2. It presents a novel representation of the FER problem using closed-loop visual learning with EC.
3. The process combines both ROI selection and feature extraction in a single evolving algorithm.
4. Using domain-independent second-order statistics, it establishes a portable solution to different pattern recognition problems.

10.5.2 Evolutionary Computation

Early in the 1960s, some researchers came to the conclusion that classical artificial intelligence (AI) approaches were an inadequate tool to comprehend complex and adaptive systems. Artificial systems based on large knowledge bases and predicate logic are extremely hard to manage and have a very rigid structure. This is one of the reasons why researchers looked elsewhere to find solutions for complex non-linear problems that were difficult to decompose, understand, and even define. The area of soft computing emerged as a paradigm to provide solutions in areas of AI where classic techniques had failed. EC is a major field in this relatively new area of research. EC is a paradigm for developing emergent solutions to different types of problems. EC-based algorithms look to emulate the adaptive and emergent properties found in naturally evolving systems. These types of algorithms have a surprising ability to search, optimize, and emulate learning in computer-based systems using concepts such as populations, individuals, fitness, and reproduction. The field of EC owes much of its current status as a viable tool to build artificial solutions to the pioneering work of John Holland [23] and his students. Holland presented the first major type of EC algorithm, the genetic plan, later to be known as a GA. A GA is a simplified model of the way in which populations of living organisms are guided by the pressure of natural selection to find optimal peaks within the species fitness landscape [24]. In a GA, a population is formed by a set of possible solutions to a given problem. Each individual in the population is represented by a coded string or chromosome that represents the individual's genotype. Using genetic operations that simulate natural reproduction and mutation, a new generation of child solutions is generated. The performance of each new solution is evaluated using a problem-dependent fitness function. The fitness function places each individual on the problem fitness landscape. The GA selects the best individuals for survival across successive generations using the fitness evaluation. When the evolutionary process terminates, the best individual found according to the fitness measure is returned as the solution.

10.5.3 *Texture Analysis and the Gray-Level Cooccurrence Matrix*

Image texture analysis has been a major research area in the field of computer vision since the 1970s. Researchers have developed different techniques and operators that describe image texture. Those approaches were driven by the hope of automating the human visual ability that seemlessly identifies and recognizes texture information in a scene. Popular techniques for describing texture information include filter banks [25], random fields [26], primitive texture elements known as textons [27], and more recently texture representations based on sparse local affine regions [28]. Historically, the most commonly used methods for describing texture information are the statistical-based approaches. First-order statistical methods use the probability distribution of image intensities approximated by the image histogram. With such statistics, it is possible to extract descriptors to describe image information. First-order statistics descriptors include entropy, kurtosis, and energy, to name but a few. Second-order statistical methods represent the joint probability density of the intensity values (gray levels) between two pixels separated by a given vector V. This information is coded using the GLCM $M(i, j)$ [29]. Statistical information derived from the GLCM has shown reliable performance in tasks such as image classification [30] and content-based image retrieval [31, 32].

Formally, the GLCM $M_{i,j}(\pi)$ defines a joint probability density function $f(i, j | V, \pi)$ where i and j are the gray levels of two pixels separated by a vector V and $\pi = \{V, R\}$ is the parameter set for $M_{i,j}(\pi)$. The GLCM identifies how often pixels that define a vector $V(d, \theta)$ and differ by a certain amount of intensity value $\Delta = i - j$ appear in a region R of a given image I where V defines the distance d and orientation θ between the two pixels. The direction of V can or cannot be taken into account when computing the GLCM.

One drawback of the GLCM is that when the amount of different gray levels in region R increases, the dimensions of the GLCM make it difficult to handle or use directly. Fortunately, the information encoded in the GLCM can be expressed by a varied set of statistically relevant numerical descriptors. This reduces the dimensionality of the information that is extracted from the image using the GLCM. Extracting each descriptor from an image effectively maps the intensity values of each pixel to a new dimension. In this work, the set Ψ of available descriptors [29] extracted from $M(i,j)$ is the following:

Entropy. A term more commonly found in thermodynamics or statistical mechanics, *entropy* is a measure of the level of disorder in a system. Images of highly homogeneous scenes have a high associated entropy, while inhomogeneous scenes possess a low-entropy measure. The GLCM entropy is obtained with the following expression:

$$H = 1 - \frac{1}{Nc \cdot ln(Nc)} \sum_i \sum_j M(i,j) \cdot ln(M(i,j)) \cdot 1_{M(i,j)} \tag{10.9}$$

where $1_{M(i,j)} = 0$ when $M(i,j) = 0$ and 1 otherwise.

Contrast. Contrast is a measure of the difference between intensity values of the neighboring pixels. It will favor contributions from pixels located away from the diagonal of $M(i,j)$.

$$C = \frac{1}{Nc(L-1)^2} \sum_k^{L-1} k^2 \sum_{|i-j|=k} M(i,j) \tag{10.10}$$

where Nc are the number of occurrences, and L is the number of gray levels.

Homogeneity. This gives a measure of how uniformly a given region is structured with respect to its gray-level variations.

$$Ho = \frac{1}{Nc^2} \sum_i \sum_j M(i,j)^2 \tag{10.11}$$

Local homogeneity.

$$G = \frac{1}{Nc} \sum_i \sum_j \frac{M(i,j)}{1 + (i-j)^2} \tag{10.12}$$

Directivity.

$$D = \frac{1}{Nc} \sum_i \sum_j M(i,j) \tag{10.13}$$

Uniformity. This is a measure of the uniformity of each gray level.

$$Ho = \frac{1}{Nc^2} \sum_i M(i,i)^2 \tag{10.14}$$

Moments. Moments express common statistical information, such as the variance that corresponds to the second moment. This descriptor increases when the majority of the values of $M(i,j)$ are not on the diagonal.

$$Mom_k = \sum_i \sum_j (i-j)^k M(i,j) \tag{10.15}$$

Inverse moments. This produces the opposite effect compared to the moments descriptor.

$$Mom_k^{-1} = \sum_i \sum_j \frac{M(i,j)}{(i-j)^k}, i \neq j \tag{10.16}$$

Maximum probability. Considering the GLCM as an approximation of the joint probability density between pixels, this operator extracts the most probable difference between pixel grayscale values.

$$max(M(i,j)) \tag{10.17}$$

Correlation. This is a measure of grayscale linear dependencies between pixels at the specified positions relative to each other.

$$S = \frac{1}{Nc\sigma_x\sigma_y} \left| \sum_i \sum_j (i-m_x)(j-m_y)M(i,j) \right| \tag{10.18}$$

$$m_x = \frac{1}{Nc} \sum_i \sum_j iM(i,j)$$
$$m_y = \frac{1}{Nc} \sum_i \sum_j jM(i,j)$$
$$\sigma_x^2 = \frac{1}{Nc} \sum_i \sum_j (i-m_x)^2 M(i,j)$$
$$\sigma_y^2 = \frac{1}{Nc} \sum_i \sum_j (j-m_y)^2 M(i,j)$$

10.5.4 Genetic Algorithm for Visual Learning

The input images for the GA are the extracted facial regions using the same face localization procedure described in [13]. The learning approach accomplishes a combined search and optimization procedure in a single step. The GA searches for the best set Ω of facial ROIs in each image and optimizes the feature extraction procedure by tuning the GLCM parameter set $\pi_i \; \forall \omega_i \in \Omega$ and selecting the best subset $\{\beta_1, \ldots, \beta_m\}$ of mean descriptor values from the set of all possible descriptors Ψ, to form a feature vector $\gamma_i = (\beta_1, \ldots, \beta_m)$ for each $\omega_i \in \Omega$. Using this representation,

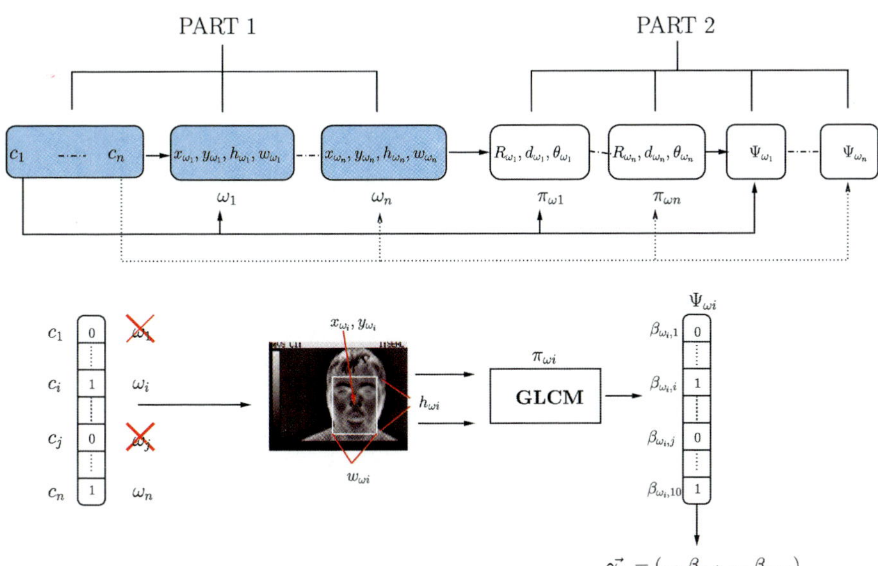

Fig. 10.10 Visual learning GA problem representation

we are tightly coupling the ROI selection step with the feature extraction process. In this way, the GA is learning the best overall structure for the FER system in a single closed-loop learning scheme. Our approach eliminates the need for a human designer, which normally combines the ROI selection and feature extraction steps. Now, this step is left up to the learning mechanism. Each possible solution is coded into a single binary string. Its graphical representation is shown in Fig. 10.10. The entire chromosome consists of 94 binary coded variables, each represented by binary strings of different sizes, from 1 to 5 bits each, depending on its use. The chromosome can be better understood by logically dividing it in two main sections. The first one encodes variables for searching the ROIs on the image, and the second is concerned with setting the GLCM parameters and choosing appropriate descriptors for each ROI.

10.5.4.1 ROI Selection

The first part of the chromosome encodes ROI selection. The GA has a hierarchical structure that includes both control and parametric variables. The section of structural or control genes c_i determines the state (on/off) of the corresponding ROI definition blocks ω_i. Each structural gene activates or deactivates one ROI in the image. Each ω_i establishes the position, size, and dimensions of the corresponding ROI. Each ROI is defined with four degrees of freedom around a rectangular region: height, width, and two coordinates indicating the central pixel. The choice of rectangular regions is not related in any way with our visual learning algorithm. It is possible to use other types of regions (e.g., elliptical regions) and keep the same overall structure of the GA. The complete structure of this part of the chromosome is coded as follows:

1. Five structural variables $\{c_1,\ldots,c_5\}$, represented by a single bit each. Each one controls the activation of one ROI definition block. These variables control which ROI will be used in the feature extraction process.
2. Five ROI definition blocks $\omega_1,\ldots,\omega_5$. Each block ω_i contains four parametric variables $\omega_i = \{x_{\omega_i}, y_{\omega_i}, h_{\omega_i}, w_{\omega_i}\}$, coded into four-bit strings each. These variables define the ROIs' center $(x_{\omega_i}, y_{\omega_i})$, height (h_{ω_i}) and width (w_{ω_i}). In essence, each ω_i establishes the position and dimension for a particular ROI.

10.5.4.2 Feature Extraction

The second part of the solution representation encodes the feature extraction variables for the visual learning algorithm. The first group is defined by the parameter set π_i of the GLCM computed at each image ROI $\omega_i \in \Omega$. The second group is defined as a string of ten decision variables that activate or deactivate the use of a particular descriptor $\beta_j \in \Psi$ for each ROI. Since each of these parametric variables is associated to a particular ROI, they are also dependent on the state of the structural variables c_i. They only enter into effect when their corresponding ROI is active (set to 1). The complete structure of this part of the chromosome is as follows:

1. A parameter set π_{ω_i} is coded $\forall \omega_i \in \Omega$, using three parametric variables. Each $\pi_{\omega_i} = \{R_{\omega_i}, d_{\omega_i}, \theta_{\omega_i}\}$ describes the size of the region R, distance d, and direction θ parameters of the GLCM computed at each ω_i. Note that R is a GLCM parameter, not to be confused with the ROI definition block ω_i (see Section 10.4.2).
2. Ten decision variables coded using a single bit activate or deactivate a descriptor $\beta_{j,\omega_i} \in \Psi$ at a given ROI. These decision variables determine the size of the feature vector γ_i, extracted at each ROI to search for the best combination of GLCM descriptors. In this representation, each β_{j,ω_i} represents the mean value of the jth descriptor computed at ROI ω_i.

10.5.4.3 Classification

Since this problem amounts to classifying every extracted region ω_i, we implement a SVM committee that uses a voting scheme for classification. The SVM committee Φ is formed by the set of all trained SVMs $\{\phi_i\}$, one for each ω_i. The compound feature set $\Gamma = \{\gamma_{\omega i}\}$ is fed to the SVM committee Φ, where each $\gamma_{\omega i}$ is the input to a corresponding ϕ_i. The SVM committee uses voting to determine the class of the corresponding image, similar to that in Sect. 10.4.4.

10.5.4.4 Fitness Evaluation

The majority of the GAs' fitness function is biased toward classification accuracy of each extracted ROI. Nevertheless, if two different solutions have the same classification accuracy, it would be preferable to select the one that uses the minimum amount of descriptors. This will promote compactness in our representation and improve computational performance. In this way, the fitness function is defined similar to [33]:

$$Fitness = 10^2 * Accuracy + 0.25 * Zeros \tag{10.19}$$

where *Accuracy* is the average accuracy of all SVMs in Φ for a given individual. In other words, $Accuracy = \frac{1}{|\Phi|} \sum_x Acc_{\phi_x}$, summed $\forall \phi_x \in \Phi$, where Acc_{ϕ_x} is the accuracy of the ϕ_j SVM. *Zeros* is the total amount of inactive descriptors in the chromosome, given by $Zeros = \sum_i \sum_j \beta_{\omega_i,j} \omega_i$ where $i = 1, \ldots, 5$ and $j = 1, \ldots, 10$. This formula is based on the work of Sun et al. [33].

10.5.4.5 GA Runtime Parameters

The rest of the parameters, as well as the GA settings, are defined as follows:

Population size and initialization. We use random initialization of the initial population. Our initial population size was set to 100 individuals.
Survival method. For population survival, we use an elitist-based strategy. The N parents of generation $t - 1$ and their offspring are combined, and the best N individuals are selected to form the parent population of generation t.
Genetic operators. Since we are using a binary-coded string, we use simple one-point crossover and single-bit binary mutation. Crossover probability was set to 0.66 and mutation probability to 0.05. Parents were selected with tournament selection. The tournament size was set to 7.

10.5.4.6 SVM Training Parameters

SVM implementation was done using libSVM [34], a C^{++} open-source library. For every $\phi \in \Phi$, the parameter setting is the same for all the population. The SVM parameters are

Kernel type. A *radial basis function* (RBF) kernel was used, given by

$$k(x,x_i) = exp\left(-\frac{\|x - x_i\|^2}{2\sigma^2}\right) \tag{10.20}$$

The RBF shows a greater performance rate for classifying nonlinear problems than other types of kernels.
Training set. The training set used was extracted from 92 different images (see Section 10.4.4).

Cross validation. To compute the accuracy of each SVM, we perform k-fold cross validation, with $k = 6$. Due to the small size of our data set, the accuracy computed with cross validation will outperform any other type of validation approach [35]. In k-fold cross validation, the data is divided into k subsets of (approximately) equal size. The SVM was trained k times, each time leaving out one of the subsets from training but using only the omitted subset to compute the classifiers' accuracy. This process was repeated until all subsets have been used for both testing and training, and the computed average accuracy was used as the performance measure for the SVM.

10.5.5 Experimental Results

This section has two objectives: to give a brief description of the database used to test the approach and to detail the obtained experimental results and comparisons. To contrast the experimental results, they are compared with classification done by human observers.

10.5.5.1 OTCBVS Data Set of Thermal Images

The OTCBVS data set [36] contains pictures from 30 different subjects taken at the UT/IRIS Lab. Each subject posed three different facial emotions; this was used as the FER classification ground truth given by the data set providers. It is apparent that the ground truth for each image was taken from the emotion that the subject "claimed" to be expressing and not by what human observers "believed" he was feeling. This is a critical point and could be considered as a shortcoming of the data set if subjects were only "acting" and were not sincerely "feeling" a given emotion. This fact is evident in Fig. 10.7.

The database contains subjects wearing glasses in some of the pictures. Despite the fact that the approach presented in the previous section showed reasonable invariance to people wearing glasses, this is not a primary goal in this section. For the purposes of testing the EC algorithm, only the best frontal views to characterize each expression class are necessary. A total of 33 surprise, 26 happy, and 33 angry images were selected, a total of 92 training/validation images to be used by the learning algorithm. The main facial area of each image was found using the same technique for automatic face localization described in [13]. This area was cropped and resized to 64×64 gray-level bitmap format (Fig. 10.11). Figure 10.3 shows sample images before cropping, corresponding to three different subjects. Each row corresponds to a different expression class. Three different poses are shown for each of the facial expressions.

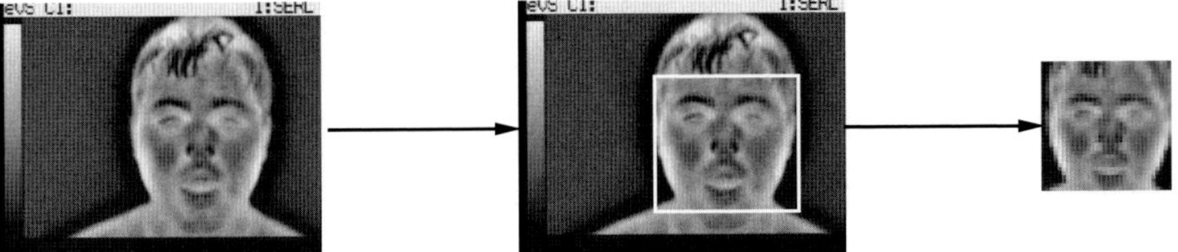

Fig. 10.11 Extracting the prominent facial region and resizing to 64 × 64 bit size

Fig. 10.12 Best individual selects four facial regions; each region shown with a white bounding box

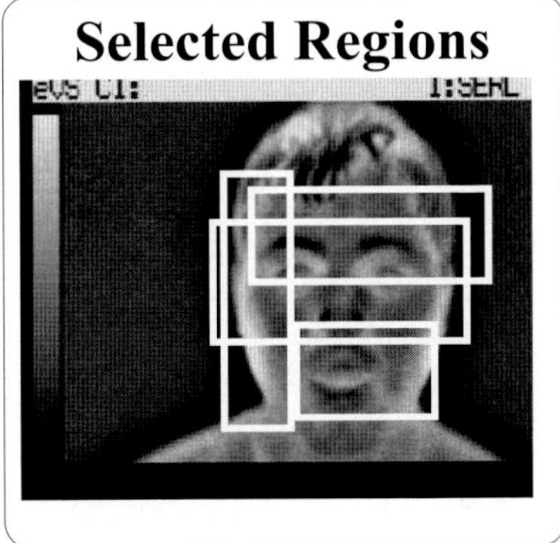

Table 10.3 Confusion matrix for cross validation

	Surprise	Happy	Angry
Surprise	77%	20%	3%
Happy	12%	70%	18%
Angry	0%	16%	84%

10.5.5.2 Approach Evaluation

As noted, due to the limited number of useful images for evaluation, cross-validation accuracy was computed. Ten different runs of the visual learning GA were carried out. To preserve compactness, we describe the results for the best of the 10 completed experiments. The superindividual or fittest individual obtained showed a cross-validation classification accuracy of 77%. The best individual extracts 4 different image regions, with some overlapping areas shown in Fig. 10.12. The complete confusion matrix for cross-validation performance is presented in Table 10.3. We can see how classification performance degrades when classifying the happy class. This is caused by two main reasons: the overlap of the expressions modeled by the human subjects (Fig. 10.7) and the questionable way in which the data set providers labeled each of the images. These two points are examples of how subjects that only act as if they experience a given emotion and are not sincerely feeling it can produce noisy data when great care is not taken by a data set designer.

Because of the lack of training data, only 10 additional images that were not used during training were used for testing. The algorithm was able to classify 8 of the 10 images correctly. The confusion matrix is shown in Table 10.4. A statistical experiment was conducted with human subjects to validate the fact that the data set used establishes an extremely hard classification problem. We randomly selected 30 images from the training set, and 100 people were asked to classify them. The experiment was repeated using the corresponding visual spectrum images to contrast the augmented difficulty of the problem when conducted for thermal imagery. This type of experimentation was also reported in [15], which is evidence of how the lack of published results and comparable data sets makes system comparison difficult. The results for the thermal and visual image experiments are shown in Table 10.5 and Table 10.6, respectively. The human classification experiments help establish the extreme difficulty of the FER problem beyond the visual spectrum. This method outperforms human classification of thermal images and is competitive with human classification of visual spectrum images, a very promising result. We argue that the low classification performance of visual spectrum images by human observers could be attributed to two primary factors: (1) cultural differences between the subjects and the people classifying them and (2) the laxness with which the data set was designed and the ground truth was established.

Table 10.4 Confusion matrix of our test results

	Surprise	Happy	Angry
Surprise	4	0	0
Happy	0	2	1
Angry	0	1	2

Table 10.5 Human classification of thermal images

	Surprise	Happy	Angry
Surprise	56%	33%	11%
Happy	23%	48%	29%
Angry	7%	14%	79%

Table 10.6 Human classification of visual spectrum images

	Surprise	Happy	Angry
Surprise	78%	21%	1%
Happy	16%	82%	2%
Angry	5%	25%	70%

The classification accuracy achieved in [13] for 30 testing images was 76.6%. This is a comparable result to the cross-validation and testing accuracy presented here. However, the previous method utilized a total of 50 eigenfeatures per region, a total that exceeds the reduced dimensionality of the feature vectors used in our proposed approach. In the example of Fig. 10.12, a total of 35 GLCM features was used for the recognition process.

10.6 Discussion and Future Work

In this chapter, we presented two different approaches for an artificial vision system that allows an acceptable FER using images of signals beyond the human visual spectrum. Our main purpose was to study how useful infrared signals could be to determine a person's emotional state from static images. To achieve this, we have presented two approaches: The first uses automatic feature localization based on interest point clustering combined with a holistic and local PCA; the second uses an evolutionary learning algorithm that searches for an optimal set of ROIs as well as a set of texture features.

Our PCA-based approach shows promising results that illustrate, through an example, how feature localization with interest points could be applied to human expression recognition to automate the eigenimage analysis. On the other hand, the EC-based approach employs a set of texture features providing comparable classification results to those obtained with PCA on the same data set. Furthermore, the technique establishes a coupled mechanism to learn ROI and feature selection, which are two of the three main tasks in common pattern recognition approaches. We believe that by combining ROI selection and feature extraction the proposed technique could be applied to a wide range of real-world problems. Interesting extensions to this technique include (1) dividing our GA chromosome into two different species and applying coevolution to construct an aggregate solution; (2) incorporating the classification mechanism into the evolving process to construct a specialized classifier for a particular problem; and (3) finally, and maybe the most straightforward, applying our technique in other problem domains, including different types of image features on the GA approach.

Acknowledgments This research was funded by CONACyT and INRIA through the LAFMI project 634-212. The bench we used is available on the Web site of IEEE OTCBVS WS Series Bench at http://www.cse.ohio-state.edu/otcbvs-bench; We would like to thank the Imaging, Robotics, and Intelligent Systems Laboratory for making the data set available online. The data set was collected under DOE University Research Program in Robotics under grant DOE-DE-FG02-86NE37968; DOD/TACOM/NAC/ARC Program under grant R01-1344-18; FAA/NSSA grant R01-1344-48/49; Office of Naval Research under grant N000143010022.

Chapter's References

1. Maja Pantic and Leon J. M. Rothkrantz. Automatic analysis of facial expressions: The state of the art. *IEEE Transactions on Pattern Analysis and Machine Intelligence*, 22(12):1424–1445, 2000

2. B. Fasel and J. Luettin. Automatic facial expression analysis: A survey. *Pattern Recognition*, 36(1):259–275, 2003

3. Fabrice Bourel, Claude C. Chibelushi, and Adrian A. Low. Robust facial expression recognition using a state-based model of spatially-localised facial dynamics. In *Proceedings of the Fifth IEEE International Conference on Automatic Face and Gesture Recognition (FGR.02)*, pages 113–118, 2002

4. Maja Pantic and Leon J. M. Rothkrantz. An expert system for recognition of facial actions and their intensity. *Image and Vision Computing*, 18:881–905, 2000

5. Ying-Li Tian, Takeo Kanade, and Jeffrey Cohn. Recognizing action units for facial expression analysis. *IEEE Transactions on Pattern Analysis and Machine Intelligence*, 23(2):97–115, February 2001

6. Severine Dubuisson, Franck Davoine, and Jean Pierre Cocquerez. Automatic facial feature extraction and facial expression recognition. In *AVBPA '01: Proceedings of the Third International Conference on Audio- and Video-Based Biometric Person Authentication*, pages 121–126, Springer-Verlag, London, 2001

7. C. Padgett and G. Cottrell. Representing face images for emotion classification. *Advances in Neural Information Processing Systems*, 9, 1997

8. Gianluca Donato, Marian Stewart Bartlett, Joseph C. Hager, Paul Ekman, and Terrence J. Sejnowski. Classifying facial actions. *IEEE Transactions on Pattern Analysis and Machine Intelligence*, 21(10):974–989, 1999

9. Michael J. Lyons, Julien Budynek, and Shigeru Akamatsu. Automatic classification of single facial images. *IEEE Transactions on Pattern Analysis and Machine Intelligence*, 21(12):1357–1362, 1999

10. Matthew Turk and Alex Paul Pentland. Eigenfaces for recognition. *Journal of Cognitive Neuroscience*, 3(1):71–86, 1991

11. Joseph Wilder, P. Jonathon Phillips, Cunhong Jiang, and Stephen Wiener. Comparison of visible and infra-red imagery for face recognition. In *2nd International Conference on Automatic Face and Gesture Recognition (FG '96)*, October 14–16, 1996, Killington, VT, pages 182–191, 1996

12. F. Prokoski. History, current status, and future of infrared identification. In *CVBVS '00: Proceedings of the IEEE Workshop on Computer Vision Beyond the Visible Spectrum: Methods and Applications (CVBVS 2000)*, page 5, IEEE Computer Society, Washington, DC, 2000

13. Leonardo Trujillo, Gustavo Olague, Riad Hammoud, and Benjamín Hernández. Automatic feature localization in thermal images for facial expression recognition. In *CVPR '05: Proceedings of the 2005 IEEE Computer Society Conference on Computer Vision and Pattern Recognition (CVPR'05) — Workshops*, page 14, IEEE Computer Society, Washington, DC, 2005

14. I. Pavlidis, J. Levine, and P. Baukol. Thermal image analysis for anxiety detection. In *International Conference on Image Processing*, volume 2, pages 315–318, 2001

15. Y. Sugimoto, Y. Yoshitomi, and S. Tomita. A method for detecting transitions of emotional states using a thermal facial image based on a synthesis of facial expressions. *Journal of Robotics and Autonomous Systems*, 31:147–160, 200

16. Y. Yoshitomi, S. Kim, T. Kawano, and T. Kitazoe. Effect of sensor fusion for recognition of emotional states using voice, face image and thermal image of face. In *Proceeding of the 2000 IEEE International Workshop on Robot and Human Interactive Communication*, Osaka. Japan – September 27–29 2000, pages 178–183

17. Seong G. Kong, Jingu Heo, Besma R. Abidi, Joonki Paik, and Mongi A. Abidi. Recent advances in visual and infrared face recognition: a review. *Computer Vision and Image Understanding*, 97(1):103–135, 2005

18. Séverine Dubuisson, Franck Davoine, and Jean Pierre Cocquerez. Automatic facial feature extraction and facial expression recognition. In Josef Bigün and Fabrizio Smeraldi, editors, *AVBPA*, volume 2091, *Lecture Notes in Computer Science*, pages 121–126, Springer, New York, 2001

19. Leonardo Trujillo and Gustavo Olague. Using evolution to learn how to perform interest point detection. In *Proceedings from the 18th International Conference on Pattern Recognition*, volume 1, pages 211–214, IEEE Computer Society, Washington, DC, 2006

20. Xin Chen, Patrick J. Flynn, and Kevin W. Bowyer. PCA-based face recognition in infrared imagery: Baseline and comparative studies. In *AMFG '03: Proceedings of the IEEE International Workshop on Analysis and Modeling of Faces and Gestures*, page 127, IEEE Computer Society, Washington, DC, 2003

21. Christopher J.C. Burges. A tutorial on support vector machines for pattern recognition. *Data Mining and Knowledge Discovery*, 2(2):121–167, 1998

22. Zhong-Qiu Zhao, De-Shuang Huang, and Bing-Yu Sun. Human face recognition based on multi-features using neural networks committee. *Pattern Recognition Letters*, 25(12):1351–1358, 2004

23. J.H. Holland. *Adaptation in Natural and Artificial Systems*. University of Michigan Press, Ann Arbor, MI, 1975

24. David E. Goldberg. *Genetic Algorithms in Search, Optimization, and Machine Learning*. Addison-Wesley Professional, Reading, MA, January 1989

25. J. Malik and P. Perona. Preattentive texture discrimination with early vision mechanisms. *Optical Society of America*, 7:923–932, May 1990

26. Jianchang Mao and Anil K. Jain. Texture classification and segmentation using multiresolution simultaneous autoregressive models. *Pattern Recognition*, 25(2):173–188, 1992

27. B. Julesz and J.R. Bergen. *Textons, the Fundamental Elements in Preattentive Vision and Perception of Textures*. Kaufmann, Los Altos, CA, 1987

28. Svetlana Lazebnik, Cordelia Schmid, and Jean Ponce. A sparse texture representation using local affine regions. *IEEE Transactions on Pattern Analysis & Machine Intelligence*, 27(8):1265–1278, 2005

29. R.M. Haralick. Statistical and structural approaches to texture. *Proceedings of the IEEE*, 67:786–804, 1979

30. J. Kjell. Comparative study of noise-tolerant texture classification. *1994 IEEE International Conference on Systems, Man, and Cybernetics. "Humans, Information and Technology"*, 3:2431–2436, October 1994

31. Peter Howarth and Stefan M. Rüger. Evaluation of texture features for content-based image retrieval. In *ACM International Conference on Image and Video Retrieval (CIVR)*, pages 326–334, 2004

32. Philippe P. Ohanian and Richard C. Dubes. Performance evaluation for four classes of textural features. *Pattern Recognition*, 25(8):819–833, 1992

33. Zehang Sun, George Bebis, and Ronald Miller. Object detection using feature subset selection. *Pattern Recognition*, 37(11):2165–2176, 2004

34. Chih-Chung Chang and Chih-Jen Lin. *LIBSVM: A Library for Support Vector Machines*, MIT Press Journals-Neural Computation, 2001

35. Cyril Goutte. Note on free lunches and cross-validation. *Neural Computation*, 9(6):1245–1249, 1997

36. IEEE OTCBVS WS Series Bench; DOE University Research Program in Robotics under grant DOE-DE-FG02-86NE37968; DOD/TACOM/NAC/ARC Program under grant R01-1344-18; FAA/NSSA grant R01-1344-48/49; Office of Naval Research under grant N000143010022.

Part V
Low-Resolution Object Detection in Airborne Infrared Videos

Chapter 11
Runway Positioning and Moving Object Detection Prior to Landing

Rida Hamza, Mohamed Ibrahim M, Dinesh Ramegowda, and Venkatagiri Rao

Abstract Safe navigation of both manned and unmanned aircraft requires a robust runway identification process and reliable obstacle detection to determine the runway status before landing. The navigation data extracted from multiple sources of current synthetic navigation sources are not adequate for positioning an aircraft, and it cannot detect moving obstacles on runways, especially during reduced visibility conditions. The enhanced vision system (EVS) described in this article can augment current synthetic vision database capabilities by providing more accurate positioning of a runway and detecting moving objects on it from an onboard infrared sensor. Our EVS is based on a two-step process. We first analyze the sensor image to identify and segment the runway coordinates. These estimates are then used to locate the runway structure and detect moving obstacles. In the segmentation process, we apply an adaptive thresholding technique to calculate the edges of a runway based on the predicted synthetic data. To match a runway template on the edges, we examine alternative fitting models. The predicted coordinates and the detected edges are then correlated to determine the location of the actual runway coordinates within the sensor image. These coordinate estimates are fed to the dynamic stabilization of the image sequence in the obstacle detection process. We also use feature points beyond the estimated runway coordinates to further improve the stabilization process. Next, we normalize the stabilized sequence to compensate for the global intensity variations caused by the gain control of the infrared sensor. We then create a background model to create an appearance model of the runway. Finally, we identify moving objects by comparing the image sequence with the background model. We have tested our EVS and reported significant improvements over the synthetic navigation data. We have been able to detect distant moving objects around the runway.

R.I. Hammoud (ed.), *Augmented Vision Perception in Infrared: Algorithms and Applied Systems,* Advances in Pattern Recognition, DOI 10.1007/978-1-84800-277-7_11,
© Springer-Verlag London Limited 2009

11.1 Introduction

The need to detect and track runways and targets from moving platforms is driving the development of sensor fusion and computer vision algorithms for next-generation situational awareness and navigation systems. For instance, safely landing an aircraft, whether an unmanned air vehicle (UAV) or manned, requires accurate information about the location of the runway. Similarly, military targeting systems for urban environments require very accurate localization to avoid collateral damage and civilian casualties. The navigation data extracted from multiple databases of current navigation data sources are not sufficient to accurately position an aircraft or a target. Furthermore, the resolution available with such sources is at least multiple meters. Low-precision guidance can be easily hampered with severe weather conditions that can present challenges for safely landing an aircraft or tracking targets of interest. At the same time, the advancement in computer vision and increases in computing power are pushing for deployment of vision sensors as a major component in navigation and target positioning.

In recent efforts, researchers have attempted to analyze sensor data using moving platforms that, with additional features, enable the commander and pilots to easily navigate or identify potential hazards, to avoid these hazards, and to obtain sufficient visual reference about the actual targets. In addition, analysis of moving platform cameras must deal with shortcomings such as competing background clutter, changing background dynamics, and artifacts of motion.

On the other hand, target detection is generally carried out using segmentation and tracking techniques based on image correlation as maintained by a static platform or static scene hypothesis; the resultant sequence of frames is then analyzed independently. When significant dynamics are available in the scene or the target is moving rapidly, this analysis usually results in jitter artifacts due to incomplete motion compensation from the sensor or the platform.

We must be able to analyze real-time sensor images from moving platforms while using the platform and sensor dynamics to compensate for the motion, thus allowing accurate estimation of the target location. This estimation provides operators with the visual references for the target required for safe navigation in detecting runways, targeting enemies, or avoiding hazards. This article describes an enhanced vision system (EVS) solution that provides accurate visual cues from sensor images that take into account the platform system dynamics and the observation measurements to discover targets and detect local obstacles.

11.2 Literature Survey

Our survey of the current literature reviewed research related to runway segmentation and how target detection and tracking are evolving from fixed to moving platforms.

Typically, researchers rely on two categories of runway segmentation techniques: feature-based and template-based analytical methods. The extraction of runway boundaries can be deduced to identify line parameters of runway contour—that is, line regression analysis. Many methods for estimating line parameters are available, but when applied to runway segmentation most of them reveal deployment problems.

Hough transform [7] can control disturbances, but the amount of computations it uses can be overwhelming when applied by brute force. Some researchers have attempted to restrict the search for a Hough fitting model to a limited range [11]; however, the final outcome still relies on a least squares (LS) fitting. This method can be easily biased even with a single outlier. Sasa et al. [10] introduced an approach for estimating aircraft orientation and pose during an approach. Their algorithm employs the Hough method on detected edges using color comparison, Sobel edge analysis, and thresholding. A linear regression is used to fit lines along the left, right, and bottom sides of a runway. The drawback of this technique is its dependency on thresholding, which can significantly affect the Sobel edge detector and color comparison limitations when applied on monochrome infrared (IR) images.

In [14], a heuristic approach was proposed to analyze the edges extracted using a Sobel-based edge detector. A combined measure was then defined to measure the strength of the edge of the line and the number of votes for line to be present in subsequent frames. The runway hypothesis is built around a few criteria that are checked through several line pairs that delineate runway boundaries. The probability of detection of this Sobel-based approach is reported to be low at nominal ranges. The algorithm is not reliable when applied to low-contrast images, and it performs poorly at longer range. Similar conclusions were deduced of the work reported in [1], in which the extraction of the runway structure was also based on a hypothesis-driven model in which multiple criteria were evaluated against the detected edges. Due to heavy compression noise artifacts, Sobel edge detection fails to provide adequate edges for analysis, and thus additional threshold manipulation was conducted to extract the actual runway edges.

Simond et al. [12] presented preliminary results estimating road boundaries that can be used for runway segmentation. They proposed a segmentation technique that uses projective properties to compute the dominant vanishing points and detect boundary lines in a captured scene. These boundary vanishing lines are computed as the highest accumulator score of a Hough transform with the detected edges. The projective invariant is used to compute the homographies between consecutive frames to match extracted features. To deal with low image quality and inaccuracy of edge detection, they based the boundary fitting on the midpoint location of each segment while allowing some margin of error on the slope and length of each segment. As suggested by the authors, this technical approach is a work in progress.

An earlier work by Tarleton et al. [16] has led to a good foundation for a reliable runway segmentation approach. They provided a general framework for analyzing the sensor images and proposed a good correlation scheme that combines navigation data and edge estimates. They used geometric hashing to correlate the extracted runway edges and their predicted coordinates. We adopted some of their concepts in our proposed architecture. An article by Gong and Abbott [2] presented an overview of additional sensor methods that are related to navigation of aircraft during landing operations. The authors provided a good survey of techniques for detecting and tracking airport runways and acknowledged the lack of a credible image analytics solution as an aid to aircraft landing.

Previous methods for detecting motion on moving platforms include an optical flow-based approach [8, 13, 15], most of which deploy Kalman filtering and a background subtraction-based approach [9, 19]. Optical flow approaches require camera motion parameters that can be deduced from the synthetic database to estimate position and velocity of the aircraft. In [13], the optical flow was first calculated for extracted features. A Kalman filter then used the optical flow to calculate the range of those features, and a range map was used to detect obstacles. In [15], the model flow field and residual flow field were initialized with the camera motion, and obstacles were detected by comparing the expected residual flow with the observed measurements. The work in [8] is a subsidiary of the [15] approach that limits the optical flow to only extracted features.

In contrast to optical flow approaches, background subtraction [9] does not need camera motion parameters. Camera motion is compensated by estimating the transformation between two images from feature points. Then, moving objects are detected by finding the frame differences between the motion-compensated pairs.

A drawback of the segmentation methods described is that most rely on edge detection, and no single edge detection method is universally applicable. In our approach, we apply blob analysis to extract the runway contour without using brute force edge detection. We implement a modified Hough and template matching to preserve the robustness of the technique while reducing the computational burden.

Our obstacle detection solution belongs in principle to the background subtraction approach and is an extension to the work of Medioni et al. [9]. We deploy Kalman filtering and make use of navigation data sets to dynamically stabilize the estimates of runway coordinates and feature points. The deployment of Kalman filtering is a prerequisite for obtaining accurate estimates of the runway coordinates at which further image stabilization is conducted.

11.3 Enhanced Vision System Framework

In image processing and scene analysis, detection of moving or static targets is usually carried out through the implementation of segmentation and tracking techniques that are based on analytical solutions maintained by the static platform hypothesis. However, detection procedures cannot rely on this analytical solution when the platform is moving. For example, motion artifacts may be introduced due to platform and sensor dynamics; dynamic stabilization techniques can compensate for the platform motion, resulting in smooth detection through the frame sequence. Our enhanced vision apparatus, illustrated in the flowchart in Fig. 11.1, dynamically detects the runway coordinates, stabilizes the estimated observation measurements by

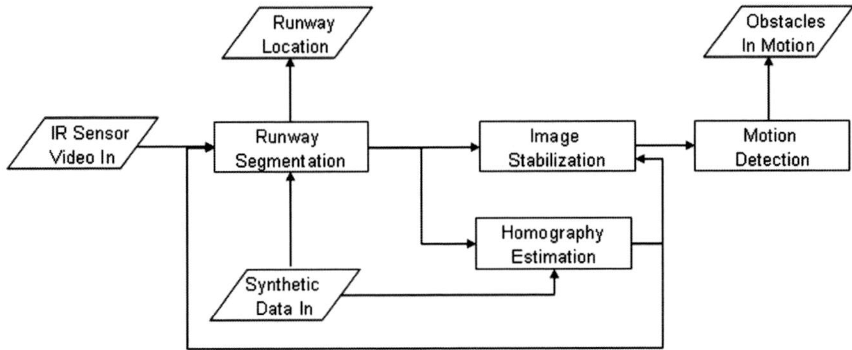

Fig. 11.1 Flowchart for Honeywell's enhanced vision system

including the system dynamics and prediction measurements based on the vehicle motion, computes image sequence homographies, and stabilizes frames accordingly to detect moving obstacles on the runway.

A runway segmentation process identifies a four-sided polygon that contains the runway location in the sensor image. These estimates are correlated with the predicted synthetic data estimates. Once the quadrilateral polygon is estimated between each consecutive pair of images, their homographies are obtained for the purpose of warping and stabilization.

We define the mathematical formulation of the homography operation. Let I_i be the ith frame in the video sequence and $H_{i,i-1}$ be the homography between I_i and I_{i-1}. Namely, $I_{i-1} = H_{i,i-1}I_i$, and $I_i = H_{i-1,i}I_{i-1}$. The homography operation is used in the runway segmentation operation to estimate the system dynamics (the subject of the next section) and in the obstacle detection process to compute the accumulated homography matrix $H_{m,Ref}$ after a lapse of m frames from a reference frame. The cumulative homography is defined as

$$H_{m,Ref} = \prod_{i=Ref+1}^{m} H_{i,i-1} \tag{11.1}$$

A reference frame could be any frame in the image sequence. The accumulated homography matrix is used to register the current frame with the reference frame. We use Ref notation to keep track of the index of the reference frame. These homographies, along with the locally stabilized image sequence, are inputs to a motion detection process that identifies moving pixels on the image.

11.4 Runway Segmentation Process

Figure 11.2 is a simplified block diagram illustrating the runway segmentation technique. A forward-looking infrared (FLIR) camera is mounted on the aircraft with a predefined field of view with an overlap of the same scene being surveyed by

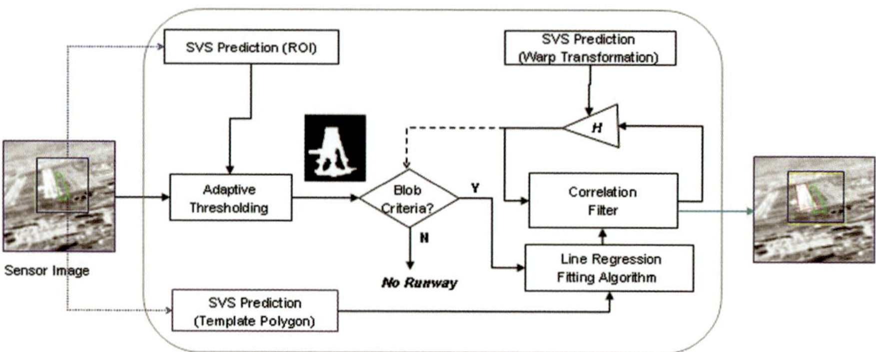

Fig. 11.2 Runway segmentation system architecture. Synthetic Vision Systems (SVS) are a set of technologies that provide pilots with clear and intuitive means of understanding their flying environment [Source wikipedia]

the navigation data from the synthetic vision system database. The navigation data provide a region of interest (ROI) where a runway is presumed to be present as well as a template of the runway perspective based on the database accuracy and the synthetic vision modeling. The sensor data are preprocessed, and the predicted ROI is analyzed for blob detection and feature extraction as shown in Fig. 11.2. The blob criteria are then evaluated to validate the presence of a runway. If a runway is present, the polygon-fitting method fits the blob contour to a runway template. We then correlate the output of the segmentation method with the dynamics of the prediction system by associating current observations with previous results to provide a stable desired output using a correlation filter. A homography matrix is computed based on the prediction estimates and is used for part of the correlation filter.

11.4.1 Adaptive Binarization

The ROI image may represent extraneous objects other than target runway. These include secondary runways, taxiways, mountainous structures near the airport, or any other structure with similar sensing characteristics within the associated IR wavelength. In our approach, the binary image used for runway estimation is generated using the adaptive threshold technique described according to the intensity density distribution method.

The adaptive binarization process depicts the probability density function of the intensity data representing the ROI to segment the foreground pixels from background. The sliding histogram estimator is used to approximate the density function. Hence, the runway blob is represented by a grouping of pixels in the histogram that makes up a certain percentage of the overall ROI data above the adaptive threshold. The percentage of the blob is determined from synthetic runway proportion estimates from data sets, based on the aircraft positioning with respect to the runway coordinates and orientation perspective. The validation processor that determines whether a runway exists within the predefined ROI may employ one or any combination of a number of validation processes of the following queries: measuring the offset of the center of moment of the runway blob with respect to the template; verifying the actual size of the estimated blob with respect to the ROI and template. Failure to validate any of these measures results in a report of no runway within the specified region.

The threshold required to segment foreground from background is derived based on the analysis of the cumulative histogram of intensities and its distribution within the ROI. First, the area that background region occupies within the ROI is estimated as

$$S_B = 1 - \left[\frac{\alpha S_p}{\omega.h}\right] \tag{11.2}$$

where ω and h represent the width and height of the associated ROI, respectively, and S_p is the area within the specified template polygon. The template size is scaled by α to account for variations in range, orientation, and synthetic noise margin.

Once the background area is estimated, the corresponding intensity level at which the cumulative histogram value equals background area will be used as the cutoff value for thresholding. If $H(g) = \partial F/\partial g$ represents ROI image normalized histogram, that is, $\int H(g)dg = 1$ and $F(g)$ represents the cumulative distribution function of H, then the threshold λ is derived such as

$$F(\lambda) = S_B \Rightarrow \lambda = F^{-1}(S_B) \tag{11.3}$$

This adaptive binarization approach strikes a good balance between providing segments that contain adequate presentation of the runway region and reducing the risk for segment-spanning multiple secondary driveways or structures that are similar and adjacent to the runway.

11.4.2 Runway Quadrilateral Fitting Algorithms

Theoretically, runway polygon fitting can be solved by finding the best line combinations that make up the runway sides. Let T be a quadrilateral polygon template and B a binary image of the blob. We want to compare the blob to the template by defining, for example, a direct Hausdorff distance that compares, in piecewise, the pixels on the side lines of the template against those of the blob contour. Thus, we define:

$$h(P(x,y),L_T) = \max_{P \in C_B} \left(\min_{L_T} |dist(P,L_T)|\right) \tag{11.4}$$

where C_B is the contour of the blob.

For instance, an estimate can be based on the maximum Hausdorff distance in both directions, that is,

$$H(L_T,\hat{L}) = max(h(L_T,\hat{L}),h(\hat{L},L_T)) \tag{11.5}$$

where most mismatched points are a measure of match.

This formulation requires that templates are well defined a priori. Because the exact orientation of the runway is unknown and blob sides can be eroded by missing edges, these factors may hamper the maximization solution provided in Eqs. (11.4) and (11.5).

In what follows, we discuss a few alternative methods for line regressions.

11.4.2.1 Adaptive Hough Fitting

Although Hough-based line detection is a well-proven concept in the literature, we proposed in [4] a configuration of the Hough method to match templates with incomplete data, especially when analyzing poor-quality IR imagery. Poor imagery contrast is a fundamental problem; multiple irregular edges mean that the Hough accumulator array will be sparsely distributed. We argue that one way to get around in matching a template using incomplete data is by clustering the detected contour of the object into subgroups: major sides (i.e., complete edges) and minor supporting segments of the contour. These major segments best characterize the target outline rather than the minor lines. The Hough method is applied specifically to the paired major sides, and then the estimated boundaries are intersected with the minor lines mapped directly from the template features. For instance, applying the concept to runway segmentation, we assume runway bottom lines slope are the same as the template's bottom edge slope and fit a line that pass through the bottommost pixel in the contour. A similar procedure is followed for fitting the top horizontal line. The horizontal lines thus obtained are intersected with already fitted vertical lines to estimate the four corners of the runway.

11.4.2.2 Vertices-Based Runway Segmentation

As an alternative to edge analysis, we propose using vertices to fit a polygon model. Because the limits of a blob can be defined as a function of its vertices, we hence propose to make use of only the blob corners to simplify the process of fitting a model to runway boundaries. From the detected blob, the contour is extracted by subjecting the blob to the Freeman chain-coding technique. The blob is initially scanned to obtain its topmost pixel, which is used as a starting point for contour tracing. Starting with this pixel and moving clockwise, each pixel is assigned a code based on its direction from the previous pixel. To extract these corner points, we define a region of support covering both sides of a boundary pixel. To qualify a boundary pixel to be a corner point, a measure called the *cornerity index*, based on statistical and geometrical properties of the pixel, is defined. Let $S_k(p) = \{P_j / j = i - k; \ldots, i, \ldots, i + k\}$ denote a small curved segment of B called the region of support of the point P_i, which is the center of the segment. The geometrical centroid of the segment is given by

$$\bar{P}_i = \left(\bar{x}_i = \frac{1}{2k+1} \sum_{i-k}^{i+k} x_j, \bar{y}_i = \frac{1}{2k+1} \sum_{i-k}^{i+k} y_j \right) \tag{11.6}$$

The computed geometrical centroid has larger shift for actual corner points and minimal shift for noncorner points. Thus, we define a corner point as a point that has a larger shift when compared to other points in the neighborhood. Therefore, the cornerity index of P_i is defined to be the Euclidean distance $d = \sqrt{(x_1 - \bar{x}_i)^2 + (y_1 - \bar{y}_i)^2}$ between the midpoints P_i and its region of support mass center point $\bar{P}_i$. The cornerity index indicates the prominence of a corner point; the larger the value of the cornerity index of a boundary point, the stronger the evidence is that the boundary point is a corner. The computed cornerity indices are subjected to thresholds such that only boundary pixels with a strong cornerity index are retained.

The corner detection estimates are sensitive to the size of the support region. While a value too large for a support region will smooth out fine corner points, a small value will generate a large number of extraneous corner points—a fundamental problem of scale.

11.4.2.3 Random Sample Consensus Line Regression for Runway Fitting

The line slope is assumed to be the same as that of the runway template. Let the line L given by $y - x\tan\alpha + \lambda = 0$ represent either side of the template. Let a $P_k(x,y) \in \Omega$ edges (right or left side) of the blob contour; thus, the distance

$$dist(P,L) = |y - x\tan\alpha + \lambda|/\|\mathbf{n}\| \tag{11.7}$$

where $\mathbf{n} = (1, -\tan\alpha)$ is the vector of the coefficients normal to the line L. Since the tangent angle is the same for all trials, this term can be dropped out of the analysis. The problem is then to find the line fit that best minimizes the error $e = \sum_{k-1}^{N} |dist(P_k, L)|^2$. In an LS sense, the solution is

$$\sum_{k-1}^{N} |y_k - x_k\tan\alpha + \lambda|^2 = 0 \Rightarrow \hat{\lambda} = \frac{1}{N} \sum_{k-1}^{N} (x_k\tan\alpha - y_k) \tag{11.8}$$

Note that $\tan\alpha$ is provided by the template slope of the associated side (left or right). Thus, the line is estimated by the two extreme points $P1$ and $P2$, where

$$\hat{P}_i = P\left(\frac{(y_1 + \hat{\lambda})}{\tan\alpha}, y_i \right); i = 1,2 \tag{11.9}$$

First, a related rationale selects only a subset of pixels within a margin distance of the predicted synthetic estimates. The RANSAC (random sample consensus) algorithm is modified to influence the selection of the random data points based on

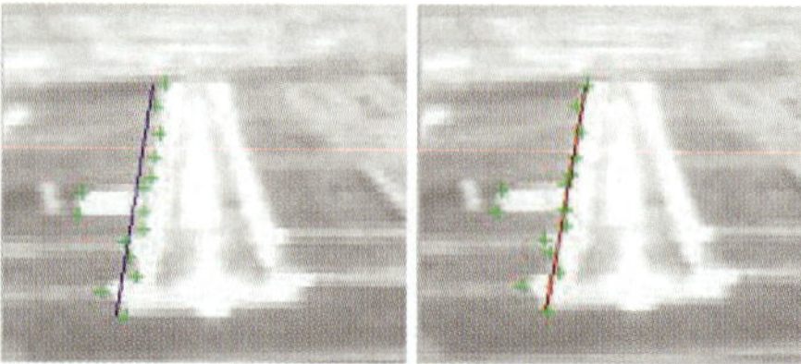

Fig. 11.3 Fitting solution: **a** least squares, **b** RANSAC

the a priori information instead of the usual random process. Second, once a suitable consensus set of edges is identified and a line instantiated, we add new points from the rest of the edges that are consistent with the line model and compute using LS a new line fitting on the basis of the larger set. Figure 11.3 illustrates the benefits of the RANSAC process when applied to the left edges of the runway.

In both Hough and RANSAC methods, we obtain estimates for the major template sides; we then connect these major segments using straight lines to piece the other polygon segments (minor segments) that are traced from the template while preserving the relative slopes from the template. The resulting quadrilateral is a best estimate that matches a given template polygon while fitting the blob structure in a nonrigid fitting model.

11.5 Dynamic Stabilization of Runway Detection

In a typical surveillance system, the detection of moving or static targets is usually carried out using segmentation and tracking techniques based on analytical solutions that are maintained by the static platform hypothesis. However, detection procedures cannot rely on this analytical solution when the platform is moving. For example, jittering artifacts could be introduced due to the platform and sensor dynamics, which motivate the need for dynamic stabilization techniques and that can compensate for this jittering. In this section, we describe an apparatus for dynamically stabilizing runway detection to reduce the jittering artifacts. We provide a general framework to extend the analysis to any target detection from moving vehicle platforms. The apparatus dynamically stabilizes the estimated observation measurements in the previous section by taking into account the system dynamics and correlating the prediction measurements based on the vehicle motion.

11.5.1 Maximum Likelihood Estimate

Before we delve into the details of the Kalman filter design, we present a simplified autoregressive and moving average model that we initially propose and provide a framework to transition from this simplified stochastic autoregressive moving average (ARMA) model to a maximum likelihood (ML) model to determine the appropriate values of the weights.

We start the analysis by formulating the runway location estimates as

$$\hat{x}_k = \alpha \hat{x}_{k-1} + (1-\alpha)z_k \tag{11.10}$$

where z_k defines the measurement of the corner location. The variables $\hat{x}_k$ and $\hat{x}_{k-1}$ are the current and previous frame estimates, and the scale α is a weighting factor that smoothes the runway location estimates (the weighting factor varies; i.e., $0 \leq \alpha \leq 1$; when $\alpha = 1$, current measurements are ignored, and when $\alpha = 0$, prior measurements are ignored). For simplicity, we define a single variable per location. In our discussion, we generalize the analysis to include both x and y coordinates and replace α with a weight vector.

If the noise distributions are known or empirically measurable, they may provide a good statistical framework to determine the scale α from the statistical parameters.

Let us assume that the inherent measurement uncertainty in the z_k estimate is defined by a standard deviation σ_k. Each type of sensor has fundamental limitations related to the associated physical medium, and when pushing the envelope of these limitations the signals are typically degraded. In addition, some amount of random electrical noise is added to the signal via the sensor and the electrical circuits. Many modeling techniques are available to estimate the statistical behavior of these typical noises. Thus, one can deduce a conditional probability of the measure $x(t)$, the location of a corner, conditioned on the measured value of being z_k. Mathematically, the conditional probability density function can be modeled as $f_{x/z}(t) \sim N(z_k, R = \sigma_k^2 I)$. For simplicity, we model the noise as white Gaussian, and thus the covariance matrix is a scalar times the identity matrix.

Similarly, let us define σ_w to be the direct measure of the system uncertainty, and thus the conditional probability measure of the estimated $f_{x/\check{x}}(t) \sim N(\check{x}_k, Q = \sigma_w^2 I)$.

At this point, we have two measurements for estimating the actual target position. The maximum likelihood estimate for the combined measurements at frame k is hence

$$\hat{x}_k = \frac{\sigma_v^2}{\sigma_v^2 + \sigma_w^2}\check{x}_k + \frac{\sigma_w^2}{\sigma_v^2 + \sigma_w^2}z_k \tag{11.11}$$

This yields a good estimate of the weight factor $\alpha = (\sigma_v^2)/(\sigma_v^2 + \sigma_w^2)$. The relative weighting between the two summations will be a function of the variance ratio. Readily conditional probability with narrower distributions will receive the higher weights.

11.5.2 Residue-Based Estimate

Ideally, α in Eq. (11.10) is chosen based on the dynamics of the vehicle and the characteristics of the sensor, including factors such as lens zoom, range to the target, and resolution of the sensor and good characterization of inherent noises as described in Eq. (11.11). Practically, this compensation is usually nonstationary due to the changing dynamics of the vehicle that is dependent on wind conditions and other uncontrollable factors. These statistical parameters in Eq. (11.11) may vary at any time. Therefore, as described in [3], we propose to dynamically adapt these parameters based on the residual of the estimate:

$$\varepsilon_k^2 = |\hat{x}_k - z_k|^2 \tag{11.12}$$

We note that, ideally, $\alpha = 1$ when $\varepsilon_k^2 = 0$, and $\alpha = 0$ when $\varepsilon_k^2 \gg 0$. In general, we express $\alpha = g(\varepsilon)$. Practically, we may want to prevent α from reaching the limits.

Next, we estimate the corner location using the transformation approximation derived from the template transition $\check{x}_k = H\hat{x}_{k-1}$ where H is the homography of Eq. (11.1). Replacing $\check{x}_k$ in Eq. (11.10) as a transformation of the previous estimate, generalizing the analysis to include both Cartesian coordinates, and using vector notations, we obtain a combined estimate for the kth measurement vector as follows:

$$\hat{\hat{\mathbf{x}}}_\mathbf{k} = \begin{bmatrix} \hat{x}_k \\ \hat{y}_k \end{bmatrix} = \overline{\alpha} H_k \hat{\hat{\mathbf{x}}}_{k-1} + (1 - \overline{\alpha}) \overline{\mathbf{z}}_k \tag{11.13}$$

The vector $\overline{\alpha} = \begin{bmatrix} \alpha_x \\ \alpha_y \end{bmatrix}$. Because Eq. (11.13) accounts for the warping effect of affine transform due to motion translation, thus the estimate compensates much better than the regression model.

11.5.3 Kalman Filter Estimate

To obtain desired estimates that are statistically optimal in the sense that they minimize recursively the mean-square estimation error and properly weigh the runway segmentation noisy estimates, we reformulate our measurement problem as a Kalman filter formulation. The recursive nature of the filter allows for efficient real-time processing to reduce the jittering effects among sequential frame reports. Kalman modeling enables the filter to account for the disparate character of the jittering due to errors in different instantaneous frames, providing an optimal integrated combination estimate while maintaining the evolution of the system in time and using all available past information.

The Kalman filter estimates a process by using a form of feedback control: The filter estimates the process state frame by frame and then obtains feedback as a noisy measurement. To account for the aircraft state dynamics, we assume that the aircraft moves before taking another measurement. Further assume that the following model approximates the corner location transition due to aircraft motion in a simplistic form:

$$\frac{dx(t)}{dt} = u(t) + w(t) \tag{11.14}$$

where $u(t)$ is a nominal velocity or change of coordinates per unit time (per frame). The variable $w(t)$ represents an additive white normal process noise that models the system uncertainty of the actual motion due to disturbances in synthetic data modeling, off nominal conditions, and errors from the source of processed information. This noise $w(t)$ is modeled as zero mean white Gaussian $f(w) \sim N(0, Q)$.

First, one must determine the appropriate statistical models of the state and measurement processes so that we can compute the proper Kalman gains. In this framework, we process each corner of the runway separately and define a single state vector (i.e., coordinates for each vertex) as unknown but assumed to be constant. The Kalman filter hence is defined to address the general problem estimating the state $\overline{\mathbf{x}}_{\mathbf{k}} = \begin{bmatrix} x_k \\ y_k \end{bmatrix} \in \mathfrak{R}^2$ of discrete time-controlled process that is governed by the linear stochastic difference equation

$$\overline{\mathbf{x}}_k = A\overline{\mathbf{x}}_{k-1} + B\overline{\mathbf{u}}_k + \overline{\mathbf{w}}_{k-1} \tag{11.15}$$

with a measurement $\overline{\mathbf{z}}_k \in \mathfrak{R}^2$ defined as

$$\overline{\mathbf{z}}_k = H\overline{\mathbf{x}}_k + \overline{\mathbf{v}}_k \tag{11.16}$$

The random variable v represents the measurement noise with the following form $f_v(v) \sim N(0, R)$. Both process and measurement noise are assumed to be independent of each other and have normal distributions. In practice, the process noise covariance matrix Q and measurement noise covariance matrix R could be changing in time. For simplicity, we will assume they are constant.

A proposed Kalman filter architecture is depicted in Fig. 11.4.

To determine the state dynamics matrix A and the control input matrix B, we refer to Eq. (11.14), at time t_k^-, just before the measurement from a runway segmentation algorithm is taken at time t_k. The a priori conditional estimate can hence be expressed mathematically as

$$\hat{x}(t_k^-) = \hat{x}(t_{(k-1)}) + u(t_k - t_{(k-1)}) \tag{11.17}$$

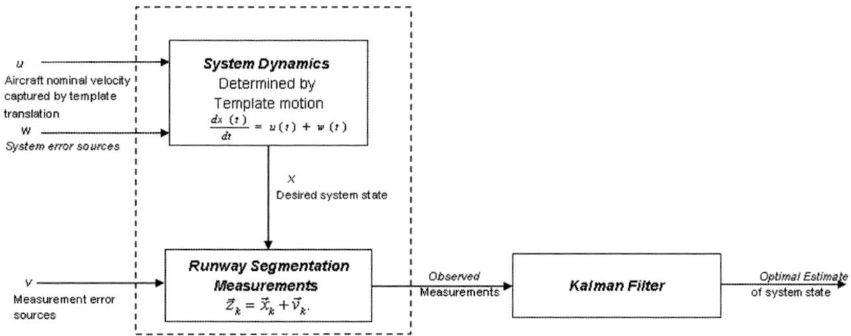

Fig. 11.4 Kalman filter architecture

This presents a state process equation form as it evolves with time. This implies that the state dynamics matrices A and B in the difference Eq. (11.15) are identity matrices. In the sequel, we shall assume H in the measurement equation to be an identity matrix as well for simplicity.

Let us now turn back to the vector presentation and define $\overline{\mathbf{x}}_k^- \in \Re^2$ to be our a priori state estimate at step k given knowledge of the process prior to step k and $\hat{\overline{\mathbf{x}}}_k \in \Re^2$ to be our a posteriori state estimate at step k given measurement $\overline{\mathbf{z}}_k \in \Re^2$. The t term is dropped out for notation convenience. We can then define a priori and a posteriori estimate errors $\breve{\overline{\mathbf{e}}}_k = \overline{\mathbf{x}}_k - \overline{\mathbf{x}}_k^-$, and $\hat{\overline{\mathbf{e}}}_k = \overline{\mathbf{x}}_k - \hat{\overline{\mathbf{x}}}_k$; the a priori estimate error covariance is then $P_k^- = E\left[\breve{\overline{\mathbf{e}}}_k \breve{\overline{\mathbf{e}}}_k^T\right]$, and the a posteriori estimate error covariance is $P_k = E\left[\overline{\mathbf{e}}_k \overline{\mathbf{e}}_k^T\right]$. With mathematical manipulations, the a priori error covariance matrix evolves as

$$P_k^- = AP_{(k-1)}A^T + Q \tag{11.18}$$

In deriving the equations for the Kalman filter, we begin with the goal of finding an equation that computes the a posteriori state estimate $\hat{x}_k$ as a linear combination of an a priori estimate $\mathbf{x}_k^-$ and a weighted measurement innovation. The innovation residual reflects the discrepancy between the predicted measurement and the actual measurement. Thus, we obtain the Kalman a posteriori state estimate equation:

$$\hat{\overline{\mathbf{x}}}_k = \hat{\overline{x}}_k^- + K_k(\overline{\mathbf{z}}_k - H\hat{\overline{\mathbf{x}}}_k^-) \tag{11.19}$$

The matrix K is the blending factor or gain that minimizes the a posteriori error covariance equation.

$$K_k = P_k^- H^T (HP_k^- H^T + R)^{-1} \tag{11.20}$$

Looking at Eq. (11.20), we deduce that as the measurement error covariance R approaches zero, that is, more reliable estimates from the segmentation algorithm, the gain K weights the residual more heavily. On the other hand, as the a priori estimate error covariance $\breve{P}_k$ approaches zero, that is, less aircraft motion, the gain K weights the residual less heavily, and the actual measurement is trusted less, while the predicted measurement is trusted more as it saturates.

The a posteriori state estimate equation is rooted in the probability of the a priori estimate conditioned on all prior measurements based on Bayes rules. The a posteriori state estimate reflects the mean (e.g., first moment), defined in Eq. (11.19) of the state distribution—it is normally distributed $p(\mathbf{x}_k/\mathbf{z}_k) \sim N(\hat{\mathbf{x}}_k, P_k)$. The a posteriori estimate error covariance equation reflects the variance of the state distribution (the second noncentral moment) and is given by the formula

$$P_k = (I - K_k H)P_k^- \tag{11.21}$$

Equations (11.17) and (11.18) represent the *predict* equations, and Eqs. (11.19), (11.20), and (11.21) represent the correction *update* equations in a typical Kalman filter formulation.

The Kalman filter estimates a process by using a form of feedback control: The filter estimates the process state at some time and then obtains feedback as a noisy measurement. The equations for the Kalman filter fall into two groups: time update equations and measurement update equations. The time update equation projects the current state and error covariance estimates forward in time to obtain the a priori estimates for the next time step. The measurement update equations are responsible for the feedback—that is, for incorporating a new measurement into the a priori estimate to obtain an improved a posteriori estimate. The time update equations can also be thought of as predictor equations, while the measurement update equations can be thought of as corrector equations. Indeed, the final estimation algorithm resembles that of a predictor-corrector algorithm. After each time and measurement update pair, the process is repeated with the previous a posteriori estimates used to project or predict the new a priori estimates. Figure 11.5 shows the customized Kalman filter design for our runway segmentation in operation.

The Kalman filter does not operate on all of the data directly for each estimate; rather, it recursively conditions the current estimate on all of the past measurements. Starting with an initial predicted state estimate and its associated covariance obtained from past information, the filter calculates the weights to be used when combining this estimate with the first measurement vector to obtain an updated "best" estimate.

The two parts of the filter's prediction and correction are akin to what made the Kalman filter so desirable. This recursive nature is one of the very appealing features of the Kalman filter—it makes practical implementations, much more feasible

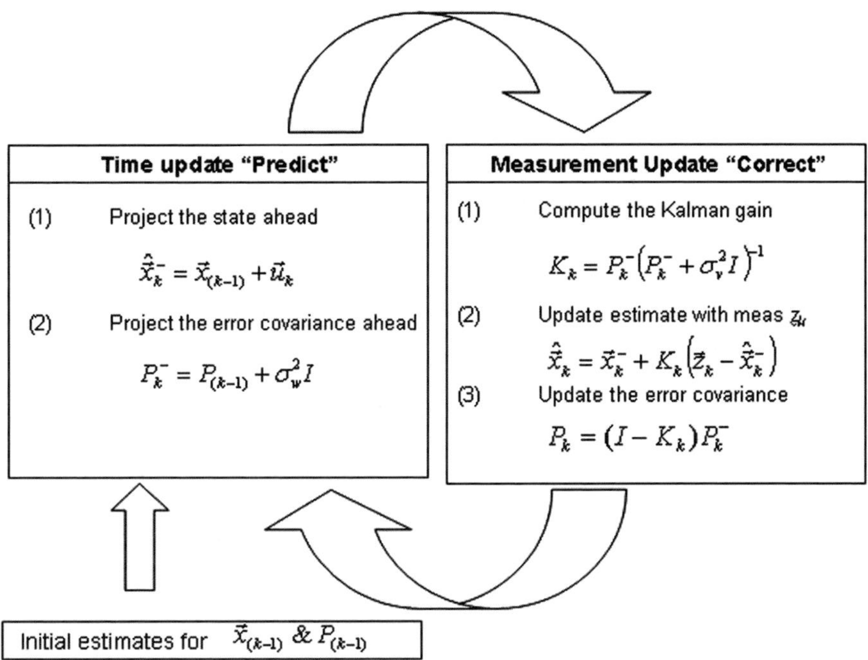

Fig. 11.5 Kalman Filter equations tailored to the runway segmentation

than other implementations, such as the Wiener filter. It is essentially a predictor-corrector type estimator that is optimal in that it minimizes the estimated error covariance—when some presumed conditions are met. It is the best possible optimal stochastic estimator from noisy sensor measurements and method of choice, while it is relatively simple in implementation and robust in nature.

11.6 Obstacle Detection Approach

We divide the problem of obstacle detection into two steps: stabilization and motion detection. The first step compensates for the camera movement by stabilizing the runway estimates using the dynamic stabilization approaches of Section 11.6. In some instances where the runway is at farther ranges, other feature points are necessary for global image stabilization, as explained in this section. Once the stabilization is obtained, we use a background model to segment moving blobs on the runway. The detected objects are filtered using a runway mask to retain the detected objects pertaining only to the runway area. The polygon vertices corresponding to the runway area are fed from the segmentation process. Homographies are used to update these vertices' coordinates in subsequent frames. The homography matrix is accumulated to warp each frame to the corresponding reference frame as defined in Eq. (11.1). All images with the same reference image are then warped to this reference image to form a locally stabilized image sequence. These homographies along with the locally stabilized image sequence are inputs to the motion detection process.

11.6.1 Stabilization Module

Since it is very hard to directly detect moving objects when the camera platform is on the move, our first goal is to stabilize the image sequence. We assume that the ground is a planar surface, which is a reasonable assumption in the neighborhood of a runway. With this assumption, the change of viewpoint between two adjacent frames could be represented by a homography as described in [9]. The four corners of a runway are sufficient for the stabilization process. However, when the runway is captured at farther distances, it represents only a fraction of the overall frame, as illustrated in Fig. 11.6, and hence additional feature points are necessary to conduct the stabilization process. We use Nobel corner points to find these additional feature points. The Nobel corner algorithm, a modification of the Harris corner detection algorithm, is reportedly more robust [5, 17] to the presence of noise in the image, and hence it has been chosen for our implementation. Scale-invariant feature transform (SIFT) features were chosen over Nobel corners in [9]. It is observed that the Nobel method achieves functional performance similar to the SIFT method but with reduced computation. In Fig. 11.6, we show our results using the Nobel technique.

Fig. 11.6 Results of Harris corner detection

11.6.1.1 Feature Point Correspondence

To establish the correspondence between feature points obtained in successive frames, a normalized correlation coefficient is used as a similarity measure. An area of 7×7 pixels centered on each feature point is considered for finding the correlation coefficient. For every feature point in the $(i-1)$th frame, the feature points present in the ith frame within a radius of 15 pixels are considered, and the feature point that yields the highest value of correlation coefficient is chosen as the match point and added to the match list. This process is repeated for all the feature points in the $(i-1)$th frame. If the correlation coefficient of the matched feature points is below a threshold value, the feature point pair is discarded from the list. The thresholding is adaptively set as follows: First, while preparing the matched list a threshold value of 0.5 is used. Then, a histogram is computed using the correlation value of all the feature points in the match list. The histogram peak represents the correlation coefficient value that is satisfied by the maximum number of feature points. A fraction of this histogram peak is used as a threshold in the next frame. This process is repeated in every frame. The matched list is sorted in the descending order of correlation coefficient value, and only the top 90 are retained for further processing.

We next compute the homography transformation. Runway images as seen by the approaching (landing) aircraft undergo oscillations in pitch, yaw, and roll as well as change in perspective. This change is captured using a projective transformation model. The projective transformation model for a given pair of point correspondences $(x, y) \leftrightarrow (X, Y)$ is given by

$$X = \frac{ax + by + c}{gx + hy + 1} \quad \text{and} \quad Y = \frac{dx + ey + f}{gx + hy + 1} \tag{11.22}$$

To estimate the unknown parameters $[a, b, c, d, e, f, g, h]$ in the above equations, a minimum of four matched pairs of feature points is required. The equations can be rewritten in standard matrix form. If we let $x = [abcdefgh]^T$, then in matrix notation, the formula reduces to $A * x = b$. The expanded matrix formula is shown in Eq. (11.23).

$$
\begin{bmatrix}
x_1 & y_1 & 1 & 0 & 0 & 0 & -X_1x_1 & -X_1y_1 \\
0 & 0 & 0 & x_1 & y_1 & 1 & -Y_1x_1 & Y_1y_1 \\
x_2 & y_2 & 1 & 0 & 0 & 0 & -X_2x_2 & -X_2y_2 \\
0 & 0 & 0 & x_2 & y_2 & 1 & -Y_2x_2 & Y_2y_2 \\
\vdots & \vdots & \vdots & \vdots & \vdots & \vdots & \vdots & \vdots \\
\vdots & \vdots & \vdots & \vdots & \vdots & \vdots & \vdots & \vdots \\
x_n & y_n & 1 & 0 & 0 & 0 & -X_nx_n & -X_ny_n \\
0 & 0 & 0 & x_n & y_n & 1 & -Y_nx_n & Y_ny_n
\end{bmatrix}
\begin{bmatrix} a \\ b \\ c \\ d \\ e \\ f \\ g \\ h \end{bmatrix}
=
\begin{bmatrix} X_2 \\ Y_1 \\ X_2 \\ Y_2 \\ \\ \\ X_n \\ Y_n \end{bmatrix}
\tag{11.23}
$$

Representing transformation in this matrix notation allows us to quickly solve for the LS solution to the problem. The set of parameters x that minimizes $\|Ax - b^2\|$ is the solution to the normal equation and is given by

$$
x = (A^T A)^{-1} * A^T * b
\tag{11.24}
$$

To be able to calculate x, we need a minimum of four matched feature pairs (8 coordinates in total). More than four matched pairs, assuming they are accurate, will result in a better estimation of the model parameters. Therefore, to find a good set of model parameters we need a good set of matched feature points. In our approach, we randomly select eight pairs (two from each quadrant in the image) of feature points of 150–300 feature points and then estimate the transformation model by applying RANSAC methodology. The RANSAC process is applied 2,000 times, and the transformation is applied to all the matched pairs to check for correctness. The transformation that is correct (one that yields minimum error) for the largest number of matched feature points is chosen as the best approximation.

In the presence of noise on input data, the accuracy of the solution of a linear system depends crucially on the condition number of the system. The lower the condition number, the less the input error is amplified and thus the system more is stable. As pointed out in [6], it is crucial that input data are properly preconditioned using a suitable coordinate change (i.e., scaling and shifting) prior to application of the RANSAC technique: Points are translated so that their centroid is at the origin and are scaled so that their average distance from the origin is $\sqrt{2}$. This preconditioning improves the condition number of the linear system that is being solved and helps to obtain a robust transformation model. The steps are explained next.

Let N be the total number of feature points. Let (xi, yi) be the coordinates of the ith feature point with reference to the top left corner of the image as the origin. We define the scale

$$
\gamma = \frac{1}{N\sqrt{2}} \sum_{i=1}^{N} \sqrt{(x_i - \bar{x})^2 + (y_i - \bar{y})^2}
\tag{11.25}
$$

where the values $\bar{x}$ and $\bar{y}$ are the mean values. Let (X_i, Y_i) be the coordinates of the same feature point after preconditioning [i.e., shifted by the mean values and scaled by γ of Eq. (11.25)]. While image warping, the transformation model computed using this set of feature points cannot be applied to the image directly because it exists in the shifted coordinate system. Hence, the model needs to be shifted back to the original coordinate system. Let S be the shifting matrix and let T be the projective transformation model in the shifted coordinate system, then the homography matrix in the original coordinate system is defined as

$$H_{i,i-1} = S * T * S^{-1} \tag{11.26}$$

where $S = \begin{bmatrix} \gamma^{-1} & 0 & 0 \\ 0 & \gamma^{-1} & 0 \\ -\bar{x}\gamma^{-1} & -\bar{y}\gamma^{-1} & 1 \end{bmatrix}$, and $T = \begin{bmatrix} a & d & g \\ b & e & h \\ c & f & 1 \end{bmatrix}$.

The homography matrix is accumulated in every frame, and the accumulated model is used to warp the current frame to register with the reference frame. We have used a bilinear interpolation scheme within our warping function.

In our implementation, the first frame is initially used as a reference frame, and all subsequent images are warped with respect to this reference frame. However, this is not a good idea for a long sequence since small errors are inevitable when doing registration; those errors may accumulate to affect later frames. To resolve this problem, we allow updating of the reference frame in the stabilization module. The decision to update could be based either on the value of the model parameters or on the shape of the warped image. We needed to define a measure that adapts to the changes in the scene (e.g., giving a longer interval when the runway is far away and more frequent updates when the runway is in close range). Since the length of warped sides (i.e., bottom or right and left sides of the warped image; see Fig. 11.7) directly reflects the amount of scene warping, it was used as a measure to decide the reference updates.

If any ratio of these lengths to their corresponding nonwarped length in the original frame is less than a preset fraction (e.g., the ratio = 90% is used in our experiments), the reference frame is updated. The length of warped image is defined

Fig. 11.7 Bottom edge length used as an indicator for reference update

as the Euclidean distance between warped image vertices. During a reference up-date, the homography matrices are initialized back to the identity matrix, and the accumulation is reset to the new reference frame as defined in Eq. (11.1).

11.6.2 Motion Detection Module

Now that we have a stabilized sequence, the background modeling can be applied to detect the moving objects on the runway or on the secondary driveways close to the runway. The approach for this module is summarized in Fig. 11.8.

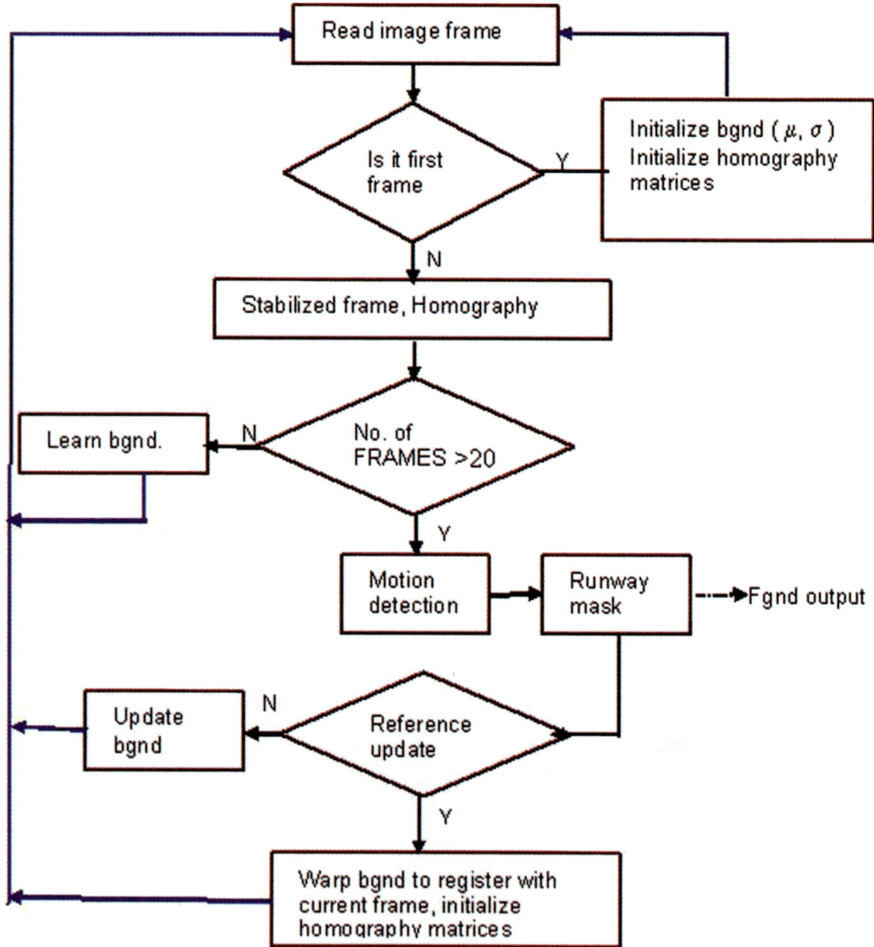

Fig. 11.8 Flowchart of motion detection module. *bgnd* background, *fgnd* foreground

A Gaussian distribution is used to model the intensity of each pixel. The mean of the Gaussian model is initialized to the value of the base image. $\mu_o = I_o$. The standard deviation σ is initialized to a constant value. Using initially K frames (e.g., $K = 20$), we build a background model (μ, σ) with reference to the first frame by adopting an exponential smoothing concept. Let I_i be the current frame. The stabilized frame is then $I_s = H_{i,Ref}I_i$, where $H_{i,Ref}$ indicates the accumulated homography matrix. Then, the background model (μ, σ) is represented by

$$\mu_i = \alpha\mu_{i-1} + (1-\alpha).I_s$$
$$\sigma_i = \alpha.\sigma_{i-1} + (1-\alpha).|I_s - \mu_i| \tag{11.27}$$

Both mean and standard deviation are updated for each new frame according to this formula during the initial learning phase (first K frames) and during the subsequent motion segmentation phase. It is assumed that a reasonably good background model is built during the first K frames. The motion detection stage is implemented from the $(K+1)$th frame onward by performing background subtraction and thresholding operations. For this purpose, we warp the background model using $H_{Ref,i}$ to register it with respect to the current frame I_i. Let μ_i^s and σ_i^s be, respectively, the warped mean and standard deviation values of the background model to adjust the orientation of our background model. Pixels having an intensity difference greater than $2\sigma_i^s(x,y)$ from $\mu_i^s(x,y)$ are marked as foreground pixels.

The foreground segmentation discussed is valid provided there is no change in intensity (either due to variation in gain or offset) between the background model and the current frame. However, in actual practice, intensity does not remain constant and must be compensated. The intensity between any two images with different gains can be modeled by an affine transformation [18].

$$\forall(x,y)I_j(x,y) = m_{i,j}I_i(x,y) + b_{i,j} + \varepsilon_{i,j} \tag{11.28}$$

By ignoring the saturated pixels, the transformation can be estimated by least mean square estimation (LMSE), and the gain can be compensated. In our implementation, the background image (μ) intensity values are compensated to account for global intensity changes by comparing it to the intensity of the current frame. After this intensity normalization, the current image is compared with the background model to identify the foreground pixels.

The four-corner-based polygon that defines a runway area is identified by the runway segmentation process. This polygon area is remapped to the current frame using the homography matrix H_{i-1i} in every frame and then used as a mask for retaining the foreground objects within the runway area. Since the area of interest is strictly the runway, other areas can be filtered out. The runway filter f is a binary image in the shape of the runway. The process simply applies an "and" operation on the image and the binary mask to single out the area of interest.

The foreground pixels that are detected due to random noise occur at random locations in the foreground image when we go from one frame to the next, whereas, the actual moving foreground objects follow a specific trajectory. Random noise can be eliminated by performing an "and" operation of the current mask with the dilated foreground mask of the previous frame:

$$\forall(x,y)C_i(x,y) = \left(\bigcup_{a=x-1,b=y-1}^{a=x+1,b=y+1} fg_{i-1}(x,y) \right) \bigcap fg_i(x,y) \qquad (11.29)$$

in which C_i refers to the output of the and operation, and fg_i, fg_{i-1} refer to the foreground mask of the previous frame and present frame, respectively.

11.7 Experimental Results

The functionalities of all components of the proposed EVS system have been tested extensively using the NASA data sets, Maxviz data sets, and simulated examples.

11.7.1 Performance Tests on Runway Segmentation

The performance of our segmentation approach is demonstrated with real data acquired by NASA & Maxviz. Figure 11.9 demonstrates the resulting performance of our adaptive Hough technique applied to runway identification at various ranges.

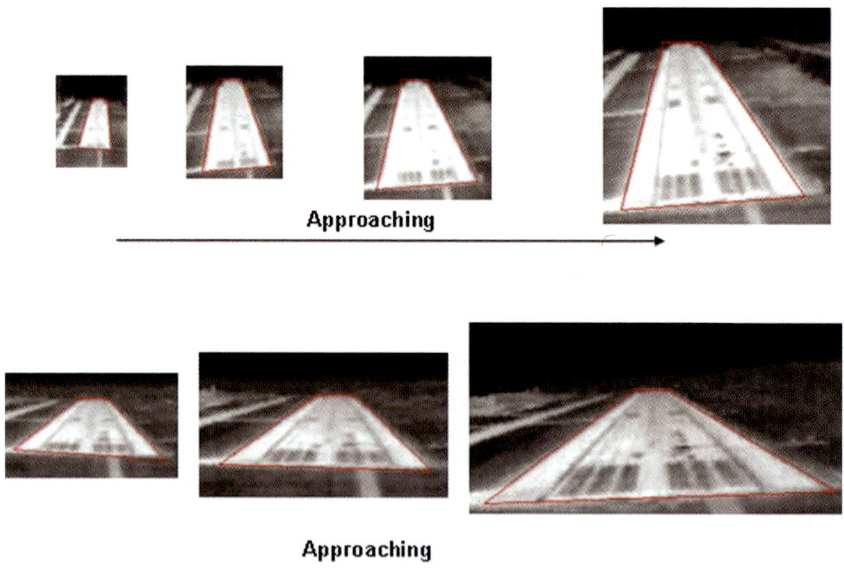

Fig. 11.9 Segmentation results using active edge method

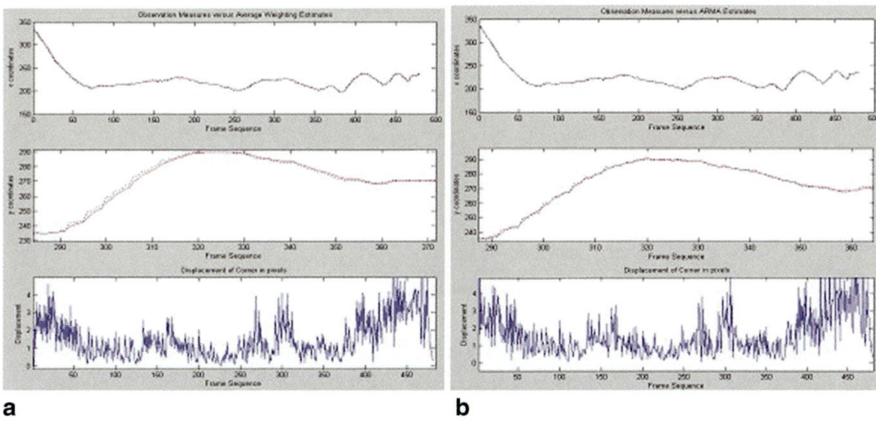

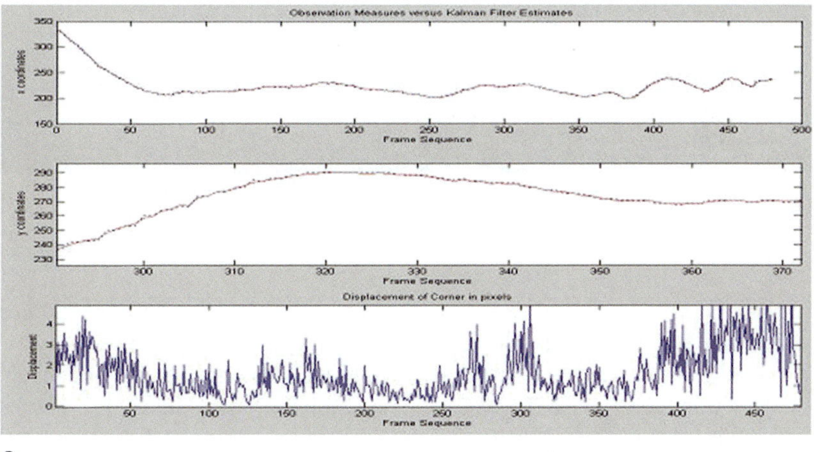

Fig. 11.10 Dynamic stabilization of runway detection: **a** Average weighting, **b** ARMA model, and **c** Kalman filter results

For detection stabilization, Fig. 11.10 shows the stabilization results for the three methods discussed in Section 11.5. Among all three methods, the Kalman process produced a more robust estimate and was simpler to design.

11.7.2 Obstacle Detection

The IR image sequences of a runway area taken by an approaching aircraft are used as test data sets. A snapshot of stabilization outcome is shown in Fig. 11.11.

The motion detection results are obtained by processing the entire frame for stabilization and background subtraction. The detected blobs are subjected to further

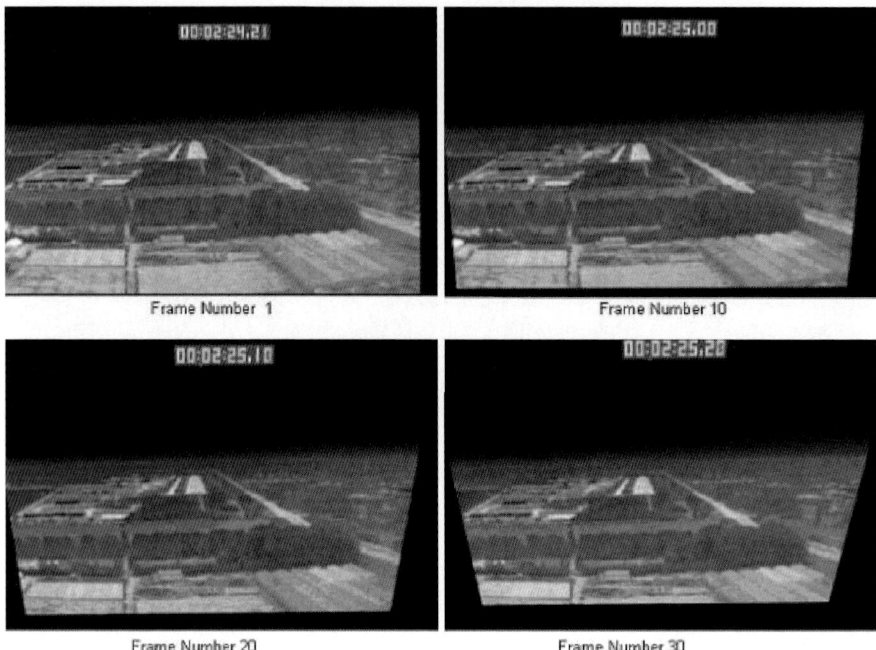

Fig. 11.11 Snapshot of image stabilization output

filtering and morphing. Since the video data sets used in our experiments did not have any moving objects in the runway area, the runway mask region was extended to the road below the runway to pick up moving vehicles on the road. The results of motion detection are shown in Fig. 11.12. The number of missed detections and spurious detections depends on object contrast, noise in the image, stabilization quality, detection threshold, and so on. Quantitative estimation of these parameters could not be carried out as a part of this effort due to resource constraints and nonavailability of ground-truthed data.

11.8 Conclusion

An EVS was designed to locate a runway and detect obstacles on it prior to landing. A sensor is mounted on an aircraft nose to capture the front view of the landscape. We capture an image of the field of view, analyze its content to detect runway borders, and correlate the information with the navigation SVS data. The sensor estimates are used to correct the navigation data to determine the accurate lateral and vertical attributes and assist in navigating the aircraft to avoid obstacles during approaching and landing operations.

We believe that this effort has laid a firm foundation for a reliable runway segmentation approach. We have implemented three new segmentation techniques to

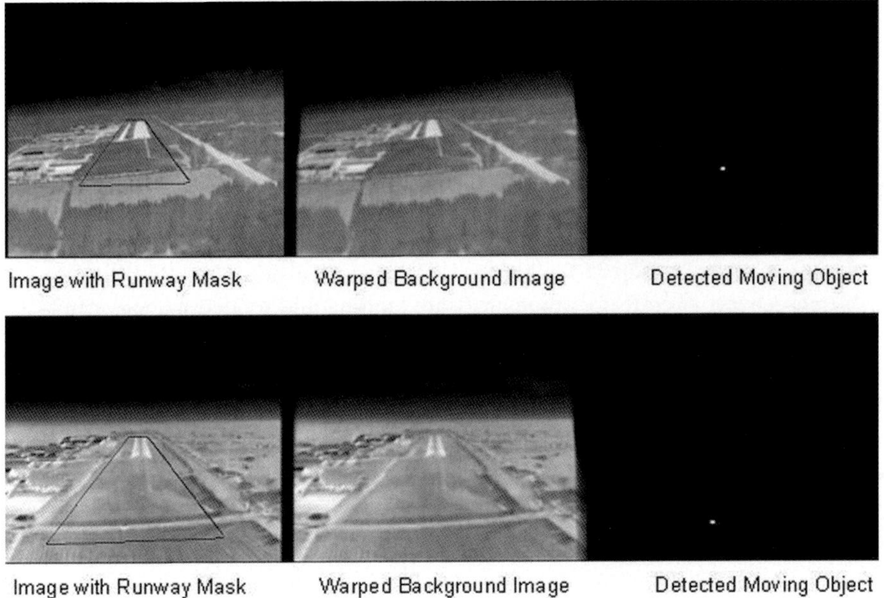

Fig. 11.12 Motion detection sample output

identify runway coordinates. All three techniques are comparable in performance based on our preliminary testing using NASA and Maxviz data sets when deployed under nominal conditions. The adaptive Hough method is more reliable when runway edges are eroded, and the template-matching method is computationally more effective than the other two techniques. In certain scenarios, extraneous objects such as roads beside the runway and similar runway look-alike structures are problematic for the corner-based segmentation method. We also noted that the Kalman correlation filter was the most effective filter for reducing jittering effects and dealt best with changing background dynamics and artifacts due to motion.

For obstacle avoidance, we deduced that a projective transformation model obtained using Nobel-Harris corner points stabilizes IR runway image sequences quite well. The results are comparable in performance and much less computationally intense than an SIFT feature-based approach [9].

Background learning using exponential smoothing of registered frames is quite effective for performing background subtraction-based motion detection. We also found that the runway mask region, which is updated in every frame, can be used to filter spurious motion blobs and restrict the motion detection to the runway area.

In the feature correspondence step, a fixed window centered on the given corner point is used to find the corresponding feature in the next frame using a normalized correlation coefficient as the similarity measure. In feature matching, we recommend adaptively changing the window size based on the size of image features and varying the search radius based on pixel displacement from frame to frame.

In the present implementation, no attempt was made to test the quality of stabilization, and since it directly affects motion detection, this quality aspect should be examined, and the stabilization module should be revisited, if required. Motion detection can be further improved by incorporating a tracker that handles both spatial and temporal consistency of the detected blobs.

Because of unavailability of data, the proposed EVS was tested on limited data sets. However, this work provides a good foundation to expand on these principles. The importance of autonomous landing and navigation has attracted much attention in recent years, and efforts to enhance navigation precision and reliability for the UAV and general aircraft operations for safe landing and related maneuverings are being promoted. The technical approaches introduced here provide a basis for expanding on these techniques, not only to process IR-based sensors, but also to be applied to different sensor modalities.

Acknowledgments We would like to acknowledge the valuable input and contributions that we have received in the obstacle detection effort from Prof. Gerard Medioni's team, Institute for Robotics and Intelligence Systems, USC. Prof. Medioni's team initiated the research effort related to detection of moving obstacles (discussed in Section 11.6). We would also like to thank Randy Bailey, Aviation Safety Program PM, for providing the NASA data sets, and Maxviz Inc. for providing additional data for proof-of-concept testing. Without their contributions, the demonstrations of these concepts would have been impossible.

Chapter's References

1. H.U. Doehler and B. Korn. *Autonomous Infrared Base Guidance System for Approach and Landing*. Technical report, Institute of Flight Guidance, German Aerospace Center, DLR, Braunschweig, Germany, 2002
2. X. Gong, L. Abbott, and G. Fleming. *A Survey of Techniques for Detection and Tracking of Airport Runways*, 44th AIAA Aerospace Sciences Meeting and Exhibit, Reno, Nevada, Jan. 9–12, 2006
3. R. Hamza and S. Martinez. Dynamic stabilization of target detection from moving vehicle. Patent application, To be published
4. R. Hamza, I. Mohammed, and D. Ramegowda. Runway positioning prior to landing using an onboard infrared camera. Tenth European Conference on Computer Vision, IEEE, Marseille France, Oct. 12–18, 2008
5. C.G. Harris and M.J. Stephens. Combined corner and edge detector. In *Proceedings of the Fourth Alvey Vision Conference*, Wiley InterScience, pages 147–151, 1988
6. Richard I. Hartley. In defense of the eight-point algorithm. *IEEE Transactions on PAMI*, 19(6):580–593
7. P.V.C. Hough. Method and means for recognizing complex patterns. Patent 1962:54-69
8. R. Kasturi, O. Camps, and S. Devadiga. Detection of obstacles on runway using ego-motion compensation and tracking of significant features. *Proceedings of the 3rd IEEE Workshop on Applications of Computer Vision (WACV '96)*, p. 168, 1996, ISBN:0-8186-7620-5
9. C-H. Pai, Y-P. Lin, G.G. Medioni, and R.R. Hamza. Moving Object Detection on a Runway Prior to Landing Using an Onboard Infrared Camera, *Computer Vision and Pattern Recognition, 2007*. CVPR apos;07. IEEE Conference on Volume, Issue, June 17–22, Page(s):1–8, DOI: 10.1109/CVPR.2007.383447

10. S. Sasa, H. Gomi, T. Ninomiya, T. Inagaki, and Y. Hamada. Position and attitude estimation using image processing of runway. In *Proceedings of the 38th AIAA Aerospace Science Meeting and Exhibit*, American Institute of Aeronautics and Astronautics, 2000
11. J. Shang and Z. Shi. Vision-based runway recognition for UAC autonomous landing. *International Journal of Computer Science and Network Security*, 7(3), 2007
12. N. Simond and P. Rives. Homography from a vanishing point in urban scenes. In *Proceedings. IEEE/RSJ International Conference on Intelligent Robots and Systems, (IROS 2003)*, 2003
13. B. Sridhar and B. Hussien. Passive range estimation for rotor-craft low altitude flight. *Machine Vision and Applications*, 6:11, 10–24, Springer, 1993
14. C. Stephan, G. Palubinskas, and R. Muller. Automatic extraction of runway structures in infrared remote sensing image sequences. In *Image and Signal Processing Remote Sensing XI, Proceedings of SPIE*, SPIE, volume 5982, 2005
15. Sanghoon Sull, Banavar Sridhar, "Runway Obstacle Detection by Controlled Spatiotemporal Image Flow Disparity," cvpr, pp. 385, 1996 IEEE Computer Society Conference on Computer Vision and Pattern Recognition (CVPR'96), 1996
16. N. Tarleton Jr., D. R. Wilkens, and P. F. Symosek. Method and apparatus for navigating an aircraft from an image of the runway. USPTO application 6,157,876
17. P. Tissainayagam and D. Suter. Assessing the performance of corner detectors for point feature tracking applications. *Image and Vision Computing*, IVC(22), No. 8, August 2004, pp. 663–679
18. R.C.H. Yalcin and M. Hebert. Background estimation under rapid gain change in thermal imagery. 2005
19. Q. Zheng and R. Chellappa. Motion detection in image sequences acquired from a moving platform. In *Proceedings of the International Conference on Acoustics, Speech, and Signal Processing*, IEEE, volume 5

Chapter 12
Moving Object Localization in Thermal Imagery by Forward-Backward Motion History Images

Zhaozheng Yin and Robert Collins

Abstract This chapter describes a moving object detection-and-localization method based on forward-backward motion history images (MHIs). Detecting moving objects automatically is a key component of an automatic visual surveillance and tracking system. In airborne thermal video especially, the moving objects may be small, color information is not available, and intensity appearance may be camouflaged. Although it is challenging for an appearance-or shape-based detector to detect the small objects in thermal images, pixel-level change detection or optical flow can provide powerful motion-based cues for detecting and localizing the objects. Previous motion detection approaches often use background subtraction, interframe difference, or three-frame difference, which are either costly or can only partially detect the object. We propose an MHI-based method that can accurately detect location and shape of moving objects for initializing a tracker or recovering from tracking failure. The effectiveness of this method is quantified using long and varied video sequences.

12.1 Introduction

Detecting moving objects in image sequences is a ubiquitous problem that plays an indispensable role in automatic surveillance and tracking. Detecting and localizing the object accurately is important for automatic tracker initialization and recovery from tracking failure. For tracker initialization, it is necessary first to localize position and shape of the object and analyze its features. Later, if the tracker fails, the moving object detection module can locate moving objects in the image, and the tracking system can associate these globally detected objects with previously tracked objects to restart the tracker.

When prior knowledge of moving object appearance and shape is not available, change detection or optical flow can still provide powerful motion-based cues for detecting and localizing objects, even when the objects move in a cluttered

R.I. Hammoud (ed.), *Augmented Vision Perception in Infrared: Algorithms and Applied Systems,* Advances in Pattern Recognition, DOI 10.1007/978-1-84800-277-7_12, 271

environment or are partially occluded. There are three main approaches to pixel-level change detection: background subtraction, interframe difference, and three-frame difference. However, these methods are either costly for moving cameras or can only partially detect the object. In this chapter, we introduce a moving object localization approach based on motion history images (MHIs) [15]. Without fine-tuning the temporal distance parameter to improve the frame difference performance, MHI integrates pixel-level change detection results over a short subsequence to find the location and shape of each moving object. Rather than referring to the edge information to get the object boundary, the forward and backward MHIs are combined to get the object boundary contour.

In Section 12.2, we review related work on pixel-level change detection methods. The details of our approach are discussed in Section 12.3. Section 12.4 presents the implementation results, which are evaluated on several video sequences. Finally, we make a brief conclusion in Section 12.5.

12.2 Related Work

Background subtraction compares the current frame with a background image to locate the moving foreground objects (Fig. 12.1). This method can extract the shape of the object well provided that the static background model is available and adapts to illumination change. Stauffer and Grimson [11] developed a probabilistic method for background subtraction. The background is adaptively updated by modeling each pixel as a Gaussian mixture model. However, in airborne video captured by a moving camera, it is costly to recover a stabilized background at every frame.

The interframe difference method computes the absolute difference between the current frame and the previous frame. If the video is captured by a moving camera, the previous frame is stabilized to the current frame coordinates before the differencing. This method easily detects motion but does a poor job of localizing the object, as shown in Fig. 12.2. If the temporal distance between two differencing frames is small, only part of the object is detected. If the temporal distance is large, two object locations are detected—one where the object is and one where it used to be.

Motivated by this problem, the three-frame difference approach uses future, current, and previous frames to localize the object in the current frame [9]. As shown

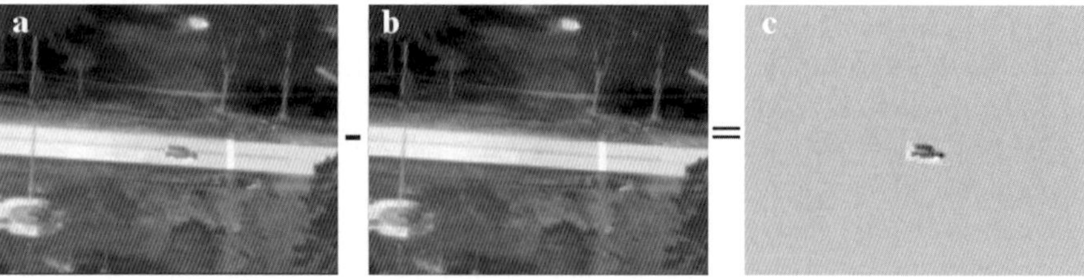

Fig. 12.1 a Current frame; **b** background; **c** background subtraction result

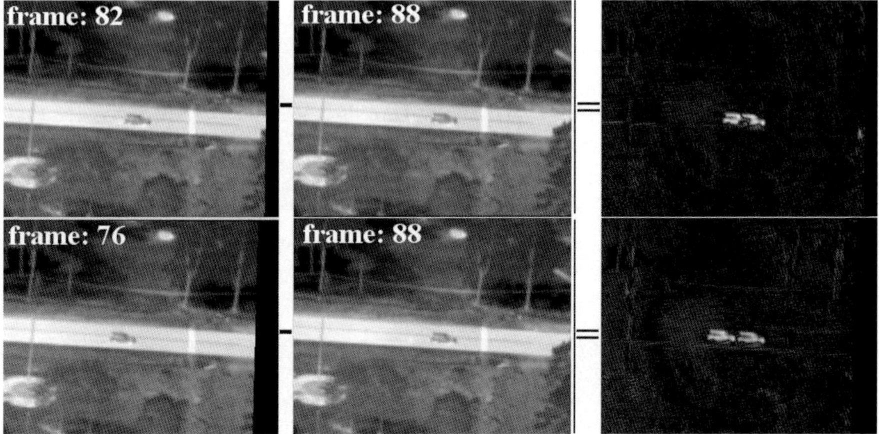

Fig. 12.2 *First row* Absolute difference between the stabilized previous frame ($I'(82)$) and the current frame ($I(88)$); *Second row* $|I'(76) - I(88)|$

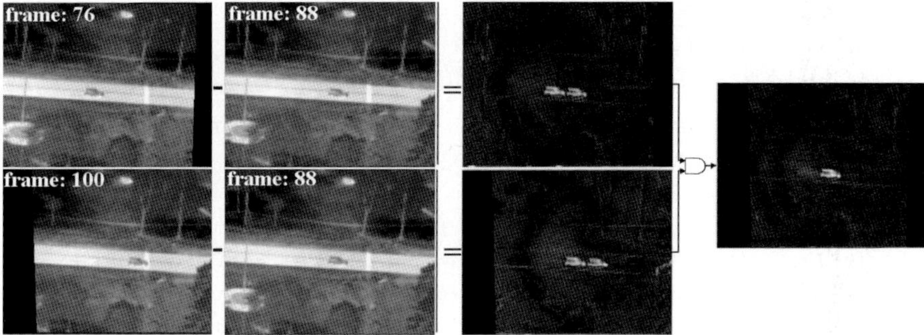

Fig. 12.3 Three-frame difference by combining $|I'(76) - I(88)|$ and $|I'(100) - I(88)|$

in Fig. 12.3, the moving object is well detected in the combined result of two inter-frame differences, provided that a suitable temporal distance is chosen (e.g., $\Delta = 12$ in Fig. 12.3). Irani and Anandan [8] provided a unified approach to moving object detection in both two-dimensional and three-dimensional (2D and 3D) scenes, in which the object is detected by three-frame difference. Using future frames intro-duces a lag in the tracking system, but this lag is acceptable if the object is far away from the camera or moves slowly relative to the high capture rate of the camera.

In the frame difference methods, the choice of temporal distance between frames is tricky. It depends on the size and speed of the moving object. Furthermore, background subtraction, interframe difference, and three-frame difference only tell us where the motion is. To segment the object from its surrounding background based on the detected motion, Strehl and Aggarwal [12] resorted to gray-level edges. Paragios and Deriche [10] presented a moving object detection-and-tracking approach using geodesic active contour and level sets. This boundary-based ap-proach applies an edge detector on the interframe difference to get the motion

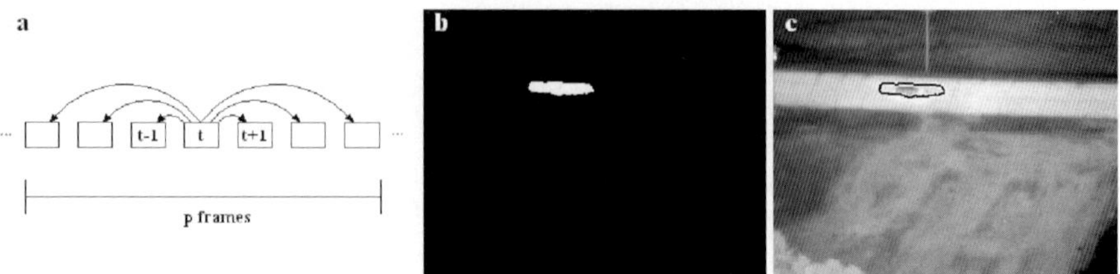

Fig. 12.4 a Each frame is analyzed with respect to its p neighbors. **b** A rough estimation of the motion region is generated by the p-frame difference. **c** The boundary of the motion mask provides a good initialization for further segmentation

Fig. 12.5 Motion history images generated from the aerobic data set. Courtesy of James W. Davis

detection boundary. Ali and Shah [1] used p-frame difference to detect object motion (Fig. 12.4). The differences between the current frame and its p neighbors are accumulated to generate a motion mask representing the region of motion. The boundary of the motion mask provides an initialization for a level set-based segmentation approach that evolves a contour that tightly detects moving objects.

In contrast to these methods, the MHI representation provides more motion properties, such as direction of motion. Bobick and Davis [2] used MHI as part of a temporal template to represent and recognize human movement (Fig. 12.5). MHI is computed as a scalar-valued image where intensity is a function of recency of motion. An extension to the original MHI framework is to compute normal optical flow (motion flow orthogonal to object boundaries) from MHI as presented by Bradski and Davis [3]. Wixson [13] presented another integration approach that integrates frame-by-frame optical flow over time. The consistency of direction is used as a filter. In the W4 system, Haritaoglu et al. [5] used a change history map to update the background model. Another related work was developed by Halevi and Weinshall [4] to track multibody nonrigid motion. Their algorithm is based on a disturbance map obtained by linearly subtracting the temporal average of the previous frames from the new frame.

We describe a moving object detection-and-localization approach based on MHI [15]. Similar to the work of Bobick and Davis [2], the motion images generated by interframe differencing are combined with a linear decay term. From previous frames to the current frame, we get the forward MHI. The trail gradient in the MHI

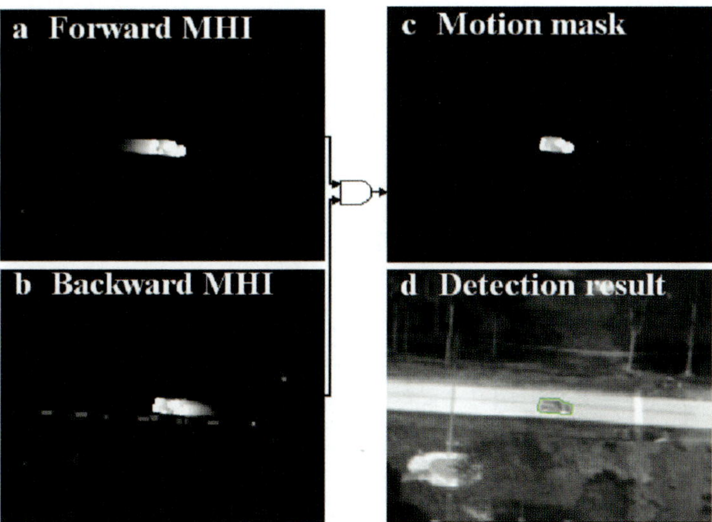

Fig. 12.6 Forward and backward MHIs combined to generate a motion mask, moving objects in the current frame localized by computing the contour of the motion mask

indicates the direction of object motion in the image. Similarly, we construct a backward MHI from the future frames to the current frame. Again, we assume the lag introduced into the tracking system is acceptable. Combining the two MHIs, we obtain the current object mask and shape. The idea is illustrated in Fig. 12.6. Compared to the previous approaches, this method does not require adaptive background reconstruction, it provides more motion information than the three-frame difference method, and it can recover the shape of the moving object without further segmentation. Our approach is also suitable for moving cameras because we do stabilization of adjacent frames in time and propagate motion information locally. This reduces the possibility of errors caused by inaccurate stabilization across large temporal distances.

12.3 Moving Object Localization by MHIs

The MHIs combine object movement information over an image subsequence. Recent object motion, as determined from frame differencing, is entered into the MHI after first reducing the previous MHI by a decay term. Motion information from multiple time periods is then represented in the same image; however, old object motion obtained from frame differences among images distant in time fades away due to the decay term. In general, MHI represents cumulative object motion as bright patches of recent motion together with a gradually fading gradient trail. Our MHI-based object detection approach is briefly described in Fig. 12.7 with three main modules:

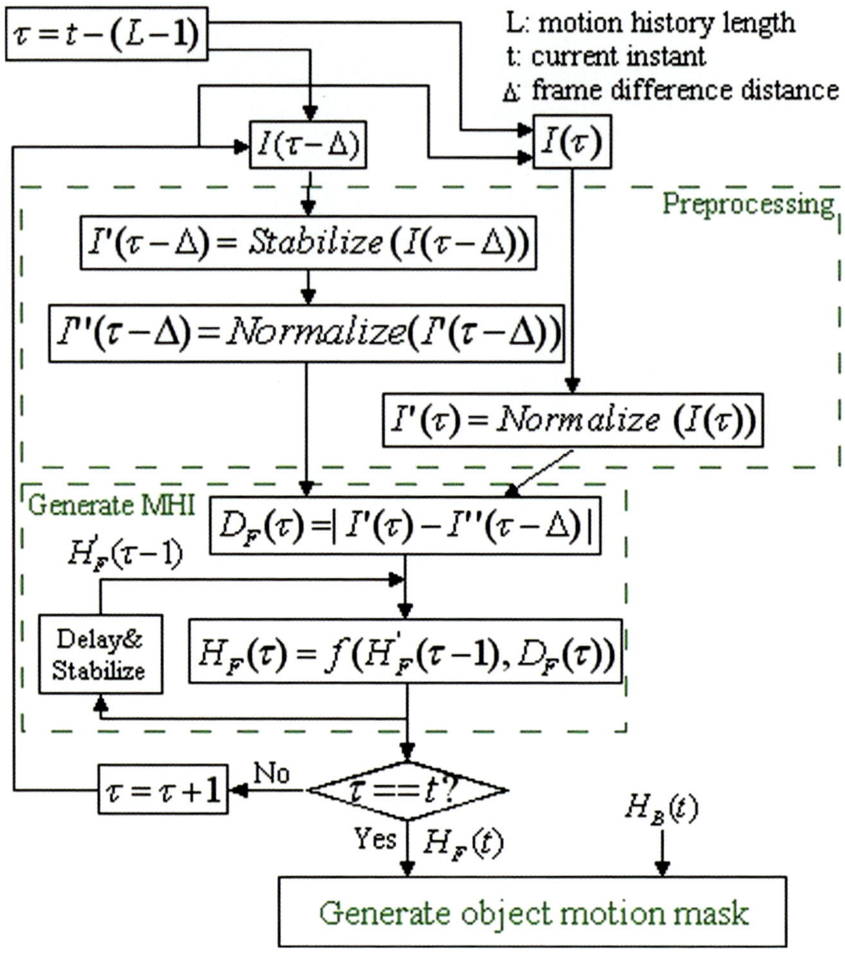

Fig. 12.7 Moving object localization system

1. *Preprocessing module.* The previous frame at time instant $\tau - \Delta$, $I(\tau - \Delta)$, is stabilized into the coordinate system of the frame at time τ, $I(\tau)$. Both of these frames are intensity normalized prior to frame differencing.
2. *MHI generation module.* The forward MHI at time τ, $H_F(\tau)$, is a function of the stabilized MHI at time $\tau - 1$, $H'_F(\tau - 1)$, and the motion image computed by frame difference at time τ, $D_F(\tau)$.
3. *Object localization module.* The forward MHI at the current instant t, $H_F(t)$, is computed recursively from previous time instants $t - (L-1)$ to t. The backward MHI, $H_B(t)$, has the same cumulative process except that τ is reduced recursively from $t + (L-1)$ to t. The forward MHI $H_F(t)$ is combined with the backward MHI $H_B(t)$ to determine the moving object mask in the current image $I(t)$.

$H_F(t)$ accumulates the motion information of L frames from time instant $t-(L-1)$ to t. When this detection system is used for tracking, subsequent forward MHIs can be computed recursively by

$$H_F(k+1) = f(H_F'(k), D_F(k+1)) \qquad (12.1)$$

where $k = t, t+1, \cdots$, and f is a function described in Section 12.3.2. $H_F(k+1)$ is determined only by motion image $D_F(k+1)$ and the single previous forward MHI $H_F'(k)$. It does not involve the whole accumulating process of L frames, hence is efficient to use in moving object tracking.

12.3.1 Preprocessing

In airborne video, the background is moving over time due to the moving camera. Before using the frame difference to get motion images, we need to stabilize the frames that is, compensate for the camera motion. If the camera is static, this step can be skipped. Two-frame background motion estimation is achieved by fitting a global parametric motion model (affine or projective) to sparse optic flow. Sparse flow is computed by matching Harris corners between frames using normalized cross correlation. Given a set of potential corner correspondences across two frames, we use a random sample consensus (RANSAC) procedure to robustly estimate global affine flow from observed displacement vectors. The largest set of inliers returned from the RANSAC procedure is then used to fit either a 6-parameter affine or 8-parameter planar projective transformation. Based on this method, subsequent frames are aligned to one another. Using $P_{\tau-\Delta}^{\tau}$ to represent the affine motion from frame $\tau - \Delta$ to frame τ, we perform the warping as

$$I'(\tau - \Delta) = P_{\tau-\Delta}^{\tau} \times I(\tau - \Delta) \qquad (12.2)$$

The transformation matrix $P_{\tau-\Delta}^{\tau}$ is cascaded as Eq. (12.3). In practice, we do not choose a large Δ since that will cause big cumulative error.

$$P_{\tau-\Delta}^{\tau} = P_{\tau-1}^{\tau} \times P_{\tau-2}^{\tau-1} \times \cdots \times P_{\tau-\Delta}^{\tau-\Delta+1} \qquad (12.3)$$

Figure 12.8 shows a by-product of the stabilization process where a mosaic image is generated by warping a 120-frame sequence to the coordinate of the first frame. Although the camera motion was large and erratic and there is no overlap between the first and last frames, all these frames can still be stabilized into the coordinate of the first frame successfully by chaining [Eq. (12.3)].

Another notorious problem in airborne video is rapid change in pixel intensities when the camera sensor has automatic gain control. Especially in thermal videos, when very hot or cold objects appear, the gray value of each pixel changes greatly as the camera rapidly adjusts its gain to avoid saturation. The changing illumination

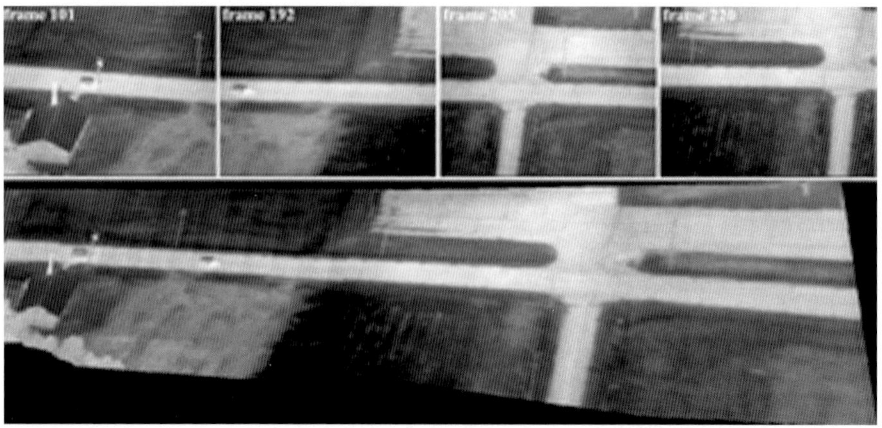

Fig. 12.8 *Top* Several sample images within a video sequence of 120 frames. The camera shook greatly between frame 193 and frame 220, and the objects were out of the view starting from frame 193. *Bottom* Mosaic result by stabilizing all the frames into the first frame coordinate system

makes the intensity-based frame difference method inadequate for obtaining accurate motion. Yalcin et al. [14] proposed an intensity-clipped affine model of camera sensor gain. Here, we use a simplified intensity normalization method:

$$I'(\tau) = \frac{I(\tau) - \overline{I(\tau)}}{std(I(\tau))} \tag{12.4}$$

where $\overline{I(\tau)}$ represents the mean value of the image, and $std(I(\tau))$ stands for the standard deviation of the image.

After the normalization step, pixel values can be negative. This will not affect the frame difference result in the next module, and we do not need to scale the pixel value into the range of 0 to 255. In the thermal video of Fig. 12.9, there is a large gain change between the 1,032nd frame and the following frames. The motion image generated by frame difference will be polluted if there is no normalization (Fig. 12.9a). Thus, the MHI at time τ will also be degraded (Fig. 12.9b). Figure 12.9c–d gives the motion image and MHI with the normalization computed from Eq. (12.4) for comparison. Figure 12.9e–f shows the final moving object localization results with the normalization.

12.3.2 Motion History Image Generation

A single motion image computed by interframe difference shows where motion (change) exists, but noise may also be above threshold. Furthermore, it is hard to choose a suitable frame difference distance Δ due to different object sizes and

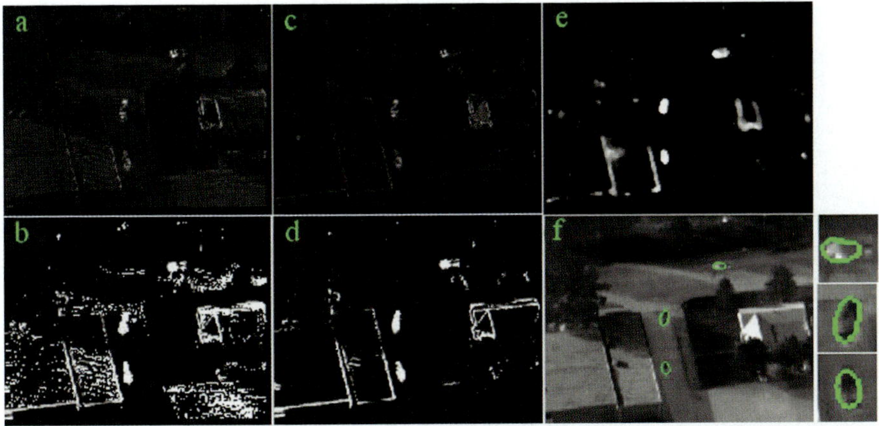

Fig. 12.9 $t = 1030, \tau = 1032, \Delta = 3$ **a** Frame difference between the τ and $\tau + \Delta$ frames without normalization, **b** $H_B(\tau)$ without normalization, **c** frame difference with normalization, **d** $H_B(\tau)$ with normalization, **e** combining $H_F(t)$ with $H_B(t)$, **f** detected object contours at current instant t

moving speed. One method of integrating motion images over time is the motion energy image (MEI), computed as[*]

$$E(t) = \sum_{t}^{t \pm (L-1)} D(\tau) \tag{12.5}$$

where $-$ means forward MEI, $+$ means backward MEI, L is the length of the time period, and $D(\tau)$ is the absolute frame difference with time distance Δ:

$$D(\tau) = |I'(\tau) - I'(\tau \pm \Delta)| \tag{12.6}$$

$I'(\tau)$ and $I'(\tau \pm \Delta)$ are stabilized and normalized images, respectively.

One drawback of MEI is that all the motion caused by noise will also be accumulated. As shown in Fig. 12.10, the MEI is blurred due to the summation of noisy motion within the time period. Thus, it is hard to distinguish the objects from the background. Instead of only showing all the existing motion during a period of time, MHI keeps a record of how the historic motion evolves with the current motion image. By incorporating a temporal decay term, the forward MHI is computed as[†]

$$H_F(x,y,\tau) = \begin{cases} max(0, P_{\tau-1}^{\tau} H_F(x,y,\tau-1) - d) & \text{if } D(x,y,\tau) < T \\ 255 & \text{if } D(x,y,\tau) \geq T \end{cases} \tag{12.7}$$

---

[*] Originally Bobick and Davis [2] used the logical 'AND' of the binary difference images to compute $E(t)$.

[†] As shorthand notation, we ignore the x, y in the parameter list and represent them as $H_F(\tau)$ and $H_B(\tau)$.

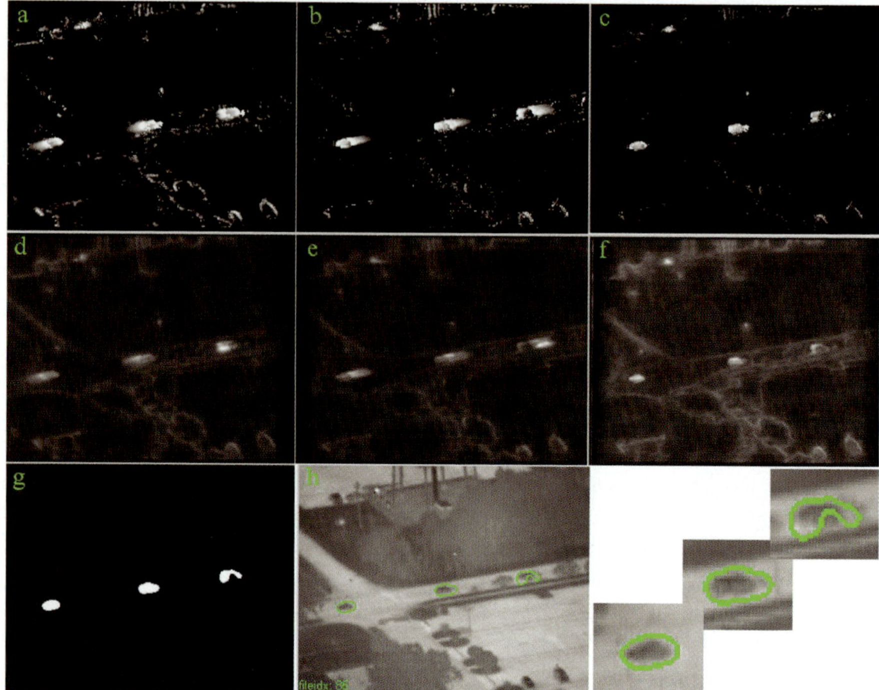

Fig. 12.10 a forward MHI, **b** backward MHI, **c** combination of **a** and **b**, **d** forward MEI, **e** backward MEI, **f** combination of **d** and **e**, **g** postprocessed **c**, **h** detected object contours. Upper right vehicle in the scene partially occluded by a tree

where $P^\tau_{\tau-1}$ is the warping matrix from frame $\tau - 1$ to frame τ, d is the decay term, and T is a threshold. The pixel value calculated is within $[0, 255]$, so the decay term d is also defined within $[0, 255]$. For example, we can define $d = 255/L$. Without loss of generality, we can also scale the pixel value into other ranges, like $[0, 1]$.

The forward MHI, $H_F(t)$, is a function of the previous forward MHI, $H_F(t-1)$ and current motion image $D_F(t)$. This satisfies the Markovian assumption that no other old motion images need to be stored. Compared to MEI, the most recent moving pixels in MHI are highlighted while the old moving pixels in MHI are darker. As a benefit, the impulse noise in the old motion image decays away while the persistent motion generated by the moving object is preserved. Similarly, we can compute the backward MHI $H_B(\tau)$. Figure 12.11 gives an example of the MHI generation process. The initial backward and forward MHIs are set to zero:

$$H_F(t - (L-1)) = 0$$

$$H_B(t + (L-1)) = 0$$

(12.8)

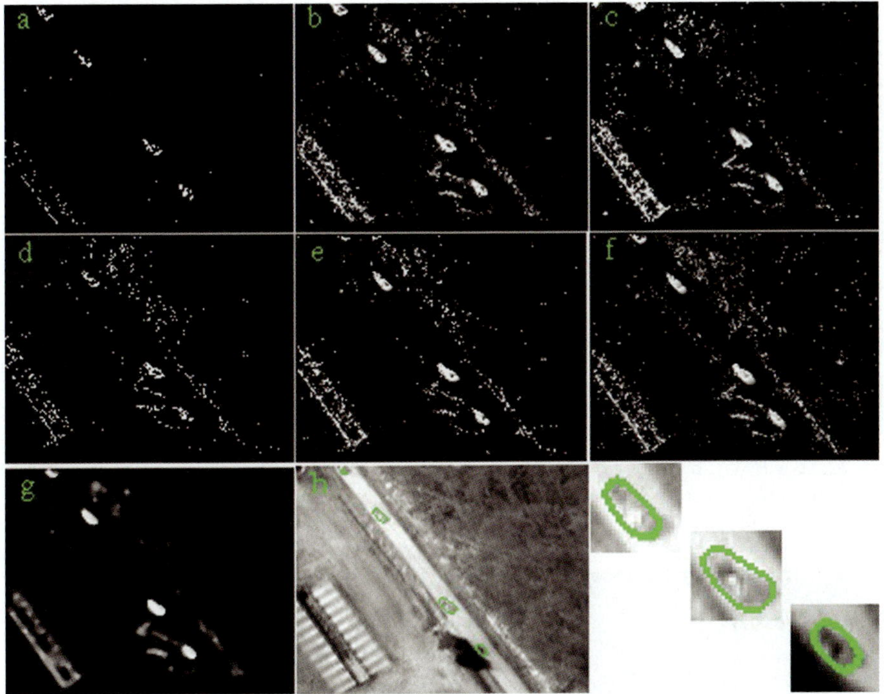

Fig. 12.11 $L = 11$ **a–c** Forward MHIs at $t - 10, t - 5, t$, respectively; **d–f** backward MHIs at $t + 10$, $t + 5, t$, respectively; **g** combination of $H_F(t)$ and $H_B(t)$; **h** detected object contours

12.3.3 Object Localization

After we get the forward and backward MHI, $H_F(t)$ and $H_B(t)$, we perform median filtering to smooth the MHIs and remove the salt-pepper noise. Alternatively, a Gaussian filter can be used. The forward-backward motion history masks are combined by

$$\text{Mask}(t) = min(med\,filt(H_F(t)), med\,filt(H_B(t))) \tag{12.9}$$

where *med filt* stands for the median smoothing filter. For objects moving in a constant direction, the "min" operator in Eq. (12.9) serves to suppress the gradient trail behind the object in the forward MHI and the gradient trail ahead of the object in the backward MHI, yielding strong response only for pixels within the current object boundary.

Figure 12.12 provides an example where two lanes of traffic move in opposite directions, while four people are seen running together in the upper left corner. The median filter smoothes the MHIs and removes the isolated salt-pepper noise. The mask generated by Eq. (12.9) is shown in Fig. 12.12e. After thresholding the mask, the final object contours are shown in Fig. 12.12f. Further morphological operations may be performed to improve the accuracy of the object mask, as shown

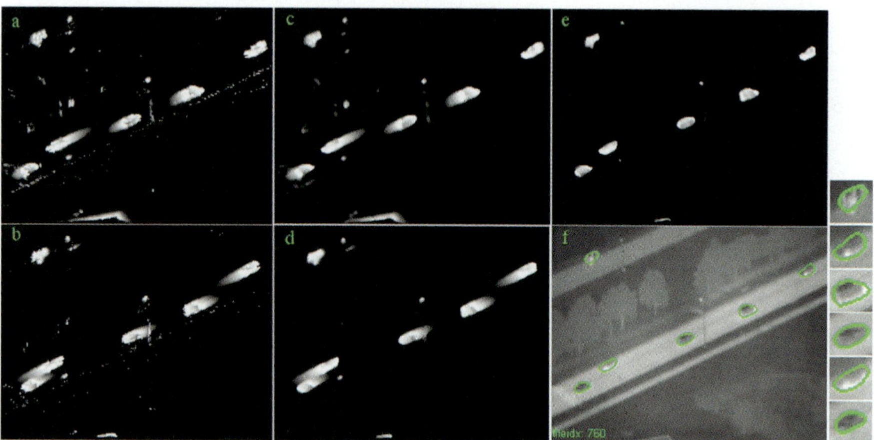

Fig. 12.12 **a** $H_F(t)$, **b** $H_B(t)$, **c** $H_F(t)$ after median filter, **d** $H_B(t)$ after median filter, **e** combined mask, **f** detected object contours. Upper left contour in the scene composed by four running people

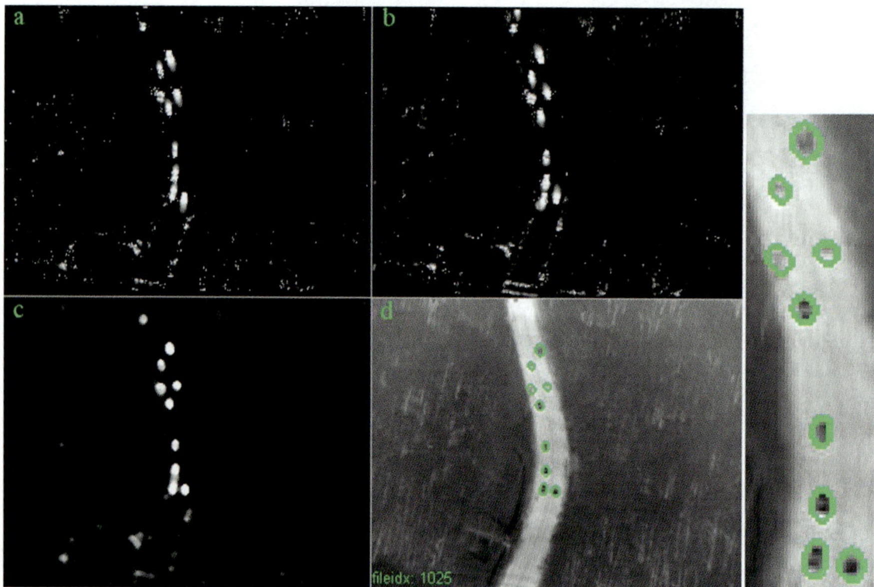

Fig. 12.13 **a** $H_F(t)$, **b** $H_B(t)$, **c** combination without morphological operation, **d** detected object contours with morphological operations

in Fig. 12.13. For example, the close or dilate operations can fill any holes or gaps within the same object, while the open or erosion operations can remove thin bridges of pixels between nearby objects as well as remove small objects caused by noise.

12.4 Experiment Analysis

12.4.1 Evaluation Metrics

Our experiment evaluation design is shown in Fig. 12.14. The ground truth object shape is labeled manually for a set of images and compared to computer-generated shape masks. Let H denote the hit area, that is, the area belonging to the object and correctly detected; M denote the miss area, that is, the area belonging to the object but incorrectly missed; and F denote the false alarm area, that is, the area not belonging to the object but incorrectly detected. The hit rate is defined as

$$\mathrm{HR} = \frac{H}{H + M} \tag{12.10}$$

The false alarm rate is

$$\mathrm{FAR} = \frac{F}{H + F} \tag{12.11}$$

Note that the miss rate is redundant with the hit rate:

$$\mathrm{MR} = \frac{M}{H + M} = 1 - \mathrm{HR} \tag{12.12}$$

so we will evaluate the detection performance based on the hit rate and false alarm rate only. A perfect detection result would have hit rate equal to one and false alarm rate equal to zero.

12.4.2 Effect of L and Δ

To achieve a good detection performance, we need to choose a suitable motion history length L. If the object has uniform intensity, s is the average moving object

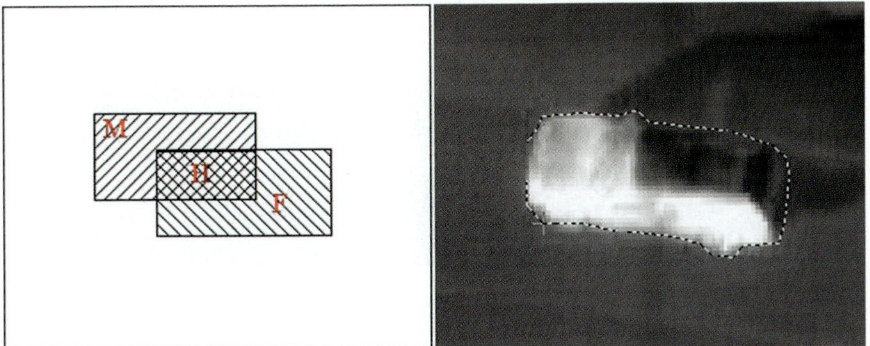

Fig. 12.14 Evaluation process and hand-labeled ground truth

speed (pixels/second) during the period L, f is the frame rate (frame/second), and l is the object length in the image (pixels), then the smallest motion history length L is constrained by

$$s\frac{L}{f} \geq l \tag{12.13}$$

Assuming that the speed and size of the moving object in the scene can be estimated when we set up the tracking system, we can calculate the minimum motion history length as before. Otherwise, we can choose a big L conservatively to guarantee that the object shape can be detected well, although big L will lengthen the lag of the tracking system. Figure 12.15 shows an example in which the hit rate increases as L increases while the false alarm rate decreases.

Another factor that affects the frame difference is the step size Δ. Normally, we choose Δ between one and four and avoid choosing larger values since that will cause more interframe stabilization error and make the MHI noisy. If the object is moving slowly and Δ is small, then only a sliver of the object can be detected at each frame; however, all the slivers will be accumulated into the final MHI with a suitable motion history length.

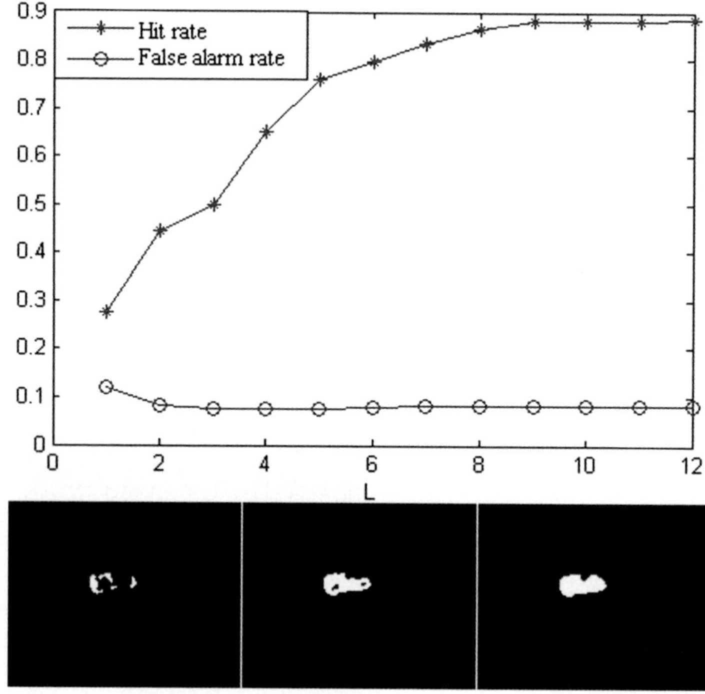

Fig. 12.15 Effect of different motion history length. *Bottom* Three motion masks related to $L = 1, 5, 9$, respectively, left to right

12.4.3 Experiment Result

Figure 12.16 shows the performance evaluation for four different thermal sequences. We randomly select 20 images from each sequence and label ground truth object shapes by hand. Note that only moving objects are labeled, and the shadow is not considered to be part of the object. If all the objects in the image are static or all the objects are totally occluded, this image is replaced by another randomly chosen image. The four sequences contain trucks and small sedans driving along a road network. The image resolution of objects in the same video sequence may be large or small due to the moving and zooming camera. Different images from the same sequence may contain single or multiple objects. Some objects may be partly occluded, and some images are blurred. Despite these challenges, among all the sampled images, the hit rate is around or above 0.8, and the false alarm rate is around or below 0.4. The detected object shape tends to match the object boundary well. Some exception cases include (1) the object slows when going around a corner so that the motion is not obvious (the 1,162nd frame of sequence 1, the 3,037th and 4,013th frames of sequence 2), which degrades the performance; (2) part of the object is just coming into the image, which cannot be detected well (the 7,537th frame of sequence 3 and the 79th frame of sequence 4); (3) the object has uniform appearance that is similar to the background. For example, the trunk of the truck in sequence 4 is dark and similar to the pavement; it is not segmented from the background perfectly.

Our approach is resistant to some uneventful background motion (Fig. 12.17). This property is desirable since distracting motion is rejected and directionally consistent motion resulting from a typical surveillance target is detected [11]. We also tested our approach on nonrigid moving objects in thermal imagery (e.g., walking people. One drawback is that it cannot handle occluded motion well. For example, the legs of a walking person are self-occluding (Fig. 12.18). More moving pedestrian detection-and-localization results are shown in Fig. 12.19 and the videos at `http://www.cse.psu.edu/~zyin/MHI.htm`. The first four rows of Fig. 12.19 show detection results on thermal videos captured at the Ohio State University campus [6]. The last two rows of Fig. 12.19 show detection results with different object sizes in outdoor motion and tracking scenarios [7]. Our approach can still detect the moving objects when they are partially occluded by trees.

To demonstrate that this approach generalizes to other kinds of scenes, we also tested it on several challenging nonthermal sequences. 40 images were chosen randomly from each sequence, and Fig. 12.20 and Fig. 12.21 provide localization results for each sequence. The upper half of Fig. 12.20 shows an airfield video with flat background. The vehicles turn around or pass by each other. The lower half of Fig. 12.20 shows two vehicles in a forest. The vehicles pass through tree shadow and become partially occluded. Figure 12.21 shows an intersection with static background in which multiple vehicles exist (Karlsruhe University) http://i21www.ira.uka.de/image.sequeces/. From the experimental results under three differernt weather conditions (normal, snow, and fog) at the same intersection, we can see that moving objects can be localized well except when the objects move close to each other in the intersection (frame 1,190 in the second row) or when

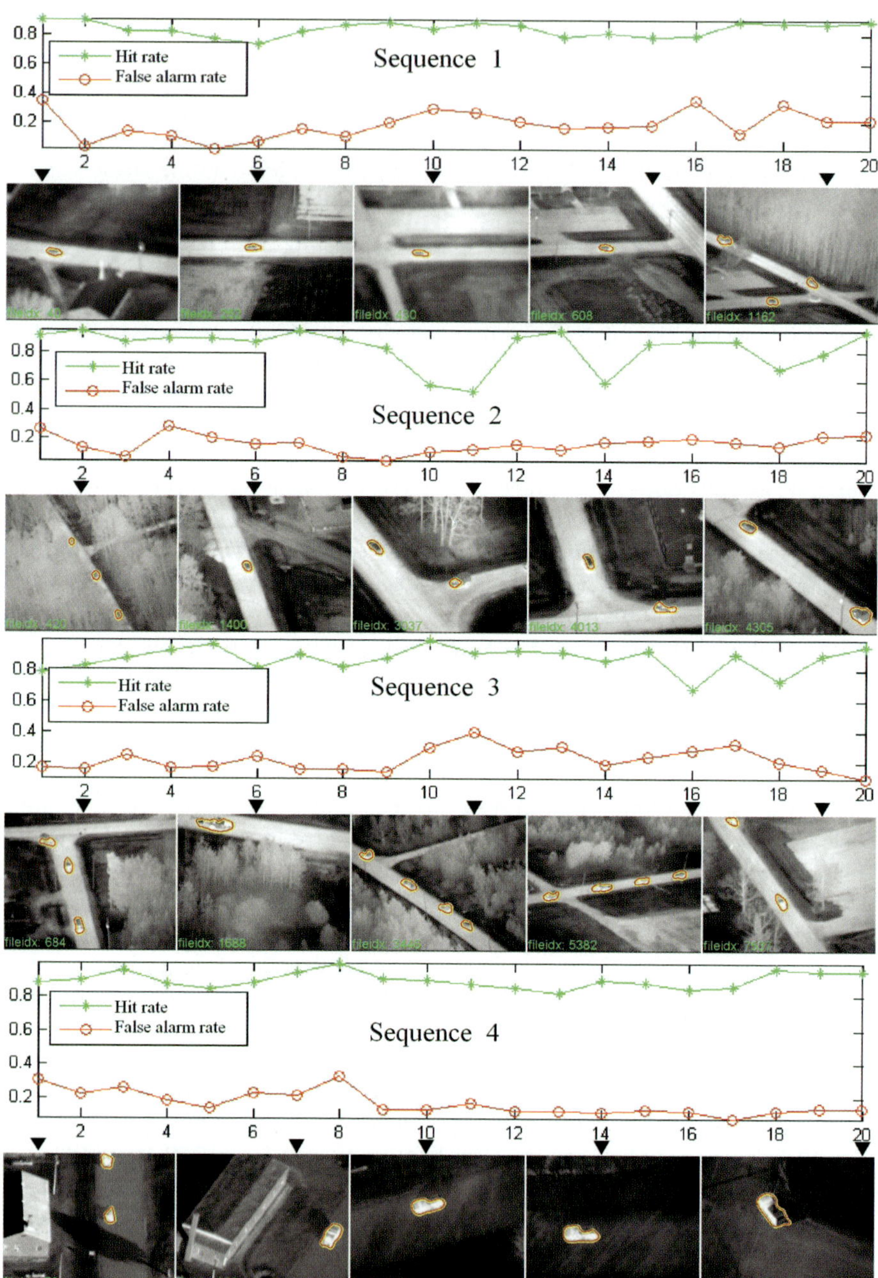

Fig. 12.16 Evaluation of the thermal videos

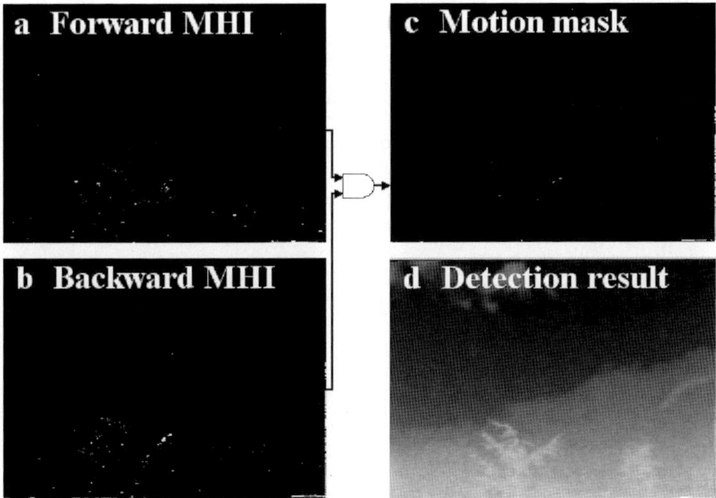

Fig. 12.17 Moving object localization in uneventful background motion scenarios. Clouds and tree branches are moving due to strong winds [7]. No salient motion detected in **d**

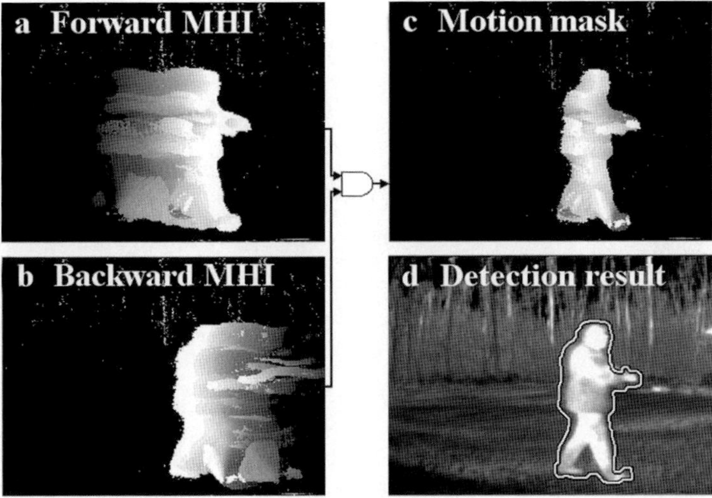

Fig. 12.18 Moving object localization by combining the forward and backward MHIs of one subject walking from left to right side of the Field-Of-View (FOV) holding an AK-47 rifle [7]

an object's intensity is very similar to the background (frame 1,250 in the second row). In addition to moving vehicles, nonrigid objects such as human bodies were also tested. As shown in Fig. 12.22, this approach can also localize some simple human body motions.

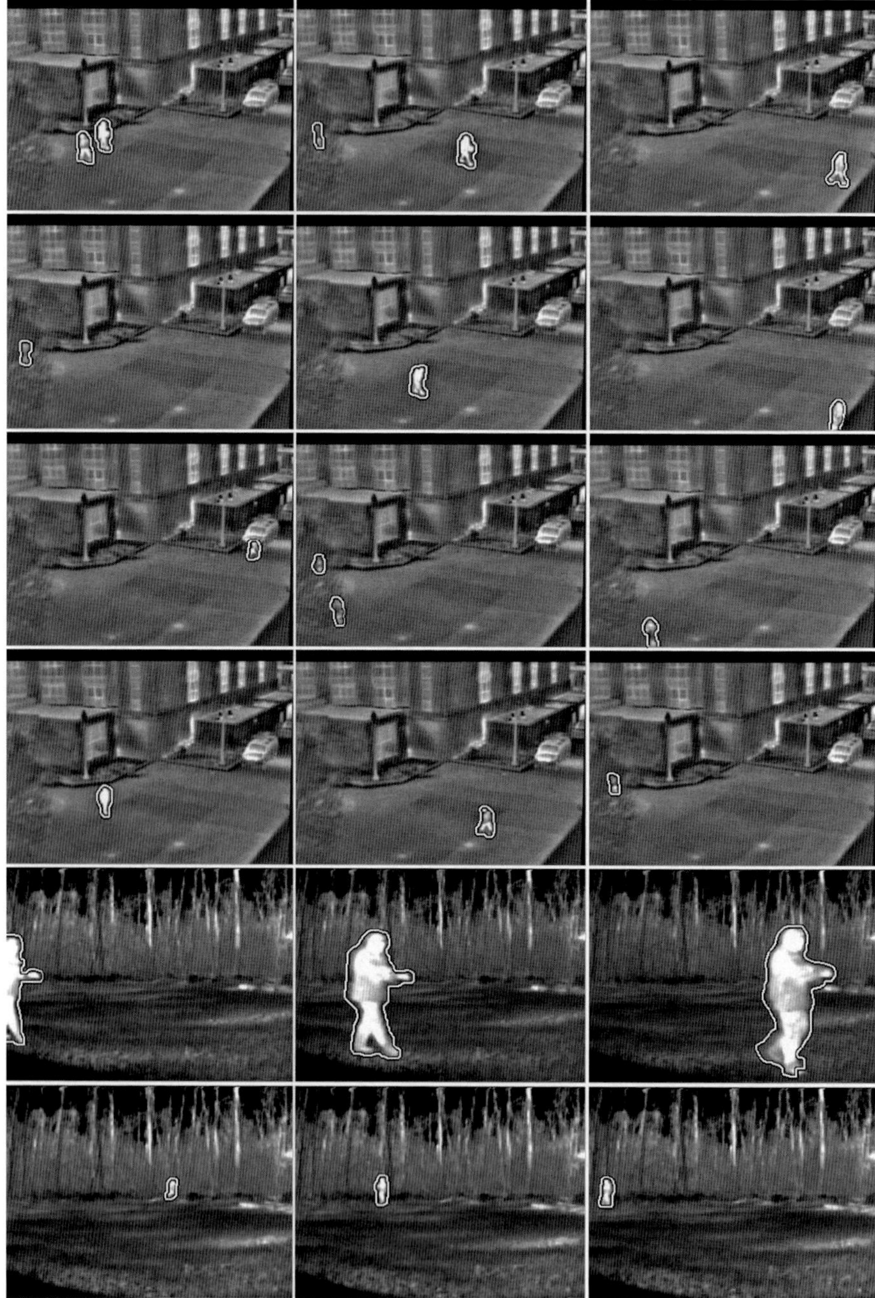

Fig. 12.19 Test of OTCVBS benchmark data set [6, 7]

Fig. 12.20 Test of airfield and forest scenes

12.5 Conclusion

The MHIs accumulate change detection results with a decay term over a short period of time. The MHI contains more motion information than a single motion image generated by frame difference. Instead of only showing where the motion is, MHI answers the questions of "what went where?"and "how did it go there?" Each moving object has a fading trail, with the trail showing the direction of movement. By combining the forward MHI and backward MHI, we can get a contour shape for the moving object at the current frame. The experiments show the effectiveness of the approach. In addition, the method is much faster to run and to implement than multiscale optical flow. There are only several subtraction and comparison operations for each pixel at each iteration. By comparison, for the optical flow method, if each pixel has a $k*k$ window, the computation cost is roughly increased by a factor of k^2.

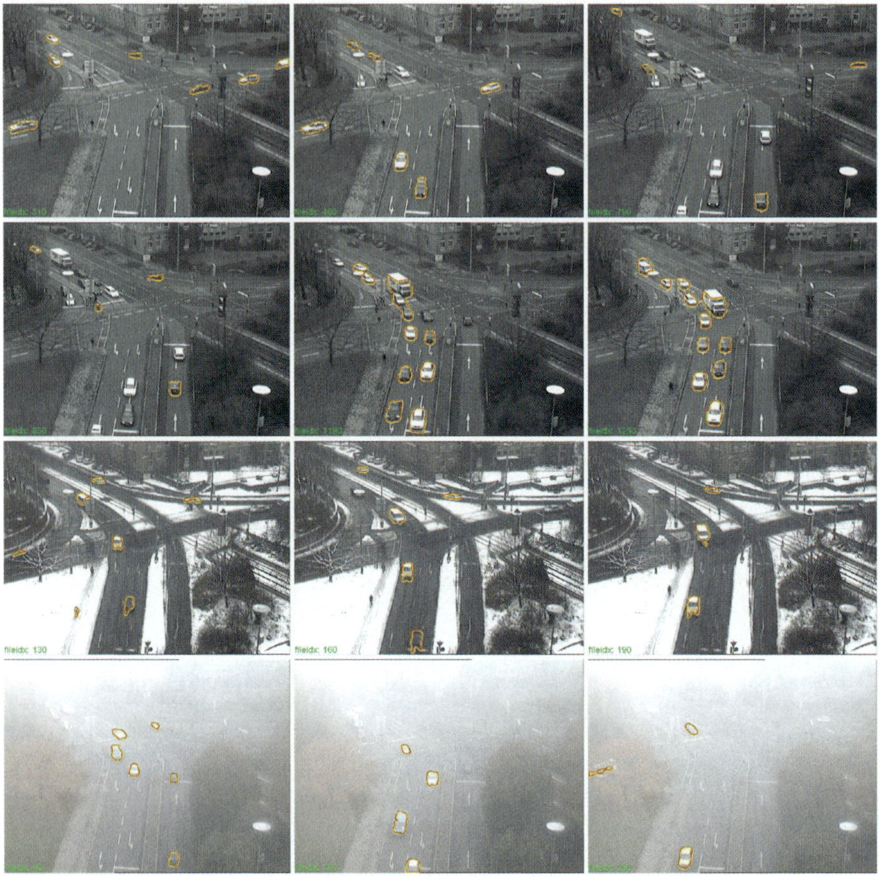

Fig. 12.21 Test of intersection scenes under different weather conditions

Fig. 12.22 Test of human body motion

Future work will implement this localization approach within a complete tracking system. The motion, shape, and appearance features of detected objects will be combined to represent and track the object. Furthermore, based on the detected object location and initial shape estimation, more accurate local segmentation methods can be performed around the object to get better layer representations of the object and background.

Acknowledgments This work was funded under the NSF Computer Vision program via grant IIS-0535324 on persistent tracking.

Chapter's References

1. S. Ali and M. Shah, COCOA—Tracking in Aerial Imagery, demo at ICCV 2005, Beijing China, October 15–21
2. A. Bobick and J. Davis, The Recognition of Human Movement Using Temporal Templates, IEEE Transactions on Pattern Analysis and Machine Intelligence, 23(3):257–267, March 2001
3. G. Bradski and J. Davis. Motion segmentation and pose recognition with motion history gradients, Fifth IEEE Workshop on Application of Computer Vision, 238–244, December 2000
4. G. Halevi and D. Weinshall. Motion of disturbances: Detection and tracking of multi-body non-rigid motion, IEEE Conference on Computer Vision and Pattern Recognition, Puerto Rico, 1997, pp. 897–902
5. I. Haritaoglu, D. Harwood, and L. Davis. W4: Real-time surveillance of people and their activities, IEEE Transactions on Pattern Analysis and Machine Intelligence, 22(8):809–830, August 2000
6. IEEE OTCBVS WS Series Bench; J. Davis and V. Sharma, Fusion-based background-subtraction using contour saliency, in Proceedings of IEEE International Workshop on Object Tracking and Classification Beyond the Visible Spectrum, June 2005
7. IEEE OTCBVS WS Series Bench; R. Miezianko, Terravic Research Infrared Database. http://www.cse.ohio-state.edu/otcbvs-bench/
8. M. Irani and P. Anandan. A unified approach to moving object detection in 2D and 3D scenes, IEEE Transactions on Pattern Analysis and Machine Intelligence, 20(6):577–589, June 1998
9. R. Kumar, H. Sawhney, et.al, Aerial video surveillance and exploitation, Proceedings of the IEEE, 89(10):1518–1539, October 2001
10. N. Paragios and R. Deriche. Geodesic active contours and level sets for the detection and tracking of moving objects, IEEE Transactions on Pattern Analysis and Machine Intelligence, 22(3):266–280, March 2000
11. C. Stauffer and W. Grimson. Learning patterns of activity using real-time tracking, IEEE Transactions on Pattern Analysis and Machine Intelligence, 22(8):747–757, August 2000
12. A. Strehl and J. Aggarwal. Detecting moving objects in airborne forward looking infrared sequences, Proceedings of the IEEE Workshop on Computer Vision Beyond the Visible Spectrum: Methods and Applications IEEE Computer Society, Washington, DC, USA, 1999, ISBN:0-7695-0050-1
13. L. Wixson. Detecting salient motion by accumulating directionally-consistent flow, IEEE Transactions on Pattern Analysis and Machine Intelligence, 22(8):774–780, August 2000
14. H. Yalcin, R. Collins, and M. Hebert. Background estimation under rapid gain change in thermal imagery, Second IEEE Workshop on Object Tracking and Classification in and Beyond the Visible Spectrum, June 20–26, 2005
15. Z. Yin and R. Collins. Moving object localization in thermal imagery by forward-backward MHI, Third IEEE Workshop on Object Tracking and Classification in and Beyond the Visible Spectrum, New York City, June 2006

Part VI
Multimodal Imagery Fusion for Human Localization and Tracking

Chapter 13
Feature-Level Fusion for Object Segmentation Using Mutual Information

Vinay Sharma and James W. Davis

Abstract A new feature-level image fusion technique for object segmentation is presented. The proposed technique approaches fusion as a feature selection problem, utilizing a selection criterion based on mutual information. Starting with object regions roughly detected from one sensor, the proposed technique aims to extract relevant information from another sensor to best complete the object segmentation. First, a contour-based feature representation is presented that implicitly captures object shape. The notion of relevance across sensor modalities is then defined using mutual information computed based on the affinity between contour features. Finally, a heuristic selection scheme is proposed to identify the set of contour features having the highest mutual information with the input object regions. The approach works directly from the input image pair without relying on a training phase. The proposed algorithm is evaluated using a typical surveillance setting. Quantitative results and comparative analysis with other potential fusion methods are presented.

13.1 Introduction

Vision applications, such as video surveillance and automatic target recognition, are increasingly making use of imaging sensors of different modality. The expectation is that a set of such sensors would benefit the system in two ways: First the complementary nature of the sensors will result in increased capability, and second the redundancy among the sensors will improve robustness. The challenge in image fusion is thus combining information from the images produced by the constituent sensors to maximize the performance benefits over using either sensor individually.

To better quantize the performance benefits and to enable the use of fusion algorithms in automatic vision systems, we adopt in our work a more "goal-oriented" view of image fusion than is traditionally used. Instead of merely improving the context (or information) present in a scene, we focus on the specific task of using image fusion to improve the estimation of the object shape (as defined by a silhouette or a boundary).

R.I. Hammoud (ed.), *Augmented Vision Perception in Infrared: Algorithms and Applied Systems,* Advances in Pattern Recognition, DOI 10.1007/978-1-84800-277-7_13,

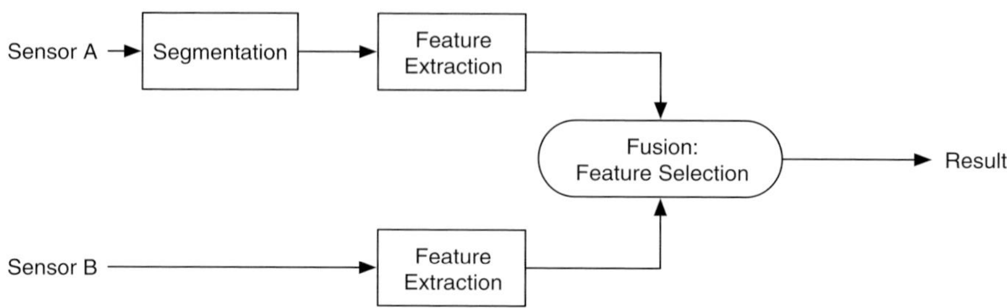

Fig. 13.1 Flowchart of proposed fusion method

We propose a novel solution to this task, one that approaches fusion as essentially a feature selection problem. Our approach is applicable to any combination of imaging sensors, provided that the sensors are colocated and registered. The processing pipeline and the main computation stages of our algorithm are shown in the flowchart of Fig. 13.1. As can be seen from the flowchart, the feature extraction stage of our algorithm is preceded by an object segmentation routine employed in only one of the input sensors (denoted by sensor A). The object segmentation routine is used only to bootstrap the feature selection process, and hence any method that provides even a rough, incomplete object segmentation can be employed at this stage. Contour-based features are then extracted from the rough object segmentation results of sensor A and from the corresponding image region of sensor B. These features are then used within a mutual information framework to extract a subset of features from sensor B that are most *relevant* (complementary and redundant) to the features obtained from the initial segmentation of sensor A.

To estimate the probability distribution required for computing the mutual information across sets of features, we present a method that relies on the regularities in shape and form found in most objects of interest. We extend the notion of affinity, originally defined to measure the smoothness of the curve joining two edge elements [37], to our contour features. Using this affinity measure, we formulate conditional probability distributions of contour features from sensor A with respect to sensor B. We then compute the mutual information between contour features from the two sensors based on these conditional distributions. Then, we identify the set of contour features from B that maximize the mutual information with the features from A. The contours from sensor A overlaid with the selected contours from sensor B form the fused result, which can then be completed and filled to create silhouettes.

Image fusion algorithms can be broadly classified into low-, mid-, and high-level techniques based on their position in the information-processing pipeline. As can be seen from Fig. 13.1, based on such a classification, the proposed algorithm can be categorized as a goal-oriented, midlevel fusion technique relying on contour-based features. Apart from the proposed algorithm, there have been several other approaches adopted for the purpose of fusing information across two imaging sensors. Next, we briefly outline two of the other popular fusion strategies that could potentially be employed with the specific aim to provide robust and accurate object segmentation.

13.1.1 Alternate Fusion Methodologies

13.1.1.1 Image Blending

Perhaps the most commonly used approach for fusion is first to create a single, fused image stream by blending images from each of the sensors [8, 18, 34]. Object location and shape are then obtained by applying relevant detection and segmentation algorithms to the fused image stream. In Fig. 13.2a we show the main processing stages typically employed by algorithms adopting this strategy. Such approaches have two potential drawbacks. First, since the fusion procedure simply blends image pairs at a low level (pixel level), the fused image is likely to contain unwanted image artifacts found in each of the two imaging domains. For example, when combining images from the thermal and visible domains, the resulting fused image can contain both shadows (from the visible domain) and thermal halos [10] (found in thermal imagery). Second, since the image characteristics of the fused stream depend on the fusion technique applied, such approaches typically require specialized object segmentation algorithms to obtain satisfactory results.

13.1.1.2 Union of Features

Another popular approach is to defer the fusion of information to a later stage in the pipeline. Such approaches typically employ complete image segmentation routines

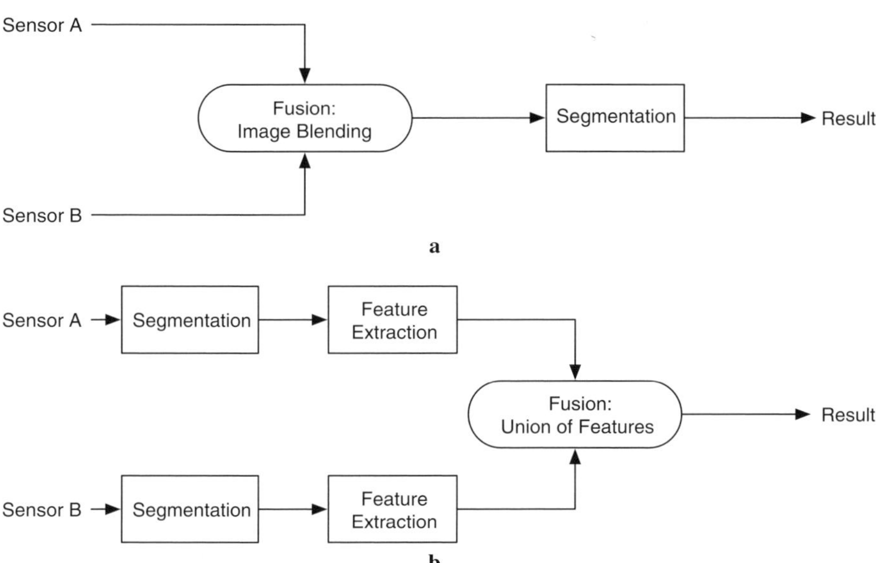

Fig. 13.2 Flowcharts of alternate fusion methods: **a** Image blending **b** Union of features

in *each* image stream and then combine the segmentation results across the sensing domains. These techniques can either be high level or midlevel, depending on whether the fusion occurs at the decision or feature level. In decision-level fusion, binary segmentation results are obtained independently in each domain, and the final result is obtained by combining the individual silhouettes [36]. In feature-level fusion, features extracted from the individual segmentation results are utilized to generate the final result. In this work, we employ for comparison a feature-level fusion approach [11]. The flowchart of this method is shown in Fig. 13.2b. As can be seen from the figure, such an approach has the undesirable property of requiring object segmentation routines to be employed independently in the different imaging domains. Since the features are extracted from the segmentation provided in each sensor, the combined result is obtained by simply performing a union of all the features obtained from both sensors. Thus, the final result is susceptible to errors in segmentation from both domains.

Based on the described flowcharts (Fig. 13.1 and Fig. 13.2), we note that the proposed approach has the potential to provide clear benefits over these alternate methods. Being a feature-level technique, it does not face the issues that hamper low-level image-blending techniques. In addition, the proposed approach completely decouples the process of segmentation from fusion. Thus, any "off-the-shelf" segmentation routine could be used in any one sensor to bootstrap the process. Compared to the feature-union method, the proposed approach provides the obvious benefit of requiring object segmentation in only one sensor. Further, given that the imaging domains employed in fusion systems are generally complementary, different segmentation algorithms are likely to be effective in each domain. While the final result in the feature-union technique will be limited by the worse of the two segmentation results, the proposed fusion technique enables the user to employ only the better of the two segmentation results to bootstrap the fusion process. Depending on the application, factors such as persistence, signal-to-noise ratio, and the availability and complexity of the segmentation scheme can influence which sensor should be chosen for providing the initial segmentation.

13.1.2 Outline

We begin by reviewing in Section 13.2 related work in image fusion. We then present the different aspects of the proposed algorithm, starting with Section 13.3, where we describe the contour-based features used in the algorithm. In Section 13.4, we describe the process of computing the mutual information between different sets of contour features, and in Section 13.5, we describe the contour selection procedure. Then, in Section 13.6, we demonstrate the approach for a video surveillance application using thermal and color cameras as the two input sensors. Based on manually segmented object regions, we show the efficacy of the proposed method by comparing segmentation performance using the fusion algorithm over using either input sensor independently. We also compare our algorithm against the two alternate fusion methods introduced and discuss the advantages of our approach over these other methods. Finally, in Section 13.7 we summarize our approach and provide directions for future work.

13.2 Related Work

Image fusion techniques have had a long history in computer vision and visualization. We categorize related work into three types, based on the processing level (low, mid, high) at which fusion is performed.

Traditionally, *low-level* techniques have been used to combine information from coregistered multisensor imagery. Improving on simple techniques such as pixel averaging, multiresolution schemes similar to the pyramid-based approaches of [4, 23, 34] were proposed. More recently, wavelet analysis has emerged as the method of choice in most multiresolution frameworks [19, 27]. Examples of other low-level techniques include the biologically motivated model based on human opponent color processing proposed in [13]. A technique based on principle component analysis (PCA) measuring pixel variances in local neighborhoods was used in [8]. Pixel-level combinations of spatial interest images using Boolean and fuzzy-logic operators were proposed in [12], and a neural network model for pixel-level classification was used in [16].

Midlevel fusion techniques have mostly relied on first- and second-order gradient information. Some of these techniques include directly combining gradients [28], determining gradients in high dimensions [33], and analyzing gradients at multiple resolutions [25, 30]. Other features, such as the texture arrays [3], have also been employed. Model-based alternatives to feature-level fusion have also been proposed, such as the adaptive model-matching approach of [5] and the model theory approach of [38]. Other midlevel fusion techniques such as the region-based methods of [17, 26, 39] make use of low-level interactions of the input domains.

High-level fusion techniques generally make use of Boolean operators or other heuristic scores (maximum vote, weighted voting, m-of-n votes) [9, 36] to combine results obtained from independently processing the input channels. Other "soft" decision techniques include Bayesian inference [1, 14] and the Dempster-Shafer method [2, 20].

Most of these fusion techniques aim at enhancing the information content of the scene to ease and improve human interpretation (visual analysis). However, the method we propose is designed specifically to enhance the capabilities of an automatic vision-based detection system. Some techniques, such as [3, 5, 12], proposed for automatic target recognition systems, have also been evaluated in terms of object detection performance. These techniques, however, are not generally applicable to the detection of nonrigid person shapes and other large, multimodal objects common in the urban environments considered in this work. Other techniques, such as those of [13], have been shown to improve recognition performance when used as inputs to separate target recognition modules.

Recently, midlevel fusion algorithm also designed with the aim of improving object segmentation was proposed [11]. However, their fusion technique was specific to thermal and color cameras and required background modeling in *both* domains (see Fig. 13.2b). Our proposed method is an improvement on both counts. First, the current algorithm is independent of the methods used for detecting the initial object regions. Thus, as long as the required contour features can be reliably extracted from the images produced by both sensors, the fusion procedure is unaffected by the modality of the sensors (far infrared, long-wavelength infrared, short-wavelength infrared, etc.). Second, the current algorithm only requires the prior ability (via method of choice) to detect object features in any *one* sensor modality.

13.3 Contour Features

Based only on the preliminary object segmentation obtained from sensor A, our goal is to be able to extract relevant information from sensor B such that the combined result is a better estimation of the object shape. The crucial step in this process is choosing the appropriate features. The importance of first-order gradient information in estimating the shape and appearance of an object is well known [7, 21]. We exploit this information by extracting features that capture the location, orientation, and magnitude of the object gradients.

We first obtain a thinned representation of the gradient magnitude image using a standard nonmaximum suppression algorithm. The thinned edges are then broken into short, nearly linear contour fragments based on changes in the gradient direction. A contour fragment is obtained by traversing along a thinned edge using a connected-components algorithm until a change from the initial edge orientation is encountered. To ensure contour fragments of reasonable size, the edge orientations are initially quantized into a smaller number of bins. We represent a contour fragment by a feature vector $c = [p_1, p_2, E_{mag}]$, where p_1 and p_2 are the coordinates of the two endpoints, and E_{mag} is the mean edge magnitude along the contour. The set of all contour features $\{c_1, \ldots, c_n\}$ forms the feature representation of the object.

In Fig. 13.3, we show an example of the feature extraction process. Figure 13.3a shows the input image, and Fig. 13.3b and c show the gradient magnitudes and the thinned binary edges, respectively. The short, linear contour features extracted for these examples are shown overlaid on the input image in Fig. 13.3d. The features were obtained by quantizing the edge orientations into 4 equal-size bins.

13.4 Estimating Feature Relevance

Having extracted the contour features, our goal now is to select features from sensor B that are relevant to the features in sensor A. Mutual information is considered to be a good indicator of the relevance of two random variables [6]. This ability to capture

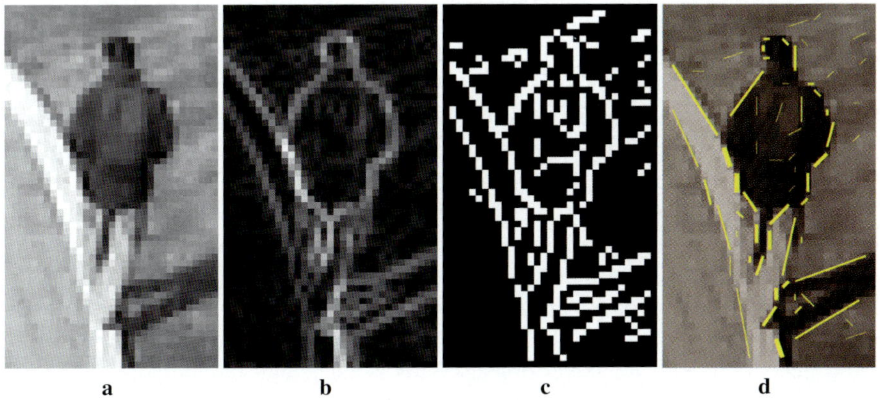

Fig. 13.3 Feature extraction. **a** Input image, **b** gradient magnitudes, **c** thinned binarized edges, **d** extracted contour features overlaid on input image

the dependence, or relevance, between random variables has recently led to several attempts at employing mutual information in feature selection schemes [15, 24, 35].

13.4.1 Preliminaries

Denoting two discrete random variables by X and Y, their mutual information can be defined in terms of their probability density functions (PDFs) $p(x)$, $p(y)$, and $p(x,y)$ as

$$I(X;Y) = \sum_{x \in X} \sum_{y \in Y} p(x,y) log \frac{p(x,y)}{p(x)p(y)} \tag{13.1}$$

Based on entropy, the mutual information between X and Y can also be expressed using the conditional probability $p(x|y)$. The entropy H of X is a measure of its randomness and is defined as $H(X) = -\sum_{x \in X} p(x) \log p(x)$. Given two variables, conditional entropy is a measure of the randomness when one of them is known. The conditional entropy of X and Y can be expressed as

$$H(X|Y) = -\sum_{y \in Y} p(y) \sum_{x \in X} p(x|y) \log p(x|y) \tag{13.2}$$

The mutual information between X and Y can be computed from the entropy terms defined by

$$I(X;Y) = H(X) - H(X|Y) \tag{13.3}$$

Let us associate random variables S_1 and S_2 with the sensors A and B, respectively. Let C_1 denote the domain of S_1 and C_2 the domain of S_2. To use either Eq. 13.1 or Eq. 13.3 to compute the mutual information between S_1 and S_2, we first need to define the domains C_1 and C_2 and then estimate the appropriate PDFs. A discretized version of the full-contour feature space of A and similarly of B are natural choices for C_1 and C_2, respectively. In general, obtaining the PDFs, especially the joint and the conditionals, of the contour features $c_i \in C_1$ and $c_j \in C_2$ is a difficult task. Indeed, it is this difficulty that primarily impedes the use of mutual information in feature selection schemes [24, 35].

Nevertheless, a typical approach would be to estimate these distributions using a large training data set consisting of manually segmented objects imaged using the sensors in question. The difficulty of generating such a data set aside, such an approach has several drawbacks. Importantly, different PDFs will need to be estimated for different object classes in the training set, and there is no guarantee that these would generalize well for novel objects. This is especially cumbersome given the enormous computation and memory requirements of nonparametric estimation techniques. Further, the well-known issue of scale (bandwidth) selection [22] in these methods becomes compounded in high-dimensional spaces such as ours.

Instead of relying on a training data set to learn the distributions of features, we propose a different approach to the problem. Drawing on the observation regarding the "natural structure" of the world, we make the assumption that objects of interest have continuous, regular boundaries. Based on this assumption, we seek to define relationships between samples from S_1 and S_2 that will enable us to identify the set of contours from sensor B with the highest relevance to sensor A.

In the context of fusion, we propose that a set of features has high relevance to another if it provides both redundant *and* complementary information. The choice of contour features (Section 13.3) enables us to further define relevance as the ability of a set of features to coincide with and complete object boundaries that have been only partially captured by another set. We now address the issue of computing contour feature relevance and folding it into a mutual information framework.

13.4.2 Contour Affinity

Assume that the pair of images shown in Fig. 13.4a and b represent the contour features of a rectangular box imaged using two sensors. Let Fig. 13.4a represent the object contours from sensor A and Fig. 13.4b the set of contours obtained from sensor B. Note that the contour features from sensor A are obtained after an initial segmentation step and hence lie along the boundary of the rectangular box. On the other hand, the feature extraction from sensor B is not preceded by any object segmentation, and hence the features are extracted directly from the entire image region.

Visualizing the contour features extracted from sensor A in image space, as in Fig. 13.4a, we see that the contour fragments form an incomplete trace of the boundary of the viewed object. As described, we desire the subset of contour features from sensor B that provides the best completion of the broken contour image formed by the features from sensor A.

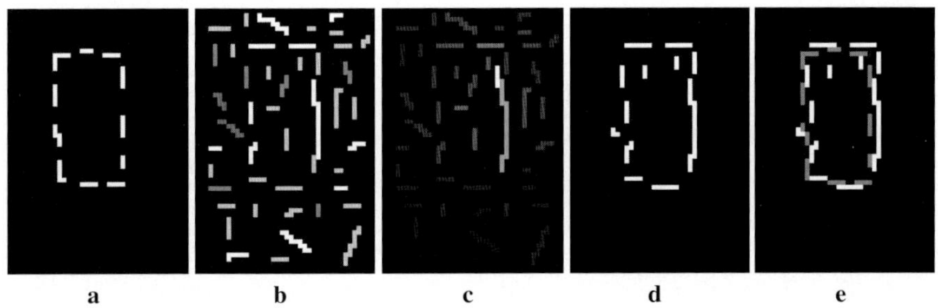

Fig. 13.4 Toy example illustrating the relevant processing stages. **a** Detected object contours from sensor *A*. **b** Contours obtained from sensor *B*. **c** Relative affinity values of contours in **b** with respect to a contour (shown in white) from **a**. **d** Set of contours selected from **b**. **e** Overlay of contours from **a**, shown in gray, with the selected contours **d**

Fig. 13.5 Computation of contour affinity

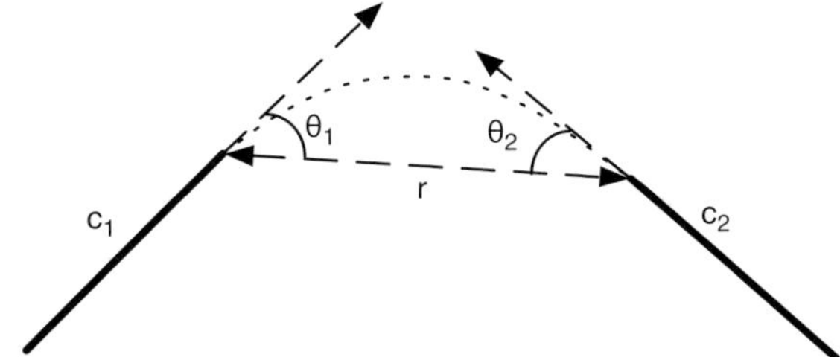

Perceptual (and computational) figure completion is a very active field of research, and several figure completion studies, such as [32, 37], have used an "affinity" measure between a pair of edge elements to compute how likely it is that they belong to the same underlying edge structure. The concept of affinity is related to the energy of the simplest curve passing through two edge elements. One popular method of determining the lowest-energy (smoothest) completion between two edge elements is minimizing the *elastica* functional, which defines energy as the integral of the square of curvature along the curve. We borrow this notion of affinity and adapt it to deal with contours of finite size instead of the dimensionless edge elements used in the literature.

Consider a pair of contours c_1 and c_2, as shown in Fig. 13.5, and hypothesize the simplest curve connecting c_1 and c_2. Any such connection would join one of the endpoints of c_1 to an endpoint of c_2. Since c_1 and c_2 have two endpoints each, all such curves fall into one of four categories based on which two endpoints are connected. Consider one such curve, shown by the dotted line in Fig. 13.5, between an endpoint of c_1 and an endpoint of c_2. Further, consider the vector joining the ends of the curve, pointing from the endpoint of c_1 to the end-point of c_2. As shown in Fig. 13.5, let θ_1 denote the angle between this vector and the unit vector at the endpoint of c_1, directed away from the contour along the tangent at c_1. Let θ_2 denote the angle from c_2, analogous to θ_1. Finally, let r denote the Euclidean distance between the two endpoints of c_1 and c_2. These quantities, θ_1, θ_2, and r, are computed for each of the four possible sets of curves between endpoints of c_1 and c_2.

We define the contour affinity $\mathrm{Aff}(c_1, c_2)$ between two contours c_1 and c_2 as the maximum affinity value over the four possible sets of curves. Following the analytical simplification for the minimization of the *elastica* functional presented in [32], the affinity for a particular curve set is defined as

$$\mathcal{A} = e^{(-r/\sigma_r)} \cdot e^{(-\beta/\sigma_t)} \cdot e^{(-\Delta/\sigma_e)} \tag{13.4}$$

where $\beta = \theta_1^2 + \theta_2^2 - \theta_1 \cdot \theta_2$ and $\Delta = |E_{mag}^{c_1} - E_{mag}^{c_2}|$ (the absolute difference in the intensity of the contours). We write the normalization factors σ_r, σ_t, and σ_e as $\sigma_r = R/f_1$, $\sigma_t = T/f_2$, and $\sigma_e = E/f_3$, respectively, where R, T, and E equal the maximum possible value of r, β, and Δ, respectively, and (f_1, f_2, f_3) are weights that can be used to change the relative influence of each term in the affinity calculation.

Contour pairs that are in close proximity, lie along a smooth curve, and have comparable intensities will have high-affinity values. Consider the pairwise affinity measurements between contour features taken one at a time from C_2 and the set of contour features C_1. If a particular contour feature $c_2 \in C_2$ lies along the object boundary, it would have very high-affinity values with neighboring contour features in C_1. If c_2 represents a nonobject contour (e.g., background edge), unless it is in close proximity to some object contour, aligns well with it, *and* has similar intensity values, we expect that it would have a low-affinity value with all the contour features in C_1.

Figure 13.4(c) shows the relative difference in affinity between the short contour shown in white (selected from Fig. 13.4a) and the other contours (from Fig. 13.4b). The brighter the contour, the higher the affinity is. For this computation of affinity, we used the weights $f_1 = 5$, $f_2 = 5$, and $f_3 = 0$ (the intensity of the contours in this example were generated randomly).

13.4.3 Estimation of Conditional Probability Using Contour Affinity

As stated, affinity captures the possibility that two contours belong to the same underlying edge structure. If we assume that one of the contours belongs to an object boundary, one can interpret the affinity between two contours to be an indication of the probability that the second contour also belongs to the object boundary. In other words, the affinity between c_1 and c_2 can be treated as an estimate of the probability that c_1 belongs to an object *given* that c_2 does.

Consider again the random variables S_1 and S_2. Let C_1, the domain of S_1, now contain contour features extracted only from the current input image from sensor A. Similarly, let C_2, the domain of S_2, contain contour features extracted from the corresponding image from sensor B. Based on the pairwise affinity between contours of C_1 and C_2, we define

$$P(c_1|c_2) = \frac{Aff(c_1,c_2)}{\sum_{c_i \in C_1} Aff(c_i,c_2)} \tag{13.5}$$

where $P(c_1|c_2) \equiv P(S_1 = c_1|S_2 = c_2)$.

13.4.4 Computing Mutual Information

The definition of the conditional probability in Eq. 13.5 enables us to measure the conditional entropy between S_1 and any contour $c_j \in C_2$. Using Eq. 13.2, this can be expressed as

$$H(S_1|c_j) = -p(c_j) \sum_{c_i \in C_1} p(c_i|c_j) \log p(c_i|c_j) \tag{13.6}$$

where the distribution $p(c_j)$ can be considered as a prior expectation of observing a given contour feature. Similarly, assuming $p(c_i)$ to be a known distribution (e.g., uniform), the entropy of S_1 can be computed as

$$H(S_1) = - \sum_{c_i \in C_1} p(c_i) \log p(c_i) \tag{13.7}$$

Using Eqs. 13.6 and 13.7 in Eq. 13.3, we can measure the mutual information $I(S_1;c_j)$. To obtain an estimate of the full joint mutual information $I(S_1;S_2)$, we consider each contour independently and use the approximation suggested in [24], which is the mean of all mutual information values between contour features $c_j \in C_2$ and S_1:

$$I(S_1;S_2) = \frac{1}{|C_2|} \sum_{c_j \in C_2} I(S_1;c_j) \tag{13.8}$$

If we assume the prior distribution of contour features $p(c_i)$ and $p(c_j)$ to be uniform, the entropy of S_1 (Eq. 13.7) is constant. Maximizing the mutual information is then equivalent to finding the set of features from S_2 that *minimizes* the conditional entropy $H(S_1|S_2)$. In other words, we seek those contour features from S_2 that minimize the randomness of the object contour features in S_1. Rewriting Eq. 13.8 using Eqs. 13.6 and 13.7 and using the assumption of uniform distributions for $p(c_i)$ and $p(c_j)$, the conditional entropy of S_1 and S_2 can be expressed as

$$H(S_1|S_2) \propto \sum_{c_j \in C_2} \left(- \sum_{c_i \in C_1} p(c_i|c_j) \log p(c_i|c_j) \right)$$

where the term in parenthesis can be interpreted as the entropy of the distribution of affinity between c_j and the contours in C_1. This is indeed the notion of relevance we wish to capture since, as described in Section 13.4.2, the entropy of affinity values is expected to be low only for c_j lying on object boundaries.

13.5 Contour Feature Selection Using Mutual Information

We now address the issue of selecting the most relevant set of contour features from S_2 based on S_1. This problem statement is very reminiscent of the feature selection problem [15, 24], and the intuition behind the solution is also similar. We seek the subset of contours from S_2 that maximizes the mutual information between S_1 and S_2.

The problem of finding the subset that maximizes the mutual information is intractable since there are an exponentially large number of subsets that would need to be compared. An alternate greedy heuristic involves a simple incremental search scheme that adds to the set of selected features one at a time. Starting from an empty set of selected features, at each iteration, the feature from S_2 that maximizes Eq. 13.8 is added to the set of selected features. This solution, as proposed in the feature selection literature [15, 24], has one drawback in that there is no fixed stopping criteria other than possibly a user-provided limit to the maximum number of features required [24]. Obviously, this is a crucial factor that would impede the use of this greedy selection scheme in most fusion applications.

We present here a modified version of the greedy algorithm that addresses the need for a reliable stopping criterion. Initially, the set C_2 contains contour features that lie along the object boundary as well as a potentially large number of irrelevant contour features due to sensor noise and scene clutter. We start by computing the mutual information I_{full} between S_1 and S_2. The algorithm is based on the observation that removing a relevant contour feature from C_2 should *reduce* the mutual information ($<I_{full}$), while removing an irrelevant feature should *increase* the mutual information ($>I_{full}$). We iterate over all the individual contours in C_2 and select only those contours that reduce the mutual information when removed from C_2. The outline of the complete feature selection algorithm is as follows:

1. Compute $I_{full} = I(S_1; S_2)$, where S_1 and S_2 are random variables defined over C_1 and C_2, respectively.
2. For each $c_j \in C_2$,

    a $C_2^j \leftarrow C_2 \setminus \{c_j\}$.
    b Compute $I^j = I(S_1; S_2^j)$, where S_2^j is defined over C_2^j.
3. Select all c_j such that $I^j \leq I_{full}$.

Figure 13.6 shows the normalized mutual information values (I^j) in descending order for the synthetic example images shown in Fig. 13.4a and b. The dashed horizontal line in the figure corresponds to I_{full} and can be considered the minimum mutual information required between S_1 and S_2. The result of the contour selection procedure for this example is shown in Fig. 13.4d. As can be seen, apart from a few internal contours, the subset of contours selected is reasonable. Figure 13.4e shows the contours from sensor A (Fig. 13.4a) overlaid in gray with the selected contours from sensor B. The slight misalignment in the contours from the two sensors was done intentionally to demonstrate the robustness of the algorithm to small errors in sensor registration.

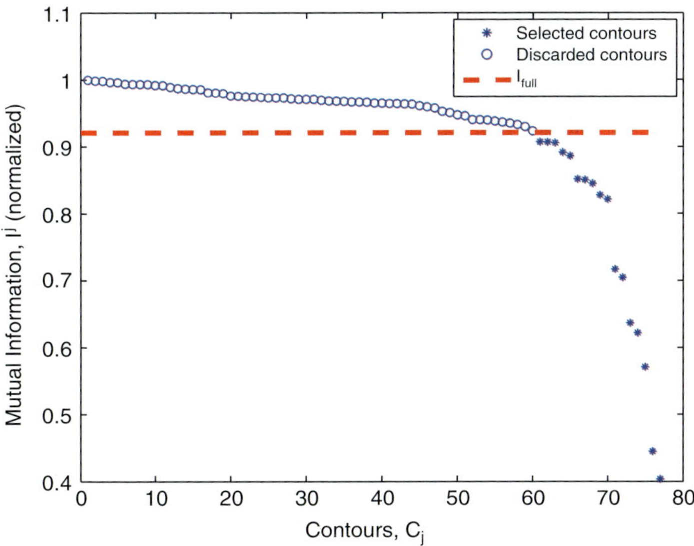

Fig. 13.6 Variation of mutual information values (I^j) for different C_2^j sorted in descending order

While using I_{full} as the threshold in the contour selection procedure is effective, it can sometimes, due to inaccuracies in the estimation of the PDFs, prove to be too strict a threshold in real-world cases. A better threshold can be obtained in practice. Observing the profile of mutual information values I_j in descending order, we often see that there is a sharp drop (corresponding to the separation of object and nonobject contours) in the mutual information at some value $I_j = I_T$ in the vicinity of I_{full} such that $I_T \geq I_{full}$. Using I_T instead of I_{full} in step 3 of the algorithm typically results in the selection of a better subset of contours.

We show two real-world examples of the contour feature selection scheme in Fig. 13.7 and also compare using I_T and I_{full} as the thresholds. Figure 13.7a and b show the contours extracted from the corresponding subimages obtained from a thermal (sensor A) and visible sensor, respectively. In Fig. 13.7c we show the set of contours selected using I_{full} as the threshold. The contours selected using I_T as the threshold are shown in Fig. 13.7d. The variation of the (normalized) mutual information values I^j for different C_2^j is shown in Fig. 13.7e. The dashed horizontal line corresponds to I_{full}. The solid line represents I_T, the point $\geq I_{full}$ in the mutual information profile with the *largest* drop. The mutual information profile corresponding to the first example shown in Fig. 13.7 shows a distinctive drop at I_T. Under conditions of high clutter, the profile of mutual information values may not contain a point with a distinctly large drop. However, as demonstrated by the second example of Fig. 13.7, the described heuristic still provides a reasonable separation of object/nonobject contours in such cases.

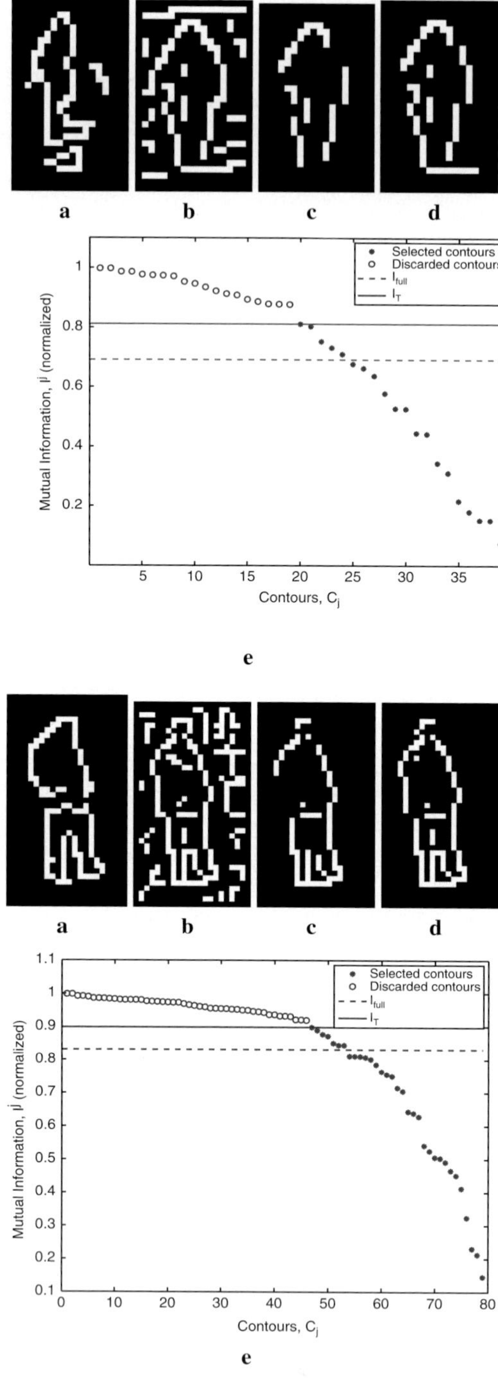

Fig. 13.7 Examples of contour feature selection based on variation of mutual information. **a** Contours from sensor A (thermal domain), **b** contours from sensor B (visible domain), **c** contours selected from sensor B using I_{full}, **d** contours selected from sensor B using I_T, **e** variation of mutual information values (I^j) for different C_2^j sorted in descending order

13.6 Experiments

To test our feature-level fusion approach, we consider a video surveillance scenario that employs a pair of colocated and registered cameras. This setting enables us to evaluate the ability of our fusion approach to improve the shape segmentation of objects found in typical urban surveillance scenarios. The two sensors used are a ferroelectric thermal camera (Raytheon 300D core) and a color camera (Sony TRV87 Handycam). We analyzed several different thermal/color video sequence pairs recorded from different locations at different times of day. The sequences were recorded on a university campus and show several people, some in groups, moving through the scene. We show an example of a typical image pair, cropped to a person region, in Fig. 13.8a and b.

We begin by describing the choices made for the internal parameters of our algorithm and providing a visual/qualitative assessment of the fusion results. Then, in Section 13.6.1, we present a detailed quantitative evaluation of the algorithm, including comparative analysis with other competing fusion techniques.

To choose the "reference" sensor (A) for our algorithm, we considered the nature of the application and the ease of obtaining an initial segmentation. The need for persistence in a surveillance application, and the ease of background modeling in the relatively stable thermal domain [10], prompted us to choose the thermal camera as sensor A.

We employ the contour-based background subtraction scheme using contour saliency maps (CSMs) [10] along with a minimal threshold to directly obtain a preliminary detection of object contours from the thermal domain. For ease of computation, we break the corresponding thermal and visible input images into subimages based on the regions obtained from background subtraction in the thermal domain. Each thermal subimage consists of contours that belong to a single object or objects that were close to each other in the input image. The matching visible subimage consists of all the thinned gradient magnitudes of the image region containing the

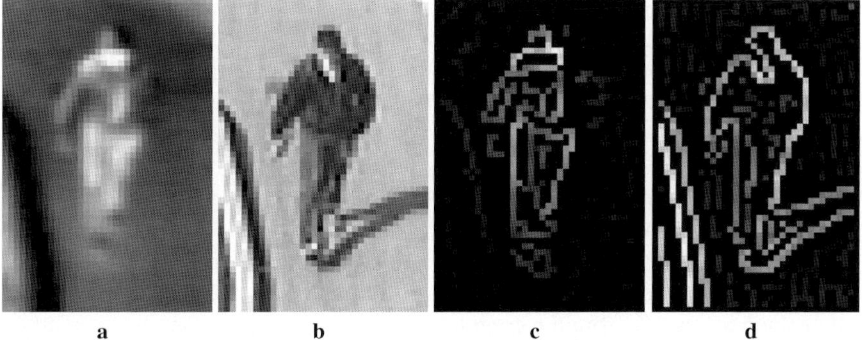

Fig. 13.8 An example input. **a** Thermal subimage; **b** visible sub-image; **c** initial object contours detected from **a**; **d** thinned gradient magnitudes from **b**

objects. In Fig. 13.8c and d, we show an example of the subimage pair corresponding to the image regions shown in Fig. 13.8a and b.

These subimage pairs form the input to our fusion algorithm. We first extract contour features from each subimage as described in Section 13.3. We used 4 orientation bins with centers at 0, 45, 90, and 135°, and a standard connected-components algorithm. For every pair of contour features from both domains, we then estimate the probability of a contour feature in the thermal domain conditioned on the occurrence of a feature from the visible domain (as described in Section 13.4.3). For the computation of contour affinity (Eq. 13.4), in all the experiments we used $f_1 = 5$, $f_2 = 5$, and $f_3 = 15$.

The set of contour features from the visible domain that are most relevant to the object contour features from the thermal domain is chosen using the steps outlined in Section 13.5. The final fused result is then obtained by overlaying these contour features selected from the visible domain with the contour features originally detected in the thermal domain. In case of misalignments that could arise due to small registration errors, standard morphological techniques are used to ensure that all contours are 1-pixel thick.

Given the contour feature sets C_1 and C_2 extracted from the two sensors, computing the mutual information involves obtaining the pairwise affinity between contours and takes order of $O(mn)$ time, where $|C_1| = m$ and $|C_2| = n$. Determining the most relevant subset of features from C_2 is an iterative procedure that has a worst-case running time of $O(m)$. In our experiments, the three stages of the algorithm, namely, extraction of features, computation of mutual information, and selection of the relevant subset of features, took an average 0.43 seconds per input subimage pair on a 2.8-GHz Intel P4 machine using software written partly in Matlab and C.

We show several examples of the fusion results in Fig. 13.12. All images have been shown in binary to improve clarity. Figure 13.12a shows the detected contours obtained from the thermal domain. Figure 13.12b shows the thinned gradients from the visible domain. The set of contours selected by our algorithm from the visible domain is shown in Fig. 13.12c. Figure 13.12d shows the final fused result obtained by overlaying Fig. 13.12c with Fig. 13.12a. Overall, the results are satisfactory. The algorithm selects contours that both strengthen and complement the set of input object contours. In general, the outer boundaries of the fused result are a reasonable approximation of the true object shape. In spite of the presence of shadows and illumination changes, the proposed fusion framework is effective in obtaining a reasonable contour segmentation in the visible domain that further improves the original segmentation acquired from the thermal sensor.

After the subimages of an image pair have been processed, the resulting fused image contains contours extracted from both domains that best represent the objects in the scene. Several different vision applications can benefit from improvements in such a result, especially those that rely on the notion of object shape. Shape could be extracted either directly from the contours or after using figure completion methods (such as [10]) on these contours. Examples of such applications include activity recognition, object classification, and tracking.

13.6.1 Quantitative Evaluation

As stated in Section 13.1, the challenge for any fusion algorithm is to utilize information from two or more sources to maximally improve the performance of the system over using either sensor individually. In this section, we analyze how our fusion algorithm stands up to this challenge for the task of shape segmentation. The quantitative evaluation is based on the manual segmentation of the object regions in 73 image pairs obtained from several thermal/color video sequences. Results of the hand segmentation (by multiple people) of each pair of images were combined using an elementwise logical-OR operation to obtain the final manually segmented images.

13.6.1.1 Experiment 1: Fusion vs Independent Sensors

Since the final result of our algorithm is a set of contours, let us assume that we have available a module that can generate a closed shape (a silhouette) from such input. For evaluation, we propose then to use this module to generate a segmentation from three different sets of contours:

- Set T: contours from the thermal sensor initially detected as lying along the object
- Set V: subset of contours from the visible sensor selected by the fusion algorithm
- Set TV: overlay of the thermal and visible contours

The comparison of the shape segmentation achieved in each of the above scenarios will provide valuable information that can be used to judge the validity of the proposed approach. Several approaches for contour-based figure completion exist. For the purpose of this evaluation, we make use of the method suggested in [10] to complete and fill the shape.

The set of 73 image pairs generated a total of 208 usable subimage pairs (a simple size criterion was used to eliminate subimages that contained person regions that were too small). For each subimage, the shape segmentation corresponding to the three sets of contours enumerated were obtained. Examples of the silhouettes obtained from set TV are shown in Fig. 13.12e. To enable a visual assessment of the segmentation result, we show in Fig. 13.12f the manual segmentation of the image regions. Corresponding to each example, we also note the F-measure value obtained by comparing the generated silhouette (Fig. 13.12e) against the manually marked ground truth (Fig. 13.12f).

To quantify the segmentation results, we compute precision and recall values using the manually segmented object regions as ground truth. *Precision* refers to the fraction of pixels segmented as belonging to the object that are in fact true object pixels, while *recall* refers to the fraction of object pixels that are correctly segmented by the algorithm. We combine these values into a single measure of performance using the F-measure [29], which is the harmonic mean of precision and recall. The higher the F-measure, the better the performance is.

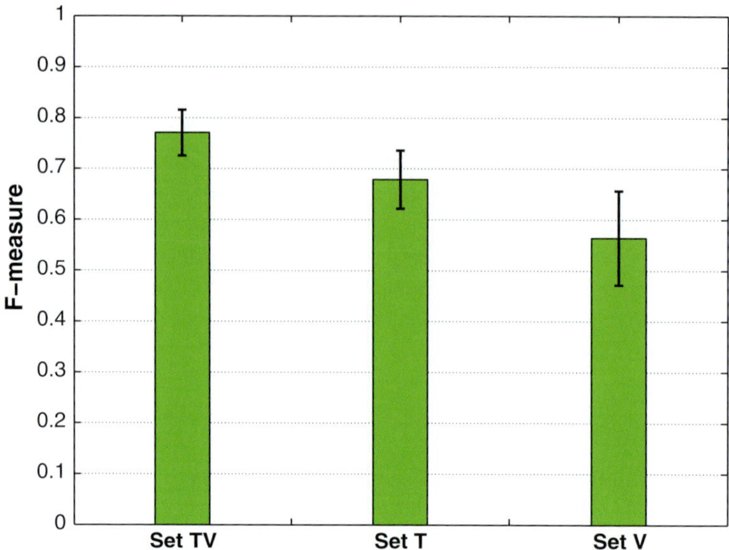

Fig. 13.9 F-measure comparison of fusion results with independent sensors

In Fig. 13.9 we present the mean F-measures evaluated for the three differ-ent scenarios over all the subimages. The error bars correspond to the variation in F-measure values obtained for each case. As can be seen from the plot, the qual-ity of the segmentation obtained using the proposed fusion method ($\overline{F}_{TV} = 0.77$) is clearly superior to that obtained from the initial object segmentation performed in the thermal domain ($\overline{F}_T = 0.67$) [two-tailed t test: $t(207) = 4.529, p < 5 \times 10^{-6}$]. The improvement in segmentation shows that the proposed fusion algorithm is in-deed able to extract relevant information from the visible domain such that the com-bination of information from the two sensors generates better results. The plot in Fig. 13.9 also shows the segmentation achieved using only the contour features cho-sen from the visible domain ($\overline{F}_V = 0.56$). While clearly lower than the segmentation performance of the thermal sensor [two-tailed t test: $t(207) = 8.346, p < 1 \times 10^{-14}$], it should be noted that the segmentation in the visible domain is not aided by any prior knowledge or background information. It is obtained purely by identifying fea-tures that best complement and support the object features detected in the thermal domain.

Overall, these numbers demonstrate the ability of the proposed algorithm to use limited cues from one sensor to extract relevant information from the other sensor. The segmentation performance obtained from the fusion results show that the al-gorithm is successful in extracting both redundant and complementary information across modalities.

We next subject our algorithm to more adverse conditions and evaluate the ability of the algorithm to identify relevant contours from the visible domain given weaker initial, detections in the thermal domain. We perform the same experiment as be-fore; however, this time we use only a subset of set T by randomly discarding $k\%$

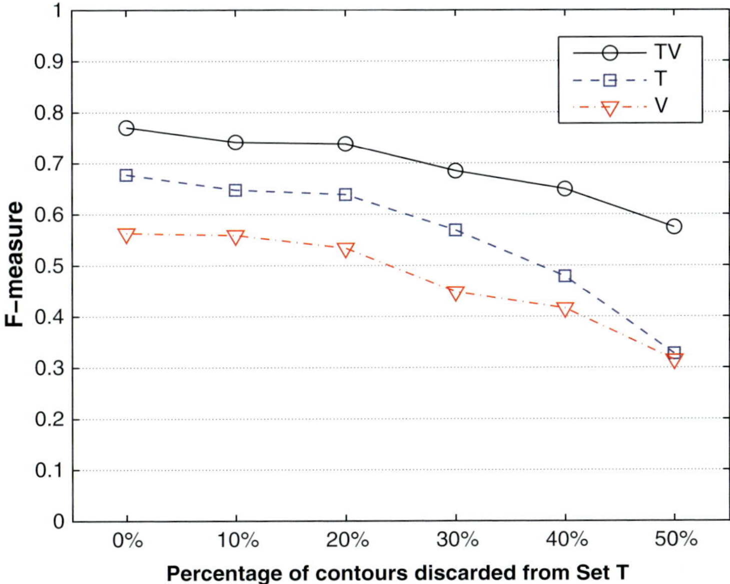

Fig. 13.10 Variation of silhouette F-measure with smaller subsets of set T

$(10 \leq k \leq 50)$ of the contours. This resulting set is then used as input into our fusion algorithm. This experiment tests if the fusion algorithm is capable of estimating the correct set of relevant features from sensor B given a more incomplete detection from sensor A. We vary the value of k, systematically from 10 to 50 at intervals of 10%. At each value of k, the experiment is repeated five times, and the results presented here are averaged over the 5 runs. In Fig. 13.10, we show the variation in segmentation performance by plotting the F-measure against the different percentages of discarded contours. As expected, the performance for each of the sets T, V, and TV decreases as the quality of the bootstrap segmentation is impoverished. However, what is interesting is the *rate* at which the performance deteriorates. It is clear from the plot that while the segmentation in the thermal domain (set T) drops sharply as k increases, the fusion results (set TV) show the most gradual decline in performance. It is also worth noting that the rate of change in the performance of the visible domain (set V) mimics closely that of set TV.

These results show that instead of being equally or perhaps worse affected by an impoverished input, the outputs of the fusion algorithm (sets V and TV) show a much more graceful degradation in performance. In fact, the drop in segmentation performance for set TV (in terms of F-measure score) is systematically lesser than set T at every value of k. Thus, as the initial bootstrap segmentation becomes weaker, the benefits of using the proposed fusion algorithm to combine information from the visible domain become increasingly apparent. These observations lead us to believe that the algorithm is indeed able to extract information from another sensor to compensate for incomplete information from one sensor.

13.6.1.2 Experiment 2: Comparison Against Other Methods

In this experiment, we compare the proposed fusion method against two other fusion approaches that could potentially be employed for object segmentation. The methods we compare against each belong to the two approaches introduced in Section 13.1.1. Here, we provide details regarding the specific algorithms employed in each case.

Image Blending

Fusion is performed by computing a regionwise weighted average of the input sensors. The weights for each circular region are determined using PCA of the pixel intensities of the input images. For each local region, this results in higher weights being assigned to the sensor that has a higher variance of pixel intensity levels (for details, see [9]). We use the method to fuse each color component of the visible domain with the thermal channel, resulting in a fused image stream with three components. As is the case with the proposed algorithm, we employ background-subtraction as the method for obtaining the object segmentation from the fused image stream. Treating the three-channel fused image as a standard color image, we construct single Gaussian background models in the normalized color and intensity spaces. Background subtraction is performed using each model separately, and the results are combined to avoid identifying shadows as foreground regions. The final foreground object regions are composed of pixels found to be statistically different from the background in the color space and statistically brighter than the background in the intensity space.

Union of Features

Binary contour fragments are obtained from the segmented foreground regions in each sensor and then combined into a single image. Since background subtraction is employed in each sensor, the combined image is formed by a union of all the extracted features. A gradient-based alignment procedure is also employed to compensate for small errors in registration (for details, see [11]). In the thermal domain, background subtraction is performed using the CSM technique [10]. In the visible domain, background subtraction is performed using separate models in the intensity and color spaces as described. Background subtraction in the visible domain is only performed within the foreground regions (or blobs) obtained from the thermal domain. This ensures that the background subtraction results in the visible domain are not adversely affected by sudden illumination changes commonly found in our data sets. The fused binary contours are completed into silhouette blobs using the same contour completion technique employed for the proposed method.

To compare the three methods, we again rely on the hand-drawn silhouettes as ground truth. We first identify bounding boxes around each hand-drawn person

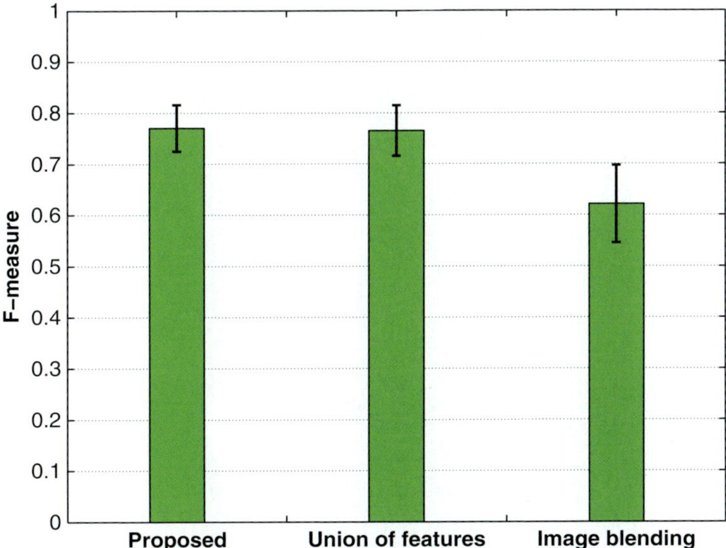

Fig. 13.11 F-measure comparison of proposed fusion method with alternate approaches

silhouette. The output silhouette images generated by each method are evaluated only within these bounding boxes so the techniques are not penalized for poor background subtraction results. The silhouette quality is measured, as before, using the F-measure of precision and recall.

The mean F-measure of the silhouettes obtained by each of the fusion methods is shown in Fig. 13.11. The error bars in the plot correspond to the variation in the F-measure score. As can be seen from the comparison, the quality of the silhouettes obtained by the two feature-level fusion methods is clearly better [two-tailed t test: $t(207) = 8.346, p < 1 \times 10^{-14}$] than that obtained from the image-blending algorithm. Among the feature-level fusion methods, the proposed technique is able to generate marginally better [two-tailed t test: $t(207) = 0.311, p < 0.756$] results than the alternate method used for comparison in spite of requiring only a rough initial segmentation from only one sensor.

13.6.2 Discussion

Both of the experiments described demonstrated several useful properties of the approach presented in this chapter. The results of the first experiment clearly show that the proposed technique adequately meets the basic requirement of any fusion algorithm, that of providing superior performance than can be achieved from using either sensor individually. More specifically, as shown in Fig. 13.9, the proposed fusion algorithm generates significantly better object silhouettes than those obtained

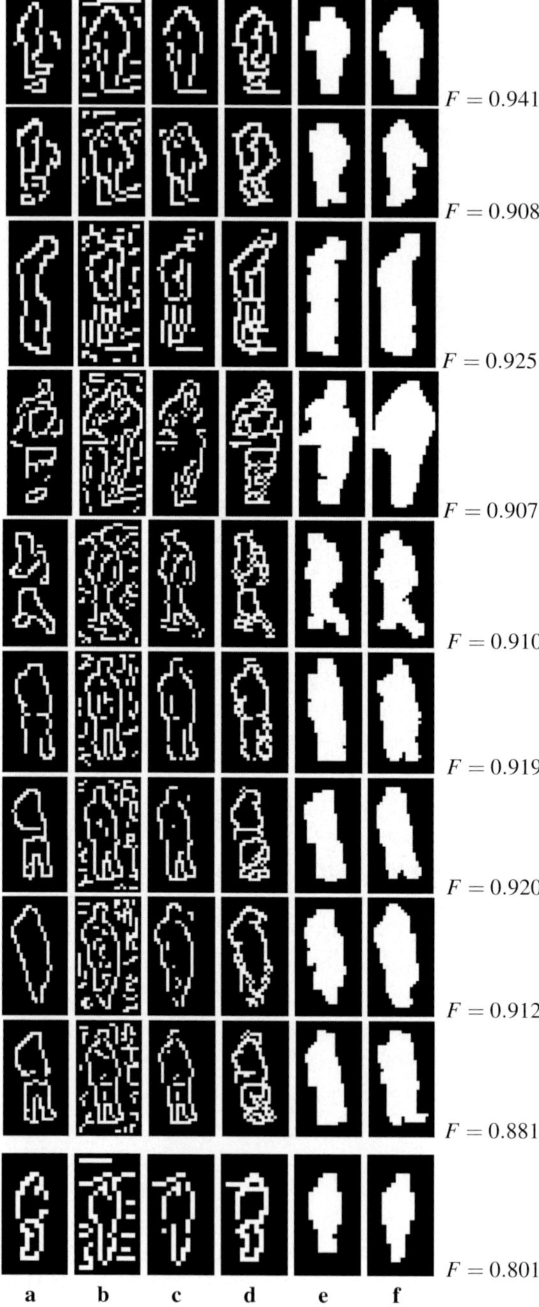

Fig. 13.12 Examples of fusion results: **a** Contours detected from thermal domain (set T). **b** Contours present in the visible domain. **c** Contours selected from **b** (set V). **d** Overlay of contours from **c** on **a** (set TV). **e** Segmentation obtained after completing and filling **d**. **f** Manually segmented object regions and corresponding F-measure values (on comparison with **e**)

from employing a standard object segmentation routine in the thermal domain. This clearly shows that, based on the available segmentation in the thermal domain, the fusion algorithm is able to extract relevant information from the visible domain to improve the overall segmentation performance.

Since the proposed algorithm requires an initial object segmentation, it is of interest to see how the quality of this initial bootstrap segmentation affects the final segmentation result. As shown in Fig. 13.10, we see that the proposed approach is able to utilize highly impoverished segmentation information and yet generate overall silhouettes of much higher quality. Thus, experiment 1 shows that the proposed algorithm is capable of significantly improving object segmentation results; further, the algorithm is able to maintain reasonable segmentation quality even as the required bootstrap segmentation is made so weak that it covers only half of the actual object region.

Next, in experiment 2, we provided a direct comparison of our approach with two other fusion strategies, one a low-level technique and the other a midlevel technique. We see that while low-level fusion techniques are useful for creating blended images for visual inspection, they are unlikely to be very effective in more "goal-oriented" scenarios, such as the object segmentation task under investigation in this work. As discussed in Section 13.1.1, the fusion process in such methods suffers from the lack of higher-level knowledge, such as which image features are likely to be useful and which detrimental to the task at hand. The fused images produced contain combinations of image features peculiar to each imaging sensor. Further, the blended image stream requires the use of segmentation routines specifically tailored to handle the unique image characteristics borne out of the fusion strategy employed.

As can be seen from the plot in Fig. 13.11, the feature-level fusion strategies fare better for goal-oriented fusion tasks. The "union-of-features" method used for comparison represents a brute force approach to fusion, wherein object segmentation is performed in each available sensor, and the final result is obtained by simply combining all the extracted features. Such an approach implicitly assumes that all of the individual object segmentation results are accurate. Any segmentation errors made in either sensor are also manifest in the final segmentation, resulting in poor-quality silhouettes. Thus, to effectively utilize such an approach, it is essential to have access to high-quality object segmentation algorithms in each of the sensors. In terms of overall performance, we see (from Fig. 13.11) that in spite of requiring object segmentation in only one of the sensors, the proposed fusion algorithm provides silhouette quality that is in fact marginally better than that provided by the brute force union-of-features technique.

Contrary to the union-of-features method, the proposed fusion approach requires only a rough, incomplete segmentation in either one of the sensors. We note that since the algorithm utilizes all of the contour features extracted from the initial segmentation, it is preferable that this segmentation be as reliable as possible, even at the cost being considerably incomplete. The intelligent feature selection process utilizes this initial segmentation to extract relevant features from the other sensor, without requiring any a priori segmentation information in that domain. Since the proposed approach can be bootstrapped using either sensor, object segmentation needs to be performed in only that sensor in which it is likely to be more reliable. For example, in our experimental setup, performing background subtraction in the single-channel thermal domain is both more reliable and computationally cheaper than performing background subtraction in the three-channel color space. While our approach requires only thermal background subtraction, the union-of-features method required additional background subtraction in the visible domain together with a shadow removal step.

13.7 Summary

We presented a new, goal-oriented, feature-level fusion technique for object segmentation based on mutual information. The proposed algorithm treats fusion as a feature selection problem. The approach utilizes the natural structure of the world within a mutual information framework to define a suitable criterion for feature selection. Starting from an initial detection of object features in one sensor, our technique extracts relevant information from the other sensor to improve the quality of the original detection.

We first defined a feature representation based on contour fragments that is rich enough to implicitly capture object shape yet simple enough to provide an easy realization of feature relevance. We then approached fusion as a variation of the mutual information feature selection problem. To avoid the pitfalls of learning the relevant probability distributions from training data, we proposed a method that generates the required probability distribution from a single pair of images. The method computes the conditional probability distribution based on the notion of contour affinity and effectively captures the expectation that objects have regular shapes and continuous boundaries. We then computed the mutual information between the features extracted from both sensors. Finally, we employed a new scheme to reliably obtain a subset of features from the secondary sensor that have the highest mutual information with the provided object contours. The final fused result is obtained by overlaying the selected contours from both domains. The final contours are then complete and filled to create silhouettes.

Our approach was tested in a video surveillance setting using colocated thermal and color cameras. The fusion algorithm improved object segmentation performance over using either sensor alone. Experiments were conducted using a set of over 200 manually segmented object regions and were evaluated using the F-measure of precision and recall. The segmentation result of the fusion algorithm yielded an F-measure of 0.77, better than those obtained from detection results of either sensor used independently. The proposed algorithm was also compared to other fusion approaches, a low-level technique [8], and another midlevel technique [11] and was shown to produce comparable (or better) results while requiring fewer computational resources.

In the future, we plan to extend the method to enable two-way information flow in our fusion pipeline. Such an approach would potentially enable the final segmentation to be built up incrementally, such that in each iteration the segmentation from one sensor would seed feature selection in the other and so on. We would also like to investigate the robustness of our feature representation to translation and rotation of the sensors. This would potentially enable our approach to withstand larger errors in image registration across the sensors.

Acknowledgments This research was supported in part by the National Science Foundation under grant 0428249. A shorter version of this chapter appeared in the 2006 *IEEE Workshop on Object Tracking and Classification in and Beyond the Visible Spectrum* [31].

Chapter's References

1. T. Bakert and P. Losiewicz. Force aggregation via bayesian nodal analysis. In *Proceedings of Information Technology Conference*, 1998
2. P. Bogler. Shafer-Dempster reasoning with applications to multisensor target identification systems. *IEEE Transactions on System, Man, and Cybernetics*, 17:968–977, 1987
3. D. Borghys, P. Verlinde, C. Perneel, and M. Acheroy. Multi-level data fusion for the detection of targets using multi-spectral image sequences. *SPIE Optical Engineering, Special Issue on Sensor Fusion*, 37:477–484, 1998
4. P.J. Burt and R.J. Kolczynski. Enhanced image capture through fusion. In *Proceedings of Computer Vision and Pattern Recognition*, 173–182, 1993
5. F. Corbett et al. Fused ATR algorithm development for ground to ground engagement. In *Proceedings of the 6th National Sensory Symposium*, volume 1, pages 143–155, 1993
6. T. Cover and J. Thomas. *Elements of Information Theory*. J. Wiley, New York, 1991
7. N. Dalal and B. Triggs. Histograms of oriented gradients for human detection. In *Proceedings of the International Conference Computer Vision*, pages 886–893, 2005
8. S. Das and W. Krebs. Sensor fusion of multi-spectral imagery. *Electronics Letters*, 36:1115–1116, 2000
9. B. Dasarathy. *Decision Fusion*. IEEE Computer Society Press, Washington, DC, 1994
10. J. Davis and V. Sharma. Background-subtraction in thermal imagery using contour saliency. *International Journal of Computer Vision*, 71(2):161–181, 2007
11. J. Davis and V. Sharma. Background-subtraction using contour-based fusion of thermal and visible imagery. *Computer Vision and Image Understanding*, 106(2–3):162–182, 2007
12. R. Delaonoy, J. Verly, and D. Dudgeon. Pixel-level fusion using interest images. In *Proceedings of the 4th National Symposium on Sensor Fusion*, volume 1, pages 29–41. IRIA (ERIM), 1991
13. D.A. Fay et al. Fusion of multi-sensor imagery for night vision: Color visualization, target learning and search. In *3rd International Conference on Information Fusion*, pages TuD3-3–TuD3-10, 2000
14. M. Hinman. Some computational approaches for situation assessment and impact assessment. In *Proceedings of the Fifth International Conference on Information Fusion*, pages 687–693, 2002
15. N. Kwak and C. Choi. Input feature selection by mutual information based on parzen window. *IEEE Transactions Pattern Analyis and Machine Intelligence*, 24(12):1667–1671, 2002
16. L. Lazofson and T. Kuzma. Scene classification and segmentation using multispectral sensor fusion implemented with neural networks. In *Proceedings of the 6th National Sensor Symposium*, volume 1, pages 135–142, 1993
17. J. Lewis, R. O'Callaghan, S. Nikolov, D. Bull, and C. Cangarajah. Region-based image fusion using complex wavelets. In *International Conference on Information Fusion*, pages 555–562, 2004

18. N. Li, S. Dettmer, and M. Shah. Visually recognizing speech using eigensequences. In *Motion-Based Recognition*, pages 345–371. Kluwer Academic, Dordrecht, 1997

19. H. Li, B.S. Manjunath, and S.K. Mitra. Multisensor image fusion using the wavelet transform. In *Graphical Model and Image Processing*, volume 57, pages 234–245, 1995

20. J. Lowrance, T. Garvey, and T. Strat. A framework for evidential reasoning system. In *Proceedings of the Fifth National Conference on Artificial Intelligence*, pages 896–901, 1986

21. K. Mikolajczyk, A. Zisserman, and C. Schmid. Shape recognition with edge-based features. In *British Machine Vision Conference*, pages 779–788, 2003

22. B. Park and J. Marron. Comparison of data-driven bandwidth selectors. *Journal of American Statistics Association*, 85(409):66–72, 1990

23. M. Pavel, J. Larimer, and A. Ahumada. Sensor fusion for synthetic vision. In *AIAA Conference on Computing in Aerospace 8*. A Collection of Technical Papers, vol. CP9110-1, pp. 164–173, AIAA, Washington, DC, 1991

24. H. Peng, F. Long, and C. Ding. Feature selection based on mutual information criteria of max-dependency, max-relevance and min-redundancy. *IEEE Transaction Pattern Analyis and Machine Intelligence*, 27(8):1226–1238, 2005

25. V. Petrovic and C. Xydeas. Gradient-based multiresolution image fusion. *IEEE Transactions on Image Processing*, 13(2):228–237, 2004

26. G. Piella. A region-based multiresolution image fusion algorithm. In *Information Fusion*, pages 1557–1564, 2002

27. C. Ramac, M. Uner, P. Varshney, M. Alford, and D. Ferris. Morphological filters and wavelet-based image fusion for concealed weapons detection. *Proceedings of SPIE*, 3376:110–119, 1998

28. R. Raskar, A. Llie, and J. Yu. Image fusion for context enhancement and video surrealism. In *Non-Photorealistic Animation and Rendering*, pages 85–94. ACM, 2004.

29. C. Van Rijsbergen. *Information Retrieval*. 2nd ed. University of Glasgow, Glasgow, 1979

30. P. Scheunders. Multiscale edge representation applied to image fusion. In *Wavelet Applications in Signal and Image Processing VIII*, pages 894–901, 2000

31. V. Sharma and J. Davis. Feature-level fusion for object segmentation using mutual information. In *IEEE International Workshop on Object Tracking and Classification Beyond the Visible Spectrum*, 2006

32. E. Sharon, A. Brandt, and R. Basri. Completion energies and scale. *IEEE Transactions on Pattern Analysis and Machine Intelligence*, 22(10):1117–1131, 2000

33. D.A. Socolinsky and L.B. Wolff. A new visualization paradigm for multispectral imagery and data fusion. In *Proceedings of Computer Vision and Pattern Recognition*, pages 319–324, 1999

34. A. Toet. Hierarchical image fusion. *Machine Vision and Applications*, 3:1–11, 1990

35. K. Torkkola. Feature extraction by non-parametric mutual information maximization. *The Journal of Machine Learning Research*, 3:1415–1438, 2003

36. P. Varshney. *Distributed Detection and Data Fusion*. Springer Verlag, New York, 1996

37. L. Williams and D. Jacobs. Stochastic completion fields: A neural model of illusory contour shape and salience. *Neural Computation*, 9(4):837–858, 1997

38. M. Kokar, and Z. Korona. Model-based fusion for multisensor target recognition. *Proceedings of SPIE*, 2755:178–189, 1996

39. Z. Zhang and R. Blum. Region-based image fusion scheme for concealed weapon detection. In *Proceedings of the 31st Annual Conference on Information Sciences and Systems*, pp. 168–173, 1997

Chapter 14
Registering Multimodal Imagery with Occluding Objects Using Mutual Information: Application to Stereo Tracking of Humans

Stephen Krotosky and Mohan Trivedi

Abstract This chapter introduces and analyzes a method for registering multimodal images with occluding objects in the scene. An analysis of multimodal image registration gives insight into the limitations of assumptions made in current approaches and motivates the methodology of the developed algorithm. Using calibrated stereo imagery, we use maximization of mutual information in sliding correspondence windows that inform a disparity voting algorithm to demonstrate successful registration of objects in color and thermal imagery where there is significant occlusion. Extensive testing of scenes with multiple objects at different depths and levels of occlusion shows high rates of successful registration. Ground truth experiments demonstrate the utility of disparity voting techniques for multimodal registration by yielding qualitative and quantitative results that outperform approaches that do not consider occlusions. A framework for tracking with the registered multimodal features is also presented and experimentally validated.

Keywords: Multimodal stereo · Person detection/tracking · Visual surveillance · Infrared imaging

14.1 Introduction

Computer vision applications are increasingly using multimodal imagery to obtain and process information about a scene. Specifically, the disparate yet complementary nature of visual and thermal imagery has been used in recent works to obtain additional information and robustness [1,2]. The use of both types of imagery yields information about the scene that is rich in color, depth, motion, and thermal detail. Such information can then be used to successfully detect, track, and analyze people and objects in the scene.

To associate the information from each modality, corresponding data in each image must be successfully registered. In long-range surveillance applications [2], the cameras are assumed to be oriented in such a way that a global alignment

R.I. Hammoud (ed.), *Augmented Vision Perception in Infrared: Algorithms and Applied Systems,* Advances in Pattern Recognition, DOI 10.1007/978-1-84800-277-7_14,

function will register all objects in the scene. However, this assumption means that the camera must be very far away from the imaged scene. When analysis of nearer scenes is desired or necessary, the global alignment will not hold.

A minimum camera solution for registering multimodal imagery in these short-range surveillance situations would be to use a single camera from each modality, arranged in a stereo pair. Unlike colocating the cameras, arranging the cameras into a stereo pair allows objects at different depths to be registered. The stereo registration would occur on a local level, similar to the way unimodal stereo camera approaches give local registration for the left and right camera pairs. However, because of the disparate nature of the imagery, conventional stereo correspondence-matching assumptions do not hold, and care needs to be taken to ensure reliable registration of objects in the scene.

One fundamental approach to multimodal stereo registration is to utilize mutual information to assign correspondence values to the scene. Egnal has shown that mutual information is a viable similarity metric for multimodal stereo registration when the mutual information window sizes are large enough to sufficiently populate the joint probability histogram of the mutual information computation [3]. An approach by Chen et al. [4] sought to obtain these large window regions by assuming that bounding boxes (BBs) could be extracted and tracked for each object in the scene. When the assumption that segmentation and tracking of each BB is perfect, the given regions can provide for accurate registration. However, in practice, it is often difficult to obtain these necessary tracking results when there are occluding objects in the scene. In these occlusion cases, segmentation often gives BBs consisting of two or more merged objects, and the BB approach will not be able to successfully register the objects in the merged BB.

This chapter introduces an approach to registering multimodal imagery that is able to accurately register occluding objects at different disparities in the scene. A disparity voting (DV) technique that uses the accumulation of disparity values from sliding correspondence windows gives reliable and robust registration results for initial segmentations that can include occluding objects. This approach requires no prior assumptions about pixel ownership or tracking. Analysis of several thousand frames demonstrated the success of our registration algorithm for complex scenes with high levels of occlusion and numbers of objects occupying the imaged space. Experiments using both ground truth and practical segmentation illustrated how the occlusion handling of the DV algorithm is an improvement over previous approaches. A framework for tracking with the registered multimodal features is also presented and experimentally validated.

14.2 Related Research

Many of the previous works in multimodal image registration have addressed the registration problem by assuming that a global transformation model exists that will register all the objects in the scene. Davis and Sharma [2], as well as O'Conaire et al. [5], used an infinite planar homography assumption to perform registration. Under this assumption, the imaged scene will be very far from the camera, so that an object's displacement from the registered ground plane will be negligible compared to the observation distance. While this is appropriate for long-distance and overhead surveillance scenes, it is not valid when objects can be at various depths and their difference is significant relative to their distance from the camera.

Other global image registration methods assume that all registered objects will lie on a single plane in the image. It is impossible to accurately register objects at different observation depths under this assumption as the displacement and scaling for each object will depend on the varying perspective effects of the camera. This means that accurate registration can only occur when there is only one observed object in the scene [6] or when all the observed objects are restricted to lie at approximately the same distance from the camera [7]. The global alignment algorithms proposed by Irani & Anandan [8] and Coiras et al. [9] do not account or experiment when there are objects at different depths or in different planes in the image. Both utilize the assumption that the colocation of the cameras and the observed distances are such that the parallax effects can be ignored.

Multiple stereo camera approaches have been investigated by Bertozzi et al. [1]. They used four cameras configured into two unimodal stereo pairs that yielded two separate disparity estimates. Registration can then occur in the disparity domain. While this approach yields redundancy and registration success, the use of four cameras can be cumbersome in physical creation, calibration, and management, as well as in data storage and processing. A registration solution using the minimum (2) number of cameras is desired.

Chen et al. [4] introduced the idea of registering partial image regions of interest instead of finding a global transformation. The main assumption of this approach is that each BB region of interest corresponds to a single object in the scene and is at a specific plane that can be individually registered with a separate homography. They proposed that the imagery can be registered using a maximization of mutual information technique on BB that correspond to detected and tracked objects in one of the modalities. The matching BB is then searched over the other modality. Each BB is independently matched so that multiple objects at different depths can be registered.

However, a limiting requirement of this approach is that BBs can always be properly segmented and tracked in one of the modalities so that the corresponding region can be identified using the maximization of mutual information technique. While [10] relaxed this somewhat by proposing an initial silhouette extraction for BB construction, the assumption that the BBs will be properly segmented will often not hold, especially when occlusions can produce BBs that contain two or more merged objects at different depths. Objects will not be registered properly when using BBs that contain multiple objects as the required assumption that a BB is contained within a single plane will not hold. In addition, Chen et al. did not actually present any registration results where there are objects that are at significantly different depths in the scene or situations where occlusions or improperly formed BBs are an issue.

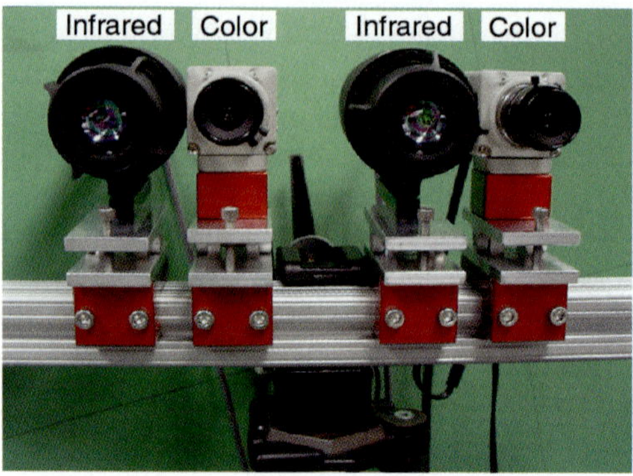

Fig. 14.1 Experimental test bed

14.3 Multimodal Test Bed

To analyze the multimodal imagery and offer a direct comparison to both unimodal color and unimodal infrared stereo setups, we have designed a test bed capable of generating the three separate, yet synchronized, stereo imageries. Utilizing a two-color, two-infrared system and a four-input frame grabber, we are able to obtain synchronized uncompressed streams from each camera. The cameras have been arranged and aligned carefully on a metal frame that supports variable baselines and easy addition, removal, and adjustment of each camera (Fig. 14.1). The cameras can be calibrated using a single calibration board to yield rectification parameters for color, thermal, and multimodal stereo pairs. Once calibrated, it is quite simple and quick to conduct experiments in a manner that can yield frame-by-frame comparison of results across individual stereo rigs.

14.4 Stereo Algorithms for Multimodal Imagery

Algorithms have been developed that utilize mutual information to solve the stereo correspondence between two images. Using mutual information to measure the similarity of potential correspondences is attractive because it is inherently robust to differences in intensities between two corresponding points. Egnal [3] is historically attributed with proposing the idea of using mutual information as a stereo correspondence-matching feature, yet results were of relatively low quality until Kim et al. [11] and subsequently Hirschmüller [12] demonstrated very successful stereo disparity generation by using mutual information in an energy minimization context. They have shown how the mutual information measure gives good results even when the images are synthetically altered by an arbitrary intensity transformation. We investigate whether these mutual information-based stereo algorithms can resolve the correspondence problem for true multimodal imagery with the same success achieved for synthetically altered imagery.

We chose to utilize the algorithm developed by Hirschmüller [12] in analyzing the use of mutual information with energy minimization for solving multimodal stereo correspondences. This choice was based on the fact that this algorithm is the mutual information-based approach that performed best on the Middlebury College Stereo Evaluation [13]. Its use of mutual information is identical to that of Kim et al. . [11], and the two algorithms differ only in how the energy function is minimized, with Kim et al. using the global optimization of *graph cuts*, while Hirschmüller utilized a faster hierarchical approach called *semiglobal matching*.

To compute mutual information in this framework, Kim et al. [11] adapted the mutual information computation to fit within the energy minimization framework. We rederive this computational framework here for convenience. The mutual information (MI) between two images I_L and I_R is defined as

$$MI_{L,R} = H_L + H_R - H_{L,R} \tag{14.1}$$

where H_L and H_R are the entropies of the two images, and $H_{L,R}$ is the joint entropy term. These entropies are defined as

$$H_L = - \int P_L(l) \log P_L(l) dl \tag{14.2}$$

$$H_{L,R} = - \int \int P_{L,R}(l,r) \log P_{L,R}(l,r) dl dr \tag{14.3}$$

where P is the probability distribution of intensities for a given image (L) or image pair (L,R), respectively. To put the entropy terms into the energy minimization framework, Kim et al. approximated the H as a sum of terms based on each pixel pair p in the imagery:

$$H_{L,R} = \sum_p h_{L,R}(L_p, R_p) \tag{14.4}$$

The joint entropy $h_{L,R}$ is computed performing Parzen estimation [twodimensional (2D) convolution with Gaussian $g(l,r)$] and approximating the probability distribution $P_{L,R}$ as the normalized 2D histogram of corresponding pixels from image pair I_L and I_R.

$$h_{L,R} = -\frac{1}{n} \log(P_{L,R}(l,r) \otimes g(l,r)) \otimes g(l,r) \tag{14.5}$$

Similarly, the entropy term is:

$$h_L = -\frac{1}{n} \log(P_L(l) \otimes g(l)) \otimes g(l) \tag{14.6}$$

From this, Kim et al. redefined mutual information as:

$$MI_{L,R} = \sum_{p} mi_{L,R}(L_p, R_p) \tag{14.7}$$

$$mi_{L,R}(l,r) = h_L(l) + h_R(r) - h_{L,R}(l,r) \tag{14.8}$$

It is this *mi* term that both Kim et al. [11] and Hirschmüller [12] used in their iterative stereo algorithm cost functions. We experimented with the stereo algorithm proposed by Hirschmüller for a variety of multimodal imagery, including color pairs, synthetically altered color pairs and paired color/infrared imagery.

Figure 14.2 shows the results of the semiglobal matching algorithm using mutual information proposed in [12] for different test images. The first row shows the results for two matched color stereo pairs. Notice how the resulting disparity image provides dense and quality estimates for the entire image. For each object in the scene, there is a silhouette of disparity that fits logically with the scene. Depth order is maintained throughout, and the overall disparity image appears similar to those reported in the stereo-matching literature [13]. These results are expected and are on par with the quality of disparity results reported in the original article. The results in the second row show the disparity image when the right image is posterized to 8 intensity levels. The results in the third row show when the right image is synthetically altered with an arbitrary transform. In this case, the transform is quite complex, and the intensity transform is not one to one, $y = 128(cos(x/15) \cdot x/255 + 1)$. Each of these disparity images gives dense and accurate estimates that are very similar to the original unaltered stereo pair. This assessment corroborates with other stereo results for synthetically altered imagery reported in [11] and [12]. In addition, the fourth row shows successful stereo matching when using two infrared images.

The final row, Fig. 14.2e, shows the results when the same algorithm is applied to multimodal stereo imagery. The resulting disparity image yields completely invalid results, and the algorithm cannot resolve any of the correct correspondences. The people in the infrared image are clearly visible, and as humans we would have no problem finding the corresponding person from the color image. The transform between color and thermal, while different from the synthetic transform, does not appear to be markedly worse, although some details, especially in the background regions, are lost. The question remains, what is fundamentally different about the infrared imagery that prevents the correct determination of correspondence values?

To try to answer this question, we need a deeper analysis of the underlying mutual information optimization scheme. At the initialization of the energy minimization algorithms, a random disparity map is chosen to initialize the probability distribution that is used to compute the mutual information terms. At this point, it is expected that the mutual information, denoted *mi* in [12] and D in [11], will appear relatively uncorrelated and give a low mutual information score. As the algorithmic iterations progress, it is desired that the *mi* values approach a maximum and the 2D *mi* plot follows the true intensity relation between the left and right images. For example, for the matched color stereo pair, the *mi* values lie along a line with negative unit slope when the correct disparity correspondences are found.

Figure 14.3 shows the *mi* plot for a pair of noncorresponding and corresponding regions for color-color (a), color-posterized (b), color-altered color (c) infrared-

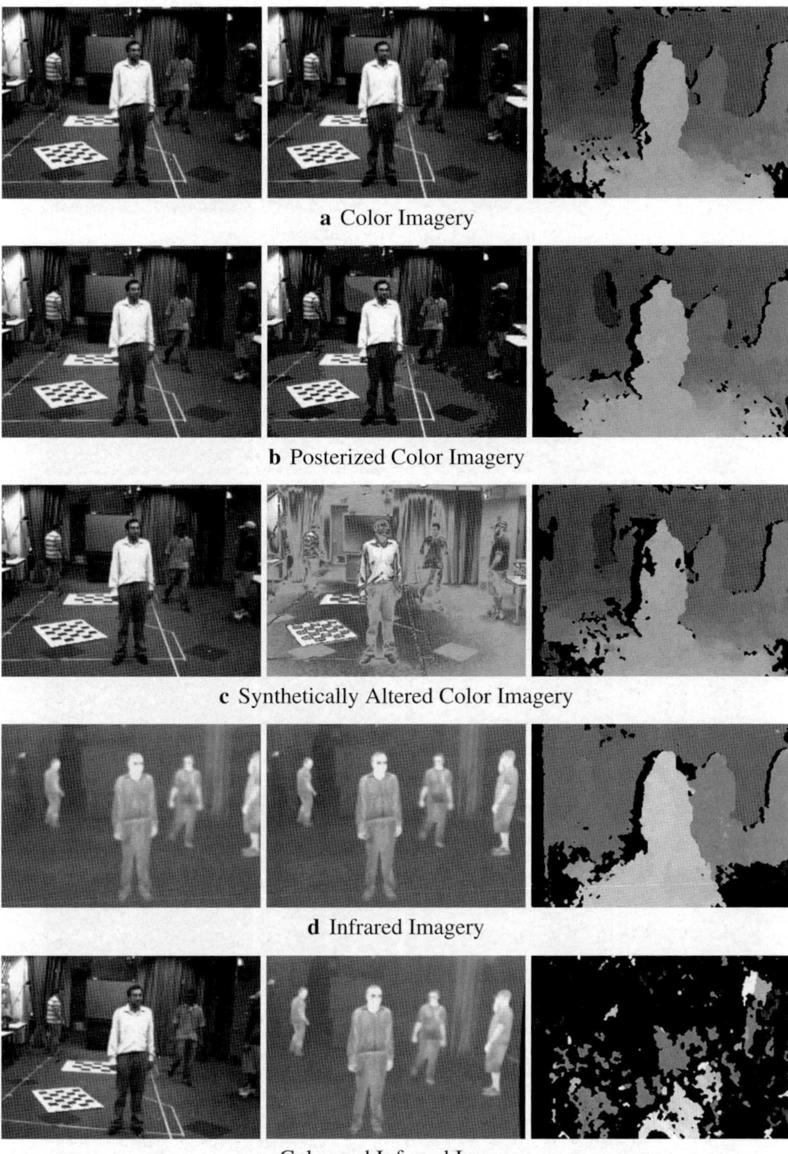

a Color Imagery

b Posterized Color Imagery

c Synthetically Altered Color Imagery

d Infrared Imagery

e Color and Infrared Imagery

Fig. 14.2 Mutual information stereo examples: Disparity results from mutual information-based stereo algorithm for different input images; disparity values are reasonable even for highly altered inputs, but the algorithm fails for natural multimodal image sets

infrared (d), and color-infrared (e) imagery. The first pair of images of each row can be thought of as starting from an initially random disparity image where most (or all) of the correspondences are incorrect. In this case, the resulting *mi* plot shows intensities that are not well correlated, as noted by its large spread across the

 S. Krotosky and M. Trivedi

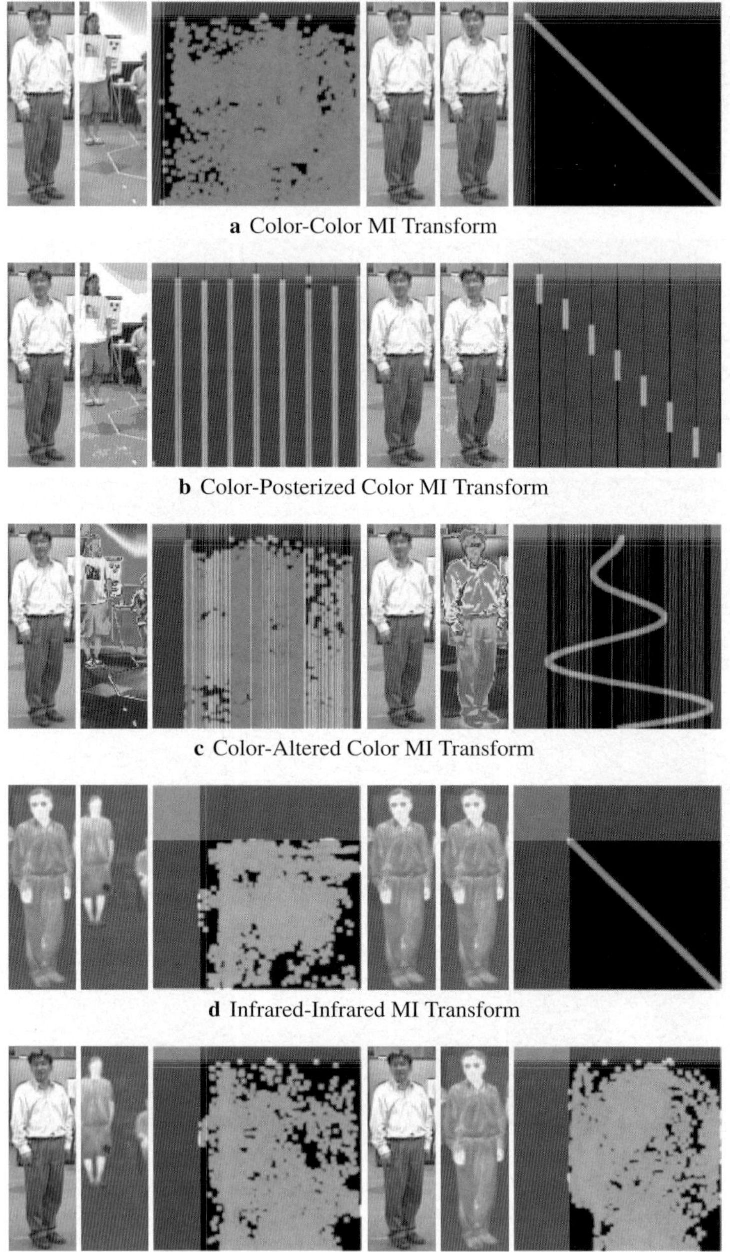

a Color-Color MI Transform

b Color-Posterized Color MI Transform

c Color-Altered Color MI Transform

d Infrared-Infrared MI Transform

e Color-Infrared MI Transform

Fig. 14.3 The *mi* plots for noncorresponding and corresponding image regions. *MI* mutual information

image 2D *mi* histogram. For the color-color, color-posterized, color-altered color, and infrared-infrared cases, when we choose corresponding image regions, the *mi* plot shows the well-correlated image intensity transform, as expected. However, for the case of corresponding color-infrared images, the *mi* value does not reduce to some easily discernable transform. In fact, the intensities for the corresponding multimodal regions appear just as uncorrelated as the intensities for the noncorresponding regions. This indicates that using these types of energy minimization algorithms is not possible with color and infrared stereo imagery. This uncorrelatedness of the color and thermal imagery means that it is difficult to predict the intensity of an infrared pixel given a corresponding color intensity. Because of this, the use of mutual information as an energy minimization term is not appropriate. The mutual information energy term (*mi* values) needs to be minimized, yet cannot be because the uncorrelation between color and thermal image intensities produces similarly large values for both good and bad matches.

14.5 Multimodal Stereo Using Primitive Matching

We have demonstrated that current state-of-the-art stereo algorithms cannot utilize mutual information to effectively solve the multimodal stereo correspondence problem. It is important to now seek out alternative features and approaches that may give some way of obtaining correspondences in the scene. To achieve any success in stereo correspondence matching with multimodal imagery, it is imperative to first identify features that are universal to both color and thermal imagery. While it is clear that there is little commonality associated with the intensities across color and thermal imagery, the example multimodal stereo pair in Fig. 14.2e suggests that there is some clear commonality on a regional (object) level and on edges associated with these region boundaries. For example, skin tone regions in the color image correspond well to bright intensity regions on the infrared image. In general, the silhouettes associated with the people in the scene have similar sizes, shapes, and edge boundaries in each modality.

Resolving stereo correspondences through regions is one of the classical approaches to utilizing image features for image matching. Traditionally, works such as those by Marapane and Trivedi [14] and Cohen et al. [15] use image segmentation to obtain regions and can achieve a coarse disparity estimate. Usually, this sort of approach is one part of a larger stereo-matching algorithm with the coarse disparity map used to guide refinements at finer detail. More recently, approaches that use the concept of oversegmentation have been applied to stereo imagery [16, 17]. By oversegmenting the image into very small regions, matching can be done in a progressive manner similar to pixel-based energy minimization functions. These oversegmentation approaches rely on the intensity similarity properties of unimodal stereo imagery and are therefore not readily extendable to the multimodal case. The challenge in applying region-based approaches to multimodal imagery lies in finding region segmentation that yields small enough regions to allow for a fine level of disparities while maintaining large enough regions to allow for reliable and robust matching.

We have developed an algorithm [18] for matching regions in a multimodal stereo context. This approach gives robust disparity estimation with statistical confidence values for each estimate for inputs that have some initial rough segmentation in each image. Currently, that segmentation is achieved through background subtraction in the color image and intensity thresholding in the infrared image. Figure 14.4 shows a flowchart outlining the algorithmic framework for our region-based stereo algorithm. Individual modules are described in the subsequent sections.

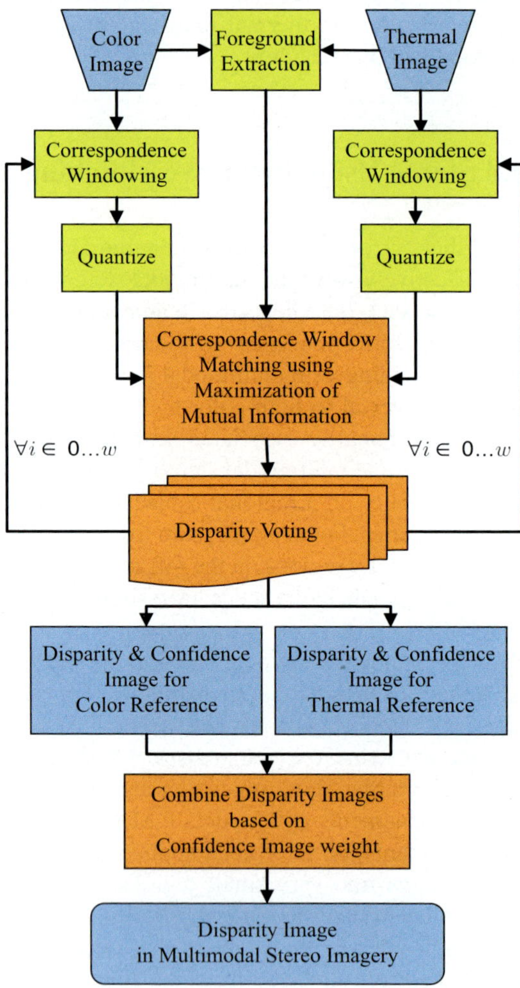

Fig. 14.4 Flowchart of disparity voting approach to region stereo for multimodal imagery

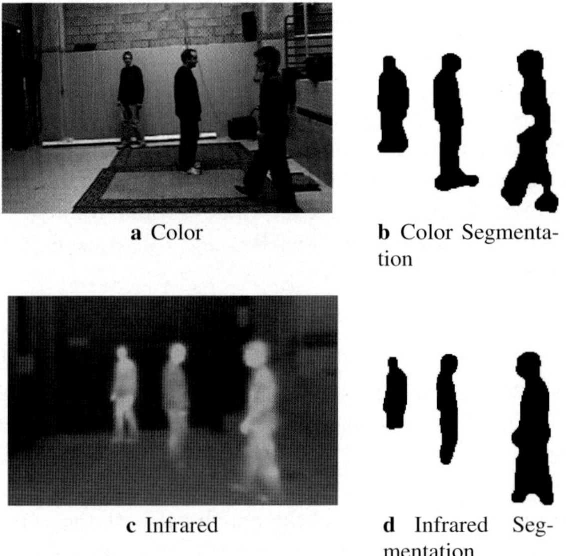

a Color

b Color Segmentation

c Infrared

d Infrared Segmentation

Fig. 14.5 Image acquisition and foreground extraction for color and thermal imagery

14.5.1 Image Acquisition and Foreground Extraction

The acquired and rectified image pairs are denoted as I_L, the left color image, and I_R, the right thermal image. Due to the high differences in imaging characteristics, it is very difficult to find correspondences for the entire scene. Instead, matching is focused on the pixels that correspond to foreground objects of interest. Naturally then, it is desirable to determine which pixels in the frame belong to the foreground. In this step, only a rough estimate of the foreground pixels is necessary, and a fair amount of false positives and negatives is acceptable. Any "good" segmentation algorithm could potentially be used with success. The corresponding foreground images are F_L and F_R, respectively. In addition, the color image is converted to gray scale for mutual information-based matching. Example input images and foreground maps are shown in Fig. 14.5.

14.5.2 Correspondence Matching Using Maximization of Mutual Information

Once the foreground regions are obtained, the correspondence matching can begin. Matching occurs by fixing a correspondence window along one reference image in the pair and sliding the window along the second image that is the best match. Let h and w be the height and width of the image, respectively. For each column

$i \in 0, \ldots, w$, let $W_{L,i}$ be a correspondence window in the left image of height h and width M centered on column i. The width M that produces the best results can be experimentally determined for a given scene. Typically, the value for M is significantly less than the width of an object in the scene. Define a correspondence window $W_{R,i,d}$ in the right image having height h^*, the largest spanning foreground distance in the correspondence window, and centered at a column $i+d$, where d is a disparity offset. For each column i, a correspondence value is found for all $d \in d_{min}, \ldots, d_{max}$.

Given the two correspondence windows $W_{L,i}$ and $W_{R,i,d}$, we first linearly quantize the image to N levels such that

$$N \approx \sqrt{Mh^*/8} \tag{14.9}$$

where Mh^* is the area of the correspondence window. The result in (14.9) comes from Thevenaz and Unser's [19] suggestion that this equation is reasonable to determine the number of levels needed to give good results for maximizing the mutual information between image regions.

Now, we can compute the quality of the match between the two correspondence windows by measuring the mutual information between them. We define the mutual information between two specific image patches as $MI_{i,d}$ where again i is the center of the reference correspondence window, and $i+d$ is the center of the second correspondence window. For each column i, we have a mutual information value $MI_{i,d}$ for $d \in d_{min}, \ldots, d_{max}$. The disparity d_i^* that best matches the two windows is the one that maximizes the mutual information:

$$d_i^* = \arg\max_{d} MI_{i,d} \tag{14.10}$$

The process of computing the mutual information for a specific correspondence window is illustrated in Fig. 14.6. An example plot of the mutual information values over the range of disparities is also shown. The red box in the color image is a visualization of a potential reference correspondence window. Candidate sliding correspondence windows for the thermal image are visualized in green boxes.

14.5.3 Disparity Voting with Sliding Correspondence Windows

We wish to assign a vote for d_i^*, the disparity that maximizes the mutual information, to all foreground pixels in the reference correspondence window. Define a DV matrix D_L of size $(h, w, d_{max} - d_{min} + 1)$, the range of disparities. Then, given a column i, for each image pixel that is in the correspondence window and foreground map, $(u,v) \in (W_{L,i} \cap F_L)$, we add to the DV matrix at $D_L(u,v,d_i^*)$.

Since the correspondence windows are M pixels wide, pixels in each column in the image will have M votes for a correspondence-matching disparity value. For each pixel (u,v) in the image, D_L can be thought of as a distribution of matching disparities from the sliding correspondence windows. Since it is assumed that all

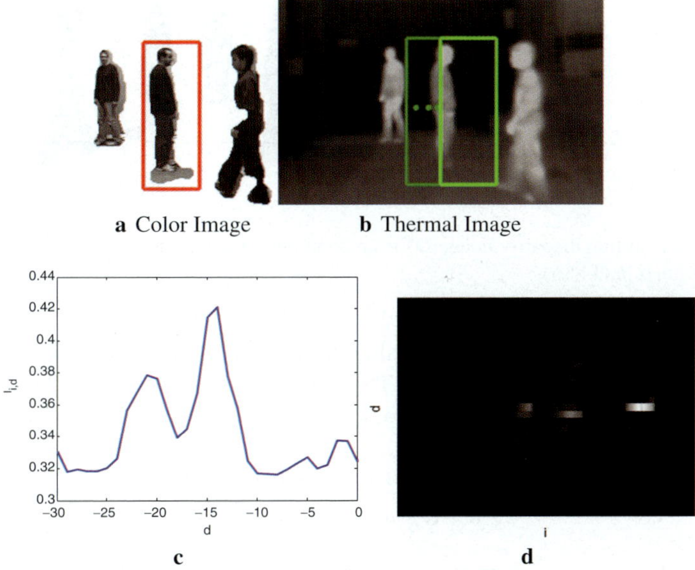

a Color Image   **b** Thermal Image

c   **d**

Fig. 14.6 Mutual information for correspondence windows

the pixels attributed to a single person are at the same distance from the camera, a good match should have a large number of votes for a single disparity value. A poor match would be widely distributed across a number of different disparity values. Figure 14.6d shows the DV matrix for a sample row in the color image. The x axis of the image is the columns i of the input image. The y axis of the image is the range of disparities $d = d_{\min}, \ldots, d_{\max}$, which can be experimentally determined based on scene structure and the areas in the scene where activity will occur. Entries in the matrix correspond to the number of votes given to a specific disparity at a specific column in the image. Brighter areas correspond to a higher vote tally.

The complementary process of correspondence window matching is also performed by keeping the right thermal infrared image fixed. The algorithm is identical to the one described, switching the left and right denotations. The corresponding disparity accumulation matrix is given as D_R.

Once the DV matrices have been evaluated for the entire image, the final disparity registration values can be determined. For both the left and right images, we determine the best disparity value and its corresponding confidence measure as

$$D_L^*(u,v) = \arg\max_d D_L(u,v,d) \tag{14.11}$$

$$C_L^*(u,v) = \max_d D_L(u,v,d) \tag{14.12}$$

For a pixel (u,v) the values of $C_L^*(u,v)$ represent the number of times the best disparity value $D_L^*(u,v)$ was voted for. A higher confidence value indicates that the disparity maximized the mutual information for a large number of correspondence

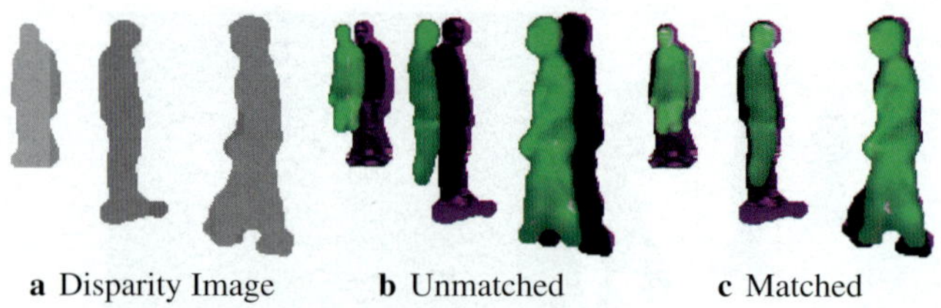

a Disparity Image   b Unmatched   c Matched

Fig. 14.7 The resulting disparity image D^* from combining the left and right disparity images D_L^* and D_S^* as defined in (14.15).

windows, and in turn, the disparity value is more likely to be accurate. Values for D_R^* and C_R^* are similarly determined. The values of D_R^* and C_R^* are also shifted by their disparities so that they align to the left image:

$$D_S^*(u + D_R^*(u,v), v) = D_R^*(u,v) \tag{14.13}$$

$$C_S^*(u + D_R^*(u,v), v) = C_R^*(u,v) \tag{14.14}$$

Once the two disparity images are aligned, they can be combined. We have chosen to combine them using an AND operation. This experimentally gives the most robust results. So, for all pixels (u,v) such that $C_L^*(u,v) > 0$ and $C_S^*(u,v) > 0$,

$$D^*(u,v) = \begin{cases} D_L^*(u,v), & C_L^*(u,v) \geq C_S^*(u,v) \\ D_S^*(u,v), & C_L^*(u,v) < C_S^*(u,v) \end{cases} \tag{14.15}$$

The resulting image $D^*(u,v)$ is the disparity image for all the overlapping foreground object pixels in the image. It can be used to register multiple objects in the image, even at very different depths from the camera. Figure 14.7 shows the result of registration for the example frame carried throughout the algorithmic derivation. Figure 14.7a shows the computed disparity image D^*, while Fig. 14.7b shows the initial alignment of the color and thermal images, and Fig. 14.7c shows the alignment after shifting the foreground pixels by the resulting disparity image. The thermal foreground pixels are overlaid (in green) on the color foreground pixels (in purple).

The resulting correspondence matching in Fig. 14.7 is successful in aligning the foreground areas associated with each of the three people in the scene. Each person in the scene lies at a different distance from the camera and yields a different disparity value that will align its corresponding image components.

14.6 Experimental Analysis and Discussion

The DV registration algorithm was tested using color and thermal data collected where the cameras were oriented in the same direction with a baseline of 10 cm. The cameras were placed so that the optical axis was approximately parallel to the

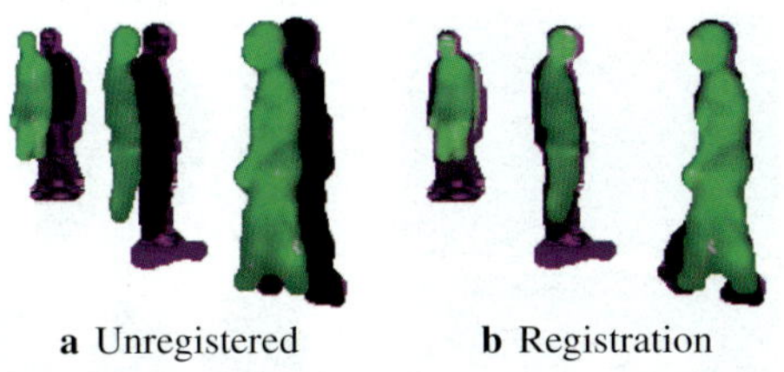

Fig. 14.8 Registration results for the disparity voting algorithm

ground. This position was used to satisfy the assumption that there would be approximately constant disparity across all pixels associated with a specific person in the frame. Placing the cameras in this sort of position is a reasonable thing to do, and such a position is appropriate for many applications. Video was captured as up to four people moved throughout an indoor environment. For these specific experiments, foreground segmentation in the visual imagery was done using the codebook model proposed by Kim et al. [20]. In the thermal imagery, the foreground is obtained using an intensity threshold under the assumption that the people in the foreground are hotter than the background. This approach provided reasonable segmentation in each image. The goal was to obtain registration results for various configurations of people, including different positions, distances from camera, and levels of occlusion.

Figure 14.8 shows the result of registration for the example frame carried throughout the algorithmic derivation. Figure 14.8a shows the initial alignment of the color and thermal images, while Fig. 14.8b shows the alignment after shifting the foreground pixels by the resulting disparity image D^* shown in Fig. 14.7. The thermal foreground pixels are overlaid (in green) on the color foreground pixels (in purple).

The resulting registration in Fig. 14.8 is successful in aligning the foreground areas associated with each of the three people in the scene. Each person in the scene lies at a different distance from the camera and yields a different disparity value that will align its corresponding image components.

Examples of successful registration for additional frames are shown in Fig. 14.9. Columns a and b show the input color and thermal images; column c illustrates the initial registration of the objects in the scene, and column d shows the resulting registration overlay after the DV has been performed. These examples show the registration success of the DV algorithm in handling occlusion and properly registering multiple objects at widely disparate depths from the camera.

14.6.1 Algorithmic Evaluation

We have analyzed the registration results of our DV algorithm for more than 2,000 frames of captured video. To evaluate the registration, we define correct registration as when the color and infrared data corresponding to each foreground object in the

S. Krotosky and M. Trivedi

a Color   **b** Infrared   **c** Unregistered   **d** Registered

Fig. 14.9 Registration results using the disparity voting algorithm for example frames

Table 14.1 Disparity voting results

Number of objects in frame	Number frames registered correctly	Total frames	Percent correct
1	55	55	100.00
2	171	172	99.42
3	1,087	1,111	97.84
4	690	720	95.83
Total	2,003	2,058	97.33

Table 14.2 Disparity voting results for frames with occlusion

Number of objects in frame	Number frames registered correctly	Total frames	Percent correct
2	51	52	98.08
3	653	677	96.45
4	581	611	95.09
Total	1,285	1,340	95.90

scene were visibly aligned. If one or more objects in the scene is not visibly aligned, then the registration is deemed incorrect for the entire frame. Table 14.1 shows the results of this evaluation. The data are broken down into groups based on the number of objects in the scene.

This analysis shows that when there was no visible occlusion in the scene, registration was correct 100% of the time. This indicates that our approach can equal the "perfect segmentation" assumption of the BB approach of Chen et al. [4]. We further break down the analysis to consider only the frames where there are occluding objects in the scene. Under these conditions, the registration success of the DV algorithm is shown in Table 14.2. The registration results for the occluded frames is still quite high, with most errors occurring during times of near-total occlusion.

14.6.2 Comparative Evaluation Using Ground Truth Disparity Values

In order to demonstrate how our DV algorithm extends and handles occlusions in an improved way over the BB approach of Chen et al. [4], we offer both a qualitative and quantitative comparison of the two approaches. It is our contention that the DV algorithm will provide good registration results during occlusions, when initial segmentation gives regions that contained merged objects. Under these circumstances, BB algorithms such as [4], which demand reliable tracking and maintenance of each object in the scene, will fail when that tracking cannot be obtained. Our DV algorithm makes no assumptions about the assignment of pixels to individual objects, only that a reasonable segmentation can be obtained. In cases when that segmentation does not include occlusion, we demonstrate successful registration on par with BB methods. In cases of occlusion, we demonstrate that the DV registration performance outperforms BB approaches and can successfully register all objects in the scene.

To demonstrate the utility of the DV algorithm for handling registration with occluding objects, we compare the BB and DV techniques when we have ground truth background segmentation. We generate the ground truth by manually segmenting the regions that correspond to foreground for each image. We then determine the ground truth disparity by individually matching each manually segmented object in the scene. This ground truth disparity image allows us to directly and quantitatively compare the registration success of the DV algorithm and the BB approach. The desire is to show that even during perfect background segmentation, BB approaches can only perform successful registration if there is no occlusion or if there is perfect tracking to assign object ownership to the pixels in the segmented foreground. By comparing the registration results to the ground truth disparities, we are able to quantify the success of each algorithm and show that the DV algorithm outperforms the BB approach for occluding object regions.

Figure 14.10 illustrates the ground truth disparity comparison tests. Column a shows the ground truth disparity, column b shows the disparity generated using the BB algorithm, and column c shows the disparity generated using the DV algorithm.

Fig. 14.10 Comparison of Region-of-interest approach [4] to the proposed disparity voting algorithm for ground truth segmentation

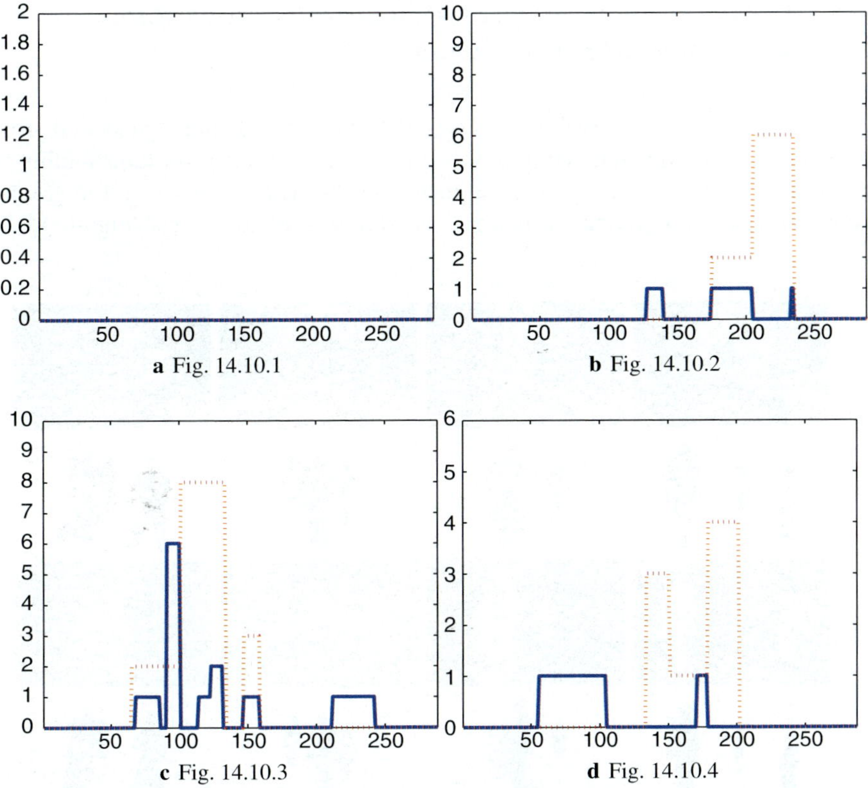

a Fig. 14.10.1 **b** Fig. 14.10.2 **c** Fig. 14.10.3 **d** Fig. 14.10.4

Fig. 14.11 Plots of $|\Delta D|$ from ground truth for each example in Fig. 14.10. Bounding box errors for an example row plotted in dotted red, errors in disparity voting registration plotted in solid blue

Figure 14.11 plots the absolute difference in disparity values from the ground truth for each corresponding row in Fig. 14.10. The BB results are plotted in dotted red, while the DV results are plotted in solid blue. Notice how the two algorithms perform identically to ground truth in the first row as there are no occlusion regions. The subsequent examples all have occlusion regions, and the DV approach more closely follows ground truth than the BB approach. The BB registration results have multiple objects registered at the same depth, although the ground truth shows that they are at separate depths. Our DV algorithm is able to determine the distinct disparities for different objects, and the $|\Delta \text{ Disparity}|$ plots show that the DV algorithm is quantitatively closer to the ground truth, with most registration errors within one pixel of ground truth, with larger errors usually occurring only in small portions of the image. On the other hand, when errors occur in the BB approach, the resulting disparity offset error is large and occurs for the entire scope of erroneously registered object.

14.6.3 Comparative Assessment of Registration Algorithms with Nonideal Segmentation

We perform a qualitative evaluation using the real segmentations generated from codebook background subtraction in the color image and intensity thresholding in the thermal image. These common segmentation algorithms only give foreground pixels and make no attempt to discern the structure of the pixels. Figure 14.12

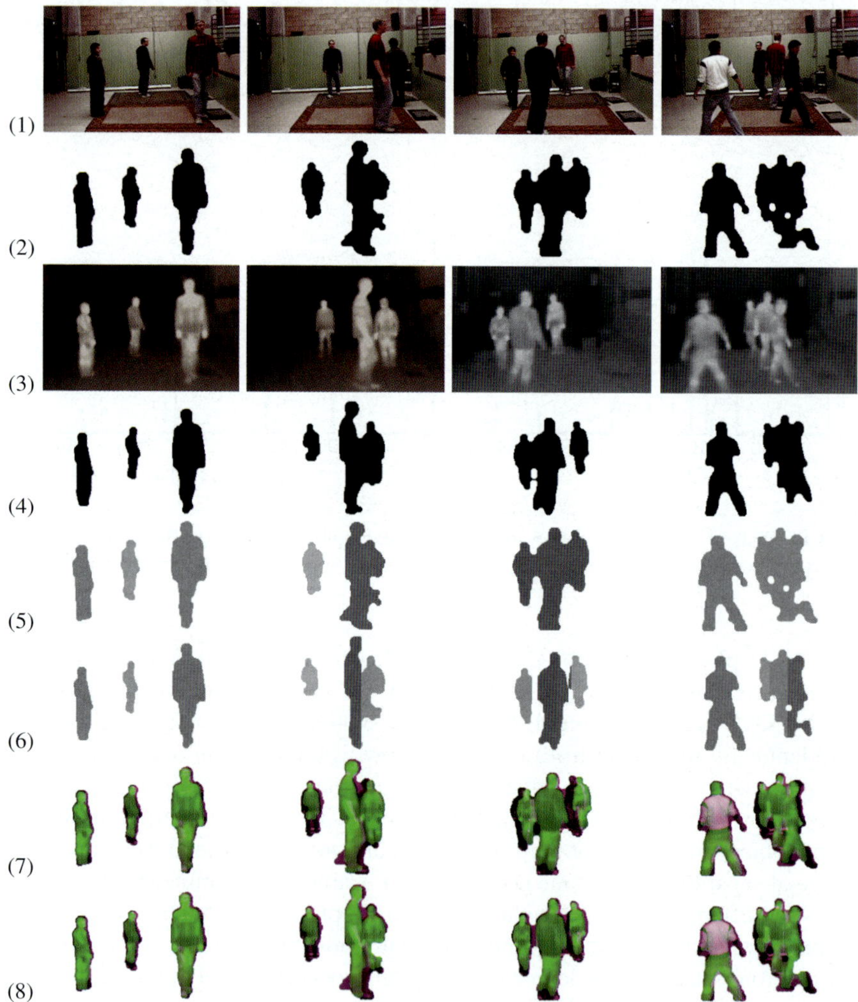

Fig. 14.12 Comparison of bounding box (BB) algorithm [4] to the proposed disparity voting (DV) algorithm for a variety of occlusion examples using nonideal segmentation: *1* the color image, *2* the color segmentation, *3* the thermal image, *4* the thermal segmentation, *5* the BB disparity image, *6* the DV disparity image, *7* the BB registration, *8* the DV registration.

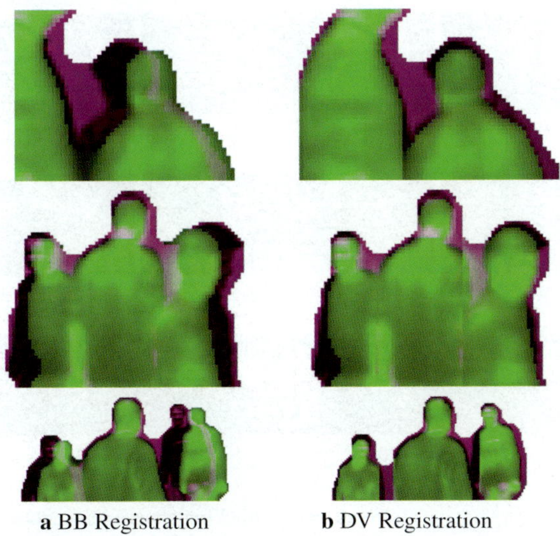

Fig. 14.13 Details of registration alignment errors in the bounding box (BB) registration approach and corresponding alignment success for the disparity voting (DV) algorithm for several occlusion examples using nonideal segmentation

illustrates several examples that compare the registration results of the DV and BB algorithms. Notice how the disparities for the BB algorithm in row 5 are constant for the entire occlusion region even though the objects are clearly at very different disparities. The disparity results for our DV algorithm in row 6 show distinct disparities in the occlusion regions that correspond to the appropriate objects in the scene. Visual inspection of rows 7 and 8 show that the resulting registered alignment from the disparity values is more accurate for the DV approach. Figure 14.13 shows the registration alignment for each algorithm in closer detail for a selection of frames. Notice how the DV approach is able to align each object in the frame, while the BB approach has alignment errors due to the fact that the segmentation of the image yielded BBs that contained more than one object. Clearly, DV is able to handle the registration in these occlusion situations, and the resulting alignment appears qualitatively better than the BB approach.

14.7 Multimodal Video Analysis for Person Tracking: Basic Framework and Experimental Study

We have shown that the DV algorithm for multimodal registration is a robust approach to estimating the alignment disparities in scenes with multiple occluding people. The disparities generated from the registration process yield values that can be used to differentiate the people in the room. It is with this in mind that we investigate the use of multimodal disparity as a feature for tracking people in a scene.

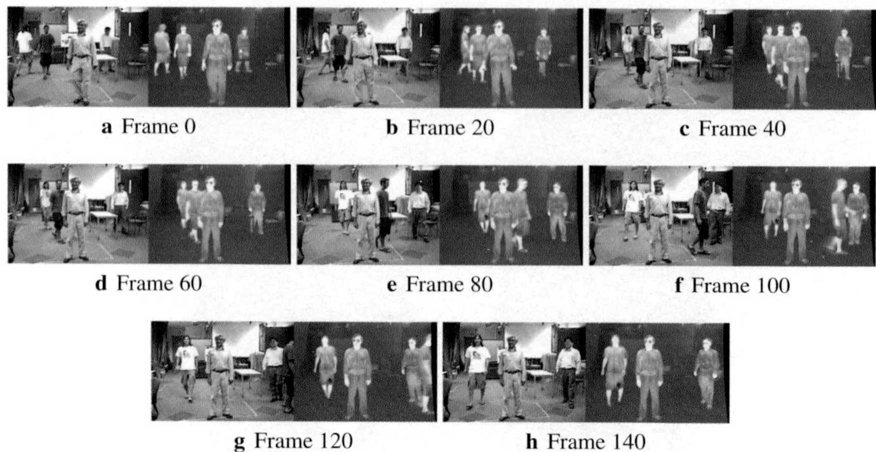

Fig. 14.14 Example input sequence for multiperson tracking experiments; notice occlusions, scale, appearance, and disparity variations

Tracking human motion using computer vision approaches is a well-studied area of research, and a good survey by Moeslund and Granum [21] gave lucid insight into the issues, assumptions, and limitations of a large variety of tracking approaches. One approach, disparity-based tracking, has been investigated for conventional color stereo cameras and has proven quite robust in localizing and maintaining tracks through occlusion as the tracking is performed in 3D space by transforming the stereo image estimates into a plan-view occupancy map of the imaged space [22]. We wish to explore the feasibility of using such approaches to tracking with the disparities generated from DV registration. An example sequence of frames in Fig. 14.14 illustrates the type of people movements we aim to track. The sequence has multiple people occupying the imaged scene. Over the sequence, the people move in a way that there is multiple occlusions of people at different depths. The registration disparities that are used to align the color and thermal images can be used as a feature for tracking people through these occlusions and maneuvers.

Figure 14.15 shows an algorithmic framework for multimodal person tracking. In tracking approaches, representative features are typically extracted from all available images in the setup [23]. Features are used to associate tracks from frame to frame, and the output of the tracker is often used to guide subsequent feature extraction. All of these algorithmic modules are imperative for reliable and robust tracking. For our initial investigations, we focus on the viability of registration disparity as a tracking feature.

To determine the accuracy of the disparity estimates for tracking, we first calibrate the scene. This is done by having a person walk around the test bed area, stopping at preset locations in the scene. At each location, we measure the disparity generated from our algorithm and use that as ground truth for analyzing the disparities generated when there are more complex scenes with multiple people

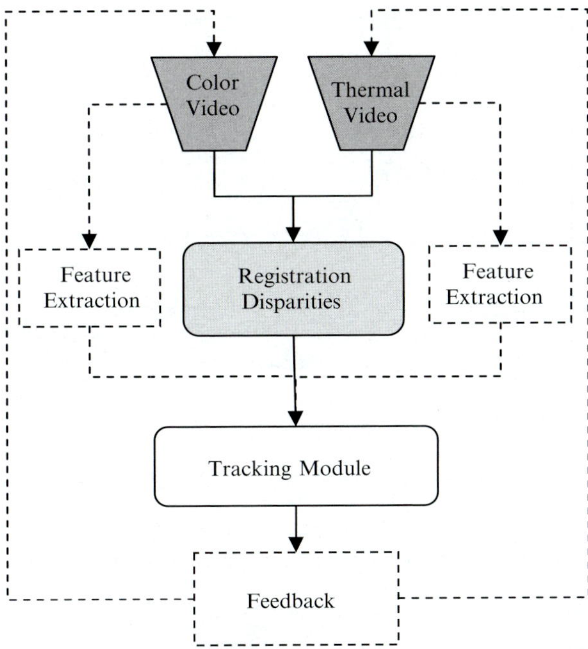

Fig. 14.15 Algorithmic flowchart for multiperson tracking

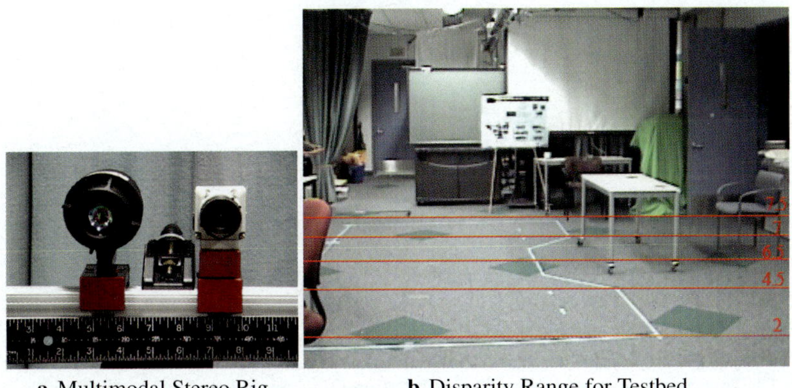

a Multimodal Stereo Rig   **b** Disparity Range for Testbed

Fig. 14.16 a Variable baseline multimodal stereo rig, **b** experimentally determined disparity range for test bed; disparities computed by determining disparities for a single person standing at predetermined points in the imaged scene

and occlusions. Figure 14.16a is the variable baseline multimodal stereo rig and Fig. 14.16b shows the ground truth disparity range for the test bed from the calibration experiments captured with this rig.

To show the viability of registration disparity as a tracking feature in a multimodal stereo context, we compare ground truth positional estimates to those

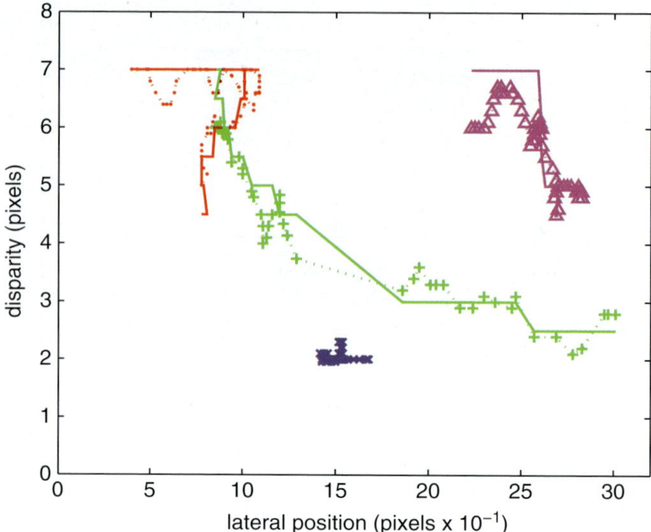

a Track patterns and ground truth for four-person tracking experiment

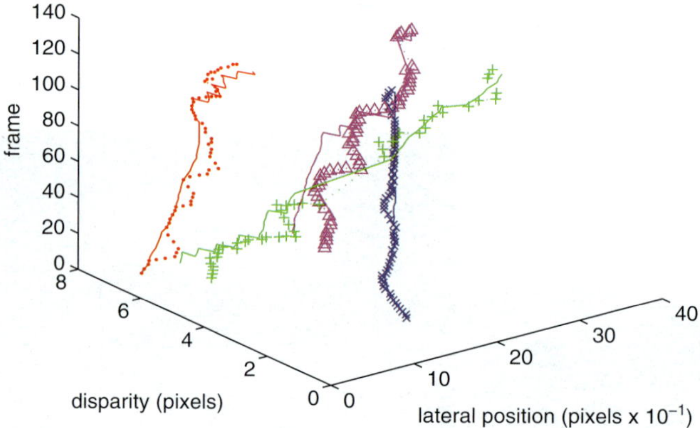

b Time-varying track patterns and ground truth for four person tracking experiment

Fig. 14.17 Tracking results showing close correlation between ground truth (in solid colors) and disparity tracked estimates (in dotted colors); each color shows the path of each person in the sequence

generated from the DV algorithm. Lateral position information for each track was hand segmented by clicking on the center point of the person's head in each image. This is a reasonable method as robust head detection algorithms could be implemented for both color and thermal imagery (skin tone, hot spots, head template matching). Approaches such as vertical projection or v-disparity could also be used to determine the locations of people in the scene. Ground truth disparity estimates were generated by visually determining the disparity based on the person's position relative to the ground truth disparity range map as shown in Fig. 14.16. Experimental disparities were generated using the DV algorithm with the disparity of each person determined from disparity values in the head region. A moving average of 150 ms was used to smooth instantaneous disparity estimates.

Figure 14.17 shows the track patterns and ground truth for the example sequence in Fig. 14.14. The ground truth is plotted in solid colors for each person in the sequence, while the disparity estimates from the DV algorithm are shown in corresponding colored symbols with dotted lines connecting the estimates. Figure 14.17a is a representation of the tracks, illustrating a "plan-view"-like representation of the movements and disparity changes of the people in the test bed. Figure 14.17b shows a time-varying version of the same data, with the frame number plotted in the third dimension.

The plots in Fig. 14.17 show that the disparities generated from the DV registration reasonably follow the ground truth tracks. As the green tracked person moves behind and becomes occluded by the blue tracked person, we see that the disparities generated when the person reemerges from the occlusion are in line with the ground truth disparities and can be used to reassociate the track after the occlusion.

Errors from ground truth are particularly apparent when people are further from the camera. This is because of the nonlinearity of the disparity distribution. There are more distinct disparities nearer to the camera. As you move deeper in the scene in Fig. 14.16, the change in disparity for the same change in distance is much less. At these distances, errors of even one disparity shift are very pronounced. Conventional stereo algorithms typically used approaches that give subpixel accuracy, but the current implementation of our DV algorithm only gives pixel-level disparity shifts. While this may be acceptable for registration alignment, refinement steps are necessary to make disparity a more robust tracking feature. Approaches that use multiple primitives [24], such as edges, shapes, silhouettes, and the like, could be used to augment the accuracy of the DV algorithm. In addition, using multiple tracking features could provide additional measurements that can be used to boost the association accuracy.

14.8 Summary and Concluding Remarks

In this chapter, we have introduced and analyzed a method for registering multimodal images with occluding objects in the scene. By using a DV approach, the algorithm has given successful and reliable registration without relying on any assumptions about the tracked ownership of pixels to object regions in the scene. An analysis of over 2,000 frames yielded a registration success rate of over 97%, with a 96% success rate when considering only occlusion examples. In addition, ground truth and segmentation comparisons illustrate how the DV algorithm improves the registration accuracy and robustness of previous BB techniques in both quantitative and qualitative evaluations. DV gives the ability to determine accurate registration disparities for occluded objects that can be used as a feature of objects in the scene for further detection, tracking, and analysis.

Multimodal imagery applications for human analysis span a variety of application domains, including medical [25], in-vehicle safety systems [26], and long-range surveillance [2]. Typically, these types of systems do not operate on data that have multiple objects and multiple depths that are significant relative to their distance from the camera. It is in this realm, including short-range surveillance [18] and pedestrian detection applications [27], that we believe DV registration techniques will prove useful.

Acknowledgment We would like to thank our research sponsors, the U.S. Department of Defense Technical Support Working Group and the U.C. Discovery Grant. In addition, we owe gratitude to the members of the Computer Vision and Robotics Research Laboratory, particularly Dr. Joel McCall and Mr. Shinko Cheng.

We would also like to express our thanks to the reviewers for their comments and assistance, which helped us improve our research.

Chapter's References

1. Bertozzi, M., Broggi, A., Felias, M., Vezzoni, G., Rose, M.D. (2006) Low-level pedestrian detection by means of visible and far infra-red tetra-vision. In: IEEE Conference on Intelligent Vehicles
2. Davis, J., Sharma, V. (2005) Fusion-based background-subtraction using contour saliency. In: IEEE CVPR Workshop on Object Tracking and Classification beyond the Visible Spectrum
3. Egnal, G. (2000) Mutual information as a stereo correspondence measure. Technical Report MS-CIS-00-20, University of Pennsylvania
4. Chen, H., Varshney, P., Slamani, M. (2003) On registration of regions of interest (ROI) in video sequences. In: IEEE International Conference on Advanced Video and Signal Based Surveillance (AVSS'03), pp. 313
5. Conaire, C.O., Cooke, E., O'Connor, N., Murphy, N., Smeaton, A. (2005) Background modeling in infrared and visible spectrum video for people tracking. In: IEEE CVPR Workshop on Object Tracking and Classification beyond the Visible Spectrum
6. Han, J., Bhanu, B. (2003) Detecting moving humans using color and infrared video. In: IEEE International Conference on Multisensor Fusion and Integration for Intelligent Systems
7. Itoh, M., Ozeki, M., Nakamura, Y., Ohta, Y. (2003) Simple and robust tracking of hands and objects for video-based multimedia production. In: IEEE Conference on Multisensor Fusion and Integration for Intelligent Systems
8. Irani, M., Anandan, P. (1998) Robust multi-sensor image alignment. In: Sixth International Conference on Computer Vision, 1998
9. Coiras, E., Santamaria, J., Miravet, C. (2000) Segment-based registration technique for visual-infrared images. Optical Engineering 39(1), 282–289
10. Chen, H., Lee, S., Rao, R., Slamani, M., Varshney, P. (2005) Imaging for concealed weapon detection. Signal Processing Magazine, IEEE Vol. 22, Issue 2, March 2005, pp. 52–61
11. Kim, J., Kolmogorov, V., Zabih, R. (2003) Visual correspondence using energy minimization and mutual information. In: Ninth IEEE International Conference on Computer Vision
12. H. Hirschmüller (2005) Accurate and efficient stereo processing by semi-global matching and mutual information. In: Computer Vision and Pattern Recognition
13. Scharstein, D., Szeliski, R. (2005) Middlebury College stereo vision research page. http://bj.middlebury.edu/~schar/stereo/web/results.php

14. Marapane, S., Trivedi, M. (1989) Region-based stereo analysis for robotic applications. IEEE Transactions on Systems, Man, and Cybernetics, Special Issue on Computer Vision 19(6), 1447–1464
15. Cohen, L., Vinet, L., Sander, P., Gagalowicz, A. (1989) Hierarchical region based stereo matching. In: Computer Vision and Pattern Recognition
16. Wei, Y., Quan, L. (2004) Region-based progressive stereo matching. In: Computer Vision and Pattern Recognition
17. Bleyer, M., Gelautz, M. (2005) Graph-based surface reconstruction from stereo pairs using image segmentation. Proc. SPIE 5665, 288–299
18. Krotosky, S.J., Trivedi, M.M. (2006) Registration of multimodal stereo images using disparity voting from correspondence windows. In: IEEE Conference on Advanced Video and Signal based Surveillance (AVSS'06)
19. Thevenaz, P., Unser, M. (2000) Optimization of mutual information for multiresolution image registration. IEEE Transactions on Image Processing 9(12), 2083–2089
20. Kim, K., Chalidabhongse, T., Harwood, D., Davis, L. (2005) Real-time foreground-background segmentation using codebook model. Real-Time Imaging 11(3), 163–256
21. Moesland, T.B., Granum, E. (2001) A survey of computer vision-based human motion capture. Computer Vision and Image Understanding 81(3), 231–268
22. Harville, M., Li, D. (2004) Fast, integrated person tracking and activity recognition with plan-view templates from a single stereo camera. In: IEEE Conference on Computer Vision and Pattern Recognition
23. Huang, K., Trivedi, M.M. (2003) Video arrays for real-time tracking of person, head, and face in an intelligent room. Machine Vision and Applications 14(2), 103–111
24. Marapane, S., Trivedi, M.M. (1994) Multi-primitive hierarchical (MPH) stereo analysis. IEEE Transactions on Pattern Analysis and Machine Intelligence 16(3), 227–240
25. Thevenaz, P., Bierlaire, M., Unser, M.: Halton sampling for image registration based on mutual information. Sampling Theory in Signal and Image Processing, May, 2008
26. Trivedi, M.M., Cheng, S.Y., Childers, E.M.C., Krotosky, S.J. (2004) Occupant posture analysis with stereo and thermal infrared video: Algorithms and experimental evaluation. IEEE Transactions on Vehicle Technology 53(6), 1968–1712
27. Krotosky, S.J., Trivedi, M.M. (2006) Multimodal stereo image registration for pedestrian detection. In: IEEE Conference on Intelligent Transportation Systems

Chapter 15
Thermal-Visible Video Fusion for Moving Target Tracking and Pedestrian Motion Analysis and Classification

Yang Ran, Alex Leykin, and Riad Hammoud

Abstract This chapter presents a novel system for pedestrian surveillance, including tasks such as detection, tracking, classification, and possibly activity analysis. The system we propose first builds a background model as a multimodal distribution of colors and temperatures. It then constructs a particle filter scheme that makes a number of informed reversible transformations to sample the model probability space to maximize posterior probability of the scene model. Observation likelihoods of moving objects account their three-dimensional locations with respect to the camera and occlusions by other tracked objects as well as static obstacles. After capturing the coordinates and dimensions of moving objects, we apply a classifier based on periodic gait analysis. To differentiate humans from other moving objects such as cars, we detect a symmetrical double-helical pattern in human gait. Such pattern can then be analyzed using the frieze group theory. The results of tracking in color and thermal sequences demonstrate that our algorithm is robust to illumination change and performs well in outdoor environments.

15.1 Introduction

Real-time pedestrian surveillance has gained a lot of attention in the machine vision community. These tasks have been identified as one of the key issues in numerous applications, ranging from collision avoidance in the automotive industry, border safety, to situation awareness in security and robotic systems [2, 12, 21, 27]. Human motion analysis based on color sensors already has been producing reliable results for indoor scenes with the constant illumination and steady backgrounds. However, outdoor scenes with significant background clutter due to illumination changes still appear to be challenging to handle using inputs from a conventional charge-coupled device (CCD) camera. Sensor fusion has become an increasingly important direction in computer vision and in particular human tracking systems in recent years.

R.I. Hammoud (ed.), *Augmented Vision Perception in Infrared: Algorithms and Applied Systems,* Advances in Pattern Recognition, DOI 10.1007/978-1-84800-277-7_15,
© Springer-Verlag London Limited 2009

A single sensory path might not provide a crucial piece of input data or might become unusable under a varying environment, such as the color camera under low-illumination conditions.

To locate pedestrians in videos, two important tasks need to be accomplished: tracking and classification. We tackle this problem with a sensor fusion approach. Human hypotheses are first generated by a shape-based detector of the foreground blobs using a human shape model. The human hypotheses are tracked with an Markov chain Monte Carlo (MCMC) filter. Hypotheses are verified while they are tracked for the first second or so. The verification is done by walking motion recognition using a novel double-helical signature (DHS) in gait.

We developed our generative tracking framework encouraged by recently found implementations for particle filtering. Random sampling was shown not only to successfully overcome singularities in articulated motion [7, 23], but also the particle-filtering approach applied to human tracking has demonstrated potential in resolving ambiguities while dealing with crowded environments [15, 35]. Working under the Bayesian framework, the particle filters can efficiently infer both the number of objects and their parameters. Another advantage is that in dealing with distributions of mostly unknown nature, particle filters do not make Gaussianity assumptions, unlike Kalman filters [16, 31]. The first contribution of the proposed method is to naturally integrate our tracker with spatial-temporal motion information by fusing color and infrared (IR) videos in a particle filter framework. The other contribution lies in a novel signature based on human gait: DHS, or double helical signature. It is concise and efficient and brings a reliable method to classify humans, associates targets across cameras, and recognize gait and activities.

We first review related work in pedestrian tracking and motion-based classification. In Section 15.3 we briefly describe the contributions of this chapter. In Section 15.4 we give an overview of our proposed system with every individual module's functionality identified. We present the tracker in Section 15.5 and the motion-based gait signature in Section 15.6. In Section 15.7, we provide preliminary results for both of the modules using color and IR sequences. Conclusions can be found in Section 15.8.

15.2 Related Work

In the areas of medical imaging, sports video, and video surveillance research, human motion analysis has become popular. Especially, human tracking and activity recognition are most commonly studied. Human tracking can be performed in two or three dimensions. Depending on the complexity of analysis, representations of the human body range from basic stick figures to volumetric models. Tracking relies on the correspondence of image features between consecutive frames, taking into consideration information such as position, color, shape, and texture. Pedestrian classification based on motion acts as a verification module. In this section, we briefly summarize the related work for these two topics, respectively.

15.2.1 Tracking Review

Substantial research has been accumulated in tracking of people. The majority of the studies addressed tracking of isolated people in well-controlled environments, but increasingly there is more attention to tracking specifically in crowded environments [3, 9, 11–13, 22]. It is worth noting that many works assume the luxury of multiple well-positioned cameras or stereo vision, which are to a certain extent not present in most establishments or do not have the desired overlapping fields of view. In contrast, cheap low-resolution digital monocular color cameras are becoming more and more readily available in stores, airports, and other public places as well as the hardware for capturing compressed real-time streams provided by these cameras. Recently, a flurry of contributions on pedestrian localization and tracking in visible and IR videos have appeared in the literature [1, 4, 10, 25, 33]. In [34], the P-tile method was developed to detect the human head first, and then the human torso and legs are included by local search. Nanda [25] built a probabilistic shape hierarchy to achieve efficient detection at different scales. In [28], a particle swarm optimization algorithm was proposed for human detection in IR imagery. Dai et al. [4] proposed a hybrid (shape plus appearance) algorithm for pedestrian detection in which shape cue is first used to eliminate nonpedestrian moving objects and an appearance cue is then used to pin down the location of pedestrians. A generalized expectation maximization algorithm has been employed by the authors to decompose IR images into background and foreground layers. These approaches rely on the assumption that the person region has a much hotter appearance than the background. Davis et al. [6] proposed to fuse thermal and color sensors in a fusion-based background subtraction framework using a contour saliency map in urban settings. Information, including object locations and contours from both synchronized sensors, is fused to extract the object silhouette. A higher performance is reported by fusing both sensors over visible-only and thermal-only imagery. This method is, however, computationally expensive as it attempts to construct a complete object contour, which does not seem a requirement in various applications like surveillance or a crash avoidance system. In [33], support vector machine and Kalman filtering were adopted for detection and tracking, respectively. In [30], two pedestrian-tracking approaches, pixel periodicity and model fitting, were proposed based on gait. The first employs computationally efficient periodicity measurements. Unlike other methods, it estimates a periodic motion frequency using two cascading hypothesis-testing steps to filter out noncyclic pixels so that it works well for both radial and lateral walking directions. The extraction of period is efficient and robust with respect to sensor noise and cluttered background. In the second method, the authors integrated shape and motion by converting the cyclic pattern into a binary sequence by maximal principal gait angle (MPGA) fitting.

15.2.2 Motion-based Classification Review

One of the most routine actions humans perform is walking. The review by Gavrila [10] categorized human motion analysis work according to whether an explicit shape model was used and the dimensionality of the model space. In another recent review work by Liang in [32], a hierarchical summary is given for research in related areas.

Many existing human motion analysis methods use a contour- or shape-based, stick figure, or volumetric model in the image domain that implicitly utilizes temporal information. Other methods, like the algorithm in this chapter, explicitly apply a temporal model. Several solutions have been proposed for characterizing the temporal periodicity, and they could be divided into two major categories. One is to analyze periodic motion at the shape or silhouette level and the other at the pixel level. Little and Boyd analyzed the shape of motion and used it for real-time target classification [19]. Niyogi and Adelson [26] presented a multiscale spatiotemporal filter bank for motion perception. In [26], the motion is represented in the X-Y-t space convolving with specific impulse responses. In the second category, Yang [29] introduced video phase-lock loop for perceiving the oscillations at the pixel level. Liu and Picard [20] found periodicity by applying Fourier analysis along pixels' trajectory. There is a major drawback in most of the algorithms in that they do not use knowledge of human kinematics. On the other hand, methods based on complicated body models require tracking of body feature points or markers, which is unreliable in many cases.

Because of the upright pose and translational global body displacement along ground surface during human walking, considering walking motion in the spatiotemporal domain is reasonable. We observe strong periodic pattern in such domain, which assemble the patterns in decorative texture or crystal structure.

15.3 Chapter's Contributions

The contribution of our tracking algorithm is twofold: First, it employs all available information to achieve the noise-free blob-map; second, subsequent it uses the blob map to perform reliable pedestrian tracking to minimize two types of tracking errors—falsely detected people and people missed by the system.

The contribution of motion-based pedestrian classification lies in the characterization of the signature generated by a walking human. To describe the computational model for this periodic helical pattern, we take the mathematical theory of symmetry groups, which is widely used in crystallographic structure research. Both observations and biometrics prove that spatiotemporal human walking patterns belong to the frieze groups because they are characterized by a repetitive motif in the direction of walking.

Despite these efforts, the challenge remains whether to use a stationary or moving imagery system. This is due to a number of key factors, like lighting changes

Fig. 15.1 Multimodal imagery of the same scene: *left* thermal frame of the scene vs *right* color frame

(shadow vs. sunny day, indoor/night vs. outdoor); cluttered backgrounds (trees, vehicles, animals); artificial appearances (clothing, portable objects); nonrigid kinematics of pedestrians; camera and object motions; depth and scale changes (child vs. adult); and low video resolution and image quality. This chapter proposes a pedestrian detection-and-tracking approach that combines both thermal and visible information (see Fig. 15.1) and subsequently models the motion in the scene using a Bayesian framework.

15.4 System Overview

The overview diagram of our approach is shown in Fig. 15.2. Our system first segments foreground regions out of each frame by using a dynamically adapting background model. We hypothesize about the number of human bodies within each such region by using the head-candidate selection algorithm. As the next step, our system constructs a Bayesian inference model based on the a priori knowledge of the human parameters and scene layout and geometry. Observations of the body appearances at each frame are a second driving force in our probabilistic scheme. Finally, the double-helical pattern detection is applied to the output object sequence for classification in both color and IR videos.

15.5 Tracking

15.5.1 Multimodal Pixel Representation

Each pixel in the image is modeled as two dynamically growing vectors of codewords. A *codeword* can be viewed as a single modality in a multimodal distribution. For the RGB (red-green-blue) input, a codeword is represented by the average pixel

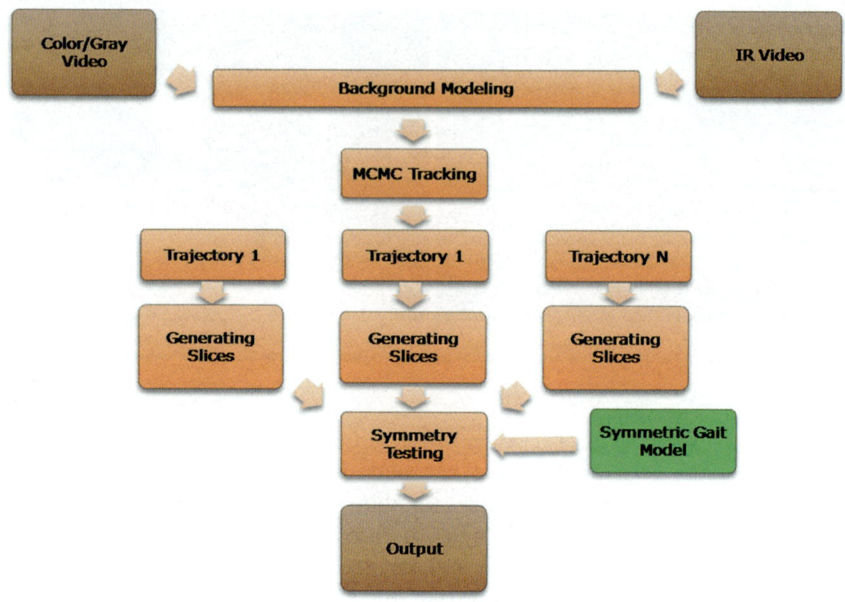

Fig. 15.2 System flow diagram

RGB value and by the luminance range I_{low} and I_{hi} allowed for this particular codeword. If an incoming pixel is within the luminance range and the dot product of p_{RGB} and *RGB* of the codeword is less than a predefined threshold, it is considered to belong to the background. For the thermal monochromatic input, a codeword is represented by intensity range T_{low} and T_{hi} occurring at the pixel location. Unlike for the color codewords, the matching of the incoming pixel temperature p_T is done by comparing the ratios of p_T/T_{low} and p_T/T_{hi} to the empirically set thresholds. This way, we can hard limit the percentage of temperature change allowed to happen at each location. By observing several thermal sequences, we have established that changes in cloud cover or shadows produced by other moving objects do not typically cause a temperature change of more than 10%.

During the model acquisition stage, the values are added to the background model at each new frame if there is no match found in the already existing vector. Otherwise, the matching codeword is updated to account for the information from the new pixel. Empirically, we have established that there is seldom an overlap between the codewords. When this is the case (i.e., more than one match has been established for the new pixel), we merge the overlapping codewords. We assume that the background changes due to compression and illumination noise are of reoccurring nature. Therefore, at the end of training we clean up the values ("stale" codewords) that have not appeared for periods of time greater than some predefined percentage of frames in the learning stage as not belonging to the background. We keep in each codeword a so-called maximum negative run length *(MNRL)*, which is the longest interval during the period that the codeword has not occurred. One additional benefit of this modeling approach is that, given a significant learning period, it is not essential that the frames be free of moving foreground objects. The background model can be learned on the fly and is helpful when tracking and model acquisition are done simultaneously. A more in-depth description of our background modeling can be found in [17, 18].

15.5.2 Bayesian Model: Observations and States

We formulate the tracking problem as the maximization of posteriori probability of the Markov chain state. To implement the Bayesian inference process efficiently, we model our system as a Markov chain $M = \{x, z, x_0\}$, where x_0 is the initial state of the chain, x is the current state, and z is the observation. To converge to the optimal state of M, we employ a variant of Metropolis-Hastings particle-filtering algorithm [8]. The state of the system at each frame is an aggregate of the state of each body $x_t = \{b_1, \ldots, b_n\}$. Each body, in order, is parametrically characterized as $b_i = \{x, y, h, w, c\}$, where x, y are coordinates of the body on the floor map; h and w represent the hight and width, respectively, measured in centimeters; and c is a two-dimensional (2D) color histogram, represented as 32 by 32 bins in hue saturation space. The body is modeled by the ellipsoid with the axes h and w. An additional implicit variable of the model state is the number of tracked bodies n.

15.5.3 Computing Posterior Probability

The goal of our tracking system is to find the candidate state x' (a set of bodies along with their parameters) that, given the last known state x, will best fit the current observation z. Therefore, at each frame we aim to maximize the posterior probability

$$P(x'|z, x) = P(z|x') \cdot P(x'|x) \tag{15.1}$$

According to Bayes rule and given (15.1), we formulate our goal as finding

$$x' = argmax_{x'} (P(z|x') \cdot P(x'|x)) \tag{15.2}$$

The right-hand side of Eq. (15.2) is comprised of the observation likelihood and the state prior probability. They are computed as joint likelihoods for all bodies present in the scene as described below.

15.5.3.1 Priors

In creating a probabilistic model of a body, we considered three types of prior probabilities: physical body parameters, body size change, and motion restrictions and floor position limitations. The first type of priors imposes physical constraints on the body parameters. Namely, body width and height are weighted $N(h_\mu, h_{\sigma^2})$ and $N(w_\mu, w_{\sigma^2})$, respectively, with the corresponding means and variances reflecting the dimensions of a normal human body. Body coordinates x, y are weighted uniformly within the rectangular region R of the floor map. Since we track bodies that are partially out of the image boundaries, R slightly exceeds the size of what corresponds to the visible part of the image to account for such cases.

The second type of priors sets the dependency between the candidate state at time t and the accepted state at time $t - 1$. First, the difference between w_t, h_t and w_{t-1}, h_{t-1} lowers the prior probability. As another factor, we use the distance between proposed body position (x_t, y_t) and $(\hat{x}_{t-1}, \hat{y}_{t-1})$—the prediction from the constant-velocity Kalman filter. The state of Kalman filter consists of the location of the body on the floor and its velocity. Although tracking the head seems like a first reasonable solution, we have established empirically that the perceived human body height varies as a result of walking; thus, the position of the feet on the floor was chosen as a more stable reference point.

The third type of priors are physical constraints with respect to other moving and static objects in the scene. First, to avoid spatial overlap between adjacent bodies (as physically improbable), we have imposed penalties on pairs of pedestrian models located closer than their corresponding body widths would allow. Second, a similar constraint was imposed on the overlap between pedestrians and stationary obstacles, which were manually marked in the frame and converted to 3D world coordinates.

When a new body is created, it does not have a correspondence; this is when we use a normally distributed prior $N(d_0, \sigma)$, where d_0 is the location of the closest door (designated on the floor plan), and σ is chosen empirically to account for image noise. The same process is taking place when one of the existing bodies is being deleted.

15.5.3.2 Likelihoods

The second component in forming proposal probability relates the observation to the model state. First, for each existing body model the color histogram c is formed by the process of weighted accumulation, with more recent realizations of c given more weight. We then compute Bhattacharya distance between proposed c'_t and corresponding c_{t-1} as part of the observation likelihood:

$$P_{color} = 1 - w_{color} * (1 - B(c'_t, c_{t-1})), \tag{15.3}$$

where w_{color} is an importance weight of the color matching

To guide the tracking process by the background map at hand, we use two more components while computing model likelihood: the amount of blob pixels not matching any body pixels P^+ and the amount of body pixels not matching blob pixels P^-. Note that we use a Z-buffer Z for these as well as for computing the color histogram of the current observation to detect occlusions. In this buffer, all the body pixels are marked according to their distance from the camera (i.e., $0 =$ background, $1 =$ furthermost body, $2 =$ next-closest body, etc.), which we obtain during the

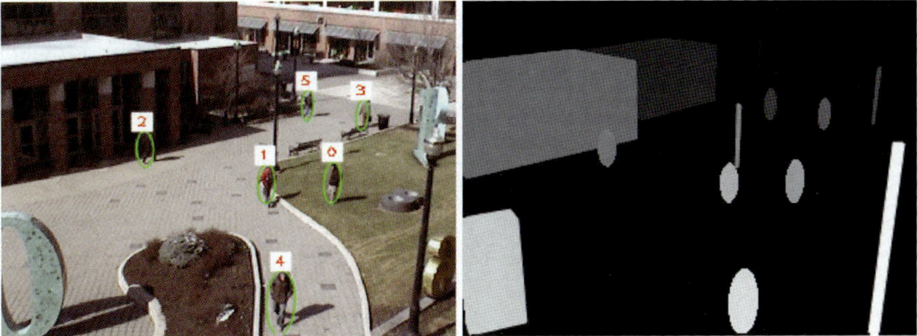

Fig. 15.3 *Left:* Original frame with tracked pedestrians. *Right:* Z-buffer (lighter shades of gray are closer to the camera)

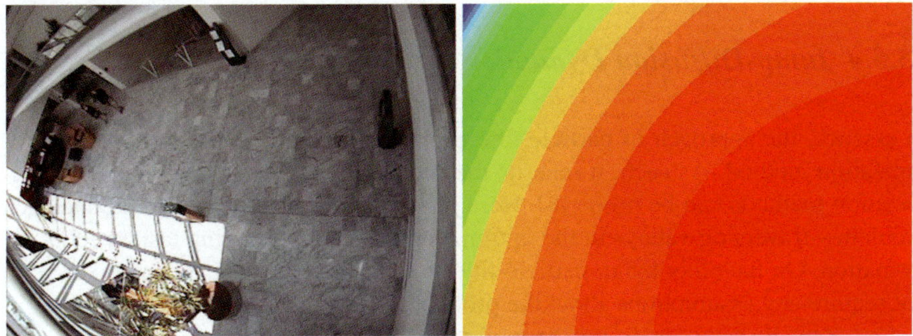

Fig. 15.4 *Left:* Original frame with tracked pedestrians. *Right:* Distance weight plane (weights increase from blue to red)

calibration process. This way, only visible pixels are considered when computing the likelihood (see Fig. 15.3). The Z-buffer is updated after each transition to reflect the new occlusion map.

In computing the likelihood as outlined, there is one major shortcoming overlooked in previous works [15,35]. If the computation is done in terms of the amounts of image pixels, it causes the bodies closer to the camera to influence the overall configuration much more, and the bodies further away are mostly neglected. This becomes particularly evident when the camera covers a large area, where pedestrian image presentations can vary from under 20 pixels of overall area in the back of the scene to more than 200 in front. In addition, such neglect makes the system absolutely tied to the current scene configuration and not portable to a different camera model.

To avoid these shortcomings, we have utilized a so-called distance weight plane D, which is the image of the same dimensions as the input frame and $D_{xy} = |P_{XYZ}, C_{XYZ}|$—the Euclidian distance between the world coordinates of the camera C_{XYZ} and the world coordinates of the hypothetical point P_{XYZ} in space located at a height $z = h_\mu/2$ and corresponding to the image coordinates (x, y). The map produced in this manner is a rough assessment of the actual-size-to-image-size ratio (see Fig. 15.4).

To summarize, the implementation of Z-buffer and distance weight plane allows to computation of multiple-body configuration with one computationally efficient step. Let I be the set of all the blob pixels and O the set of all the pixels corresponding to bodies currently modeled, then,

$$P^+ = \sum \frac{\left(I - O \cap Z_{(Z_{xy} > 0)}\right) \cdot D}{|I|}$$

$$P^- = \sum \frac{\left(O \cap Z_{(Z_{xy} > 0)} - I\right) \cdot D}{|O|}$$

where $\cap$ is set intersection, $\cdot$ is elementwise multiplication, $-$ is set difference, and $||$ is set size (number of pixels).

15.5.4 Jump-Diffusion Dynamics

In essence, the approach of particle filtering is a nondeterministic multivariate optimization method. As such, it inherits the problems to which other, classical optimization methods can be prone [8]. Here, we present a way to overcome one such problem—traversing valleys in the optimization space by utilizing task-specific information. On the other hand, particle-filtering methods are robust because they do not require any assumptions about the probability distributions of the data.

Our joint distribution is not known explicitly, so we have chosen to use the Metropolis-Hastings sampling algorithm.

$$\alpha(x, x') = min\left(1, \frac{P(x')}{P(x_t)} \cdot \frac{m_t(x|x')}{m_t(x'|x)}\right). \tag{15.4}$$

where x' is the candidate state, $P(x)$ is the stationary distribution of our Markov chain, and m_t is the proposal distribution. In Eq. (15.4), the first part is the likelihood ratio between the proposed sample x' and the previous sample x_t. The second part is the ratio of the proposal density in both directions (1 if the proposal density is symmetric).

This proposal density would generate samples centered around the current state. We draw a new proposal state x' with probability $m_t(x'|x)$ and then accept it with the probability $\alpha(x, x')$. Notice that the proposal distribution is a time function; that is, at each frame it will be formed based on the rules outlined in this chapter.

To form the proposal distribution, we have implemented a number of reversible operators. There are two types of jump transitions and five types of diffuse transitions implemented in our system: adding a body, deleting a body, recovering a recently deleted body, changing body dimensions, changing body position, moving a body, resizing a body. Notice that we use a set of controllable weight probabilities to add more emphasis to one or another transition type. In our application, normally around 100 jump-diffuse iterations are required for each frame to reach convergence.

15.6 Symmetry-Based Pedestrian Classification

15.6.1 Symmetry in Gait

Symmetry is a fundamental concept for understanding repetitive patterns in art decoration, crystallography, and more. This has been a primary motivation for developing the branch of mathematics known as geometric group theory. A geometric figure is said to be symmetric if there exist isometries that permute its parts while leaving the object as a whole unchanged. An isometry of this kind is called a *symmetry*. The symmetries of an object form a group called the *symmetry group* of the object. A symmetrical group spanning in one dimension (1D) is defined as a *frieze group* and is defined as a *wallpaper group* in 2D space. Because the human walking motion generates translation along planes parallel to the direction of global body translation, we are more interested in planar symmetries such as reflections and half turn. There are seven distinct subgroups (up to scaling) in the discrete frieze group generated by translation, reflection (along the same axis or a vertical line), and a half turn (180° rotation). Therefore, we use the study of symmetry as pioneered in frieze group theory to analyze and extract human motion-based signatures regarding questions such as whether the individual is carrying a load.

As a human walks, the swing of the limbs generates a symmetrical double-helical pattern that can then be analyzed using frieze group theory. To capture and analyze the double-helical gait signature of an individual, we first track a moving individual by putting a bounding box around the person being tracked. Then, we stack all the bounding boxes of all the frames to create a $x - y - t$ volume in which the DHS resides. Note that the DHS actually resides in the $x - t$ slice of the $x - y - t$ volume. Therefore, we obtain the $x - t$ slice of this volume. Figure 15.4 shows an $x - t$ slice of the volume. The double-helical gait signature is clearly seen in this slice. To segment and extract the helix from the background, we divide the helix into four quadrants and fit 1D curve models for each of the quadrants separately. These model parameters are then refined by incorporating consistency and continuity constraints.

15.6.2 Double-Helical Signature

In the symmetrical twin-pendulum model describing hip-to-toe motion as shown in Fig. 15.5 as a classification module, each leg is modeled as a hand of the twin pendulum with equal length and a uniform angular swing speed $d\theta$ and period T. The generated gait signatures are shown in the second row. The slices generated by the "legs" of that model do contain twisted DNA-like patterns. When investigating the relationship between those twisted DNA structures, we realize that they contain different symmetries, such as reflection symmetry in horizontal and vertical axis, 180° rotation:

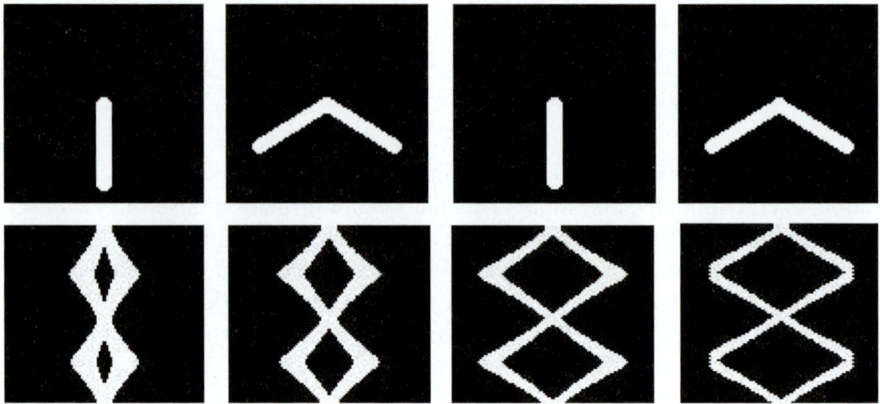

Fig. 15.5 An example to illustrate motion signature inspired by twin-pendulum model. *Top*: *X-Y* image at different gait phase; *bottom*: selected slices containing helical structure

Fig. 15.6 An example to illustrate real motion signature. *First column*: selected slices containing helical structure; *second column*: frames

Hence, any gait pattern S_z at height z is represented as:

$$
\begin{aligned}
S_z(t) &= \{P_1, P_2\} \\
&= \{(z/\sin\theta * \cos\theta(t), t), (z/\sin\theta * \cos\theta'(t), t)\} \\
&= \{(z\tan\theta(t), t), (z\tan\theta'(t), t)\}
\end{aligned}
\tag{15.5}
$$

Such signature is used as a cue for gait-based classification in the proposed system as shown in Fig. 15.6. Most of the current approaches are based on $x - y$ domain (frames) and depend on accurate silhouette, while what we presented based on DHS brings a real-time solution for both static and moving platforms since no segmentation is required.

15.7 Experimental Results

In this section, we present experimental results for every building block of the proposed pedestrian and tracking system.

15.7.1 Tracking

For testing and validation purposes, we used a thermal and color data set from OTCBVS [5]. The set contains short outdoor pedestrian sequences in two locations. Each scene is filmed both with RGB and thermal cameras at the identical resolution, thus providing a pixel-to-pixel correspondence between two types of sensors. The operation of our color-thermal background model significantly reduces and in most cases fully eliminates two types of false foreground regions: shadows as the result of a moving cloud cover and shadows cast by moving pedestrians.

We performed a preliminary evaluation of our tracking system for the presence of three major types of inconsistencies: misses, false hits, and identity switches. A *miss* is when the body is not detected or detected but tracked for an insignificant portion of its path ($<30\%$). A *false hit* is when a new body is created where there is no actual person present. Most of the false hits are a result of more than one body in the model being assigned to a single body in the scene. An *identity switch* is when two or more bodies exchange their IDs once within close proximity. By counting the number of each type of errors on a number of sequences of 6,000 frames overall, we obtained the results summarized in Table 15.1.

Table 15.1 Tracking results based on the manually observed ground truth

Sequence	Frames	People	People missed	Frames missed[a]	False hits	Frames in false hits[a]	Identity switches
1	1,054	15	3	20	1	1	3
2	0601	8	0	0	3	2	4
3	1,700	16[b]	5	10	14	5	15
4	1,506	3[c]	0	0	0	0	1
5	2,031	2	0	0	0	0	2
6	1,652	4	0	0	0	0	1
All	8,544	48	8	5	3	1.3	4.3
%%	100	100	16.6	N/A	6.2	N/A	8.9

[a] - %% percentage of frames missed/false hits averaged over all missed/false hit pedestrians
[b] - Two infants, below the tracked height limit, not counted
[c] - Two pedestrians covered by trees not counted
N/A - not available or not applicable

The most common mistakes made by the tracker were false hits. We observed that the majority of false hits (more than 50%) were short lived (i.e., typically last for only several frames). False hits are typically created when the system recognizes a blob as containing more people than the actual number present in the scene. This can be explained by the distortion in shape of the blob caused by pixel-level noise. These cases can be further postprocessed by temporal filtering to remove insignificantly short paths. Sometimes, however, false detections are accompanied by ID switches when the body tracked for a long time is substituted for a false hit. This presents a more complicated case and deserves further study.

Misses typically occur due to partial or total occlusions by the scene objects or due to the clothing color being too close in color to the background. The first type of misses is usually promptly recovered by the tracker, but if the recovery took place in a different location, a new ID is assigned, resulting in a "switch."

Overall performance of the tracker shows space for improvement; it produces satisfactory detection and prolonged tracking in crowded scenes (Fig. 15.7), although the ratio of pedestrian ID switches has to be reduced. It becomes apparent that the complexity of the scene (i.e., the number of pedestrians) decreases the performance of the tracker.

15.7.2 Classification

Since human motion is naturally different than other types of motions, the sequence within bounding boxes can be further analyzed to verify whether or not the tracked subject is indeed a human (Fig. 15.8). We used the human motion analyzer proposed by Ran et al. [30]. The algorithm proposed tests the spatiotemporal pixels to prove or disprove the null hypothesis that the signal being tested is periodic. Briefly, for a given pixel location (x, y) at some slices in the bounding box, the algorithm computes the periodogram of the color value of (x, y) across some slices of the subject. A peak in the periodogram proves that the spatiotemporal signal is periodic at that pixel location. On the other hand, a flat periodogram means that the spatiotemporal signal is white noise. The periodicity test is repeated for all pixel locations in the slices within the bounding box. If the histogram of periodic pixels has a peak that is higher than a certain threshold, then the subject undergoing the test is a human. Otherwise, the null hypothesis that the subject is a human is rejected. For more details on the algorithm, refer to [30]. The results are shown in Fig. 15.9.

15.7.3 Activity Matching

After locating and classifying pedestrians by the proposed tracking and motion-based recognition methods, we now present some preliminary results for activity

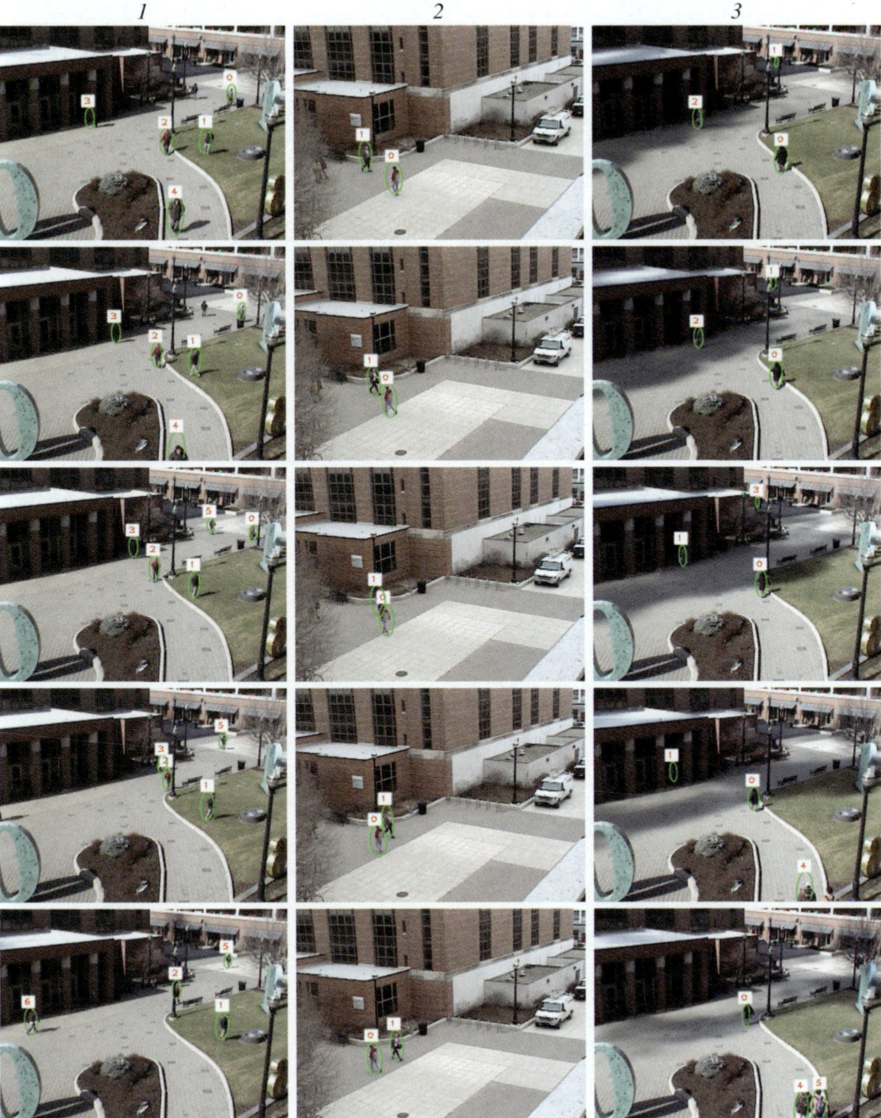

Fig. 15.7 Sample frames showing tracking in three sequences (for tracking videos, see http://www.cs.indiana.edu/cgi-pub/oleykin/otcbvs.html)

matching. We use pedestrian gait as examples. Matching using shape and appearance plays a key role in gait recognition as well as cross-camera target association.

Since we already extracted the slices containing helical signature, they can be directly used for matching and hence for gait recognition and other related areas.

Fig. 15.8 An example to illustrate periodic motion signature verification. *Bottom* tracked target; *top* DHS formed by slicing the video cube

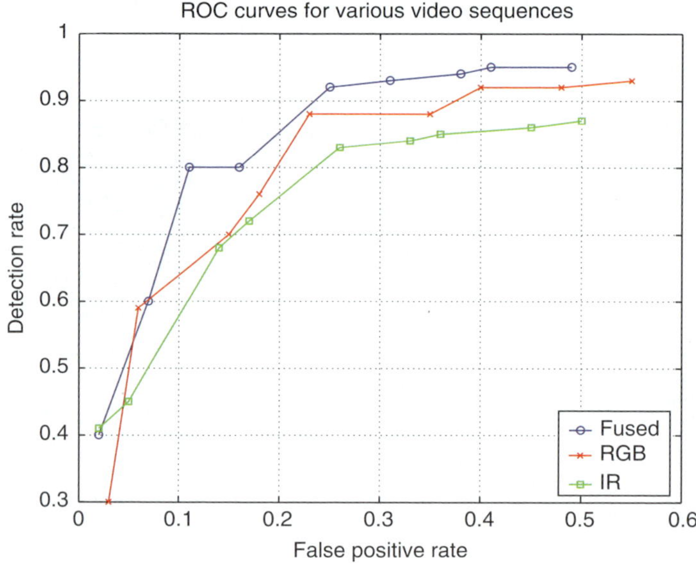

Fig. 15.9 Classification ROC curve results. RGB and IR are the ROC curves for RGB and IR sensors, respectively. Fused is the curve created by fusing them through a majority voting

15.7.3.1 Across Cameras

Because of the equivalency of viewing angle and pose change in generating image sequences, we studied them in a unified framework using a homography constraint reported in [14]:

Table 15.2 USF data: DHS matching across cameras

Error%	02463G2AR	02539G1AR	03500G0AR	03507C0AR	03509C0AR
02463G2AL	1.3	5.1	7.2	4.0	5.0
02539G1AL	3.7	1.0	5.2	7.5	5.0
03500G0AL	7.2	2.7	1.1	10.0	5.0
03507C0AL	4.1	5.3	7.9	1.4	5.4
03509C0AL	5.5	7.1	8.4	5.0	1.5

$$s'(y_0) = H \, s(y_0) \tag{15.6}$$

As in (15.6), the normalized $x - t$ DHSs in various views (poses) are related to a specific transformation. If we can estimate such an H and measure the quality of restored geometry, we obtain a similarity measure between two activity sequences. There are quite a few methods available for solving (15.6) given point coordinates. We use the method reported in [14] to estimate H by using a set of linear equations. After calculating the homography H, we obtain the error as the overall average pointwise difference for a set of DHS slices $s = \{\bar{s}_0, \bar{s}_1, \ldots, \bar{s}_{N-1}\}$ at $y = \{\bar{y}_0, \bar{y}_1, \ldots, \bar{y}_{N-1}\}$ for a pair of gait sequences g and g' assuming that they are from the same subject:

$$Error(g, g') = \frac{1}{\sum N_i} \sum_{s_i \in s} \sum_{p \in s_i} |x_p - \hat{F} x'_p|^2 \tag{15.7}$$

where $\hat{F}$ is the estimated transformation, and N_i is the number of points in slice $S_{\bar{y}_i}$. To validate the proposed method, we present the results for matching targets from a subset of USF (University of South Florida, gait dataset) data (02463G2AR, 02539G1AR, 03500G0AR, 03507C0AR, 03509C0AR and 02463G2AL, 02539G1AL, 03500G0AL, 03507C0AL, 03509C0AL) captured by two cameras. The confusion matrix (not symmetrical) is given in Table 15.2. In this experiment, we divided each sequence into clips containing DHS for one stride. Each entry in the table presents the average *Error*. It shows that the proposed method consistently matched the DHS across cameras and did not confuse with other subjects.

15.7.3.2 Across Time

When the two clips to be matched are from similar viewing directions but have different speeds, the DHS from the same individual will have different lengths for a cycle. This may also be due to videos captured at various time instants or the like. The matching considers translation and scale change in both space and time, which works for DHS from multiple views for the same individual at the same time. However, if the two DHSs are captured at different times, the matching may become nonlinear. Equation (15.6) is not enough to incorporate the subtle change in gait.

Temporal alignment deals with matching two sequences with different lengths and plays a key role in human identification by gait. Among many existing methods, dynamic time warping (DTW) is the most-used technique for nonlinear alignment. It uses dynamic programming to find the optimal alignment of two sequences of different lengths. But, a major challenge is accurately determining the location of the start and endpoints in samples. In our case, the complete DHSs are partitioned into strides, which makes DTW [24] an effective tool. DTW yields an accumulated distance (AD) matrix representing the best possible match between the input pattern and the template. The DHS pattern giving the lowest AD is the best match for the input pattern.

A challenge in DTW for temporal multidimensional signal warping is the cost involved to compute the minimum-cost assignment path, which also holds for multidimensional activity space. But, since we decompose g into slices and each DHS contains two spirals for left and right limbs, we can view them as two temporal signals $s(t)_l, s(t)_r$. To compare two DHSs at the same height from two clips, the cost function is given as

$$D(s,s') = \min\{d(s(t)_l, s'(t)_l) + d(s(t)_r, s'(t)_r), \; d(s(t)_l, s'(t)_r) + d(s(t)_r, s'(t)_l)\}$$

$$(15.8)$$

where $d(s(t), s'(t))$ is the DTW cost function for two single spirals, and $D(s,s')$ is the overall cost for a DHS containing multiple spirals. Each DHS contains two spirals for left and right limbs. We compare the two possible permutations and choose the one with minimum error. When a set of DHSs is used to represent an activity g, we average the cost for each and write the cost function as

$$D(g,g') = \frac{1}{N} \sum_{i=1,s_i \in g, s'_i \in g'}^{N} D(s_i, s'_i)$$

$$(15.9)$$

Comparing two sequences by measuring similarity in DHS is a major advantage in our framework. The shape sequence is represented by a small set of temporal slices. An example of the proposed method is given in Fig. 15.10. In our experiment, only 3 DHSs were used. They were at heights $(0.4, 0.6, 0.8)$ in the bounding

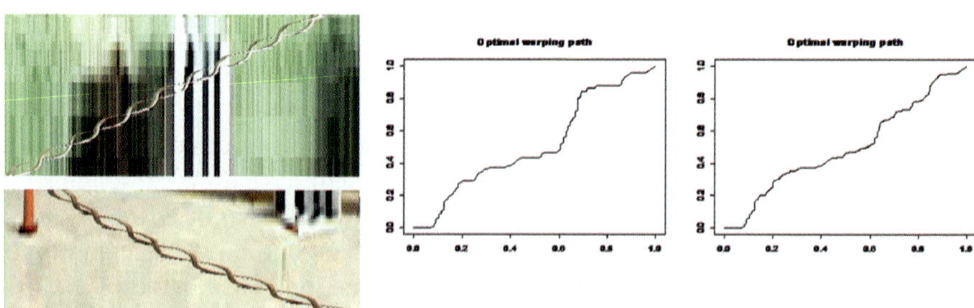

Fig. 15.10 DHS matching using DTW for gait activities with different lengths

Table 15.3 USF data: DHS matching across time

$D(g,g')$	02463G2AR	02539G1AR	03508G0BR	03521G0AR	03603G0AR
02463G0AR	2.1	6.2	6.3	5.7	7.0
02539G0AR	6.2	1.7	5.7	6.7	5.0
03508G1AR	6.3	5.7	3.1	8.0	6.7
03521C0AR	5.7	6.7	8.0	2.3	6.4
03603C0AR	7.0	5.0	6.7	6.4	1.9

boxes, corresponding to arms, hip, and legs. The confusion matrix (symmetrical) is given in Table 15.3. This experiment suggests the potential of using DHS for human identification and activity recognition.

15.8 Conclusion

In this chapter, we presented a vision system combining tracking and motion analysis. This area has drawn the attention of many researchers, and the major challenge lies in the fact that human motion is nonrigid and articulated. The proposed method naturally integrates a tracker with spatial-temporal motion information by fusing color and IR videos in an MCMC framework. By fusing the information from both sensors, it provides more robust tracking results. Moreover, we have implemented a pedestrian classification module capable of recognizing human body motion and matching across various cameras and time. It is based on the DHS, a human gait signature found in decomposed horizontal temporal slices. The advantage of using DHS is twofold. First, no accurate segmentation is required, and hence it is more robust to noise and camera motion. Second, only a small number of slices can have strong classification capability. The overall system can process the detected moving targets, track their motion, and classify them into pedestrian and other objects. It also has the power to do cross-camera association for identification and clustering for event classification.

Chapter's References

1. B. Bhanu and J. Han. Kinematic-based human motion analysis in infrared sequences. In *IEEE Workshop on Applications of Computer Vision*, pages 208–212, 2002
2. R. Collins, A. Lipton, and T. Kanade. Introduction to the special section on video surveillance. *IEEE Transactions on Pattern Analysis and Machine Intelligence*, 22(8):745–746, 2000
3. R. Collins, A. Lipton, T. Kanade, H. Fujiyoshi, D. Duggins, Y. Tsin, D. Tolliver, N. Enomoto, O. Hasegawa, P. Burt, and L. Wixson. A system for video surveillance and monitoring. Technical Report CMU-RI-TR-00-12, Carnegie Mellon University, Pittsburgh, May 2000
4. C. Dai, Y. Zheng, and X. Li. Layered representation for pedestrian detection and tracking in infrared imagery. In *IEEE CVPR Workshop on OTCBVS*, 2005

5. J. Davis and V. Sharma. Fusion-based background-subtraction using contour saliency. In *IEEE International Workshop on Object Tracking and Classification beyond the Visible Spectrum*, IEEE OTCBVS WS Series Bench, 2005

6. J. W. Davis and V. Sharma. Fusion-based background-subtraction using contour saliency. In *IEEE CVPR WS on Object Tracking and Classification Beyond the Visible Spectrum*, pages 19–26, 2005

7. J. Deutscher, B. North, B. Bascle, and A.B. Blake. Tracking through singularities and discontinuities by random sampling. In *International Conference on Computer Vision*, 1999

8. A. Doucet, N. de Freitas, and N. Gordon. *Sequential Monte Carlo Methods in Practice*. Springer-Verlag, New York, 2001

9. A. Elgammal and L. Davis. Probabilistic framework for segmenting people under occlusion. In *International Conference on Computer Vision*, 2001

10. D. Gavrila. The visual analysis of human movement: A survey. *Computer Vision and Image Understanding*, 73(1):82–98, 1999

11. I. Haritaoglu and M. Flickner. Detection and tracking of shopping groups in stores. In *International Conference on Computer Vision and Pattern Recognition*, 2001

12. I. Haritaoglu, D. Harwood, and L. Davis. W-4: Real-time surveillance of people and their activities. *IEEE Transactions on Pattern Analysis and Machine Intelligence*, 22(8):809–830, 2000

13. L. Havasi and T. Sziranyi. Motion tracking through grouped transient feature points. In *Advanced Concepts for Intelligent Vision Systems*, 2004

14. A.Z.R. Hartley. *Multiple View Geometry in Computer Vision. Part 2*. Cambridge University Press, Cambridge, UK, 2004

15. M. Isard and J. MacCormick. Bramble: A Bayesian multiple-blob tracker. In *International Conference on Computer Vision*, 2:34–41, 2001

16. C. Kemp and T. Drummond. Multi-modal tracking using texture changes. Image and Vision Computing Volume 26, Issue 3 (March 2008). Pages: 442–450

17. A. Leykin. Visual Human Tracking and Group Activity Analysis: A Video Mining System for Retail Marketing. Ph.D. thesis, Indiana University, 2007

18. A. Leykin and R. Hammoud. Robust multi-pedestrian tracking in thermal-visible surveillance videos. *Computer Vision and Pattern Recognition Workshop (cvprw)*, Volume, Issue 17–22, 136–136,, 2006

19. J. Little and J. Boyd. Recognizing people by their gait: The shape of motion. *Journal of Computer Vision Research*, 1(2):1–32, 1998

20. F. Liu and R. Picard. Finding periodicity in space and time. *Proceedings of the Sixth International Conference on Computer Vision*, pages 376–382, 1998

21. S. Maybank and T. Tan. Introduction to special section on visual surveillance. *International Journal of Computer Vision*, 37(2):173–173, 2000

22. A. Mittal and L. Davis. M2tracker: A multi-view approach to segmenting and tracking people in a cluttered scene using region-based stereo. In *European Conference on Computer Vision*, 2002

23. D. Morris and J. Rehg. Singularity analysis for articulated object tracking. In *International Conference on Computer Vision and Pattern Recognition*, 1998

24. C.S. Myers and L.R. Rabiner. A comparative study of several dynamic time-warping algorithms for connected word recognition. *Bell System Technical Journal*, 60(7):1389–1409, 1981

25. H. Nanda and L. Davis. Probabilistic template based pedestrian detection in infrared videos. In *IEEE Intelligent Vehicles Symposium*, 2002

26. S. Niyogi and E. Adelson. Analyzing and recognizing walking figures in xyt. *Proceedings of IEEE Conference on Computer Vision and Pattern Recognition*, pages 469–474, 1994

27. N. Oliver, B. Rosario, and A. Pentland. A Bayesian computer vision system for modeling human interactions. *IEEE Transactions on Pattern Analysis and Machine Intelligence*, 22(8):831–843, 2000

28. Y. Owechko, S. Medasani, and N. Srinivasa. Classifier swarms for human detection in infrared imagery. In *IEEE CVPR on Object Tracking and Classification Beyond the Visible Spectrum*, 2004

29. I.Y. Ran, Q. Zheng and L.S. Davis. Pedestrian classifcation from moving platforms using cyclic motion pattern. *International Conference of Image Processing*, pages 854–857, 2005

30. Y. Ran, Q.Z.I. Weiss, and L.S. Davis. Pedestrian detection via periodic motion analysis. *Interenational Journal of computer Vision*, 2006

31. C. Sminchisescu and B. Triggs. Kinematic jump processes for monocular 3D human tracking. In *International Conference on Computer Vision and Pattern Recognition*, 2003

32. W.H.L. Wang and T. Tan. Recent developments in human motion analysis. *Pattern Recognition*, 36(3):585–601, 2003

33. F. Xu and K. Fujimura. Pedestrian detection and tracking with night vision. In *IEEE Intelligent Vehicles Symposium*, 2002

34. M. Yasuno, N. Yasuda, and M. Aoki. Pedestrian detection and tracking in far infrared images. In *IEEE CVPR Workshop on Object Tracking and Classification Beyond the Visible Spectrum*, 2004

35. T. Zhao and R. Nevatia. Tracking multiple humans in crowded environment. In *International Conference on Computer Vision and Pattern Recognition*, 2004.

Chapter 16
Multi Stereo-Based Pedestrian Detection by Daylight and Far-Infrared Cameras

Massimo Bertozzi, Alberto Broggi, Mirko Felisa, Stefano Ghidoni,
Paolo Grisleri, Guido Vezzoni, Cristina Hilario Gómez, and Mike Del Rose

Abstract This chapter presents a tetravision (4-camera) system for the detection of pedestrians by means of the simultaneous use of two far infrared and visible camera stereo pairs. The main idea is to exploit the advantages of both far infrared and visible cameras to develop a system that combines the advantages of using far infrared or daylight technologies. Different approaches are used to process the two stereo flows in an independent fashion to produce a list of areas of attention that potentially contain pedestrians. Then, four different following approaches are used to refine and filter this list and to validate the presence of a pedestrian. Preliminary results show that the combined use of two vision systems as well as the use of different and independent validation steps enable the system to effectively detect pedestrians in different conditions of illumination and background.

16.1 Introduction

Pedestrian detection is an important field of research for commercial or governmental organizations. The automatic identification of people or obstacles can improve safety for regular vehicles. In particular, the U.S. Army is actively developing obstacle detection for mule operations, path following, and intent-based antitamper surveillance systems for its robotic vehicle safety [11, 12, 18].

Nevertheless, the potential field of use of such technology is larger than U.S. Army necessities: surveillance systems, driver assistance systems, and intelligent systems for autonomous or semiautonomous driving represent a few of the many areas that need to detect the presence of persons to take appropriate actions, like avoid or enabling safety countermeasures.

Unfortunately, the detection of pedestrians is a challenging task, and the difficulty of the problem increases when the movement of the sensors, uncontrolled outdoor environments, and variations in pedestrian appearance and pose have to be taken into account.

R.I. Hammoud (ed.), *Augmented Vision Perception in Infrared: Algorithms and
Applied Systems,* Advances in Pattern Recognition, DOI 10.1007/978-1-84800-277-7_16,
© Springer-Verlag London Limited 2009

Many different approaches to solve this problem have been tested. Some use LADAR or laser scanners to retrieve a three-dimensional (3D) map of the terrain and detect pedestrians [13–15]; others use ultrasonic sensors to determine the reflection of pedestrians [1]. Radar sensors can be used in a similar way: Pedestrian detection is based on the analysis of the reflections on the targets to discriminate whether they are pedestrians [20, 23].

Vision is a natural choice for a pedestrian detection sensor because it is based on how people perceive humans (through visual cues). In particular, for the U.S. Army, vision-based detection is important since cameras are nonevasive sensors. A number of systems that approach this task have been developed based both on monocular [27, 29, 34] or stereo [4, 7, 16, 25, 28, 30] vision.

Recently, infrared [FIR] also infrared technologies (both far, infrared [FIR] and near infrared based) came into in the pedestrian detection arena [2, 5, 10, 24, 33], thanks to the decreasing cost of infrared technology.

In many scenarios, FIR cameras are more suited than daylight ones for detecting pedestrians, especially when the background is colder than the human beings or in night or low-illumination conditions. Moreover, FIR images generally show less-noisy details, easing the initial steps of a detection process. FIR cameras fail the detection of pedestrians in hot or sunny weather, namely, when pedestrians are not warmer than the background.

The following presents a pedestrian detection system based on the simultaneous use of a visible and FIR stereo systems to exploit the benefits of both approaches.

Section 16.2 summarizes the whole algorithm; Section 16.3 introduces the initial phase of the detection of areas of attention. In Section 16.4, a refinement step based on the use of different symmetry operators is explained; the result is fed to three different validators that are explained in Section 16.5. Section 16.6 sketches the acquisition system developed to deal with four video streams. Section 16.7 shows the results of each validation process, while Section 16.8 summarizes and concludes the chapter.

16.2 The Approach

Visible cameras have been widely used for surveillance and detection systems; more recently, thanks to decreasing costs, also FIR and near infrared devices became a viable option for such tasks. Generally, the choice between infrared and daylight technologies depends on the specific use and on cost-vs-benefit analysis.

An object in a FIR domain shows a pattern that depends on its thermal features, namely, on the amount of heat it emits. Moreover, this pattern is barely affected by illumination changes. FIR images are also generally less noisy thanks to the relative absence of shadows and textures due to colors.

Therefore, FIR images ease the detection of objects warmer (or colder) than the background (like pedestrians, moving vehicles, etc.) since they are sufficiently contrasted with respect to the background. Moreover, FIR cameras are suitable for night or low-illumination scenarios since they are barely affected by illumination.

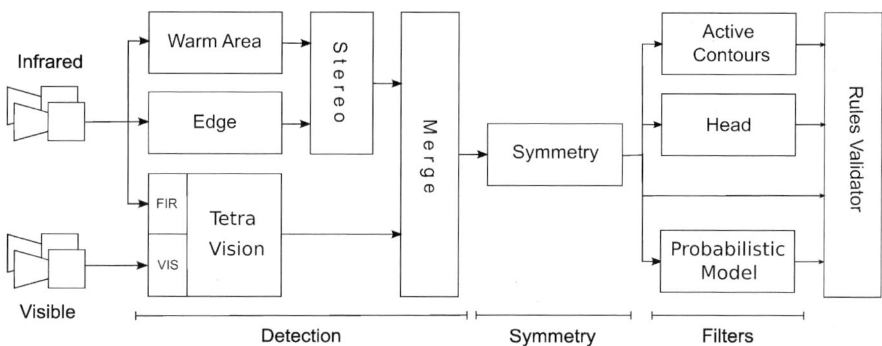

Fig. 16.1 Overall algorithm flow

Unfortunately, strong sun heating or high-temperature conditions can increase the background thermal impact in the FIR images and can also introduce additional textures due to the different thermal behavior of different materials. Moreover, for pedestrian detection, clothes affect the thermal footprint of a human shape, making detection more challenging. Conversely, images acquired using daylight cameras show a number of small details and shadows. Visible cameras are also highly affected by changes in the illumination conditions.

Often, in the first stages of a detection process, the shape is used as a preliminary detector pattern. In such a case, the presence of a big number of many details is cumbersome and makes the detection more difficult. Nevertheless, it is of paramount importance for distinguishing between pedestrians and objects.

In the following, a system based on the simultaneous use of two infrared and visible camera stereo pairs is described. The main idea is to exploit the advantages of FIR and visible cameras, trying at the same time to cope with the deficiencies of each system. Different approaches have been developed for pedestrian detection in the two different image domains: warm area detection, vertical edge detection, and an approach based on the simultaneous computation of disparity space images (DSIs) in the two domains.

These first stages of detection output a list of bounding boxes that enclose potential pedestrians. A symmetry-based approach is further used to refine this rough result. Then, a number of validators are used to evaluate the presence of a human shape inside each bounding box. The validation is performed searching for human shape characteristics: head detection, shape detection, and an active contour-based approach. Figure 16.1 sketches the overall algorithm flow.

16.3 Detection of Areas of Attention

Different approaches are used for computing a list of areas of the image that potentially contain a pedestrian.

16.3.1 Run-Time Calibration

The knowledge of correct calibration parameters is of paramount importance for machine vision since it is mandatory to resolve the relationship between image matrix and real-world coordinates.

Many vision-based systems for intelligent vehicles rely on the assumption of moving on a flat road or—more realistically—on a road with a smoothly varying slope. Anyway, this assumption can lead to errors in computing scene parameters or even to misdetections, not only when the road is not really flat, but also due to the fact that a camera installed on a moving vehicle continuously varies its yaw, pitch, and roll.

A V-disparity approach is used for computing the actual road slope, exploiting the presence of two stereo vision systems [6, 21]. The knowledge of the road slope allows computation of the actual pitch angle that is widely used in the following detection steps.

A Sobel filter is used to enhance image features, especially edges. Then, a correlation is computed for different offset values (*disparities*), for each pair (left and right) of rows of the Sobel images. Thanks to the specific setup of the stereo pair, the same rows in the stereo images are epipolar lines. This process has been tested both for visible and infrared images; the result is a new image, the *V-disparity image*, encoding correlation values (see Fig. 16.2). The brighter the pixel, the higher the match quality is.

It can be noticed that in visible images, the ground components produce a slanted line that can give information about the road cross section. Conversely, due to the

Fig. 16.2 V-disparity computation in visible and infrared domains: *left* original images, *center* Sobel images, and *right* V-disparity image

bare presence of ground edges in infrared images, the FIR V-disparity image does not permit recovery road data.

Therefore, only the visible domain is used to compute the ground slope; anyway, this information can be used in the following process for the infrared domain thanks to the knowledge of relative calibration of all four cameras.

16.3.2 FIR-Only Processing

Two different processings are dedicated to analyze FIR images for detecting a list of areas in the image that potentially contain pedestrians: the detection of warm areas and an edge-based filtering.

16.3.2.1 Warm Area Detection

The warm area detection approach is based on the fact that pedestrians are usually warmer than the background, and therefore they are rendered as bright regions in FIR images.

Therefore, as a first step of the algorithm, high-intensity areas on the input image are searched for. This is obtained using two different threshold values: Initially, a high threshold is applied on all pixels to get rid of cold or barely warm areas, selecting pixels corresponding to very warm objects only. Then, pixels featuring a gray level higher than a lower threshold are selected if they are contiguous to other already selected pixels in a region-growing fashion. The resulting image contains only warm contiguous areas that present hot spots (Fig. 16.3).

To select vertical stripes containing hot regions, a columnwise histogram is computed on the resulting image (see Fig. 16.4a). The histogram is filtered with an adaptive threshold whose value is a fraction of the average value of the whole histogram. Obviously, multiple hot objects may be vertically aligned in the image so

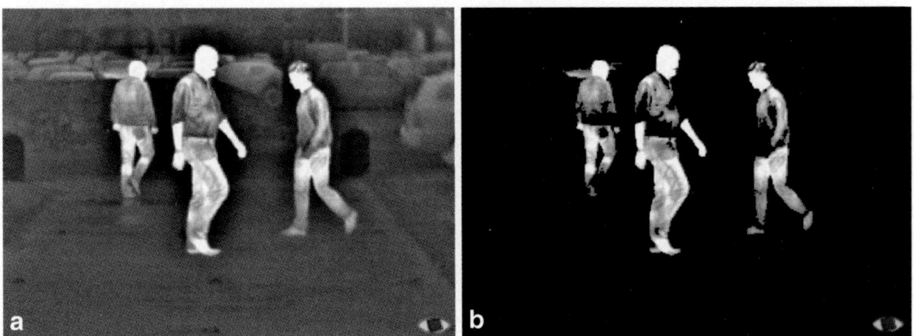

Fig. 16.3 Preprocessing phase: **a** original input image, **b** focus of attention

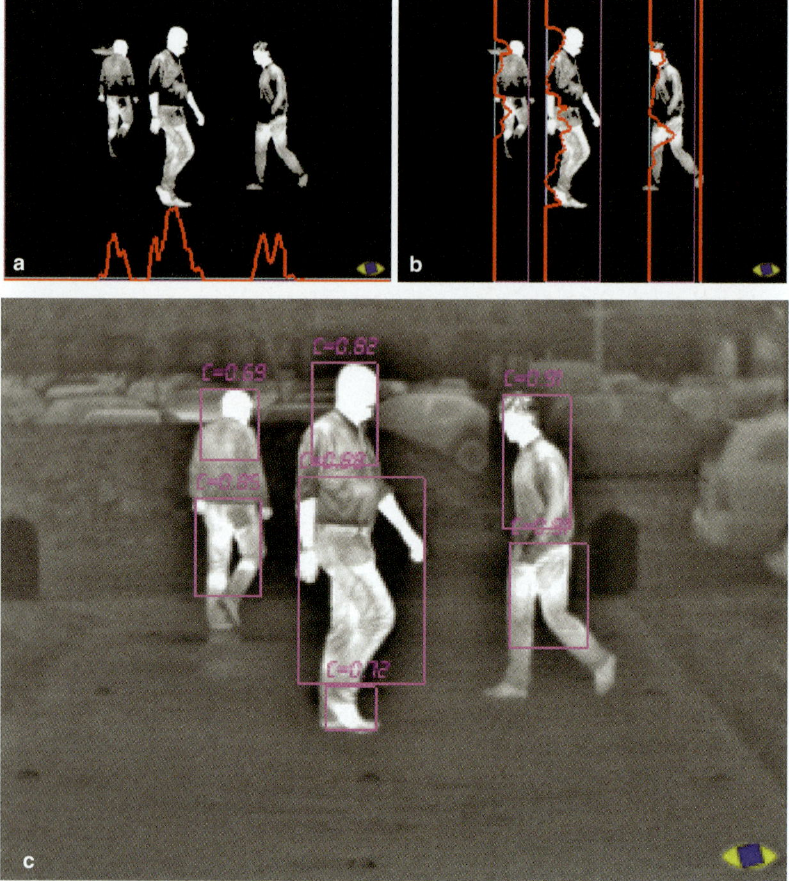

Fig. 16.4 Warm element detection: **a** a columnwise histogram detects columns with warm object, **b** following-row histograms for each column detect initial areas of attention, **c** final results on the original image

that their contributions add up in the histogram. Nevertheless, warm areas belonging to the same horizontal stripe can be distinguished by computing a new rowwise histogram of the gray levels for each stripe (Fig. 16.4b). This procedure yields rectangular bounding boxes framing areas where pedestrians may be located. To refine these bounding boxes, the columnwise and rowwise histogram procedure is iteratively applied to each rectangular box until its size is no longer reduced. Results are shown in Fig. 16.4c.

In this phase, small bounding boxes are removed as they represent nuisance elements.

16.3.2.2 Edge Detection

Another intrinsic characteristic of FIR images is the relative absence of image textures. This is due to the higher homogeneity of thermal patterns with respect to color features. Therefore, FIR images present fewer edges than their daylight counterparts; in addition, these edges are much more likely due to the object borders, which reduces the problem of noise in an approach based on edge detection.

Usually, a human shape features more vertical edges than the background or than other objects. Therefore, this approach is based on the detection of areas that contain a high amount of vertical edges.

Initially, acquired images are filtered using a Sobel operator and an adaptive threshold to detect nearly vertical edges (Fig. 16.5a and 16.5b). Unfortunately, also some objects other than pedestrians contain a high amount of vertical edges: buildings, cars, poles, and so on, but vertical edges of such objects are generally longer and more uniform than the ones that belong to human shapes. Therefore, a filtering phase devoted to the removal of regular vertical edges that are longer than a given threshold is performed. Also, isolated pixels are considered as noise and removed as well (Fig. 16.5c).

To further enhance vertical edges and to group edges of a pedestrian in the same cluster, a morphological expansion is performed using a 3×7 operator. In the final result (Fig. 16.5d), only a few clusters are present.

A labeling approach is used to compute connected clusters of pixels; the resulting image is analyzed, and a list of bounding boxes containing connected clusters of pixels is built (Fig. 16.5e). Generally, in complex scenarios, a large number of small clusters are detected, affecting both the effectiveness and the efficiency of the subsequent processing steps. Thus, a filtering phase is used to remove bounding boxes that are too small and contain cold areas only; therefore, boxes sufficiently large or containing bright pixels survive this step (Fig. 16.5f).

16.3.2.3 Stereo Match

Bounding boxes detected during the previous two steps are then matched against the other image to compute their size and position in the real world. Thanks to the knowledge of vision system calibration computed in Section 16.3.1, a search area in the other image for each bounding box can be estimated.

The following Pearson's correlation function is used to evaluate the result of the match:

$$r = \frac{\sum a_{xy}b_{xy} - \frac{\sum a_{xy}\sum b_{xy}}{N}}{\sqrt{\left(\sum a_{xy}{}^2 - \frac{(\sum a_{xy})^2}{N}\right)\left(\sum b_{xy}{}^2 - \frac{(\sum b_{xy})^2}{N}\right)}} \qquad (16.1)$$

where N is the number of pixels in the considered bounding box, a_{xy} and b_{xy} are gray-level values of corresponding pixels of the two images. The bounding box in

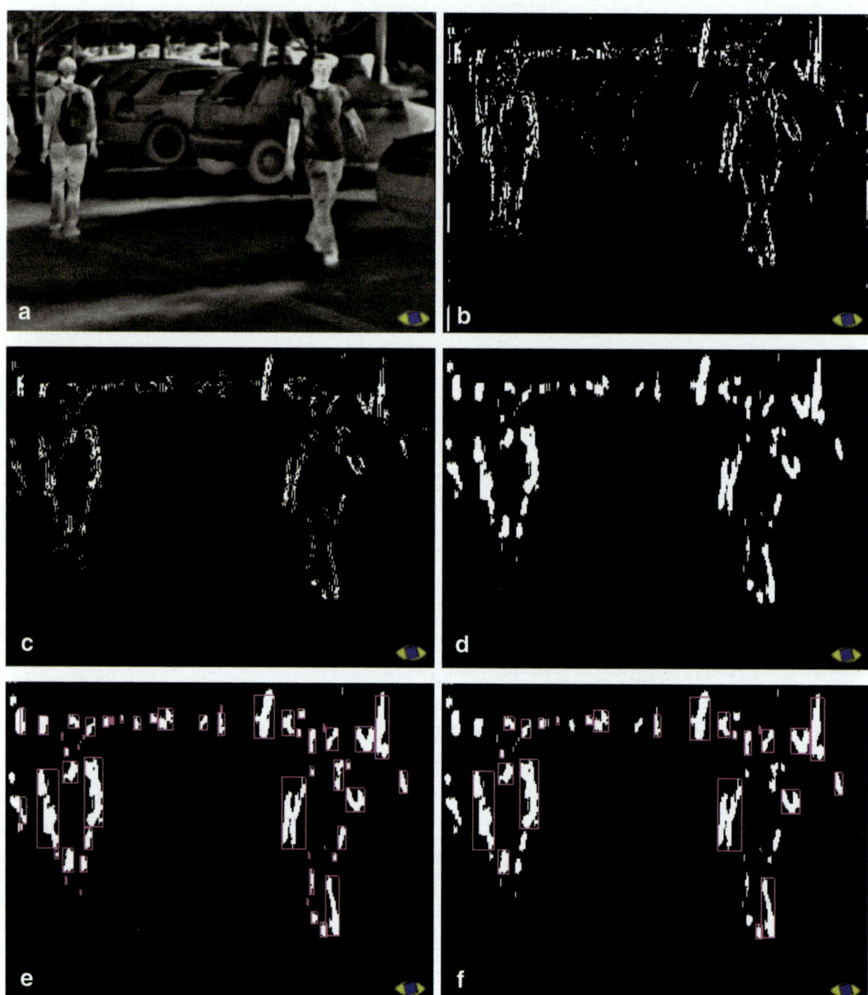

Fig. 16.5 Edge detection: **a** original image, **b** vertical edge detected using the Sobel operator and a threshold, **c** edge after the removal of strong and regular edges, **d** expanded edges, **e** bounding boxes containing connected clusters of pixels, and **f** final resulting list of bounding boxes

the other image featuring the maximum value for the correlation is selected as the best match. The algorithm discards the matches whose correlation values are lower than a given threshold (Fig. 16.6).

A triangulation technique is used to estimate the distance between each object and the vision system.

Fig. 16.6 Stereo match: homologous bounding boxes in the left image

16.3.3 Independent Tetravision Obstacle Detection

In the following a multidomain approach to detect the region of interest is explained. Dealing with two stereo systems that work in different domains is a bit tricky since the two stereo systems feature different points of view, different angular apertures, and—generally—different frame rates and resolutions. In addition, the whole system cannot rely on a search for homologous points since it is very difficult to match features across different spectral domains.

Many image registration strategies have been proposed [17], but unfortunately they require annotation of homologous points or an a priori knowledge of the environment. Conversely, the proposed system independently processes the two stereo flows and then fuses the results coming from the two different domains.

Initially, a *disparity space image* (DSI) computation is used to detect the presence of obstacles: The right images of each flow are subdivided into 3×8 pixel regions, and their corresponding regions are searched for into the left images. The reason for a 3×8 region size is that a pedestrian shape is generally characterized by strong vertical features. The search for a homologous region is limited to the same row of the left image since the optical axis of the two stereo systems are parallel; thus, two corresponding rows in the two images are epipolar lines. Moreover, calibration information permits further bounding of the search area, reducing both the required computational load and the probability of wrong matches. For each region in the image, the best-matching area in the other image is considered, and the disparity between their coordinates is computed.

The match is performed using the following correlation formula:

$$C = \frac{\sum_{i=1}^{3} \sum_{j=1}^{8} \left(L_{i,j} \times R_{i,j} \right)}{max \left(\sum_{i=1}^{3} \sum_{j=1}^{8} (L_{i,j})^2, \sum_{i=1}^{3} \sum_{j=1}^{8} (R_{i,j})^2 \right)} \tag{16.2}$$

where (i, j) are the coordinates of each pixel into the 3×8 regions. The result is a new image, the DSI, in which each pixel encodes the disparity value. Figure 16.7a

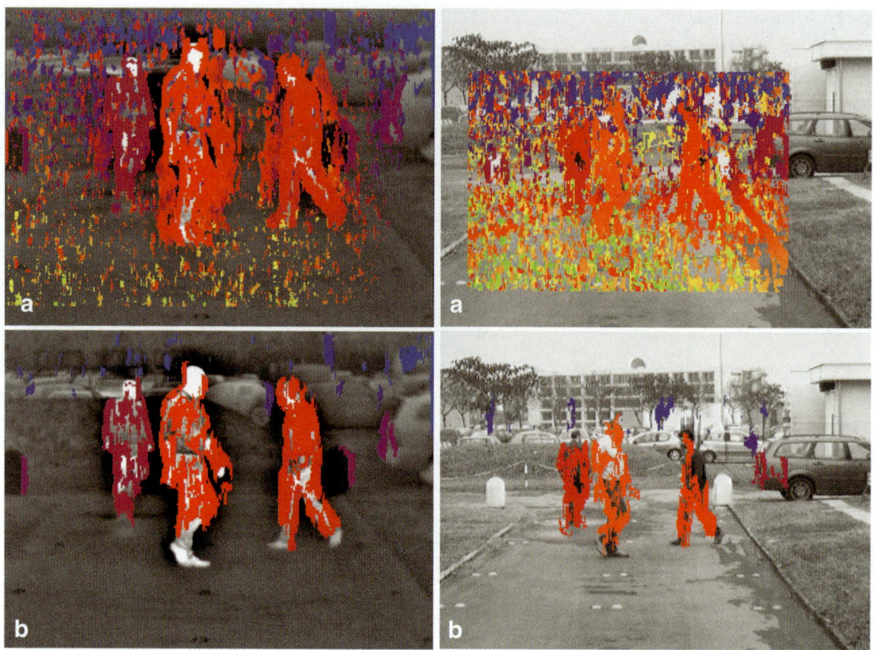

Fig. 16.7 DSI: **a** the disparity images for visible and infrared domains—the darker the color, the smaller the disparity; **b** detected obstacle regions labeled using different colors

shows an example of DSIs computed for both the visible and infrared domains; the colors encode the disparity value: Bright colors correspond to high disparities, while dark colors correspond to small disparity values. It can be noticed that only obstacles feature large clusters of pixels that have similar disparity values; conversely, background features present colors that evolve from bright to dark when moving from the bottom part of the image toward the top.

An aggregation step on the DSI is therefore used to detect the presence of obstacles: Large connected regions featuring a similar disparity are marked as obstacles. DSI pixels that have the same value as the ground disparity are removed. Then, the DSI images are clustered to detect the connected areas having similar disparity. Not all clusters of pixels are considered obstacles; in fact, size, distance, and height constraints are used to further reduce their number. Figure 16.7b shows resulting clusters with different colors.

16.3.4 Merge and Rough Filtering Step

Each area of attention detected by the segmentation process described is marked using a bounding box (see Fig. 16.8a). Unfortunately, since different approaches are used to process the same scene, different bounding boxes often belong to the

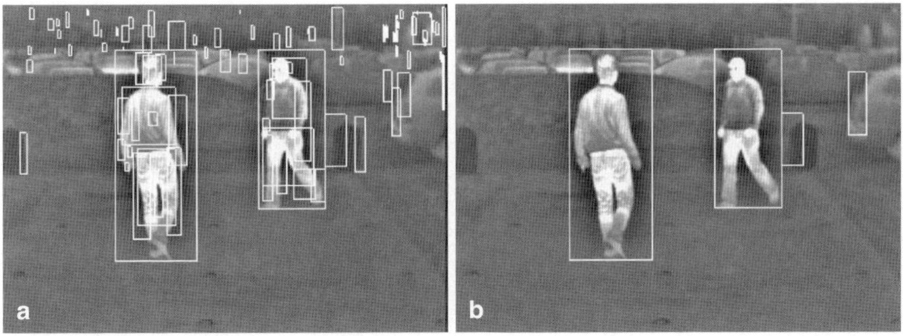

Fig. 16.8 Bounding box generation and merging: **a** detected regions enclosed by bounding boxes and **b** resulting bounding boxes after fusion

same obstacle; therefore, a merging process is mandatory, following the rule that two bounding boxes are fused to compose a larger bounding box when they are sufficiently close in real-world coordinates.

Figure 16.8b shows an example of the merging process in the infrared domain; thanks to the use of world coordinates for the merging, bounding boxes belonging to obstacles that partially overlap are not merged.

After the merging step, another filtering phase takes place. Bounding boxes that are too small, too big, or featuring an aspect ratio incompatible with a human shape are discarded.

Finally, the size of resulting bounding boxes is refined. In fact, thanks to the knowledge of the road slope computed as described in Section 16.3.1, it is possible to compute the point of contact between each object framed by a bounding box and the ground. Thus, the bounding boxes' bottoms are stretched to the ground [2] (see Fig. 16.9). This allows including legs when they are misdetected and, at the same time, removing portions of the background below the human shape when it has been incorrectly included in a bounding box.

16.3.4.1 Bounding Box Registration

At the end of the previous stages, two lists of bounding boxes were produced: one for the obstacles detected in the visible domain and another for the obstacles detected in the infrared images.

Due to the different extrinsic and intrinsic calibration parameters of the two stereo systems, a registration phase is required. In fact, even bounding boxes that correspond to the same obstacle in the two lists are differently sized and positioned. Therefore, to fuse the results obtained working on the two spectral domains, it is necessary to compute the position and size of bounding boxes of each list in the other domain. Moreover, the different field of view of the two stereo systems has to be taken into account; in this case, visible cameras have a larger field of view, so not all detected objects in such a domain have a correspondent in the infrared images.

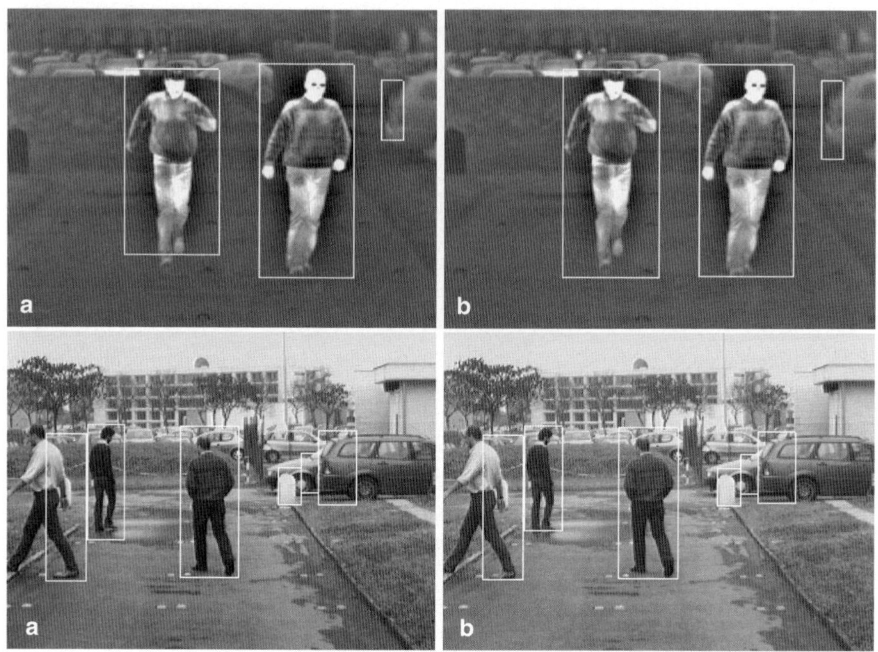

Fig. 16.9 Two different cases of bounding box refinement for the two stereo flows: bounding boxes **a** before and **b** after bottom refinement

Thanks to the knowledge of the calibration data and road slope, the bounding boxes computed in the visible domain are resized, repositioned according to the infrared vision system setup, and finally added to the bounding boxes detected in the infrared domain. Bounding boxes that are completely or partially out of the field of view of the infrared stereo system are removed or cropped.

The same process is performed to compute size and position of bounding boxes detected in the infrared images toward the visible domain.

16.3.4.2 Cross-Domain Fusion of Results

Since most of the obstacles in front of the vision system are correctly detected both in the visible and in the infrared domains, the two separate processings can produce two different bounding boxes for the same obstacle. Moreover, during the previous step, the two lists of bounding boxes have been merged; therefore, in each list two bounding boxes can refer to the same object so that another merge step is required. The merging is operated using a bounding box that encloses the two bounding boxes to be fused together. To determine if two bounding boxes refer to the same object, the percentage of overlapping in the images and their distance in world coordinates are considered (see Fig. 16.10).

Fig. 16.10 Bounding box registration and fusion: bounding boxes computed in the **a** visible and **b** infrared domains and their union in the **c** infrared and **d** visible domains; box fusion in **e** infrared and **f** visible images

At the end of this stage, the two lists of bounding boxes contain the same number of bounding boxes, and each bounding box in a list have a homologous one in the other list.

16.4 Symmetry-Based Refinement

The previously discussed low-level obstacle detectors generate a bounding box list. Unfortunately, the same bounding box sometimes may enclose different pedestrians, especially if they are at a similar distance from the cameras (Fig. 16.12a). In addition, both objects and pedestrians can be detected.

Therefore, a refinement and filtering process has been developed to get one bounding box for each object and to get rid of nonpedestrian obstacles. More specifically, a symmetry-based processing is used to split, refine, and preliminary validate each bounding box. In fact, pedestrians are generally symmetrical; then, a symmetry computation can be used both as a validator and as a refinement step.

Experimental results demonstrated that infrared images are more suitable than daylight images for a symmetry computation since they contain a lower number of small details. Therefore, this phase is performed on the right infrared images only; the whole process is depicted in Fig. 16.11. This step produces an updated bounding boxes list for both infrared and visible domains.

Two different vertical symmetry measures are performed: one on the gray-level values and one on the vertical edge values. The density of edges is also considered to remove homogeneous edgeless areas.

For each bounding box, all the possible positions of a symmetry axis are considered. For each axis, different symmetry window widths are evaluated consider-

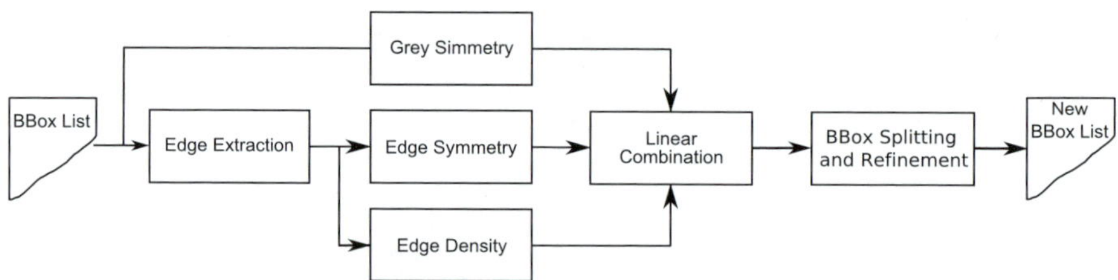

Fig. 16.11 Flow of symmetry processing *BBox* bounding box

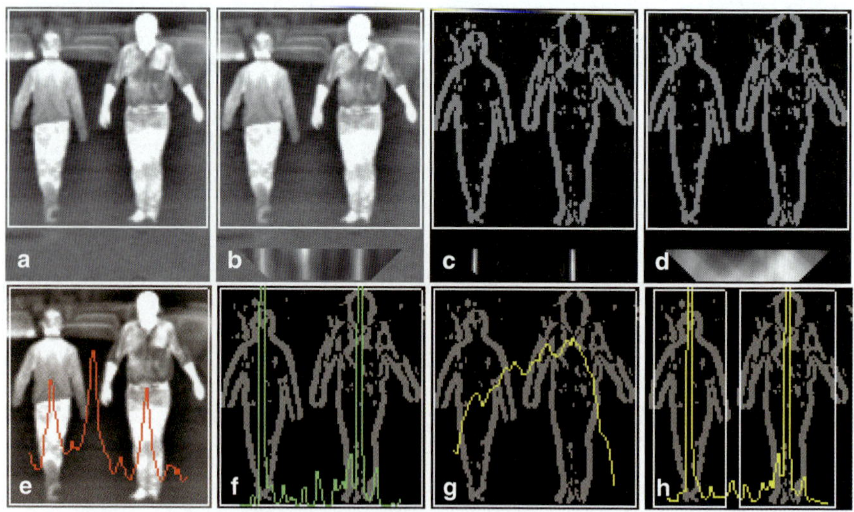

Fig. 16.12 Symmetry computation steps: **a** original bounding box, **b** grey and **c** edge symmetry maps, **d** edge density map, **e** gray and **f** edge symmetry histograms, **g** edge density histograms, **h** overall histogram and new split bounding boxes

ing the size of a human shape at the bounding box distance. Two symmetry maps are computed: one for gray levels (Fig. 16.12b) and one for edges (Fig. 16.12c). In these maps, the horizontal axis encodes the symmetry axis horizontal position within the bounding box; the vertical axis represents the symmetry window width. Figure 16.12d shows the density map that is computed in a similar way to the one used for the symmetry map. The triangular shape of the maps is due to the limitation in scanning large windows for peripheral vertical axis.

Bright points encode the presence of a high symmetric content: The value of each symmetry map pixel depends on the number of symmetric points in the related symmetry window. In the gray-level symmetry computation, two points are regarded as symmetric if they are equidistant from symmetry axis and have about the same gray value. Conversely, in the edge symmetry computation, they must also feature an opposite edge sign to match edges belonging to the same object (Fig. 16.13).

Let $A = (x_a, y_a)$ and $B = (x_b, y_b)$ be two symmetric edge points with $x_a < x_b$ and let $\rho_a, \rho_b \in \pm 1$ be, respectively, their edge sign. A and B are regarded as *warm symmetric edge points* if $\rho_a = 1$ and $\rho_b = -1$, namely, if A is a transition from dark-to-bright and B is a bright-to-dark one. The edge symmetry computation mainly exploits the warm symmetric edge points. This particular approach is aimed to regard as symmetric the objects that are brighter than background. Also, the match between bright-to-dark and dark-to-bright transitions is considered for computing the symmetry maps; in fact, when given an axis position and a symmetry window, the edge symmetry is higher than a given threshold, this contribution is also computed and added to the symmetry map. The underlying ratio of this approach is to further enhance the symmetry weight related to warm objects.

Gray-level symmetry (Fig. 16.12e), edge symmetry (Fig. 16.12f), and edge density (Fig. 16.12g) histograms are computed scanning the related maps and detecting

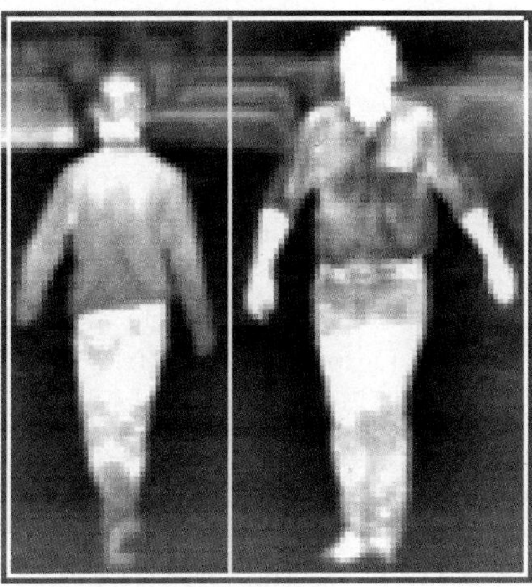

Fig. 16.13 Wrong axis not detected thanks to the use of edge sign

the highest values for each column. Symmetry histograms and edge density are then combined to compute the overall histogram (Fig. 16.12h), as shown in the following formula:

$$TotSym = (EdgeSym + k \times GraySym) \times EdgeDens \qquad (16.3)$$

where k is a constant value used to weight gray-level symmetry contribution. A new bounding boxes list is then generated by thresholding the resulting histogram. Therefore, for each bounding box, the overthreshold peaks generate new bounding boxes that contain the shape of a potential pedestrian. This step allows both refinement of bounding box width and splitting bounding boxes that contain more than one object.

The new bounding box width is, in fact, computed maximizing the edge density. Unfortunately, the so-computed bounding boxes tend to be narrower than the object they contain (Fig. 16.14a). Therefore, the bounding boxes are further expanded using a vertical histogram of edges to better match the object's width (Fig. 16.14b).

Also, the symmetry content of each bounding box is added to the bounding box list. In fact, the list of refined and filtered bounding boxes is used by following validator stages and by the final decision step, as depicted in Fig. 16.1.

16.5 Human Shape Detection

After the symmetry step, the list of bounding boxes that contains potential pedestrians is further investigated to validate the presence of a human shape. Three different validation processes are currently used to evaluate the boxes content. One is based on the search of the most evident feature of a human shape, namely, the presence of a head. The two others are based on the evaluation of the overall shape of the object: active contour-based shape detection and the use of a probabilistic model.

A final rules validator is used to compute the probability that a box contains a human shape and to discard boxes whose probability is too low.

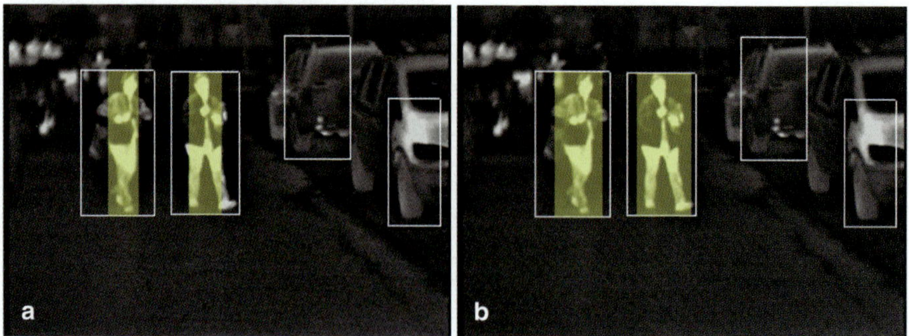

Fig. 16.14 The bounding boxes before **a** and after **b** the expansion. The yellow areas show the refined bounding box; the white border depicts the bounding box before the symmetry-based step

16.5.1 Active Contour Models

Active contour models, also known as *snakes*, are a widely used method in pattern recognition for extracting an object's shape when it is already roughly known. Initial stages of the system provide a bounding box for each detected object. The active contour models technique can be effectively exploited to obtain the object's precise contour starting from the bounding box enclosing it.

First introduced by Kass et al. [19], this topic has been extensively explored in the last years. Basically, a snake, in its simplest version, is a curve described by the parametric equation $\mathbf{v}(s) = (x(s), y(s))$, where s is the normalized length, assuming values in the range $[0, 1]$. Such a curve consists of points feeling the effect of a field generated by some forces that are the unique control over point movements; for this reason, the snake capability of following the desired contour in an image only derives from the careful choice of such forces. All snake points change their position, reaching a location where a minimum of the energy associated with the field is found.

Forces, and associated energies, can be divided into two different categories: internal and external. Internal energy only depends on the topology of the snake and controls the continuity of the curve derivatives; it is evaluated by the equation:

$$E_{int} = \alpha(s)|\mathbf{v}_s(s)|^2 + \beta(s)|\mathbf{v}_{ss}(s)|^2 \tag{16.4}$$

where $\mathbf{v}_s(s)$ and $\mathbf{v}_{ss}(s)$ are, respectively, the first and second derivatives of $\mathbf{v}(s)$ with respect to s. The first contribution appearing in the sum is responsible for the tension of the snake and makes it behave like an elastic band; the second one gives the snake resistance to bending; $\alpha(s)$ and $\beta(s)$ are weights. Internal energy therefore causes the snake to have certain mechanical properties but is independent of the image; external energy, on the contrary, is a function of the image and can therefore attract the snake to the desired features. The snake will try to minimize the whole energy balance, given by the equation:

$$E_{snake} = \int_0^1 (E_{int}(\mathbf{v}(s)) + E_{ext}(\mathbf{v}(s))) \, ds \tag{16.5}$$

The main task, when using active contour models, is to properly choose external energies because these determine how the snake will be attracted by desired features. In particular, the image energy must decrease in the regions where the snake should be attracted. The following discussion thus focuses on the techniques adopted to obtain the contour of recognized pedestrians.

When applying the snake method, an initialization algorithm should be chosen, that is, a way of placing the snake in its initial position. As discussed, the shape of the object should already be roughly known. The natural and simplest choice for the system being presented was to place the snake along the contour of the bounding box that contains each object.

Moreover, in this system both visible and FIR images are available, but the latter seem much more convenient for extracting pedestrian contours. In particular, the edge image obtained applying a Sobel filter has been selected as a starting point for computing the image energy. Moreover, such image has already been computed because it is needed by previous steps of the recognition algorithm. The original edge image is smoothed using a Gaussian filter to enlarge the edges, namely, the region where the active contour should be attracted; then, an energy level is associated to each pixel, depending on the gray level of the smoothed edge image: the brighter the pixel, the lower is the energy associated. Therefore, *snaxels* (the points into which the active contour is sampled) are attracted by bright (i.e., strong), edges.

Once the active contour has been placed, its energy is minimized by iteratively moving each snaxel, trying to find a local minimum. Many solutions to the snake minimization problem have been proposed in the literature [19, 31, 32]. The greedy snake algorithm [31], applied on 5×5 neighborhood, has been used in this work.

Because of the tension of the snake, it tends to shrink until some other kind of energy impresses an opposite behavior; this can be, for instance, the presence of edges in the image; in general, the snake will lower its length while adapting to the object's shape, and the mean distance between two adjacent snaxels will decrease. This does not necessarily happen in a uniform way; when the detected object is a pedestrian, the snake contracts a lot toward the head and less along the body, so snaxel concentration can vary in different portions of the contour. For example, where a large number of snaxel are present in a small region, some artifacts may appear (Fig. 16.15a). To avoid this situation, the snake is periodically resampled while it is contracting, keeping controlled the mean distance among snaxels; in this way the mechanical characteristics are preserved even if the active contour length undergoes sensible changes.

To improve the shape extraction, also a double-snake approach was explored. Substantially, two snakes were used to extract the shape: One works like the one already described, while the other one is initialized inside the bounding box and

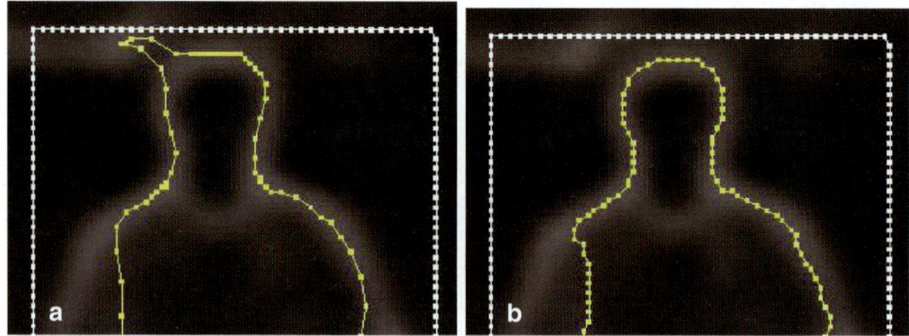

Fig. 16.15 Final configuration of the snake without resampling **a** and with resampling **b**. White dots represent snaxels of the snake in its initial position; yellow dots compose the snake in its final configuration. In **b** the correct snaxel disposition and the absence of the artifact toward the head are evident

Fig. 16.16 Example of a
pedestrian contour extracted
by double-snake technique

expands until it reaches the pedestrian contour. This technique should improve the
extraction of concave contours, like the inner part of a pedestrian's legs.

When dealing with double snakes, it is important to care about how each snake is
attracted by the other one. The easiest way is to add a force that moves each snaxel
toward the nearest one of the other snake. This solution, however, turned out not
to be effective enough because the new force modifies the behavior of the snake in
a few cases, so that the two snakes do not get sufficiently close. A more effective
solution was then adopted; it consisted of associating each snaxel of the inner snake
with the nearest snaxel of the outer one. Then, one of the snaxel is moved toward
the other, depending on which movement minimizes the overall energy of the two
snakes. By adopting this approach, the two snakes tend to collapse into a single
snake. A shape extraction result adopting the double-snake technique is shown in
Fig. 16.16.

Even if the double-snakes technique improves the contour extraction in some
cases, it does not substantially enhance contour extraction. At the moment, the
single-snake approach is preferred because of the lower computational cost.

Results given by contour extraction can be used in the object classification phase,
whose task is to decide whether a detected obstacle is a pedestrian. Such phase is
based on the use of multiple classification algorithms, each one providing a vote;
the final decision is derived from all votes. Therefore, the extracted contours should
be further processed to derive a single vote. Some shape-matching algorithms [8, 9]
were used to compare the final configuration of each snake with some models of
pedestrian shape in a database. These approaches have proven to feature not only a
low number of false negatives, but also an unacceptable number of false positives
were obtained; this led to the conclusion that shape matching is unsuitable to classify
pedestrians in this application. Another approach, based on neural networks, is currently under development: Every snake is resampled so that it is composed of a given
number of snaxels, which are fed to a classification network that evaluates the vote.

16.5.2 Probabilistic Models

A match against probabilistic models of a human shape is also used to validate bounding boxes [22, 24, 26]. To make this validation more reliable and to detect the pedestrian pose, four different probabilistic models that comprise a number of pedestrian poses and orientation are used for the match (Fig. 16.17). Currently, the models are classified considering the position of the legs during the walking process: open, almost open, almost closed, and fully closed legs. Four different training sets of images that contain human shapes in these different poses have been used to obtain the four probabilistic models. The training sets also show people who walk in front or laterally with respect to the camera axis to take into account different angles of view for human shapes.

Figure 16.18 shows the resulting models used for the recognition step. These models are created offline; therefore, their computation is not time consuming.

Initially, a mask model (Fig. 16.18a) is used to remove the background and to enhance the foreground of the bounding boxes. The mask is computed as the average of the four probabilistic models. Since sometimes the bounding boxes are not sufficiently accurate, the search region is enlarged proportionally to each bounding box size, and the mask model is scaled according to this search region (Fig. 16.19).

To enhance the foreground and remove the background, the following formula is used:

$$I(x,y) = \begin{cases} 255 & \text{if } p(x,y) \geq 0.75 \text{ AND } I(x,y) > 127 \\ 0 & \text{if } p(x,y) \leq 0.20 \text{ AND } I(x,y) \leq 127 \\ I(x,y) & \text{otherwise} \end{cases} \qquad (16.6)$$

where $I(x,y)$ are the values of search region pixels, and $p(x,y)$ are the values of the mask pixels.

Fig. 16.17 A few examples of training data set images

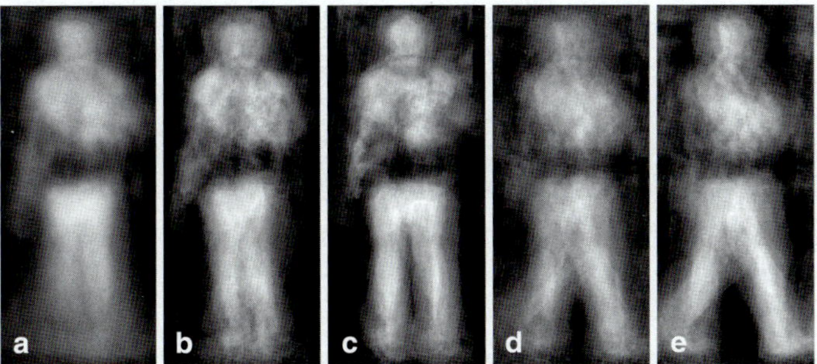

Fig. 16.18 **a** Mask model obtained considering all kind of poses and models used for the recognition step: **b** closed, **c** almost closed, **d** almost open, **e** open legs

Fig. 16.19 Foreground enhancement and background removal: **a** original bounding boxes, **b** search areas (blue) and rectangles (red) that correspond to the size of the mask model, and **c** final result

Empirically, it was found that regions where $p(x,y) \geq 0.75$ corresponds to head and torso, and $p(x,y)$ is lower than 0.20 in correspondence to the background. Intermediate situations are generally due to legs or arms.

After this phase, the different models are matched against the extracted foreground. Each model is resized to fit the corresponding search region. For each point (x,y) within the search area, a correlation probability $cp(x,y)$ is computed using the formula:

$$cp(x,y) = \sum_{i=1}^{m} \sum_{j=1}^{n} (th_{xy}(i,j) - 127) \times (p(i,j) - 0.5) \qquad (16.7)$$

where m and n are the model width and height, and $th_{xy}(i,j)$ is the image after the enhancement operation. The function $cp(x,y)$ encodes the probability that the area $m \times n$ centered in (x,y) contains a pedestrian.

Since four different models are used, the classification is solved as a maximization problem. The model that gets the higher probability value is considered the best match, and the probability itself is considered as the final vote and fed to the validator. Bounding boxes featuring a probability lower than a given threshold are discarded (Fig. 16.20).

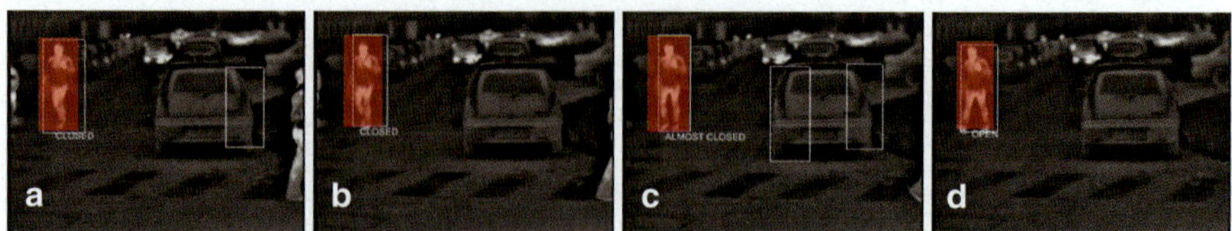

Fig. 16.20 Probabilistic model results: pedestrians **a** and **b** are labeled as having Closed legs, in **c** the pedestrian is labeled as having the legs almost closed, and in **d** as open legs

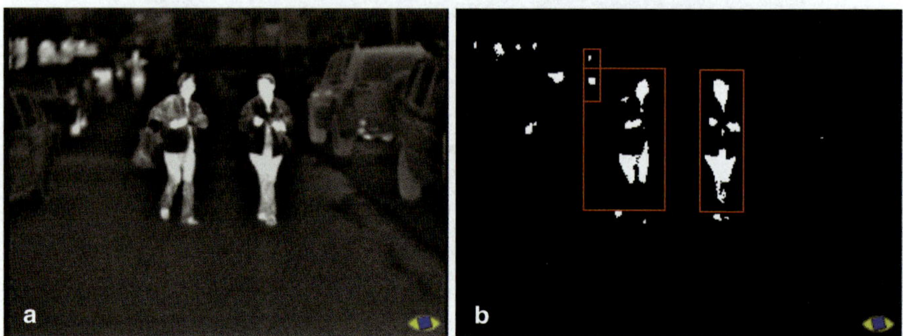

Fig. 16.21 Head detection by pattern matching: **a** original image and **b** binarized image

16.5.3 Head Detection

For each potential pedestrian, the presence of the head is evaluated. This step is performed on the FIR domain only because the head is the most evident feature of a human shape in the FIR domain due to the high heat dispersion, which makes pixels brighter than the background. In addition, the position of the head is barely affected by pedestrian pose, being always in the upper part of the bounding box.

The head detection system exploits three different approaches; two of them are based on a model match, the last one on the search of warm areas. The use of a number of different approaches has been selected to minimize the risk of false negatives.

16.5.3.1 Pattern Matching

The first approach relies on a pattern-matching technique. Two different models of a head are used to perform different matching operations.

The first model encodes thermal characteristics of a head warmer than the background. To ease the match, the areas of attention are binarized using an adaptive threshold (Fig. 16.21). This model is a binary mask showing a white head on a black background (Fig. 16.22a).

The model is scaled according to the bounding box size and assuming that a head measures nearly one sixth of a human shape height. The match is performed on an

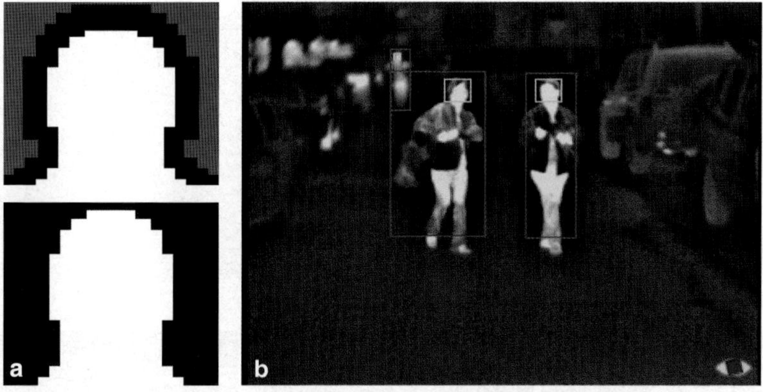

Fig. 16.22 Head search: **a** the two models used for head matching, **b** final result

area centered around the top of the bounding box using Eq. (16.1). The highest correlation value obtained is considered the match quality (P_w).

Unfortunately, the head is not always warmer than the background. Environmental conditions, hats, helmets, or hair may mask heat radiation. To cope with this problem, an additional head model is used. This model encodes the head shape (Fig. 16.22a) and is used to perform another match in the top area of each bounding box. In this case, the areas of attention are not binarized. For each position of the model, the average values of pixels that correspond to the internal (white) or external (black) part of the model are computed. The quality of the match is considered the absolute value of the difference between these two averages. A higher difference is obtained in correspondence to objects that feature a shape similar to the model. The shape-matching quality (P_s) is computed as the highest of such differences.

The final match parameter (P_m) is computed as $P_m = 1 - ((1 - P_w) \times (1 - P_s))$ and fed to the rules validator.

16.5.3.2 Probabilistic Model

The probabilistic approach defines a matching method between a head image and a suitable model [24]. The underlying idea is the same as that described in Section 16.5.2, but the implementation is slightly different.

The region of interest focuses on the head; as discussed, the model is generated using a number of head images. The *training set* images show pedestrians in several positions and movements. Then, for each pixel the probability that it belongs to a pedestrian head is computed. The probabilistic head model (Fig. 16.23) is then resized according to the bounding box width to perform the match.

For each bounding box, the area of attention for the matching is located in the upper part. The height of the searching area is, again, about one sixth of the box height. To ease the match, a threshold is applied to remove the background. Moreover,

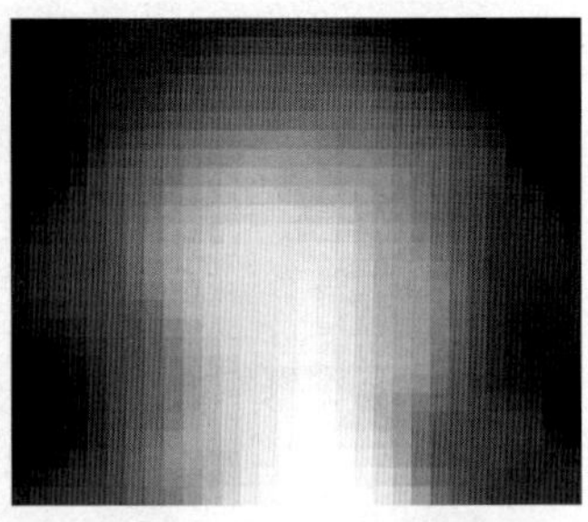

Fig. 16.23 Head model for probabilistic search

Fig. 16.24 Head search by probabilistic model: **a** detected obstacles, **b** preprocessed image, and **c** final result

bright zones are enhanced using a morphological filter (Fig. 16.24); the match is performed using Eq. (16.7) and the result is a probability map.

This map contains all the probability values greater than a threshold. Then, the local maximum is found and considered as the best candidate for the head position in the bounding box; the correspondent correlation probability value is provided to the rules validator (Fig. 16.24c).

16.5.3.3 Warm Area Search

The last approach is based on warm area analysis. This search criterion exploits, as a preprocessing phase, the double-threshold algorithm described in Section 16.3.2. Different operations are then performed due to the different aims of the search.

The upper side of the bounding box is considered as a search area. Moreover, since sometimes bounding boxes are not sufficiently accurate, this search area is slightly enlarged, proportional to the bounding box size.

After the preprocessing phase, warm areas are labeled to detect contiguous areas. These areas are then enclosed in smaller bounding boxes, which become candidates for the head position.

The biggest warm area that satisfies headlike criteria is chosen as the best match for the head. Criteria are based on head aspect ratio and on the ratio between the head area and the bounding box area. Figure 16.25 shows the different process steps.

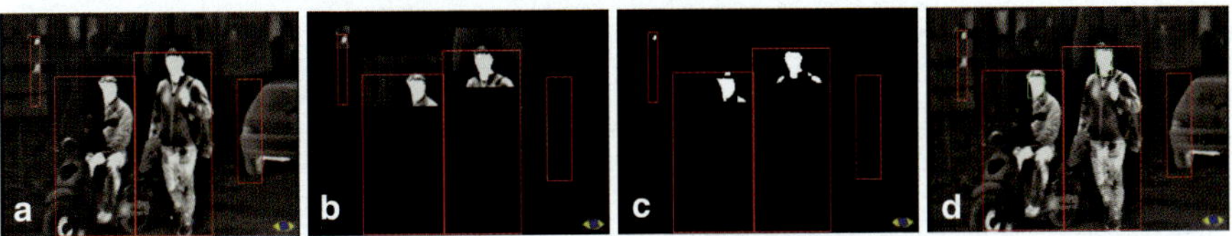

Fig. 16.25 Head search by warm area search: **a** detected obstacle, **b** searching area, **c** preprocessed image, and **d** final result

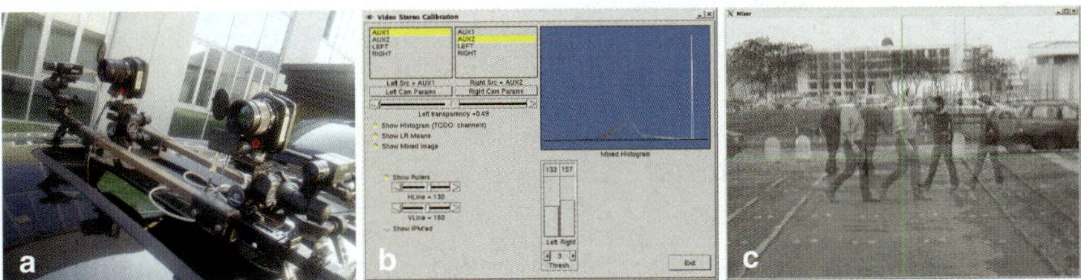

Fig. 16.26 Acquisition system: **a** The four cameras used to acquire the images installed on the vehicle roof, **b** the user interface of the video calibration tool, and **c** the mixed image used to adjust the camera orientation

16.6 Acquisition

This section describes the main aspects of the video acquisition system and the procedure used to perform image calibration and camera orientation.

16.6.1 System Description

The hardware used to capture the images consists of two rigid metal bars connected through flanges forming a frame. On this frame, four 3-axis, geared camera mounts have been mounted; two of them hold the FIR cameras, and the others hold the visible cameras. Particular care is used to avoid movements due to vibrations even when carrying a heavy load such as the FIR cameras (Fig. 16.26a).

Table 16.1 below reports the main parameters of the two cameras.

Each FIR camera is connected to a dedicated BT878-based frame grabber, while the visible images are acquired through the IEEE 1394 bus.

Neither FIR cameras nor visible ones have an external synchronization system; therefore, small differences among the acquisition timings of the four cameras are experienced. A newer vision system capable of synchronizing all the cameras using an external trigger is currently under development.

Table 16.1 Camera parameters

Characteristic	FIR	Visible
Wavelength	7–14 µm	0.4–0.7 µm
Sensor type	Uncooled FPA	CCD
Sensor size	320 × 240 pixel	640 × 480 pixel
Image depth	8 bit	8 bit
Horizontal FOV	0.1535 rad	0.2216 rad
Vertical FOV	0.1182 rad	0.1680 rad
Baseline	0.500 m	1.000 m
Standard	National Television System Committee (NTSC)	IIDC (Inquirer Inter-collegiate Debating Championship)

16.6.2 Cameras Calibration

Calibration is an important issue because it is needed by both visible and FIR stereo pairs.

Cameras are adjusted using a software tool specifically developed for this application; a graphical interface allows selecting, overlapping, and mixing of two acquisition streams, therefore easing the tuning of camera positions. The same interface displays color and luminance histograms of the acquired images, allowing the tuning of camera apertures and gains (Fig. 16.26b).

Specific regions of the mixed streams can be magnified to perform a fine-tuning for focus and orientation, especially when framing faraway objects. Two rulers can be used during the orientation adjustment operations, showing horizontal and vertical lines that can be moved as markers to keep references in a specific position. This is particularly helpful when the vehicle is positioned on a calibration grid (Fig. 16.26c).

16.7 Validation and Results

The results of each validation step are used to filter out false positives. In the current system, the vote fed by each validator is linearly combined to compute a final *fitness* value. This simple approach has proven to be insufficiently effective since it does not take into account the different behavior of each validator. In fact, each validator features different perception characteristics that are hard to join together. For example, a validator can feature a low false-positive rate as well as a medium false-negative percentage; in such a case, when a human shape is detected, the probability it is a false positive is low; at the same time, when a bounding box is discarded, the risk it contains a pedestrian is not negligible. Therefore, a single vote for each validator is not sufficient to evaluate the presence of a pedestrian.

Currently, a neural network approach is under development as a rules validator, to take into account the behavior and characteristics of each validator.

Figure 16.27 shows a few results of the validator based on the probabilistic model. If the confidence of the detection is above a threshold, the corresponding box is drawn in red, meaning that the region has a high probability of containing a human shape, or in blue when the confidence is low.

Figure 16.28 shows the results of the symmetry computation step. This allows refining bounding box width, splitting bounding boxes that contain more than one object, and validating them, filtering out edgeless or asymmetric bounding boxes.

Some symmetrical objects, like cars or trees, are validated as well. Moreover, some validation problems are encountered when the FIR images are not optimal, like those acquired in summer under heavy direct sunlight; in these conditions, many

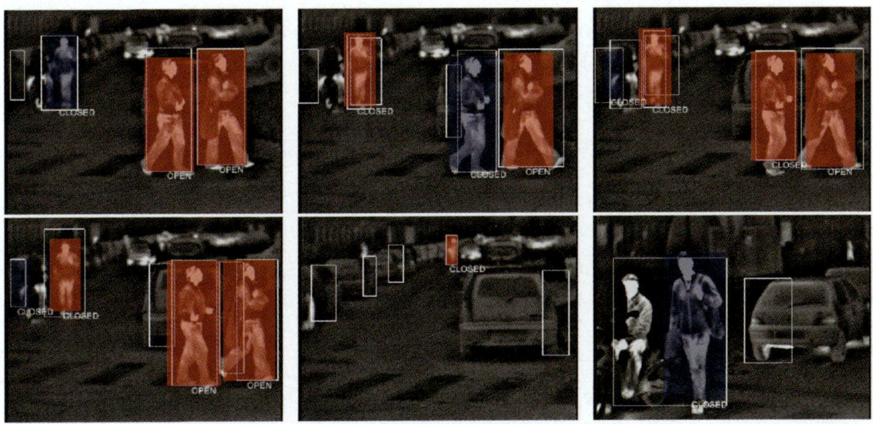

Fig. 16.27 Results of the probabilistic model validators: detected pedestrians shown using a superimposed red box when the confidence on the classification is high or blue when the confidence is low

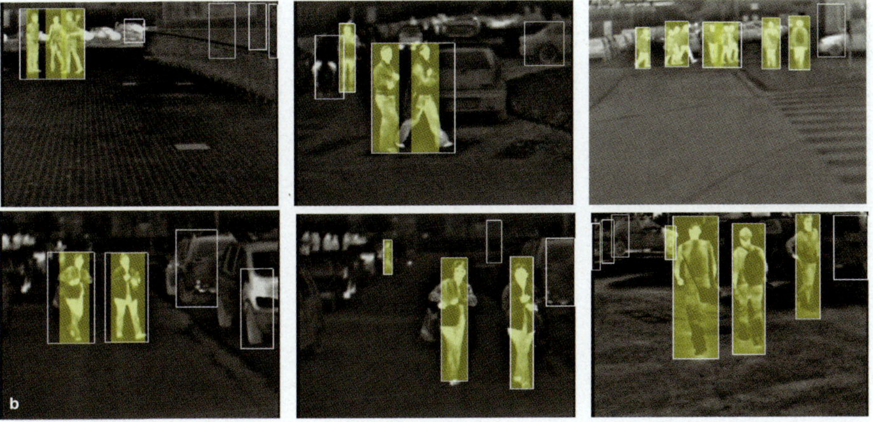

Fig. 16.28 Symmetry analysis results: validated pedestrians shown using a superimposed yellow box; white rectangles are the bounding boxes generated by previous steps before the validation, splitting, or resizing

objects in the background become warm, and the assumption that a pedestrian features a higher temperature than the background is not satisfied. This causes some problems in the edge and symmetry computation, and therefore in the results.

The result of active contour-based validation is shown in Fig. 16.29. It can be noticed that computed contours fit well with pedestrian shapes. A match with a model has been tested to validate the result before passing it to the rules validator; nevertheless, a more sophisticate filter is currently under development.

Figure 16.30 shows the results of the three different approaches for head detection. The use of different approaches allows reducing the number of false negatives, especially when the head is not much warmer than the background.

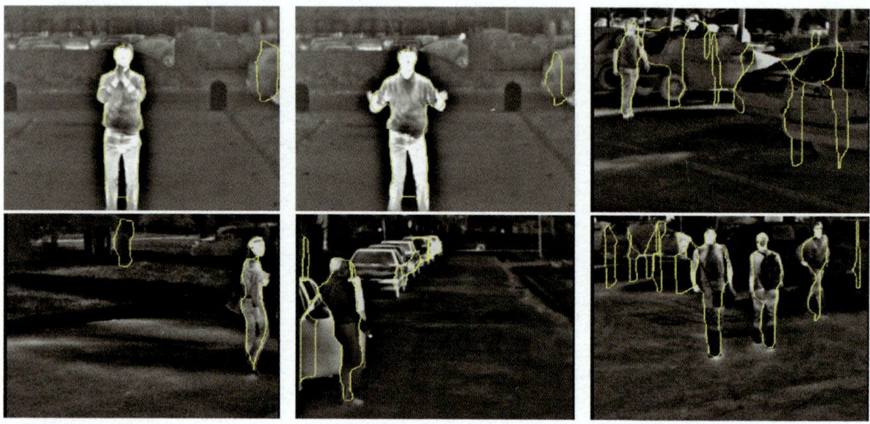

Fig. 16.29 Examples of contour extraction using snakes: the active contour model is applied to each detected obstacle. Contour extraction gives good results also when some background details are present

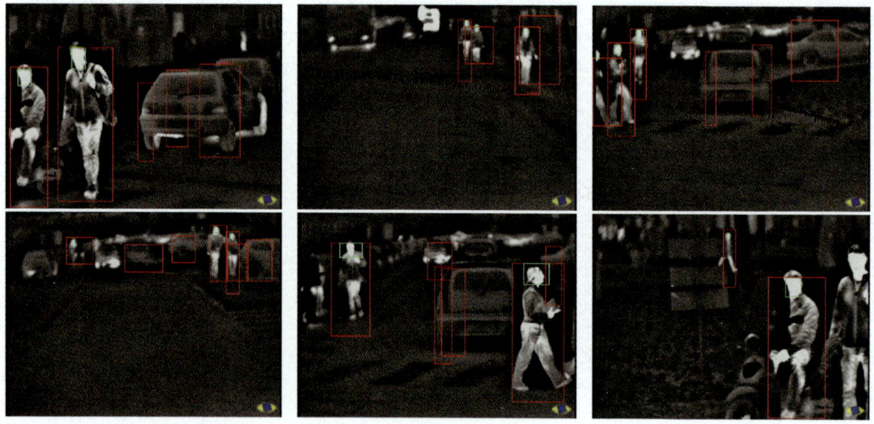

Fig. 16.30 Results of the head detector validator: green bounding boxes show heads detected inside the original areas of attention in red

16.8 Conclusions

In this chapter, a system aimed at the detection of pedestrians exploiting both a FIR and a visible daylight system has been presented. It has been tested in urban and rural environments using an experimental vehicle equipped with four cameras, two infrared cameras that work in the 7 to 14μm spectrum range and two CMOS-based CCD cameras.

The algorithm is based on the use of different approaches for detecting an initial list of areas of attention: the warm area and vertical edge detection in the FIR domain and the DSI-based computation in both domains. These different approaches independently produce different lists of areas of attention; therefore, a merging step is used for coalescing these lists into a single one. Thanks to a precise knowledge of the calibration parameters and to the stereo-based computation of the road slope, distance, size, and position of each area are computed, and areas of attention that feature parameters not compatible with the presence of a pedestrian (or a small group of pedestrians) are discarded.

Thanks to the assumption that a human shape is mostly symmetrical, a following symmetry-based process is used to refine and further filter out the areas of attention.

Finally, to validate the presence of pedestrians in the surviving areas, a number of validators are used: a head presence detector, active contour-based processing, and a match with a probabilistic model of human shape.

Preliminary results demonstrated that this approach is promising. The use of sensors that work in different domains allowed increasing the detection rate with respect to previous similar systems [2] and detecting pedestrians even if they were not warmer than the background or in low-illumination conditions. Moreover, the symmetry-based refinement step and the use of some validators allowed the system to detect pedestrians where they were not detected by a previous system version [3].

A neural network-based rules validator is currently under development. Neither temporal correlation nor motion cues are used for the processing.

Chapter's References

1. NIT Phase II: Evaluation of Non-Intrusive Technologies for Traffic Detection. Technical Report SRF No. 3683, MN DOT Research Report, 2002

2. M. Bertozzi, A. Broggi, M. Del Rose, and A. Lasagni. Infrared Stereo Vision-Based Human Shape Detection. In *Proceedings of IEEE Intelligent Vehicles Symposium 2005*, pages 23–28, Las Vegas, June 2005

3. M. Bertozzi, A. Broggi, M. Felisa, G. Vezzoni, and M. Del Rose. Low-Level Pedestrian Detection by Means of Visible and Far Infra-red Tetra-vision. In *Proceedings of IEEE Intelligent Vehicles Symposium 2006*, pages 231–236, Tokyo, Japan, June 2006

4. D. Beymer and K. Konolige. Real-time Tracking of Multiple People Using Continuous Detection. In *Proceedings of the International Conference on Computer Vision*, Kerkyra, 1999

5. B. Bhanu and J. Han. Kinematics-Based Human Motion Analysis in Infrared Sequences. In *Proceedings of IEEE International Workshop on Applications of Computer Vision*, Orlando, 2002

6. A. Broggi, C. Caraffi, R. I. Fedriga, and P. Grisleri. Obstacle Detection with Stereo Vision for Off-Road Vehicle Navigation. In *Proceedings of International IEEE Workshop on Machine Vision for Intelligent Vehicles*, San Diego, June 2005

7. R. Cutler and L. S. Davis. Robust Real-time Periodic Motion Detection, Analysis and Applications. *IEEE Transaction on Pattern Analysis and Machine Intelligence*, 22(8):781–796, August 2000

8. M.-S. Dao, F.G.B.D. Natale, and A. Massa. Edge Potential Functions and Genetic Algorithms for Shape-Based Image Retrieval. In *Proceedings of IEEE International Conference on Image Processing (ICIP'03)*, volume 2, pages 729–732, Barcelona, Spain, September 2003

9. M.-S. Dao, F.G.B.D. Natale, and A. Massa. Efficient Shape Matching Using Weighted Edge Potential Function. In *Proceedings of 13th International Conference on Image Analysis and Processing (ICIAP'05)*, Cagliari, Italy, September 2005

10. J.W. Davis and V. Sharma. Robust Background-Subtraction for Person Detection in Thermal Imagery. In *Proceedings of International IEEE Workshop on Object Tracking and Classification Beyond the Visible Spectrum*, Washington, D.C, 2004

11. M. Del Rose and P. Frederick. Pedestrian Detection. In *Proceedings of Intelligent Vehicle Systems Symposium*, Traverse City, MI, 2005

12. M. Del Rose, P. Frederick, and J. Reed. Pedestrian Detection for Anti-Tamper Vehicle Protection. In *Proceedings of Ground Vehicle Survivability Symposium*, Monterey, CA, 2005

13. A. Fod, A. Howard, and M.J. Mataric. Laser-Based People Tracking. In *Proceedings of IEEE International Conference on Robotics and Automation*, Washington, DC, 2002

14. K.C. Frerstenberg, J. Dietmayer, and V. Willhoeft. Pedestrian Recognition in Urban Traffic Using Vehicle Based Multilayer Laserscanner. In *Proceedings of Automobile Engineers Cooperation International Conference*, Paris, 2001

15. K.C. Frerstenberg and U. Lages. Pedestrian Detection and Classification by Laser-Scanners. In *Proceedings of IEEE Intelligent Vehicles Symposium 2002*, Paris, June 2002

16. K. Grauman and T. Darrell. Fast Contour Matching Using Approximate Earth Mover's Distance. Technical Report AI Memo, AIM-2003-026, MIT, Cambridege, MA, 2003

17. G.D. Hines, Z. Rahman, D.J. Jobson, and G.A. Woodell. Multi-image registration for an enhanced vision system, *Proceedings - SPIE The International Society for Optical Engineering*, pp. 231–241, April 2003

18. R. Kania, M. Del Rose, and P. Frederick. Autonomous Robotic Following Using Vision Based Techniques. In *Proceedings of Ground Vehicle Survivability Symposium*, Monterey, CA, 2005

19. M. Kass, A. Witkin, and D. Terzopoulos. Snakes: Active Contour Models. *International Journal of Computer Vision*, 1(4):321–331, 1988

20. M. Koltz and H. Rohling. 24 GHz Radar Sensors for Automotive Applications. In *Proceedings of International Conference on Microwaves and Radar*, Warsaw, Poland, 2000

21. R. Labayrade and D. Aubert. A Single Framework for Vehicle Roll, Pitch, Yaw Estimation and Obstacles Detection by Stereovision. In *Proceedings of IEEE Intelligent Vehicles Symposium 2003*, pages 31–36, Columbus, OH, June 2003

22. L. Lee, G. Dalley, and K. Tieu. Learning Pedestrian Models for Silhouette Refinement. In *Proceedings IEEE International Conference on Computer Vision*, Nice, France, 2003

23. S. Milch and M. Behrens. Pedestrian Detection with Radar and Computer Vision. In *Proceedings of Conference on Progress in Automobile Lighting*, Darmstadt, Germany, 2001

24. H. Nanda and L. Davis. Probabilistic Template Based Pedestrian Detection in Infrared Videos. In *Proceedings of IEEE Intelligent Vehicles Symposium 2002*, Paris, June 2002

25. C. Papageorgiou, T. Evgeniou, and T. Poggio. A Trainable Pedestrian Detection System. 38: 15–33, June 2000

26. A. Senior. Tracking People with Probabilistic Appearance Models. In *ECCV Workshop on Performance Evaluation of Tracking and Surveillance Systems*, 2002

27. A. Shashua, Y. Gdalyahu, and G. Hayun. Pedestrian Detection for Driving Assistance Systems: Single-frame Classification and System Level Performance. In *Proceedings of IEEE Intelligent Vehicles Symposium 2004*, Parma, Italy, June 2004

28. H. Shimizu and T. Poggie. Direction Estimation of Pedestrian from Multiple Still Images. In *Proceedings of IEEE Intelligent Vehicles Symposium 2004*, Parma, Italy, June 2004

29. G.P. Stein, O. Mano, and A. Shashua. Vision Based ACC with a Single Camera: Bounds on Range and Range Rate Accuracy. In *Proceedings of IEEE Intelligent Vehicles Symposium 2003*, Columbus, OH, June 2003
30. S. Tate and Y. Takefuji. Video-Based Human Shape Detection Deformable Templates and Neural Network. In *Proceedings of Knowledge Engineering System Conference*, Crema, Italy, 2002
31. D.J. Williams and M. Shah. A Fast Algorithm for Active Contours and Curvature Estimation. *CVGIP: Image Understanding*, 55(1):14–26, 1992
32. C. Xu and J. Prince. Snakes, Shapes, and Gradient Vector Flow. *IEEE Transactions on Image Processing*, 7(3):359–369, 1998
33. F. Xu and K. Fujimura. Pedestrian Detection and Tracking with Night Vision. In *Proceedings of IEEE Intelligent Vehicles Symposium 2002*, Paris, June 2002
34. L. Zhao. Dressed Human Modeling, Detection, and Parts Localization. Ph.D. dissertation, Carnegie Mellon University, 2001

Part VII
Multitarget Tracking Using Laser and Infrared Sensors

Chapter 17
Real-Time Detection and Tracking of Multiple People in Laser Scan Frames

J. Cui, X. Song, H. Zhao, H. Zha, and R. Shibasaki

Abstract This chapter presents an approach to detect and track multiple people robustly in real time using laser scan frames. The detection and tracking of people in real time is a problem that arises in a variety of different contexts. Examples include intelligent surveillance for security purposes, scene analysis for service robot, and crowd behavior analysis for human behavior study. Over the last several years, an increasing number of laser-based people-tracking systems have been developed in both mobile robotics platforms and fixed platforms using one or multiple laser scanners. It has been proved that processing on laser scanner data makes the tracker much faster and more robust than a vision-only based one in complex situations. In this chapter, we present a novel robust tracker to detect and track multiple people in a crowded and open area in real time. First, raw data are obtained that measures two legs for each people at a height of 16 cm from horizontal ground with multiple registered laser scanners. A stable feature is extracted using accumulated distribution of successive laser frames. In this way, the noise that generates split and merged measurements is smoothed well, and the pattern of rhythmic swinging legs is utilized to extract each leg. Second, a probabilistic tracking model is presented, and then a sequential inference process using a Bayesian rule is described. A sequential inference process is difficult to compute analytically, so two strategies are presented to simplify the computation. In the case of independent tracking, the Kalman filter is used with a more efficient measurement likelihood model based on a region coherency property. Finally, to deal with trajectory fragments we present a concise approach to fuse just a little visual information from synchronized video camera to laser data. Evaluation with real data shows that the proposed method is robust and effective. It achieves a significant improvement compared with existing laser-based trackers.

R.I. Hammoud (ed.), *Augmented Vision Perception in Infrared: Algorithms and Applied Systems*, Advances in Pattern Recognition, DOI 10.1007/978-1-84800-277-7_17,
© Springer-Verlag London Limited 2009

17.1 Introduction

The detection and tracking of people in crowds is a problem that arises in a variety of different contexts. Examples include intelligent surveillance for security purposes, scene analysis for a service robot, crowd behavior analysis for human behavior study, traffic flow analysis for intelligent transportation, and many others.

Over the last several years, an increasing number of laser-based people-tracking systems have been developed in both mobile robotics platforms [1–6] and fixed platforms [7–10] using one or multiple laser scanners. It has been proved that processing of laser scanner data makes the tracker much faster and more robust than a vision-based one in complex situations with varied weather or light conditions.

However, all these systems are based on a basic assumption that laser points that belong to the same person can easily be clustered or grouped as one feature point. Then, data associations can be processed for multiple people tracking. In real experiments with an unknown number of people, especially when dealing with a crowded environment, such systems will greatly suffer from poor features provided by a laser scene. Laser points of different objects are often interlaced and undistinguishable and cannot provide reliable features.

The same problem arose in our previous work [7]. The experimental result showed that, in some cases, a simple clustering method will fail in detecting a person due to clutter from other objects that move with people, such as nearby people or luggage. And, when two people are walking or their legs are too close together, it across each other's path likely causes tracking error or broken trajectory. To ease the understanding of laser scan data, we show a registered laser scan image (see Fig. 17.1) with four laser scanners scanning at a height of 16 cm from horizontal ground. In Fig. 17.1), nearly 30 people are separately distributed in an open area. Red circles denote laser scanners. White points are foreground laser points, mainly human legs. Green points are background laser points, including walls, chairs, and

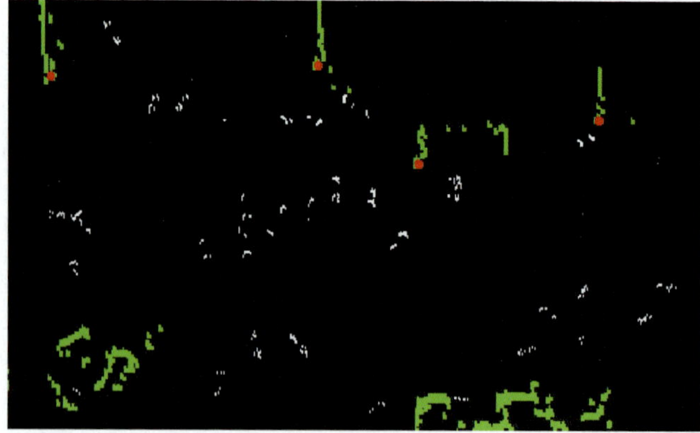

Fig. 17.1 A laser scan image: *red circles* laser scanners; *white points* foreground laser points, mainly human legs, *green points* background laser points

Fig. 17.2 One sample of fusion of image frame and laser scan frame. Each cluster of laser points that belong to human leg has been manually circled

and more. Figure 17.2 is one sample of fusion of image frame and laser scan frame. This image is just used to illustrate what the laser points mean. Each cluster of laser points that belong to a human leg has been manually circled.

In this chapter, we present a robust tracker to detect and track multiple people in a crowded and open area. We first obtain raw data that measure two legs for each people at a height of 16 cm from horizontal ground with multiple registered laser scanners. Then, a kind of stable feature is extracted using accumulated distribution of successive laser frames. In this way, the noise that generates split and merged measurements is smoothed well. A region coherency property is utilized to construct an efficient measurement likelihood model. Then, a tracker based on the combination of independent Kalman filter and a Rao-Blackwellized Monte Carlo data association filter (RBMC-DAF) is introduced. Finally, to deal with broken fragments caused by complex situations as well as long-term occlusions, a visual tracking and matching approach is introduced. Evaluation with real data showed that the proposed method is robust and effective and deals with most well-known difficulties encountered by conventional laser-based trackers very well (e.g., measurement split/merge and temporal occlusion).

The remainder of this chapter is organized as follows: After discussion of related work in the following section, the principle of single-row laser scanner, system architecture, and data collection are briefly introduced in Section 17.3. And then, feature extraction approach is described in Section 17.4. In Section 17.5. and Section 17.6, we will present the laser-based tracking framework and visual-assisted processing respectively. Last, evaluations of detection and tracking results are presented in Section 17.7, followed by some discussions and conclusions in the final section.

17.2 Related Work

Research on laser-based people tracking originated from Prassler's work [5]. In recent years, laser scanner became much cheaper, and the scan rate became higher than before (from 3 fps used in [2, 3, 5, 6] to 30 fps used in [7]). In the context of robotic technology, a lase-based people tracker [1–4, 6] has been a fundamental part of a mobile robotic system. These trackers mainly focus on how to correctly detect moving people and distinction of people from static objects with a mobile platform, then tracking one or few moving persons surrounding the mobile robot within the successive laser scan images. On the other hand, in the context of intelligent monitoring and surveillance, multiple laser scanners [7, 9] are deployed to cover a wide area. In this case, the task is to effectively extract an individual person from cluttered and scattered laser points and then simultaneously track a large number of people robustly and reliably. Thus, we can see that, for both kinds of laser trackers, no matter whether with a static or mobile platform, there are two fundamental aspects: people extraction (i.e., people detection), and data association.

Clustering or grouping in each scan image is the most commonly used and almost the only people extraction strategy for existing laser-based trackers [1–9]. In [5], a grid map representation is used for detection of moving cells. One group of nearby cells is considered as one person. Then, a trajectory of moving targets is obtained by a nearest neighbor criterion between groups of cells marked as moving in consecutive scan samples.

In [2], at each time step, the laser scan image was segmented and further split into point sets representing objects. At first, the scan was segmented into densely sampled parts. In the second step, these parts were split into subsequences describing "almost convex" objects using the assumption that there are distance gaps between distinct objects. A threshold value was used to find distinct objects. For tracking, the motion of object shapes in successive scan images is represented as flows in bipartite graphs. Then by network optimization techniques in graph theory, plausible assignments of objects from successive scans were obtained.

In [3], they found violation points (corresponding to moving objects) from each range scan at first, then all detected violations were viewed as a Gaussian impulse living on the two-dimensional world map. New hypotheses were created at violation points that had a function value over a certain threshold. In fact, this is a continuous version of a point-clustering process instead of a discrete one. Previous hypotheses were propagated of moving objects by a gradient ascent method on the function. In essence, this is a local searching method, similar to nearest neighbor searching.

In [6], a mobile platform was equipped with two laser-range scanners mounted at a height of 40 cm. Each local minimum in the range profile of the laser range scan was considered as a feature that represented an object. Moving objects such as a person were distinguished from static objects by computing local occupancy grid maps. In [1] and [4], a simple clustering method was used for object extraction similar to the first step described in [3]. For tracking, in [1, 4, 6], several novel data association and tracking algorithms were proposed that incorporate particle filters and a Joint probabilistic data association filter (JPDA). This sampling-based approach has the advantage that there are no restrictions for the analytic form of model, although the required number of particles for a given accuracy can be very high. Due to the weakness of the measurement likelihood model, the tracking performance is greatly dependent on the performance of the filter.

The studies mentioned mainly focused on tracking one or a few moving persons surrounding the mobile robot, and up to now, only a few works [7, 9] aimed at tracking a large number of people with fixed laser scanners. In [9], the authors used multiple laser scanners waist height. By subtraction from the background model, the foreground was obtained. They defined a *blob* as a grouping of adjacent foreground readings that appeared to be on a continuous surface and assumed that measurements that were spatially separated by less than 10 cm belonged to the same blob. Scanning at the height of the waist will suffer greatly from occlusions by nearby people and the unpredictable range reflections from swinging arms, handbags, coats, and so on, which are difficult to model for accurate tracking. For tracking, they associated a Kalman filter with each object to alleviate the consequences of occlusions and to reduce the impact of occlusions and model inaccuracies.

Compared with other works, the system described in [7] gives the most promising result for tracking a large number of people simultaneously. The laser scanners were on the ground at a height of 20 cm. At this height, one person generates two point clusters, one for each foot. Simple clustering was also used to extract the moving feet. Then, a given distance range was used to group two nearby feet as one step. The following conditions were used for data association. First, two step candidates in successive frames overlapped at the position of at least one foot candidate. Second, the motion vector decided by the other pair of nonoverlapping foot candidates changed smoothly along the frame sequence. The experimental result showed that, in some cases, the simple clustering will fail in detecting a person due to clutter from other objects that moved with people, such as nearby people or luggage. And, when two people were walking or their feet were too close together, it likely caused across each other's path tracking error or broken trajectory.

Finally, as for related work on a combination of visual data and laser data, several methods have been proposed in the robotics area [10–12]. They tried to detect people on a mobile platform based on a combination of distance information obtained from laser range data and visual information obtained from a camera. Blauco et al. [10] obtained laser range data first to detect moving objects, then the position information was used to perform face detection on the subimage using a face detection algorithm. In [12], a simple face-tracking module was used to verify whether the nonconfirmed foreground objects belonged to people. Byers et al. [11] used laser range data to verify that skin-colored pixels, which were isolated by a color blob detection algorithm as possibly belonging to faces, did indeed belong to a face. All these methods can only detect one or a few people. They all used one sensor for coarse detection and another sensor for verification. Furthermore, a combination of two modes was studied only for detection.

17.3 Sensor, System Architecture, and Data Collection

17.3.1 Sensor: Single-Row Laser Scanner

Laser scan frames are generated by single-row laser scanners. Single-row-type laser range scanners produced by SICK Corp. are employed (LMS-200, see Fig. 17.3). It measures range distances from the sensor to surrounding objects using the method of time of flight. That is, a single laser pulse is sent out and reflected by an object surface within the range of the sensor, and the elapsed time between emission and reception of the laser pulse serves to calculate the distance between object and laser scanner. Via an integrated rotating mirror, the laser pulses sweep a radial range in front of the scanner. In this way, it sends pulses of light in plane within a certain degree (typically 100° or 180°; see Fig. 17.4) and measures the time it takes the pulse to return. Thus, it can reconstruct a map of its surroundings. Fig. 17.5 shows a laser scan frame captured in a small room about 3×3 m in size.

A laser scanner has the advantages of direct measurement, high accuracy (average distance error is 4 cm), wide viewing angle (180°) and long-range distance (maximum range distance of 30 m). Moreover, it has a high-angle resolution of 0.25° because of little diffusion of the laser beam. The frequency is 37.5 Hz, and wave length of the laser beam is 905 nm (class 1A, near infrared spectrum, safe to eye and human skin). Ambient operating temperature 0°C to +50°C. It can be bought at a rather low price on market (about U.S. $5,000). Recently, it increasing attention in the field of moving object detection and tracking.

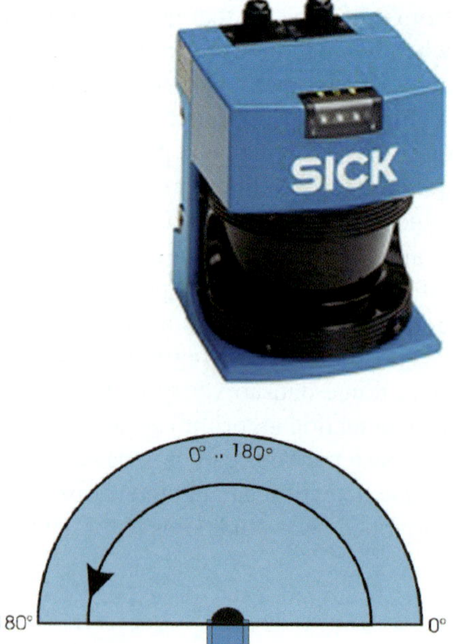

Fig. 17.3 Single-row laser scanner: LMS-200 by SICK optics

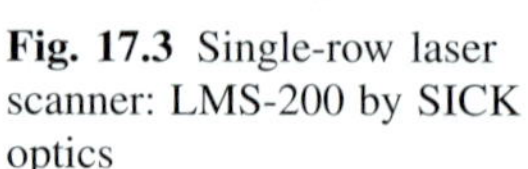

Fig. 17.4 Measurement range 0° to 180° (view from above, scan from right to left)

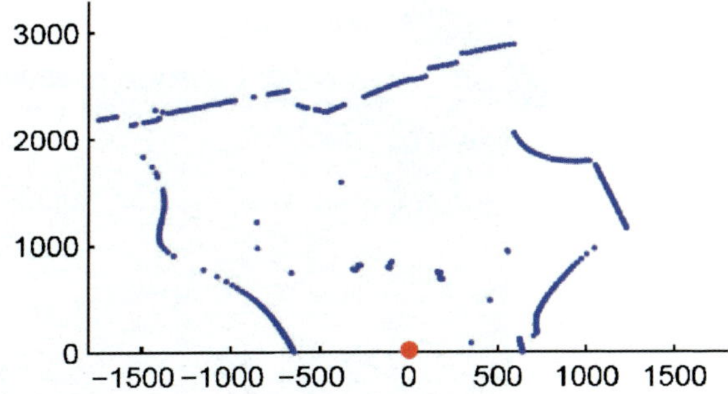

Fig. 17.5 One laser scan frame captured in a small room about 3 × 3 m in size; *red point* laser scanner

17.3.2 System Architecture and Data Collection

Multiple single-row laser-range scanners were exploited in our experiments. For each laser scan, one laser scanner profiles 180° on the scanning plane at a frequency of 37.5 fps. The range data can be easily converted to rectangular coordinates (laser points) in the sensor's local coordinate system [7]. Scanners are set doing horizontal scanning at ground level, so that cross sections at the same horizontal level of about 16 cm containing the data of moving (e.g., human legs) as well as still objects (e.g., building walls, desks, chairs, and so on) are obtained in a rectangular coordinate system of real dimension. Those laser points of moving objects are obtained from background image subtraction, then moving points from multiple laser scanners are temporally and spatially integrated into a global coordinate system (see Fig. 17.1).

For registration, laser scans keep a degree of overlay between each other. Relative transformations between the local coordinate system of neighboring laser scanners are calculated by pairwise matching their background images using the measurements of common objects. If common features in overlapping area are too few for automated registration, an initial value is first assigned through manual operation, followed by automated fine-tuning. Assigning an initial value to laser scanners' relative pose is not a tough task as two-dimensional laser scans are assumed to coincide in the same horizontal plane; operators can shift and rotate one laser scan on the other one to find the best matching between them. Specifying one local coordinate system as the global one, transformations from each local coordinate system to the global one are calculated by sequentially aligning the relative transformations, followed by a least-square-based adjustment to solve the error accumulation problem. A detailed address registering multiple laser scanners can be found in [13].

One of the major differences of our system compared with other research efforts is that we put laser scanners on the ground level (about 16 cm above the ground surface), scan pedestrians' feet, and track the pattern of rhythmically swinging feet. There are two reasons to target pedestrians' feet. The swinging feet of a normal pedestrian, no matter a child or an adult, no matter a tall or short person, can be scanned on the ground level with the least occlusion. In addition, the data of swinging feet can be modeled as the same pattern and tracked simply and uniformly.

Several real-world data sets are collected based on two sensor configurations. The first one is shown in Fig. 17.6 and Fig. 17.7 with a rather limited field of view

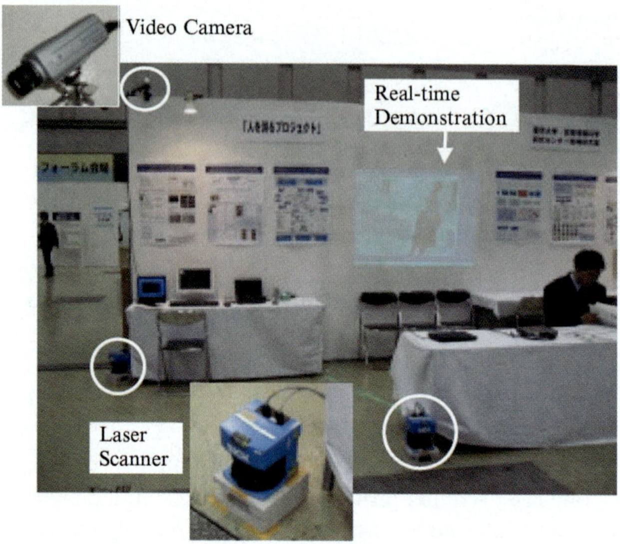

Fig. 17.6 A picture of the demonstration site in Fig. 17.7) with two laser scanners located in the middle

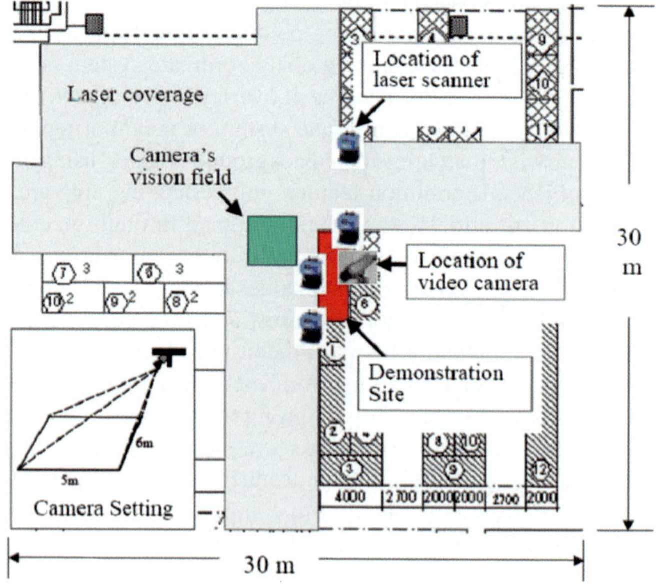

Fig. 17.7 A map of sensor's location and measurement coverage in the experiment

Fig. 17.8 An image frame captured by the camera in Fig. 17.6) with very limited FOV

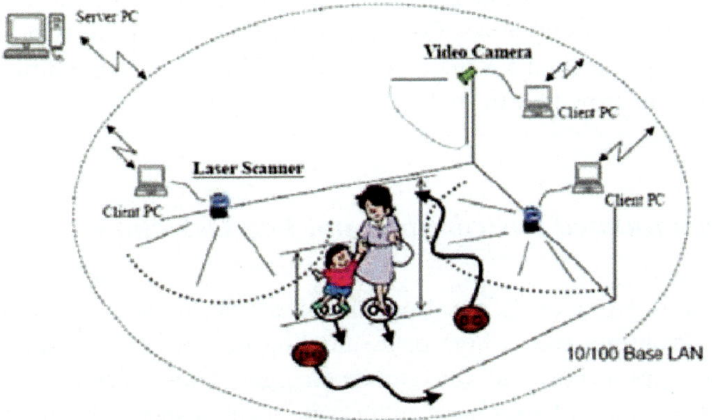

Fig. 17.9 An illustration of system architecture. only two laser scanners are figured out

(FOV) due to physical limitation the of camera set, (see Fig. 17.8). The second one is introduced in Section 17.6 with a larger FOV that is better for visual processing.

In Fig. 17.6, four laser scanners are located on the floor. Laser scans cover an area of about $30*30\,\mathrm{m}^2$ around our demonstration corner. On the other hand, a video camera is set on the top of a booth, about 3.0 m high from the floor, monitoring visitors at a slant angle and covering a floor area of about $5*6\,\mathrm{m}^2$. A map of the sensor's location and measurement coverage is shown in Fig. 17.7

In addition, the illustration of the system architecture is shown in Fig. 17.9. The laser scanner used in the experiment LMS200 by SICK. Each sensor controlled by an IBM ThinkPad X30 or X31. They were connected through 10/100 Base local-area network (LAN) to a server personal computer (PC).

17.4 Feature Extraction and People Detection

17.4.1 Single-Frame Clustering

People detection from a laser scan frame suffers greatly from the poor features provided by the laser points. Existing clustering-based people extraction methods use the assumption that there are distance gaps between distinct objects. Then, a threshold value is used for clustering to find distinct objects. However, in real scenes with crowds people, the clustering-based detection maps do not reflect real positions of legs. Occlusions make the points that belong to one leg split to multiple-point clusters, or there are no points that belong to the leg. In addition, mutual interactions make the points belonging to two different legs merge into one cluster.

In Fig. 17.10 and Fig. 17.11, raw data of one single laser scan frame and the clustering result from these data are shown, respectively. In the raw image, laser points of each person are manually circled. There were a total of four persons in the scene, and in the result, only one person correctly detected, was both two legs that is extracted correctly with no noise. Only one leg was extracted for two persons, and there was a noisy detection for the other person. Thus, it is rather difficult to track multiple people in crowds with such a clustering result.

17.4.2 Accumulated Distribution and Leg Detection

In this section, we propose a novel people extraction method called *accumulated distribution*. Profiting from a high data-sampling rate, it is reasonable to get successive range scan images with only subtle changes. Time- accumulation means to accumulate the count of laser points at the same pixel of successive multiple frames

Fig. 17.10 An illustration of system architecture. Only two laser scanners are figured out

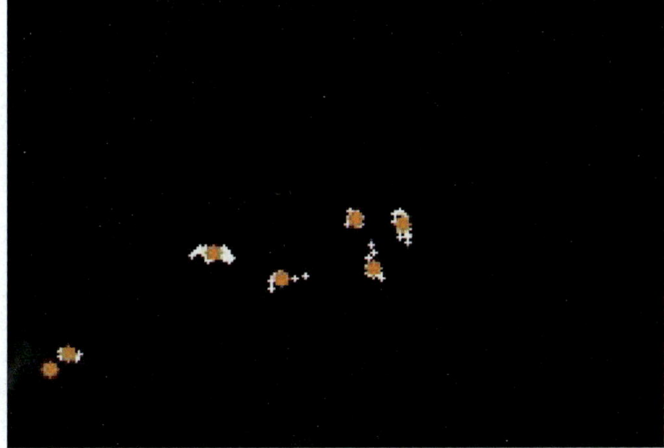

Fig. 17.11 System architecture; only two laser scanners are figured out

Fig. 17.12 Accumulated image at frame 364; data represented as a set of discrete points

as the intensity of that pixel (Fig. 17.12 shows a time accumulation image). If an object stops at a position for a while, the laser points belonging to it will accumulate at a nearby position (considering the measurement noise), thus in the final time accumulation image, the intensity of the pixel corresponding to the object position will be much higher than other pixels and appear to be a maximum in a neighborhood.

According to the usual walking model of humans, when a normal walking person steps forward, one of the typical appearances is, at any moment, that one foot swings by pivoting on the other one, as shown in Fig. 17.13. Two feet interchange their motion by landing and moving shifts in a rhythmic pattern. It was reported that [14], muscles act only to establish an the initial position, and the velocity of the feet at the beginning half of the swinging phase then remains inactive throughout the other half

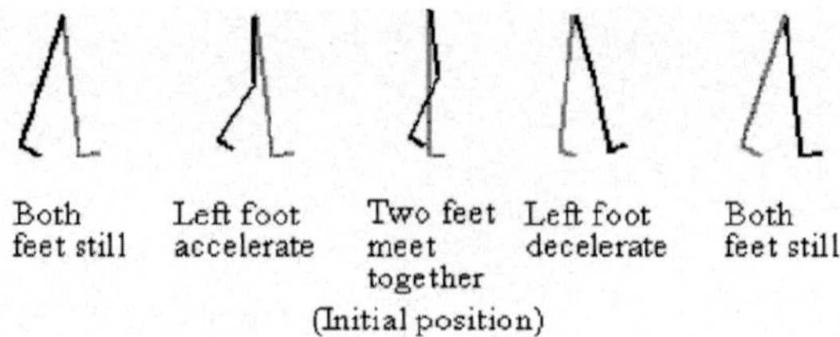

Fig. 17.13 Walking model

of the swinging phase. Actually, from Fig. 17.12 we can see that the brighter points
in the accumulated image are directly related to the inactive feet of persons. If we
can accurately locate these points, they can provide us very direct and stable cues to
infer the trajectories of walking people. The distribution of laser points in the image
is discrete, and it is hard to directly locate the points with maximal intensity. Parzen
window density estimation [15] is a well-known nonparametric method to estimate
distribution from sample data, and we utilize it to convert discrete sample points to
a continuous density function. The general form of the density is then

$$\hat{p}(x) = \frac{1}{n} \sum_{i=1}^{n} \phi(x - x_i, h) \tag{17.1}$$

in which $\{x_1, \ldots, x_n\}$ is a set of d-dimensional samples ($d = 2$ in this case), $\hat{p}(x)$
is the window function, and h is the window width parameter. Parzen showed that
$\hat{p}(x)$ converges to the true density if $\phi(\cdot)$ and h are selected properly [15].

The most popular window function is the Gaussian distribution, and we also
chose it because of its good features:

$$\phi(z, h) = \frac{1}{(2\pi)^{\frac{d}{2}} h^d |\Sigma|^{\frac{1}{2}}} \exp(-\frac{z^T \Sigma^{-1} z}{2h^2}) \tag{17.2}$$

where Σ is a covariance, and we use the identity matrix simply, considering its
property of isotropy in two dimensions. The selection of h here depends on the size
of the foot region. One result of the accumulated image after the Parzen window is
shown in Fig. 17.14.

For the search of local maxima in a kernel-based density, mean shift appears to
be a common method and should be able to give a promising result. Considering
the real-time requirement of tracking and the low dimensionality of laser data, we
chose a simple local search strategy, which was proven very effective and fast by
our experiments.

In summary, we process the laser data with accumulation, Parzen window filtering, and local maximum search at every time step. A number of measurements
will be obtained, each of which represents one foot that remains static for a while
in a small region so that its intensity is one local maximum in the current accumulated image. Since one local maximum might appear in several successive frames,
only newly appearing ones are considered as measurements at the current time
(Fig. 17.15).

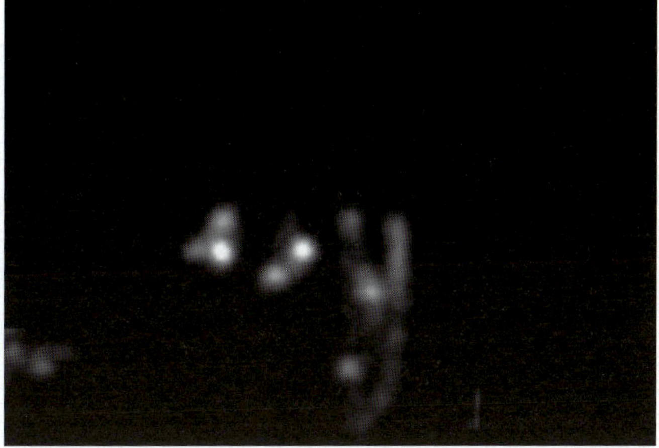

Fig. 17.14 After Parzen window in Fig. 17.12, with continuous intensity distribution

Fig. 17.15 Leg extraction result in Fig. 17.14, each circle denotes an inactive leg

17.4.3 Evaluations of Detection Algorithm

We evaluated our detection algorithm through a sequence of 1,000 frames (about 3 s). The detection performance was evaluated by comparing the true count and the estimated count of persons in the scene, and two different situations were considered separately. One was to consider the whole laser coverage of the sensors (situation 1). *Laser coverage* here means the area that was measured by at least one laser scanner, considering only the occlusions from the physical layout of the environment, and the area was about 30 * 30 m. The other situation (situation 2) considered the central area of laser coverage measured by at least two laser scanners and not too far away from the locations of the laser scanners, considering occlusions from moving people and the physical layout, and the area was about 20 * 18 m.

As a result of the difficulty and workload in obtaining a ground truth count of multiple people, we evaluated our detection algorithm with sampling results at an interval of 50 frames. In Fig. 17.16, the count of correctly detected persons is compared with the true count at every 50 frames.

In Table 17.1, the detection ratios are listed. The detection ratio in situation 1 was much lower than situation 2. because in situation 1, persons with few points were also included, as shown in Fig. 17.17 In some remote corners, persons were almost occluded by the environment, and thus only two or three points of one leg were visible. In situation 2, only the central area was considered, and these areas have been taken out. Consequently, the detection ratios greatly increase.

The errors arise from two reasons: occlusion and noise. Most of the detection failures from occlusions can be recovered with time accumulation computation within the following 1 to 10 frames. Noisy measurements mainly come from luggage dragged by the persons or measurement split by partial occlusion.

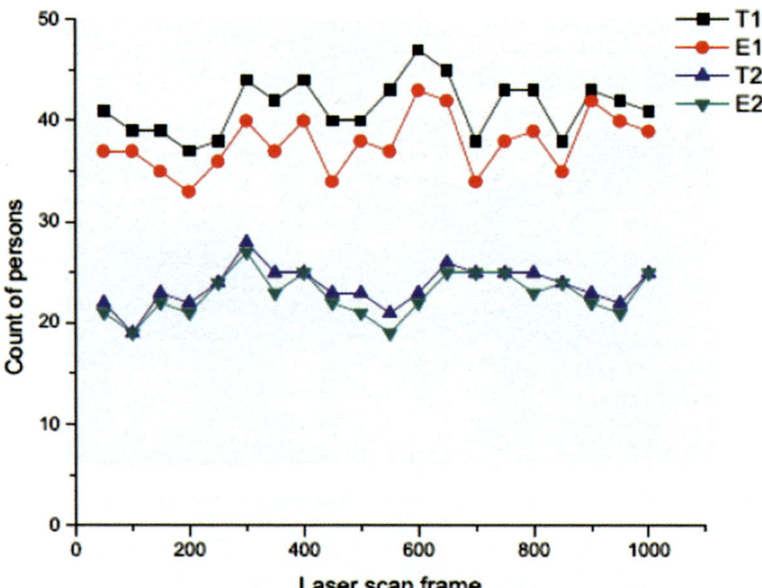

Fig. 17.16 Evaluation of the detection results in two situations. $T1$ true count of persons in the whole area (situation 1). $E1$ detection result in the whole area. $T2$ true count of persons in the central area (situation 2). $E2$ detection result in the central area

Table 17.1 Detection ratios

Situation	Highest ratio	Lowest ratio	Average ratio
Situation 1	97.67%	85%	91.41%
Situation 2	100%	90.48%	96.41%

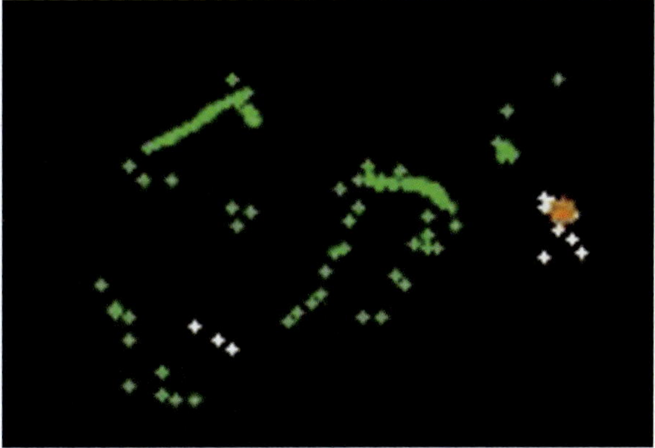

Fig. 17.17 Difficulty in people detection for situation 1: only three points are measured for one leg

In Fig. 17.16 and Table 17.1, only the detection result at one frame is considered. Another evaluation can be done considering the detection ratio of persons through the whole sequence of 1,000 frames. There appeared a total of 96 persons, and only 1 person was missed in the whole sequences because of severe occlusion. 93 persons were successfully detected within 5 frames, and 95 persons were detected within 10 frames after their first appearance in the scene. In addition, there were 6 noisy measurements.

17.5 Bayesian Tracking and Data Association

After the inactive legs are detected, they are stored as measurements at current time. As there are multiple targets and multiple measurements, the direct estimation of the target states is difficult due to the unknown data associations. So-called data association is to associate these measurements to the trajectories in previous frames. To address the problem, the target states can be augmented with the unknown associations, and the joint distribution of states and associations are estimated sequentially.

In the following, the probabilistic tracking model is introduced first, and then the sequential inference process using Bayes' rule is described, which is difficult to compute analytically. Finally, two strategies are presented to simplify the computation, respectively, for the case of independent tracking and the case of joint tracking of multiple targets.

17.5.1 Probabilistic Tracking Model

Here, we describe a probabilistic model for tracking that addresses the problem of multiple measurements and multiple targets. We assume that there are T targets, where T is fixed, and write their joint state as X_k. At each time step we have M measurements Y_k, where M can change at each time step. The data association set is denoted as θ_k. In this chapter, we assume one target can generate up to one measurement, and one measurement can be generated from up to one target.

First, we specify the joint distribution $P(X_{0:k}, Y_{1:k}, \theta_{1:k})$ over the actual measurements $Y_{1:k}$ data associations $\theta_{1:k}$, and states $X_{0:k}$ of the targets between time steps 0 and k,

$$P(X_{0:K}, Y_{1:K}, \theta_{1:K}) = P(X_0) \prod_{k=1}^{K} P(X_k|X_{k-1}) P(Y_k|\theta_k, X_k) P(\theta_k) \tag{17.3}$$

where we assumed that the target motion is Markov, each measurement set Y_k is conditionally independent given the current state X_k, and X_k depends only on the previous time step. Since the actual state X_k of the targets does not provide us with any information on the data association. Consequently, we also assume that the prior over data associations $P(\theta_k)$ does not depend on the target state:

$$P(Y_k, \theta_k|X_k) = P(Y_k|\theta_k, X_k) P(\theta_k) \tag{17.4}$$

It is convenient to write inference in this model recursively via the Bayes filter. The objective is to infer the current position X_k of the targets given all of the measurements $Y_{1:k}$ observed so far. In particular, the posterior distribution $P(X_k|Y_{1:k})$ over the joint state X_k of all present targets given all observations $Y_{1:k} = \{Y_1, \ldots, Y_k\}$ up to and including time k is updated according to the recursive formula

$$P(X_k|Y_{1:k}) = c \sum_{\theta_k} P(X_k, \theta_k|Y_{1:k}) \tag{17.5}$$

$$= c \sum_{\theta_k} P(Y_k|X_k, \theta_k) P(\theta_k) \int_{X_{k-1}} P(X_k|X_{k-1}) P(X_{k-1}|Y_{k-1})$$

where c is a normalizing constant.

Usually, this expression of the sequential update equation cannot be solved analytically. Further assumptions are required to simplify this model, which is introduced in Sections 17.5.2 and 17.5.3. In the following sections, we concentrate on deriving an expression for the posterior $P(X_k|Y_{1:k})$ on both X_k and the data association θ_k by providing further details of the motion model $(X_k|X_{k-1})$ and the measurement model $P(Y_k|X_k, \theta_k)$.

17.5.1.1 State Space and Observation Space

The state space for each target includes both position and velocity

$$X_{k,i} = \left[x_{k,i}, y_{k,i}, vx_{k,i}, vy_{k,i} \right]^{\mathrm{T}}, \quad i = 1, \ldots, T$$

at time step k. Measurements were simply 2-d positions

$$Y_{k,j} = \left[u_{k,j}, v_{k,j}\right]^{\mathrm{T}}, \quad j = 1\dots,M.$$

17.5.1.2 The Motion Model

For the motion model, we assume a standard linear Gaussian model. That is, we assume that the initial joint state is Gaussian

$$P(X_0) = N(X_0; m_0, V_0) \tag{17.6}$$
$$m_0 = \{m_{0,1}, \dots, m_{0,T}\}, V_0 = \{V_{0,1}, \dots, V_{0,T}\}$$

where m_0 is the mean, and V_0 is the corresponding covariance matrix. In addition, we assume that targets move according to a linear model with additive Gaussian noise,

$$P(X_k|X_{k-1}) = N(X_k; AX_{k-1}, Q_{k-1}) \tag{17.7}$$

where Q_{k-1} is the prediction covariance, and A is a linear prediction matrix. We model the motion of each target independently with a constant velocity model, that is,

$$A = diag\left\{A_1, \dots, A_T\right\}, A_i = \begin{bmatrix} I_{2\times2} & I_{2\times2} \\ 0 & I_{2\times2} \end{bmatrix}, i = 1, \dots, T \tag{17.8}$$

where $I_{2\times2}$ denotes a 2×2 identity matrix.

17.5.1.3 The Measurement Model

We represent a data association set θ_k by a

$$\theta_k = \{(i,j)|\theta_{k,j} = i\}, (i,j) \in \{0, \dots, T\} \times \{1, \dots, M\} \tag{17.9}$$

where $\theta_{k,j} = i$ denotes that the jth measurement is generated by ith target, and $i = 0$ implies that the measurement is clutter. Given the data association θ_k, we can divide the measurements into clutter and observations, respectively [16],

$$P(Y_k|X_k, \theta_k) = P(Y_{c,k}|\theta_k)P(Y_{o,k}|X_k, \theta_k) \tag{17.10}$$

Then, we assume that each clutter measurement (i.e., an unassigned measurement) is independently and uniformly generated over the FOV. Consequently, the clutter model is a constant C proportional to the number of clutter measurements $|Y_{c,k}|$:

$$P(Y_{c,k}|\theta_k) = |Y_{c,k}|/C \tag{17.11}$$

The constant C is related to the size of the FOV, in a 720×480 image $C = 720 \cdot 480$.

To model the observations, we map the data association in a Gaussian observation model

$$P(Y_{o,k}|\theta_k, X_k) = N(Y_{o,k}; HX_k, R_k) \tag{17.12}$$

where R_k is the measurement covariance.

We assume that each measurement is generated independently, and we once again obtain a block-diagonal structure:

$$H = diag\{H_1, \ldots, H_M\}, H_j = \left[I_{2\times2}\ 0_{2\times2}\right], j = 1, \ldots, M \tag{17.13}$$

17.5.2 Independent Tracking Using Kalman Filters

Once the models are specified, the joint distribution of data association and state can be estimated recursively using Eq. (17.5). Usually, this expression of the sequential update equation cannot be solved analytically. Further assumptions are required to simplify this equation.

In this section, we introduce the first simplification. Now, we assume that all the targets move independently, and the targets' states are not correlated or mutually independent. The possible measurements of target data associations are determined by a simple gating strategy [6]. That is, targets are only assigned to measurements within standard deviations of the predicted position of the target.

Then, we can construct their measurement likelihood independently and compute the MAP (maximum a posteriori probability) estimation for each target, respectively:

$$P\left(x_{j,k}, \theta_{j,k}\right) = P(\theta_{j,k})P(Y_k|x_{j,k}, \theta_{j,k})\int P(x_{j,k}|x_{j,k-1})P(x_{j,k-1}|Y_{1:k-1})dx_{j,k-1} \tag{17.14}$$

The assignment that provides MAP is chosen as the potential association for each target.

The dynamic model for each target is one component of Eq. (17.7) that is linear with constant velocity.

$$P(x_{j,k}|x_{j,k-1}) = N(x_{j,k}|Ax_{j,k-1}, Q_{j,k-1})$$
$$P(x_{j,k-1}|Y_{1:k-1}) = \sum_{\theta_{j,k-1}} P(x_{j,k-1}, \theta_{j,k-1}|Y_{1:k-1}) \tag{17.15}$$

Assume the initial prior distributions of the target states are Gaussian.

$$P(x_{j,0}) = N(x_{j,0}; m_{j,0}V_{j,0}) \tag{17.16}$$

For the target originated measurement likelihood, we use an additional coherency cue. Then, with the independent assumption, the measurement likelihood is

$$P(Y_k|x_{j,1:k}, \theta_{j,k}) = Z_{j,k}^{position} \times Z_{j,k}^{coherency} \tag{17.17}$$

$Z^{position}_{j,k}$ is the cue derived from the distance between measured position and predicted position as shown in (17.13):

$$Z^{position}_{j,k} = P(y^{position}_{s,k}|X_k, \theta_{jk} = s) = N(y_{s,k}|Hx_{j,k}, R_{j,k}) \qquad (17.18)$$

where $\theta_{j,k} = s$ means $\theta_{j,k} = s$ measurement $y_{s,k}$ is from target j, and $y^{position}_{s,k} = Hx_{j,k} + r_{j,k}$, $r_{j,k} \sim N(0, R_{j,k})$, $H = \begin{bmatrix} I_{2\times2} & 0_{2\times2} \end{bmatrix}$. $Z^{coherency}_{j,k}$ is the cue derived from the region membership of measured position belonging to target trajectory (see Fig. 17.18). It's formulated based on the observation that two successive measurements of the same person belong to the same coherent region in accumulated image. Some advanced region segmentation and analysis methods should be effective, but also slow. We chose a quite practical and effective approach that measures the intensity of points on the line linking the measured position and the last position of target trajectory:

$$
\begin{aligned}
Z^{coherency}_{j,k} &= \sqrt[|E|]{\prod_{p\in E} intensity(p)} = \exp\left(\ln \sqrt[|E|]{\prod_{p\in E} intensity(p)} \right) \\
&= \exp\left(\frac{1}{|E|} \sum_{p\in E} \ln(intensity(p)) \right) \qquad (17.19) \\
&= \exp\left(- \sum_{i\in \text{hist}(E)} \text{hist}(i)\ln(1/i) \right)
\end{aligned}
$$

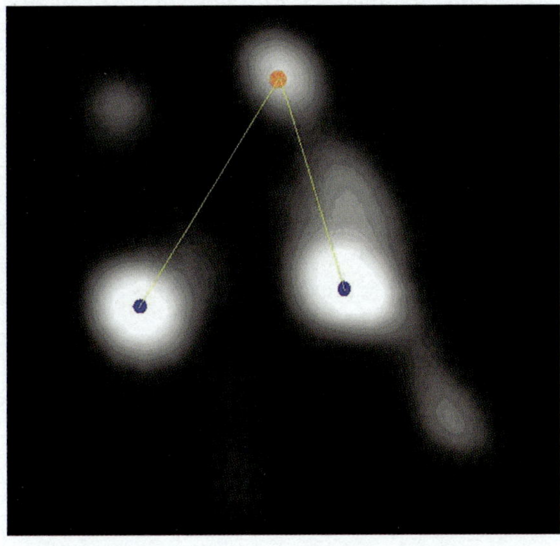

Fig. 17.18 Region coherency likelihood on accumulated image: *Top point* a new detected measurement. The points linked with this new measurement are locations of two trajectories in the previous frame. Lines denote possible data associations

where E is the line linking measured position and last trajectory position, p is the pixel on the line, hist(E) is the histogram of line pixels, and hist(E)$\in [0, 1]$ is the histogram value of a specific intensity value $i \in [0, 1]$.

This coherency likelihood was demonstrated with great robustness and effectiveness in our experiments and could uniquely make data associations correctly in most cases. Even if a trajectory is new and thus the position is difficult to predict or a person changes walking direction and the predicted position is wrong, the coherency likelihood could also assign measurement to the correct trajectory.

17.5.3 Joint Tracking of Multiple Targets Using RBMC-DAF

While using independent filters is computationally tractable, the result is prone to failure. In a typical failure mode, two targets walk close, and the measurement of one target "hijacks" the filter of another nearby target with a high likelihood score.

The JPDAF [2,3] can address these situations. However, the JPDAF represents the belief of state of the targets as Gaussian and may not accurately capture the multi modal distribution over the target states.

On the other hand, in [16,17], the RBMC-DAF algorithm was introduced to estimate data associations with a SIR filter and the other parts with a Kalman filter. This idea originated from Rao-Blackwellized particle filtering (RBPF) [18]. That is, sometimes it is possible to evaluate part of the filtering equations analytically and the other part by Monte Carlo sampling instead of computing everything by pure sampling. In this way, multimodal distribution of the target state can be considered with limited computation.

In the following, we first provide a Monte Carlo strategy sampling of the data association, and then a Rao-Blackwellized data association algorithm is presented as a practical strategy to improve the computation.

17.5.3.1 Monte Carlo Sampling of Data Association

A Monte Carlo sampling method approximates a probability distribution by a set of samples drawn from the distribution. In a typical Monte Carlo sampling method, one starts by inductively assuming that the posterior distribution over the joint state of the targets at the previous time step is approximated by a set of S samples:

$$P(X_{k-1}|Y_{1:k-1}) \approx \left\{ X_{k-1}^{(s)} \right\}_{s=1}^{S} \tag{17.20}$$

Given this representation, we obtain the following Monte Carlo approximation of the Bayes filter:

$$P(X_k|Y_{1:k}) \approx c \sum_{\theta_k} P(Y_k|X_k, \theta_k)P(\theta_k) \sum_{s=1}^{S} P(X_k|X_{k-1}^{(s)}) \tag{17.21}$$

A straightforward implementation of this equation is intractable due to the large summation over the space of data associations θ_k combined with the summation over the indicator s. To address this problem, a second Monte Carlo approximation can be introduced.

$$P(X_k|Y_{1:k}) \approx c \sum_{w=1}^{W} P(Y_k|X_k, \theta_k^{(w)}) P(\theta_k^{(w)}) \sum_{s=1}^{S} P(X_k|X_{k-1}^{(s)}) \tag{17.22}$$

Then, the evaluation of this equation can be achieved with Gaussian assumption of state distribution in a RBM(=1)AF framework.

17.5.3.2 Rao-Blackwellized Monte Carlo Data Association

At each time step, we run the tracking and data association process as in the following procedure.

Initialization

We assume that we can approximate the posterior $P(X_{k-1}|Y_{1:k-1})$ by the following mixture of Gaussians:

$$P(X_{k-1}|Y_{1:k-1}) \approx \frac{1}{S} \sum_{s=1}^{S} N(X_{k-1}; m_{k-1}^{(s)}, V_{k-1}^{(s)}) \tag{17.23}$$

Prediction

Because the target motion model is linear Gaussian, the predictive density over X_k for each value of the mixture indicator s can be calculated analytically:

$$\int_{X_{k-1}} P(X_k|X_{k-1}) N(X_{k-1}; m_{k-1}^{(s)}, V_{k-1}^{(s)}) = N(X_k; Am_{k-1}^{(s)}, Q_{k-1}^{(s)}) \tag{17.24}$$

Hence, the predictive prior $P(X_k|Y_{1:k-1})$ on the current state is also a mixture of Gaussians:

$$P(X_k|Y_{1:k-1}) \approx \frac{1}{S} \sum_{s=1}^{S} N(X_k; Am_{k-1}^{(s)}, Q_{k-1}^{(s)}) \tag{17.25}$$

Evaluation

The sequential Monte Carlo approximation of Eq. (17.21) to the target posterior using the Bayes filter becomes

$$P(X_k|Y_{1:k}) \approx c \sum_{\theta_k} P(Y_k|X_k,\theta_k)P(\theta_k)\frac{1}{S}\sum_{s=1}^{S} N(X_k;Am_{k-1}^{(s)},Q_{k-1}^{(s)})$$

$$\approx c \sum_{w=1}^{W} P(Y_k|X_k^{(w)},\theta_k^{(w)})P(\theta_k^{(w)})N(X_k^{(w)};Am_{k-1}^{(s')},Q_{k-1}^{(s')}) \quad (17.26)$$

using a set of sampled states, data associations, and mixture indicators:

$$\left\{ X_k^{(w)},\theta_k^{(w)},s^{(w)} \right\}_{w=1}^{W}$$

where $s' = s^{(w)}$ is the wth sampled mixture indicator drawn from the following target density:

$$\tilde{\pi}(X_k,\theta_k,s) = cP(Y_k|X_k,\theta_k)P(\theta_k)N\left(X_k;Am_{k-1}^{(s)},Q_{k-1}^{(s)}\right) \quad (17.27)$$

Now, we can analytically marginalize out the current state X_k based on Eq. (17.27), and obtain a Rao-Blackwellized target density:

$$\pi(\theta_k,s) = P(Y_{c,k}|\theta_k)P(\theta_k)\int_{X_k} N(Y_{o,k};HX_k,R_k)N\left(X_k;Am_{k-1}^{(s)},Q_{k-1}^{(s)}\right) \quad (17.28)$$

The key observation here is that the product of the likelihood and the predictive prior

$$N(Y_{o,k};HX_k,R_k)N\left(X_k;Am_{k-1}^{(s)},Q_{k-1}^{(s)}\right)$$

is proportional to a Gaussian. $P(\theta_k)$ is assumed to be uniformly distributed. As a result, the integral over X_k is analytically tractable and is also Gaussian.

Sampling

Finally, samples $\left\{ \theta_k^{(w)},s^{(w)} \right\}_{w=1}^{W}$ drawn from the Rao-Blackwellized target density $\pi(\theta_k,s)$ based on Eq. (17.28) are used to construct a new mixture of Gaussians over the current state

$$P(X_k|Y_{1:k}) = \frac{1}{W}\sum_{w=1}^{W} N(X_k;m_k^{(w)},V_k^{(w)}) \quad (17.29)$$

where $m_k^{(w)}$ is the mean, and $V_k^{(w)}$ is the covariance of the target state at the current time step.

Practical Heuristics

We apply two heuristics to obtain some additional gains in efficiency. First, we gate the measurements based on a covariance ellipse around each target. Targets are only assigned to measurements within standard deviations of the predicted position of the target. Second, the components of the association set are sampled sequentially conditional on the components sampled earlier in the sequence. We make use of this property to ensure that measurements associated with targets earlier in the sequence are not considered as candidates to be associated with the current target. In this way, the algorithm is guaranteed to generate only valid association hypotheses. In addition, sampling is done using roulette wheel selection. That is, states with high density would be selected with high probability.

17.5.3.3 Mutual Correlation Detection and Modeling

For detection of mutual correlation between multiple targets, a graph is used, with the nodes representing the targets, and the edges representing that there is a correlation or interaction between corresponding nodes. Targets within a certain distance (e.g., 20 cm, 15 pixels) of one another are linked by an edge. The absence of edges between two targets provides the intuition that targets far away will not influence each other's motion. At each time step, the correlation graph is updated. The targets with no edge are tracked with independent filters, and targets with edges are tracked with RBMC-DAF. In this work, up to two targets were considered and jointly tracked using RB-MCDAF.

17.5.4 Evaluations of Tracking Results

Four single-row laser-range scanners were exploited in our experiments and covered a corner of an exhibition hall as shown in Fig. 17.7 Each laser scanner profiles 360° range distances equally in the scanning plane with a frequency of 30 fps. Scanners were set for horizontal scanning at a height of 16 cm from horizontal ground. We used a sequence with 9,035 frames as the experimental data.

Figure 17.19 is a screen copy of generated trajectories; red circles denote the location of laser scanners; green points represent background image; white points represent moving legs. Color lines are trajectories. Extreme points of trajectories are locations of inactive legs of people at the current time. Laser points of one person were manually circled to ease observing. Two close persons (rectangle in the right of the figure) were tracked with RBMC-DAF, and other persons were tracked with independent Kalman Filter (KF).

With conventional laser-based tracking, tracking often fails in three kinds of situations: (1) If people walk too close; (2) if people walk a cross each others path and their feet are too close together at the intersection point; (3) if there is a temporal occlusion. Our method could handle all these cases well.

To make quantitative analysis of the tracking performance, we compared the count of correctly tracked trajectories and the true number of trajectories in Fig. 17.20. Again, we sampled the results at an interval of 50 frames. There were 96

Fig. 17.19 A screen copy of tracking result

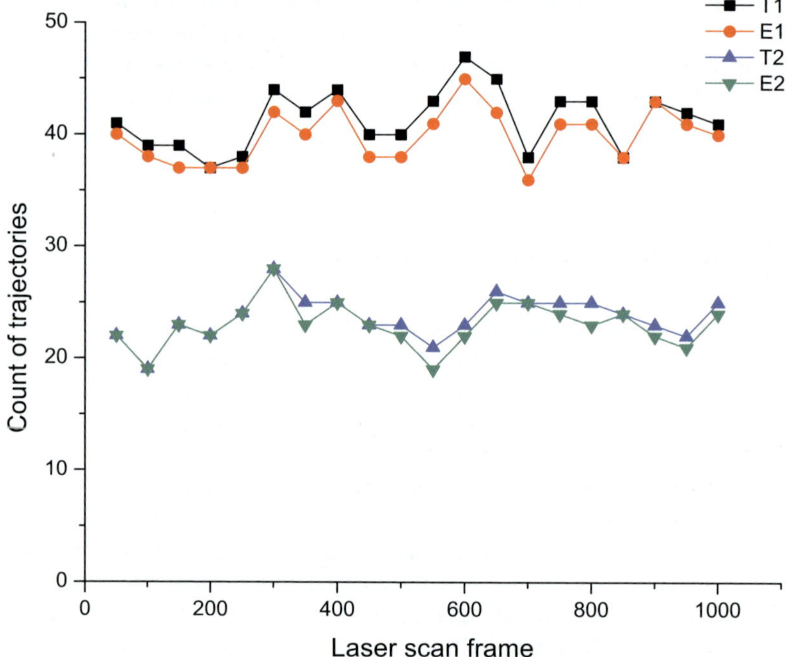

Fig. 17.20 Evaluation of the tracking results in two situations. T1 true count of trajectories in the whole area (situation) E1 correctly tracked trajectories in the whole area T2 true count of trajectories in the central area (situation) E2 correctly tracked trajectories in the central area

trajectories in the whole area, and 14 of them had one or several failures through 200 time steps (1,000 frames). Thus, the success ratio of the tracker was 85.42%. The reasons and corresponding number of the failures are listed in Table 17.2.

Table 17.2 Reasons for tracking failures

Reason for failures in frame 0–1000	Count of failures
Nondetection results in nontrajectory.	1
Few laser points with low detection ratio of one person through the sequence results in a broken trajectory	3
Walking too fast makes data association within nearby range fail	3
Noisy measurement generates a nonexistent trajectory	2
Noisy measurement disturbs an existing trajectory nearby and results in a broken trajectory	1
Noisy measurement attracts an existing trajectory, and results in a tracking error, but then the tracker can recover from this error	2
When a new object is detected, an existing trajectory is attracted by the new measurement, which results in errors in both trackers of two persons	2
Tracking error caused from mixed data of two closely situated persons	0

17.6 Data Association with Assistance of Visual Data

From the discussion, it can be concluded that compared to tracking methods using normal video cameras, the method using single-row laser-range scanners can achieve robust and real-time tracking performance. However, the limitations of the laser-based methods are inherent and obvious. They cannot provide information such as color of objects, so it is difficult to obtain a set of features that uniquely distinguish one object from another. If a trajectory is broken due to occlusion, it is difficult to connect the fragments. In addition, if people walk in a group or cross each other's path and their feet are too close together, the data will be mixed, and the feet positions will be lost in extraction. Thus, visual computation is strongly required in these laser-based methods.

In our previous work of laser and vision fusion [8], we proposed a decision-level fusion framework. People detection was achieved by feet trajectory detection from laser data. Laser-based detection results were used to initialize a mean-shift visual tracker using camera calibration parameters. Visual color tracking by the mean shift method with a color histogram and laser tracking that uses a typical pedestrian model is independently processed. Then, a Bayesian method is used to fuse these two sets of tracking information. Compared with other research on two-modal combination, our method can overcome the major tracking errors that cannot be tackled with only laser data or video data, such as the laser data missing due to occlusion or data mixed when two people walk closely or walk across each other's path, and the severe occlusion of video data in complex visual situations. However, due to frame-by-frame fusion of laser data and vision data, the computational cost is high.

In this section, we use a rather simple but effective approach with little computation consumption. The basic idea is to extract and attach visual information to each trajectory for the learning the appearance of the corresponding human body. When

a laser-based tracker fails, a mean shift-based visual tracker is triggered to continue
tracking until the laser tracker redetects this person and track. The latter trajectory's
ID is reassigned to the original one if visual-based matching succeeds.

17.6.1 Sensor Configuration and Data Collection

In Section 17.3, we showed one sensor configuration with a limited FOV (see
Fig. 17.6 and Fig. 17.7). Here, we show the other sensor configuration with two
laser scanners and one camera that can provide a larger FOV with a higher camera position; see Fig. 17.21 for the real site at a corner of an exhibition hall, and
Fig. 17.22 is a sample image captured by the camera.

17.6.2 Calibration and Body Localization

To extract and attach the visual representation to each detected trajectory from a
laser-based tracker, a calibration matrix between the laser scan frame coordinates
and image frame coordinates is necessary. This matrix can be used to localize the
human body in the image from the leg position in the laser scan frame.

The video camera was set 4.5 with a mm lens with a diagonal angle of about 98°,
and a slant angle of 30–45° toward the ground overhead of the laser scanners. The
video camera was calibrated independently of the global coordinate system using
Tsai's model [19], by which both internal and external parameters are calculated

Fig. 17.21 A real experimental site with two laser scanners and one camera

Fig. 17.22 An image frame captured by the camera in Fig. 17.21 with a larger FOV

using at least 11 control points. A global coordinate system is defined with its xy axes coincident with those of the integrated coordinate system of laser scanners, with its z-axis vertically upward and its origin on the ground surface. The elevation of the laser scanning plane is detected using the sensor chip developed in [20], so that the z-coordinate is associated with each laser point. Control points are obtained by putting markers on the vertical edges of the wall, desks, chairs, boxes, and so on, with themselves visible on the video image and the vertical edges measured by laser scanners. The z-coordinate of each marker is its elevation from the ground surface, which is physically measured previously. The xy-coordinates of each marker come from the laser scans measurement of the vertical edge.

With the pinhole model, the project transformation of the global coordinates to the image plane is

$$
Z_c \begin{bmatrix} u \\ v \\ 1 \end{bmatrix} = \mathbf{M} \begin{bmatrix} X_w \\ Y_w \\ Z_w \\ 1 \end{bmatrix} = \begin{bmatrix} m_{11} & m_{12} & m_{13} & m_{14} \\ m_{21} & m_{22} & m_{23} & m_{24} \\ m_{31} & m_{32} & m_{33} & m_{34} \end{bmatrix} \begin{bmatrix} X_w \\ Y_w \\ Z_w \\ 1 \end{bmatrix}
\tag{17.30}
$$

where $\begin{bmatrix} u & v & 1 \end{bmatrix}^T$ is the homogeneous coordinates in the image plane (in pixels), $\begin{bmatrix} X_w & Y_w & Z_w & 1 \end{bmatrix}^T$ is the homogeneous coordinates in the global (world) coordinate system, $\mathbf{M}$ is the project matrix, and Z_c is a constant factor for a fixed point (actually, the z-coordinate in the camera coordinate system).

With this model, laser points can be projected to the visual image plane. In Fig. 17.2, we show one sample image of registered laser data and vision data. White points denote moving laser points of human feet.

Once a person is detected and tracked in the laser scan frames, the corresponding body region in the synchronized video image frames can be determined using the project model. Laser scanners can tell the xy-coordinates of the position center of two feet $\left(X_w \; Y_w \; Z_w \right)$ in the world coordinate system, and the z-coordinate is the elevation of the laser scanning plane. Thus, the image coordinates of the feet center position can be calculated using the project transformation without any difficulty.

An ellipse is utilized to model the body the region, which is described by a 5-dimensional vector (x, y, h_x, h_y, θ), where x and y are the location of the ellipse center, h_x, h_y are the lengths of axes, and θ is the rotation angle. However, it impossible to calculate the body ellipse only using the feet position in the world coordinate system, and thus we introduce two constraints on body region. It is assumed that a human has an average height (e.g., 170 cm) and width (e.g., 50 cm) and always stands vertically on the ground. With such an assumption, body of the ellipse can be calculated from feet position easily since the world coordinate of the human head is also known (with a fixed z-coordinate). Actually, only (x, y) is enough to describe a body region (other parameters will depend on it). With a fixed z-coordinate, the project transformation could be rewritten as

$$
Z_c \begin{bmatrix} u \\ v \\ 1 \end{bmatrix} = \begin{bmatrix} m_{11} & m_{12} & m_{13}Z_w + m_{14} \\ m_{21} & m_{22} & m_{23}Z_w + m_{24} \\ m_{31} & m_{32} & m_{33}Z_w + m_{34} \end{bmatrix} \begin{bmatrix} X_w \\ Y_w \\ 1 \end{bmatrix} = \mathbf{M}' \begin{bmatrix} X_w \\ Y_w \\ 1 \end{bmatrix} \tag{17.31}
$$

In Fig. 17.23, it is shown that body regions are calculated from given feet positions effectively.

17.6.3 Visual Representation and Similarity Distance

After body region localization in video image frames, corresponding visual color information is extracted frame by frame. The utilization of visual information is

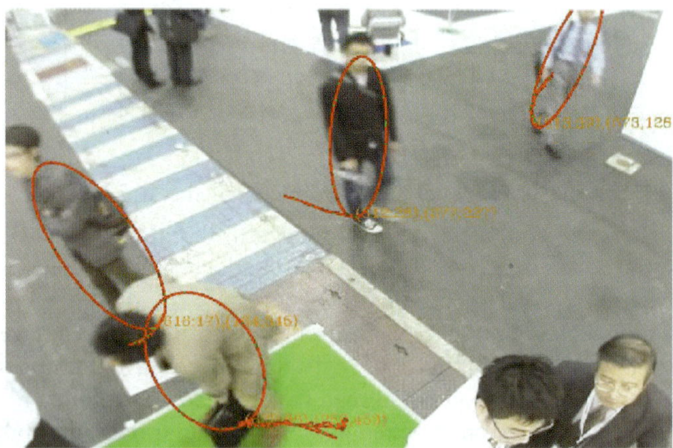

Fig. 17.23 The calculated human body regions using feet positions obtained from laser-based tracker

twofold. First, when some trajectory generated from the laser-based tracker is broken, the visual tracker is processed to continue tracking. Meanwhile, the broken trajectory is labeled as the to-be-linked (TBL) trajectory. Second, when a new target is detected by the laser-based tracker, the visual similarity distance is calculated between the new target and each TBL trajectory. The TBL trajectory with the highest similarity score that is also higher than a predefined threshold is linked with the newly detected trajectory and removed from the TBL trajectory list.

The visual representation, similarity distance measurement used here is motivated from a region-based algorithm named mean shift, which is a nonparametric statistical method that has recently been widely adopted as an efficient technique for visual appearance-based object region tracking [21]. It employs mean shift iterations to find the most similar region compared with a given region in terms of color distribution. The similarity of two color distributions is expressed by a metric based on the Bhattacharyya coefficient. The main advantages of this method are that it can achieve real-time performance, and it is robust to partial occlusion and rotations in depth.

The color histograms are used to represent the color distribution in an image. Mainly, the color histogram approach counts the number of occurrences of each unique color on a sample image. Since an image is composed of pixels and each pixel has a color, the color histogram of an image can be computed easily by visiting every pixel once. By examining the color histogram of an image, the colors existing in the image can be identified with their corresponding areas as the number of pixels. One possible way of storing the color information is to use three different color histograms for each color channel. Another possible method is to have a single-color histogram for all of the color channels. In the latter approach, the color histogram is simply a compact combination of three histograms, and the empty slots can be discarded easily. We used the latter one.

We denote by $\{\mathbf{x}_i^* = (x_i^*, y_i^*)\}_{i=1,\ldots,n_h}$ the pixel locations of the target model, centered at 0, representing the body region ellipse. Let b be the function that associates to the pixel at location $\mathbf{x}_i$ the index $b(\mathbf{x}_i)$ of the histogram bin corresponding to the color of that pixel in the RGB space with $16 \times 16 \times 16$ bins. The probability of the color r in the target model is derived by employing a convex and monotonic decreasing function K that assigns a smaller weight to the locations that are farther from the center of the target. The weighting increases the robustness of the estimation since the peripheral pixels are the least reliable, being often affected by occlusions (clutter) or background. By assuming that the generic coordinates x and y are normalized with the height h_x and width h_y, respectively, we can write

$$\hat{q}_r = C \sum_{i=1}^{n} K(\mathbf{x}_i^*)\delta[b(\mathbf{x}_i^*) - r] \tag{17.32}$$

where δ is the Kronecker delta function. C is the normalization constant $C = 1/\sum_{i=1}^{n} K(\mathbf{y} - \mathbf{x}_i^*)^2 \delta[b(\mathbf{x}_i^*) - r]$.

Let us denote by $\{\mathbf{x}_i = (x_i, y_i)\}_{i=1,\ldots,n_h}$ the pixel locations of the target candidate, centered at $\mathbf{y}$ in the current frame. Employing the same weighting function K, the probability of the color r in the target candidate is given by

$$\hat{p}_r(\mathbf{y}) = C_h \sum_{i=1}^{n} K(\mathbf{y} - \mathbf{x}_i)\delta[b(\mathbf{x}_i) - r] \tag{17.33}$$

where $C_h = 1/\sum_{i=1}^{n} K(\mathbf{y} - \mathbf{x}_i^*)^2 \delta[b(\mathbf{x}_i^*) - r]$ is the normalization constant.

The Bhattacharyya coefficient is used to measure the similarity between two histograms:

$$\hat{\rho}(\mathbf{y}) \equiv \rho[\hat{p}(\mathbf{y}), \hat{q}] = \sum_{r=1}^{m} \sqrt{\hat{p}_r(\mathbf{y})\hat{q}_r} \tag{17.34}$$

Based on that equation, the similarity distance between two histograms can be defined as

$$D(\mathbf{y}) = \sqrt{1 - \rho[\hat{p}(\mathbf{y}), \hat{q}]} \tag{17.35}$$

The minimization of the similarity distance can be efficiently achieved based on the mean shift iterations [3] via two steps. First, derive the weights

$$\omega_i = \sum_{i=1}^{n} \delta[b(\mathbf{x}_i - r)]\sqrt{\hat{q}_r/\hat{p}_r(\mathbf{y}_0)} \tag{17.36}$$

where $\mathbf{y}_0$ is the predicted location. Then, derive the new location:

$$\hat{\mathbf{y}}_1 = \sum_{i-=1}^{n} \mathbf{x}_i \omega_i K\left(\|\hat{\mathbf{y}}_0 - \mathbf{x}_i\|^2\right) \Big/ \sum_{i=1}^{n} \omega_i K\left(\|\hat{\mathbf{y}}_0 - \mathbf{x}_i\|^2\right) \tag{17.37}$$

Now, update the histogram and evaluate the similarity distance. If the distance becomes smaller, then shift to the new location $\hat{\mathbf{y}}_0 \leftarrow \hat{\mathbf{y}}_1$. At last, mean shift iterations converge to position $\hat{\mathbf{y}}_0$ and the final similarity distance is

$$D[\mathbf{y}] = (1 - \hat{\rho}(\hat{\mathbf{y}}_o))^{1/2} \tag{17.38}$$

17.6.4 Approach Summary

The proposed visual-assisted and laser-based tracking algorithm can be summarized as follows:

Detection and visual representation calculation:

1. At each time t, human detection is done by a laser-based feet detection algorithm based on laser points that are not associated with trajectories at time $t - 1$.
2. For each newly detected person, the corresponding body region in the video image is localized from average Person's height, feet position, and calibrated camera model.

3. Then an appearance model for each person based on a color histogram is calculated, followed by mean shift matching with the appearance model of objects whose trajectories have broken.
4. If matching succeeds, then this is a fragment of one existing object; otherwise, it belongs to a new object, that is, a newly entered person.

Laser-based tracking:

1. At each time t, for each person, laser-based foot tracking operates with the method just described.
2. Laser-based tracking results are transformed to the position of the center of two feet in image coordinates.
3. Visual representation for each target is updated time by time.

Visual-assisted data association for linking broken trajectories:

1. At each time t, for each broken trajectory, mean shift-based visual tracking operates and tries to match with newly detected targets.
2. If matching succeeds, then the broken trajectory is linked with the newly detected trajectory; otherwise, it continues to track and match at the next time.

17.6.5 Evaluations of Visual-Assisted Tracking Results

Visual assistance to the existing laser-based tracking system is specifically important when the configuration of laser number and locations is limited, which is quite a real-world situation. In Fig. 17.21, we show a real scene at a corner of an exhibition hall with two laser scanners and one camera. We could only put the laser scanners around our booth. Feet were measured only from one side, and thus occlusions occurred very often. The height of the camera set was restricted as well. We evaluated our method with 10-min data and present results by showing that broken trajectories were linked successfully with visual assistance.

Detection performance of our approach was very reliable, only 6 of 167 people were missed due to severe occlusion; that is, 96.4% of people who appeared in the scene were captured. We ran our laser-based tracker; 358 trajectories were generated, which implied that many trajectories were broken, and these broken fragments were difficult to connect with laser data. The totally laser-based tracker failed 201 times, including 197 breaks and 4 errors, and 177 of them (i.e.) 88% were recovered by assistance of visual information.

In Fig. 17.24, Fig. 17.25, Fig. 17.26 and Fig. 17.27 we show some multiple people tracking results with visual assistance. The laser-based tracker made the trajectory of each person fragmented into several parts. Each part was denoted with a colored line. Here, the introduction of visual information has two advantages. One is to extend the trajectory when there are no laser data. In the figures, white

Fig. 17.24 When laser-based tracker fails, visual data could help to extend the trajectory or connect the trajectory fragments that belong to the same person. Tracking result of frame 27

Fig. 17.25 Tracking result of frame 38

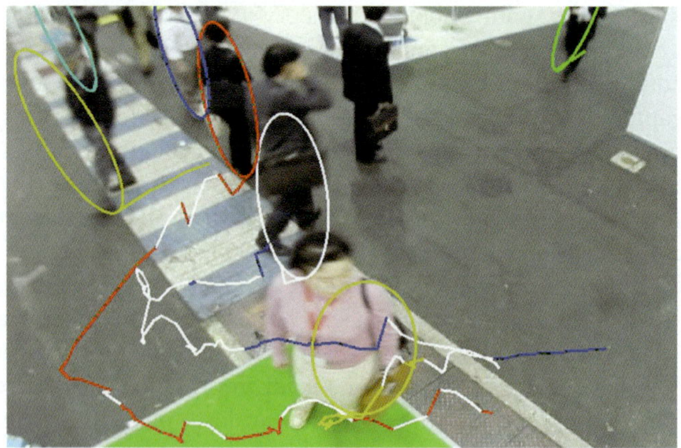

Fig. 17.26 Tracking result of frame 98

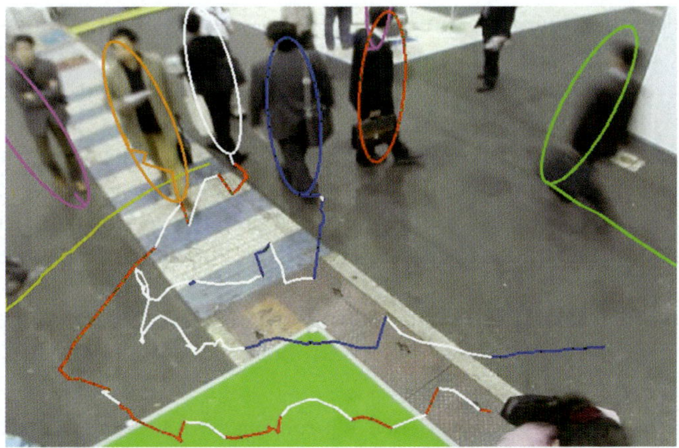

Fig. 17.27 Tracking result of frame 114

lines and ellipses denote trajectories and regions generated by visual tracking. The other advantage is to connect the trajectory fragments belonging to the same person. Through this real experiment, the effectiveness and feasibility of our approach was examined.

17.7 Conclusions

There are two main issues for laser-based multiple-people tracking. One is the difficulty in effective feature extraction. The other is joint estimation of target states and data associations. In this chapter, a novel method was presented of tracking multiple people in a wide and open area, such as a shopping mall and an exhibition hall, by scanning the feet of pedestrians using a number of single-row laser-range scanners.

In our experiment, multiple laser scanners and one camera were set in an exhibition hall, monitoring visitors' flow during exhibition day. About 50 visitors were tracked simultaneously during a peak hour with near real-time performance, which is much faster than our previous work [10, 22]. Compared with existing laser-based trackers, our method has three significant advantages: The extracted feature is very stable and deals with the measurement noise very well; the measurement likelihood is very strong and could uniquely make data associations correctly in most cases. Additional RBMC-DAF is used for tracking two correlated targets. Finally with very limited assistance from visual data, the system not only can maintain correct tracking in various complex situations, but also does not consume more time on video processing. The experimental results showed that our proposed method is very effective and robust.

There are still several problems that have not yet been solved. If one person moves very fast (jogs, for example), the accumulated image might not provide a significant local maximum for some static foot positions. We might miss that position and get a broken trajectory sometimes. This could be improved by a finer search strategy of local maximum or using a slide window to consider simultaneously several successive scan images.

For people carrying luggage, we can correctly track the person in most cases. But some times, the person and luggage together will generate two trajectories that mutually cross because we do not use a specific model for luggage. This problem could be tackled by learning patterns, respectively, for human and luggage in our future work.

In addition, a tracking algorithm will be developed for monitoring not only pedestrians, but also shopping carts, baby Strollers, bicycles, motor cars, and so on.

Acknowledgments This work was supported in part by the NKBRPC (2006CB303100), NSFC grant (60333010), and NSFC grant (60605001)

Chapter's References

1. Frank, O., Nieto, J. , Guivant, J., Scheding, S.: Multiple target tracking using sequential monte carlo methods and statistical data association. In: Proc. 2003 IEEE/RSJ Int. Conf. Intell. Robots Syst. (2003)
2. Kluge, B., Koehler, C., Prassler, E.: Fast and robust tracking of multiple moving objects with a laser range finder. In: Proc. IEEE Int. Conf. Robot. Automation 1683–1688 (2001)
3. Lindstrom, M., Eklundh, J.-O.: Detecting and tracking moving objects from a mobile platform using a laser range scanner. In: Proc. 2001 IEEE/RSJ Int. Conf. Intell. Robots Syst. 1364–1369 (2001)
4. Montemerlo, M., Thrun, S., Whittaker, W.: Conditional particle filters for simultaneous mobile robot localization and people-tracking. In: Proc. IEEE Int. Conf. Robot. Automation (2002)
5. Prassler, E., Scholz, J., Schuster, M., Schwammkrug, D.: Tracking a large number of moving objects in a crowded environment. In: IEEE Workshop Perception Mobile Agents (1998)
6. Schulz, D., Burgard, W., Fox, D., Cremers, A.: Tracking multiple moving targets with a mobile robot. In: Proc. IEEE Comput. Soci. Conf. Comput. Vis. Pattern Recogn. (2001)
7. Zhao. H., Shibasaki, R.: A novel system for tracking pedestrians using multiple single-row laser range scanners. IEEE Trans. SMC. Part A: Syst. Humans. 35, 283–291 (2005)
8. Cui, J., Zha, H., Zhao, H., Shibasaki, R.: Tracking multiple people using laser and vision. In: Proc. 2005 IEEE/RSJ Int. Conf. Intell. Robots Syst. 1301-1306 (2005)
9. Fod, A., Howard, A., Mataric, M.J.: A laser-based people tracker. In: Proc. IEEE Int. Conf. Robot Automation 3024-3029 (2002)
10. Blanco, J., Burgard, W., Sanz, R., Fernandez, J.L.: Fast face detection for mobile robots by integrating laser range data with vision. In: Conf. Advanced Robotics (2003)
11. Byers, Z., Dixon, M., Goodier, K., Grimm, C.M., Smart, W.D.: An autonomous robot photographer. In: IEEE Conf. Robots Syst. (2003)
12. Scheutz, M., McRaven, J., Cserey, G.: Fast, reliable, adaptive bimodal people tracking for indoor environments. In: Proc. 2004 IEEE/RSJ Int. Conf. Intell. Robots Syst. (2004)
13. Zhao, H., Shibasaki, R.: A robust method for registering ground-based laser range images of urban outdoor environment. Photogram. Eng. Remote Sens. 67, 1143–1153 (2001)
14. Mochon, S., McMahon, T. A. Ballistic walking. J. Biomech. 13, 49–57 (1980)
15. Parzen, E.: On estimation of a probability density function and mode. Ann. Math. Statistics. 33, 1065–1076 (1962)

16. Khan, Z., Balch, T., Dellaert, F.: Multitarget tracking with split and merged measurements. In: IEEE Conf. Comput. Vis. Pattern Recog. (2005)
17. Sarkka, S., Vehtari, A., Lampinen, J.: Rao-Blackwellized Monte Carlo data association for multiple target tracking. In: The 7th Int. Conf. Inform. Fusion (2004)
18. Doucet, A., De Freitas, J.F.G., Gordon, N.J. (eds.) Sequential Monte Carlo Methods in Practice. Springer-Verlag, New York, 2001
19. Tsai, R.Y.: A versatile camera calibration technique for high-accuracy 3D machine vision metrology using off-the-shelf TV cameras and lenses. IEEE J. Robotics Automation 4(3), 323–344 (1987)
20. Nishimura, T., Hideo, I., Yoshiyuki N., Yoshinobu, Y., Hideyuki, N.: A compact battery-less information terminal (CoBIT) for location-based support systems. SPIE, Vol. 3001, 124–139 (2004)
21. Comaniciu, D., Ramesh, V., Meer, P. Kernel based object tracking. IEEE Trans. Pattern Anal. Mach. Intell. 25 (2003)
22. Reprinted from: Computer Vision and Image Understanding, Vol. 106, no 2 , J. Cui, H. Zha, H. Zhao, R. Shibasaki, Laser-based detection and tracking of multiple people in crowds, 300–312, Copyright (2007), with permission from Elsevier

Chapter 18
On Boosted and Adaptive Particle Filters for Affine-Invariant Target Tracking in Infrared Imagery

Guoliang Fan, Vijay Venkataraman, Li Tang, and Joseph P. Havlicek

Abstract We generalize the usual white noise acceleration target model by introducing an affine transformation to model the target aspect. This transformation is parameterized by scalar variables that describe the target shear, scale, and rotation and obey a first-order Markov chain. Our primary interest is in achieving robust Monte Carlo estimation in a complex state space under low signal-to-noise ratios, where a low-entropy unimodal likelihood function contributes to the failure of the conventional particle-filtering algorithms. Motivated by recently developed particle filter consistency checks, we develop a new track quality indicator that monitors tracking performance, triggering one of two actions as needed to improve the quality of the track. The first action is a boosting step, by which a local detector is defined based on the most recent tracker output to induce additional high-quality boosting particles. The original idea of boosting is extended here by encouraging positive interaction between the detector and the tracker. The second action is an adaptation step in which the system model self-adjusts to enhance tracking performance. In the context of affine-invariant target tracking, we compare these two techniques with respect to their effectiveness in improving the particle quality. We present experimental results that show that both techniques can improve the tracking performance by balancing the focus and the diversity of the particle distribution and by improving the particle filter consistency.

18.1 Introduction

Target tracking is usually formulated as a problem of estimating a dynamic state-space system, in which the state comprises the target kinematics and distinct models for the state evolution and the observation process are explicitly involved. Given a probabilistic state-space formulation, the tracking problem is well suited for the recursive Bayesian approach, which attempts to construct the posterior probability

R.I. Hammoud (ed.), *Augmented Vision Perception in Infrared: Algorithms and Applied Systems,* Advances in Pattern Recognition, DOI 10.1007/978-1-84800-277-7_18, © Springer-Verlag London Limited 2009

density function (PDF) of the state based on all state and observation information available. Recently, particle filtering has received increasing attention because of its flexibility and its ability to deal with nonlinear and non-Gaussian estimation problems by representing a continuous density with a discrete approximation [1]. The key idea is to represent the required PDF by a set of weighted particles and to estimate the state based on these weighted particles. A recursive Bayesian filter can be implemented by Monte Carlo simulations to continuously update the particles and their associated weights. To ensure success of the approach, it is important to control the quality of the particles, that is, to maintain a balance between diversity (having multiple distinct samples) and focus (having multiple copies of samples with large weights) [2]. Moreover, from a practical standpoint, it is desirable to implement a consistency check of the particle filter to provide continuous insight into the quality of the tracking process and improve the confidence level of the obtained tracks [3].

State estimation in a high-dimensional space requires an exponentially increasing number of particles. There are two different but relevant methodologies to improve the quality and efficiency of the particles: the "bottom-up" and the "top-down" approaches. The bottom-up methods control the particle quality via direct particle modifications, including particle reweighting and resampling [2], the kernel-based particle filter [4, 5], and the annealed particle filter [6]. By contrast, the top-down methods focus on the problem formulation, that is, the system model, or *prior*, and the observation model, or *likelihood*. On the one hand, the target motion is assumed to be governed by a certain motion model such as a white noise acceleration model [7] or a random walk [8]. Adaptive motion models have also been developed that can be learned on the fly on the arrival of new observations [9]. The Adaboost particle filter [10] incorporates the detection hypothesis in the proposal distribution in which the detector is involved as an independent process to track new targets. On the other hand, the observation model mainly depends on target descriptions or sensor models. For a static target template, the sum of squared distance (SSD) in intensity is minimized to determine the likelihood. Other features have also been derived to describe targets and objects, including edges or histograms in a region of ellipses [4] or rectangles [5] with a varying scale and a fixed aspect ratio. When a geometric contour is used [11], the boundary of an object evolves as an active contour, and the likelihood function is derived by minimizing an image-based energy function. Moreover, the affine model was used in many trackers to support robust and accurate tracking for deforming targets [9, 11]. Most of the aforementioned approaches are mainly for visible wavelength optical images in which the objects of interest usually have relatively high visibility and large size. In the context of small target tracking in infrared image sequences or other remotely sensed imagery, low Signal-to-noise ratio (SNR), poor target visibility, and unknown dynamics of the target aspect render the tracking problem more challenging. In such cases, additional prior knowledge about the system and observation models is needed to ensure robust tracking performance. For example, the technique proposed in [8, 12] assumes that the clutter is characterized by a first-order Gauss-Markov random field (GMRF) and that a finite set of multiaspect target signatures is represented by discrete-valued indices in the state vector. Excellent tracking performance was reported in [12] on simulated infrared image sequences under low SNR, ranging from -5.7 to 7 dB. Two key elements of the techniques given in [8, 12] are the highly selective likelihood function derived from the GMRF and the predefined multiaspect target templates as well as their transition probabilities.

Motivated by these results, we here consider a more general formulation of the multiaspect tracking problem applicable to infrared image sequences by introducing a continuous-valued affine model for the dynamics of the target aspect, including rotation and scale. Specifically, we assume that the affine model is characterized by a first-order Markov chain and is incorporated into the state and observation models. This new formulation is capable of describing more realistic target motions and aspect dynamics. However, due to the complexity of the state space, which involves additional continuous-state variables, traditional particle-filtering algorithms such as sequential importance resampling (SIR) and the auxiliary particle filter (APF) discussed in [12] fail to provide satisfactory results under the new formulation. In SIR, the transition prior is chosen as the proposal distribution to simplify the computation of the importance weights. Since the present observation is not taken into consideration in the prediction stage, the particles may not be placed in the modal regions of the posterior; this can be especially problematic in high-dimensional state spaces. Although a large number of particles may be introduced to increase the chance of overlap with the underlying true posterior, doing so significantly increases the computational load. Alternatively, with the APF the current observation is considered in the prediction stage by drawing particles twice. The second sampling places increased emphasis on those promising particles with large weights. This improves the efficiency of the prediction to some extent. However, for high-dimensional particle filtering such as the affine-invariant target tracking formulation discussed in this chapter, new strategies are needed to make the algorithm practically feasible. Two approaches that are typically exploited to overcome problems of this type are to draw particles near the modes of the posterior and to devise an appropriate means of evaluating the importance of these particles. If the new strategies can succeed in placing particles in modal regions of the posterior or if the importance function is close to the state posterior, then a set of particles with low variance will be obtained, implying that the particle impoverishment problem is effectively mitigated [13].

In this chapter, we revisit two recent improvements to particle filters, namely, boosting and adaptation. Boosting usually involves an independent process that adds new particles [10], whereas adaptation is a strategy for adjusting the system and observation models on the arrival of new observations [9]. Both of these techniques have been proven efficient for object tracking in optical images with relatively high object visibility. Our interest here is in how to achieve robust and accurate Monte Carlo estimation in a high-dimensional state space with poor target visibility. Central to this work is a track quality indicator that estimates the tracking performance based on the observation model and can trigger boosting or adaptation as needed. In the boosting step, a detector is constructed based on the most recent output of the track filter to induce more promising particles. The original notion of boosting is extended here by encouraging positive interaction between the detector and the tracker. In the adaptation step, the system model self-adjusts to enhance the tracking performance. In the context of affine-invariant target tracking, we compare these two methods with respect to their contributions toward improving the particle quality.

18.2 Problem Formulation

In this section, we briefly review the state transition and observation models that were given in [7]. We then extend these models by introducing a new, continuous-valued affine-invariant model capable of representing highly realistic evolution of the target aspect parameters, including scale and rotation.

18.2.1 System and Observation Models

Let Δ denote the time interval between two consecutive observed frames. For $k \in \mathbb{N}$, the state vector at the instant $t = k\Delta$ consists of the position (x_k, y_k) and velocity $(\dot{x}_k, \dot{y}_k)$ of the target centroid expressed in two-dimensional (2D) Cartesian image plane coordinates, such that $\mathbf{x}_k = [x_k\ \dot{x}_k\ y_k\ \dot{y}_k]^T$. The target position and velocity in the x and y directions are assumed to be independent and to evolve over time according to the white noise acceleration model

$$\mathbf{x}_k = \mathbf{F}\mathbf{x}_{k-1} + \mathbf{w}_{k-1} , \tag{18.1}$$

where the process noise $\mathbf{w}_k$ is zero mean, white, and Gaussian. Let the state transition matrices for the x and y directions be

$$\mathbf{F_x} = \mathbf{F_y} = \begin{bmatrix} 1 & \Delta \\ 0 & 1 \end{bmatrix} .$$

Then, the state transition matrix in (18.1) is given by

$$\mathbf{F} = \begin{bmatrix} \mathbf{F_x} & \mathbf{0} \\ \mathbf{0} & \mathbf{F_y} \end{bmatrix} .$$

The observed frames are acquired from an imaging sensor and are of size $L \times M$ pixel sites $\{(i, j)\,|\,1 \leq i \leq L, 1 \leq j \leq M\}$. The observation model for the frame $\mathbf{z}_k$ acquired at the time instant $t = k\Delta$ (time step k) is given by

$$\mathbf{z}_k = \mathbf{H}(\mathbf{x}_k) + \mathbf{v}_k , \tag{18.2}$$

where $\mathbf{v}_k$ is a GMRF clutter field with a template of a specific intensity distribution, and $\mathbf{H}$ is a function of the state vector that produces a clutter-free frame containing a template of the target at the position specified by $\mathbf{x}_k$.

We assume that $\mathbf{w}_k$ in the state model (18.1) is statistically independent from $\mathbf{v}_k$ in (18.2) [8]. The clutter frames $\{\mathbf{v}_k | k \in \mathbb{N}\}$ are assumed to be iid GMRF sequences with zero mean and nonsingular covariance matrices. Each clutter frame is described by the first-order GMRF given in [14]:

$$\mathbf{v}_k(i,j) = \beta_v^c[v_k(i-1,j) + v_k(i+1,j)] + \beta_h^c[v_k(i,j-1) + v_k(i,j+1)] + \varepsilon_k(i,j) \,, \tag{18.3}$$

where the parameters β_v^c and β_h^c are, respectively, the vertical and horizontal predictor coefficients, and ε_k is the prediction error such that [7]

$$E[v_k(i,j)\varepsilon_k(l,r)] = \sigma_{c,k}^2 \delta_{i-l,j-r} \,. \tag{18.4}$$

One can estimate the GMRF parameters β_h, β_v, and σ_c^2 for each frame $\mathbf{z}_k$ via the suboptimal approximate maximum likelihood (AML) algorithm given in [15].

18.2.2 Affine-Invariant Target Model

In [12], a discrete set of target templates with different aspects was used for multi-aspect tracking, and the state vector was augmented with an aspect index s to account for the aspect variability, so that $\mathbf{x}_k = [x_k \ \dot{x}_k \ y_k \ \dot{y}_k \ s]^T$. Our objective in this section is to generalize this multiaspect tracking approach by introducing a continuous-valued affine model that is often used to track deformable objects [9,11].

The new formulation, which we refer to as *affine-invariant target tracking*, models the temporal evolution of the target signature by applying an affine transformation to the base target template at every time step. This affine transform consists of two parts: a translation that models the motion of the target centroid and an affine transformation $\mathbf{T}_a$, given in (18.7) below, that models the dynamics of the target aspect. The specific transformation $\mathbf{T}_a$ applied in any given time step is determined by a set of stochastic scalar aspect parameters, including a shearing parameter α, an x-axis scaling parameter s_x, a y-axis scaling parameter s_y, and a rotation parameter θ. The aspect parameters are modeled as continuous-valued random variables that follow a first-order Markov chain with equal transition probabilities (*i.e.*, 1/3) of increasing or decreasing by a fixed quantization step (Δ_α, Δ_{s_x}, Δ_{s_y}, Δ_θ) or of maintaining the value from the previous time step. In addition, we add uniform white process noises γ_α, γ_{s_x}, γ_{s_y}, and γ_θ to the aspect parameters at each state transition to obtain increased random variability in the model. Thus, for example, the target rotation angle θ is modeled according to

$$\theta_k = \begin{cases} \theta_{k-1} - \Delta_\theta + \gamma_\theta & \text{with prob. } p(\theta_k|\theta_{k-1}) = 1/3, \\ \theta_{k-1} + \gamma_\theta & \text{with prob. } p(\theta_k|\theta_{k-1}) = 1/3, \\ \theta_{k-1} + \Delta_\theta + \gamma_\theta & \text{with prob. } p(\theta_k|\theta_{k-1}) = 1/3, \end{cases} \tag{18.5}$$

where γ_θ is a zero-mean uniform white noise with a maximum value proportional to the quantization step size Δ_θ. For affine-invariant target tracking, the aspect parameters are incorporated into the state model by defining a new augmented state vector:

$$\mathbf{x}_k = [x_k \; \dot{x}_k \; y_k \; \dot{y}_k \; s_x^k \; s_y^k \; \alpha_k \; \theta_k]^T. \tag{18.6}$$

The clutter-free target frame $\mathbf{H}(\mathbf{x}_k)$ in the observation model (18.2) is then obtained by injecting the base target template into a zero-valued frame at the location (x_k, y_k) and applying the affine transformation

$$\mathbf{T}_a = \begin{bmatrix} 1 & \alpha_k & -\alpha_k y_k \\ 0 & 1 & 0 \\ 0 & 0 & 1 \end{bmatrix} \begin{bmatrix} s_x^k & 0 & (1-s_x^k)x_k \\ 0 & s_y^k & (1-s_y^k)y_k \\ 0 & 0 & 1 \end{bmatrix} \times \begin{bmatrix} \cos\theta_k & \sin\theta_k & (1-\cos\theta_k)x_k - (\sin\theta_k)y_k \\ -\sin\theta_k & \cos\theta_k & (1-\cos\theta_k)y_k + (\sin\theta_k)x_k \\ 0 & 0 & 1 \end{bmatrix}$$
$$\tag{18.7}$$

to every pixel. This transformation is calculated in homogeneous coordinates [16], where the intensity value originally located at pixel site (i, j) is mapped to a transformed site (i', j') given by $[i' \; j' \; 1]^T = \mathbf{T}_a [i \; j \; 1]^T$. It should be noted that this implies that the computation of $\mathbf{H}(\mathbf{x}_k)$ requires gray-level interpolation of the original target template under the affine transformation since $i', j' \notin \mathbb{Z}$ in general.

While this new formulation is conceptually a straightforward extension of the one given in [12], the inherent complexity of the resulting high-dimensional state space introduces significant difficulties, as discussed in detail in Section 18.3. In particular, the continuous-valued aspect parameters make the target detection and tracking problems more challenging at low-to-moderate SNR. Because of these problems, conventional particle filters will generally fail to provide satisfactory performance in practice, and we propose new techniques to overcome the limitations of traditional particle filters for this extended problem formulation. To close this section, we note that (18.5–18.7) provide a general model for the dynamics of the target aspect as projected on the image plane. This generality is needed to accommodate scenarios such as aerial combat, where the target and sensor may be in complex motion relative to one another. In other applications, such generality may be unnecessary and even undesirable. If the sensor is stationary and the target is a rigid ground vehicle, for example, the shearing parameter α is probably not needed; including it in the formulation would only lead to wasted particles corresponding to highly improbable and impossible states. In the case of a slow-moving ground vehicle, it may be unnecessary to implement separate scaling parameters for the x and y directions. In the interest of simplicity, we assume throughout the remainder of this chapter that the shearing parameter $\alpha = 0$, and that there is only a single scaling parameter $s = s_x = s_y$.

18.2.3 Likelihood Function of the Observations

The form of the likelihood function is rooted in the structured nature of the background clutter typical of infrared imagery, which may be effectively modeled as a first-order GMRF [7, 12, 14]. Let Z_k and $h(\mathbf{x}_k)$ be one-dimensional (1D) column vector representations of the observed frame and the clutter-free target frame, obtained by concatenating the rows of $\mathbf{z}_k$ and of $\mathbf{H}(\mathbf{x}_k)$, respectively. Similarly, let V_k denote the 1D vector representation of the clutter frame $\mathbf{v}_k$ defined in (18.3). Then, the likelihood function is given by [8]

$$p(Z_k|\mathbf{x}_k) \propto \exp\left[\frac{2\lambda(Z_k) - \rho(\mathbf{x}_k)}{2\sigma_{c,k}^2}\right], \tag{18.8}$$

where $\rho(\mathbf{x}_k)$ is an *energy term* associated with the clutter-free target frame, and $\lambda(Z_k)$ is a *data term* that quantifies the agreement between the target signature in the observed frame and that in the clutter-free target frame. The energy term depends on the current target template (with a particular set of aspect parameters) and on the correlation structure of the GMRF clutter model according to

$$\rho(\mathbf{x}_k) = h^T(\mathbf{x}_k)(\sigma_{c,k}^2 \Sigma_v^{-1})h(\mathbf{x}_k), \tag{18.9}$$

while the data term depends on the observed frame and also implicitly on the current state according to

$$\lambda(Z_k) = Z_k^T(\sigma_{c,k}^2 \Sigma_v^{-1})h(\mathbf{x}_k). \tag{18.10}$$

One may regard $\lambda(Z_k)$ as the result of a matched filtering or correlation operation between the observed frame $\mathbf{z}_k$ and the clutter-free target frame for a particular choice $\mathbf{x}_k$ of the state vector.

Given a particle (a hypothesis), its likelihood computation involves both the energy and data terms. The energy term $\rho(\mathbf{x}_k)$ depends only on the current target state and clutter statistics and is independent of the current observation. Therefore, we focus on the likelihood $p(Z_k|\mathbf{x}_k)$ and the data term $\lambda(Z_k)$. Suppose that the observed frame $\mathbf{z}_k$ contains an instance of the target with state vector $\mathbf{x}_k$, centroid $\mathbf{x}_k^c = [x_k, y_k]^T$, scale parameter s_k, and rotation parameter θ_k. In addition, let $\hat{\mathbf{x}}_k$ be an estimate of the state vector with centroid $\hat{\mathbf{x}}_k^c = [\hat{x}_k, \hat{y}_k]^T$, scale parameter $\hat{s}_k$, and rotation parameter $\hat{\theta}_k$. Then, the signed scalar $\varepsilon_k^c = \text{sgn}(\hat{x}_k - x_k)\text{sgn}(\hat{y}_k - y_k)||\hat{\mathbf{x}}_k^c - \mathbf{x}_k^c||_{\ell^2}$ quantifies the centroid position error associated with the estimate $\hat{\mathbf{x}}_k$. The (signed) errors in the aspect parameters are given by $\varepsilon_k^s = 1 - (\hat{s}_k/s_k)$ and $\varepsilon_k^\theta = \hat{\theta}_k - \theta_k$. For the moment, let us consider the likelihood function $p(Z_k|\mathbf{x}_k)$ and the data term $\lambda(Z_k)$ as functions of the three scalar variables ε_k^c, ε_k^s, and ε_k^θ. Figure 18.1 shows the variations in the likelihood function and data term with respect to all pairwise combinations of these three variables. Our objective in Fig. 18.1 is to study the selectivity and sensitivity of $p(Z_k|\mathbf{x}_k)$ and $\lambda(Z_k)$ as functions of the state estimation errors. It is clearly shown that the likelihood function is both significantly more sensitive and significantly more selective by comparison.

It was suggested in [17] that an ideal likelihood function should admit three main characteristics: (1) It should have a flat area at the mode to provide sufficient robustness against minor modeling inaccuracies and to ensure that there are only small differences between the weights of multiple particles that are all close to the true state. (2) It should provide a significant difference between good particles that are close to

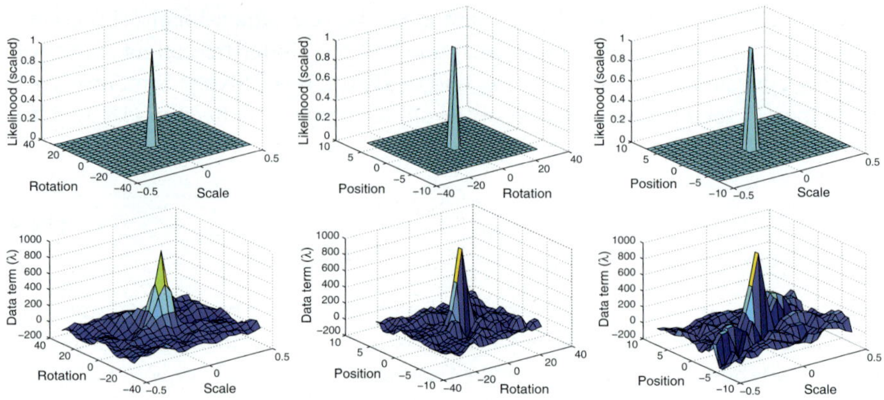

Fig. 18.1 Sensitivity and selectivity of the likelihood function (18.8) (*top row*) and the data term (18.10) (*bottom*) with respect state estimation errors in the target position, scale, and orientation

the true state and bad particles that are far from it. (3) Bad particles that are far from the true state should be similarly weighted so that the filter avoids locking onto local maxima due to clutter if the target is momentarily lost. As shown in Fig. 18.1, the likelihood function given by (18.8) is highly selective with a very sharp peak at the true state. This provides both high accuracy and high sensitivity for detection and tracking at low SNR [7].

These same characteristics can prove detrimental in a high-dimensional state space with mixed parameters, however, because the Monte Carlo simulation may fail to place any particles near the true target state. Indeed, the situation may be likened to the proverbial "person (i.e., the Monte Carlo simulation) searching for a needle (the true target state) in a haystack at night (the high-dimensional state space) while looking through a drinking straw (the highly selective likelihood function)." In devising a solution for this problem, we do not want to change the highly selective nature of the likelihood function because of its potential to deliver high-precision tracking. Rather, what is needed are new strategies to improve the robustness and efficiency of the sampling process.

18.3 Particle-Filtering Theory

18.3.1 Recursive Bayesian Estimation

With the affine-invariant target model given in Section 18.2.2, we have cast the target-tracking problem as one of estimating the state of a linear system driven by white noise, where the state $\mathbf{x}_k$ is first-order Markov. The available observations $\mathbf{z}_{1:k}$ are nonlinearly related to the augmented state vector (18.6) through the observation model (18.2). The Bayesian estimation framework provides a well-known theoretical solution for this problem by which the posterior density $p(\mathbf{x}_k|\mathbf{z}_{1:k})$ is obtained recursively by alternating prediction and refinement of the state vector estimate. The prior $p(\mathbf{x}_k|\mathbf{z}_{1:k-1})$ at time step k is obtained from the posterior $p(\mathbf{x}_{k-1}|\mathbf{z}_{1:k-1})$ at time step $k-1$ by

$$p(\mathbf{x}_k|\mathbf{z}_{1:k-1}) = \int p(\mathbf{x}_k|\mathbf{x}_{k-1})p(\mathbf{x}_{k-1}|\mathbf{z}_{1:k-1})d\mathbf{x}_{k-1}, \tag{18.11}$$

where $p(\mathbf{x}_k|\mathbf{x}_{k-1})$ is the transition prior. When the observation $\mathbf{z}_k$ arrives, (18.11) is refined using Bayes' rule to obtain the updated posterior for time step k according to

$$p(\mathbf{x}_k|\mathbf{z}_{1:k}) \propto p(\mathbf{z}_k|\mathbf{x}_k)p(\mathbf{x}_k|\mathbf{z}_{1:k-1}). \tag{18.12}$$

Since this theoretical solution can almost never be realized in practice, numerical techniques such as particle filtering have been devised to approximate (18.11) and (18.12).

18.3.2 Basic Particle Filtering

A particle filter approximates the posterior $p(\mathbf{x}_k|\mathbf{z}_{1:k})$ by a set of N_p weighted particles $\{\mathbf{x}_k^i, w_k^i\}_{i=1}^{N_p}$. The weights are chosen using the principle of *importance sampling*, which involves an *importance density* $q(x)$. If the samples $\mathbf{x}_{1:k}^i$ are drawn from an importance density $q(\mathbf{x}_{0:k}^i|\mathbf{z}_{1:k})$, then the weights are computed by [18]

$$w_k^i \propto \frac{p(\mathbf{x}_{0:k}|\mathbf{z}_{1:k})}{q(\mathbf{x}_{0:k}|\mathbf{z}_{1:k})}, \tag{18.13}$$

and these weights are recursively updated based on new observations $\mathbf{z}_k$ according to

$$w_k^i \propto w_{k-1}^i \frac{p(\mathbf{z}_k|\mathbf{x}_k^i)p(\mathbf{x}_k^i|\mathbf{x}_{k-1}^i)}{q(\mathbf{x}_k^i|\mathbf{x}_{k-1}^i, \mathbf{z}_k)}. \tag{18.14}$$

SIR is the basic particle-filtering algorithm, in which the transition prior $p(\mathbf{x}_k^i|\mathbf{x}_{k-1}^i)$ is used as the importance density, and *resampling* is applied at every time step to inhibit degeneracy of the particle set [18]. After resampling, all of the new particles are assigned equal weights, which corresponds to an idealized case of (18.13) in which the importance density exactly coincides with the posterior $p(\mathbf{x}_{0:k}|\mathbf{z}_{1:k})$. A key feature of SIR is that the importance density fails to incorporate the most recent observation $\mathbf{z}_k$ and thus may differ substantially from the posterior in practice. SIR is inherently suboptimal in this regard.

The APF was proposed to improve the SIR by incorporating a two-stage sampling process by which the current observation $\mathbf{z}_k$ is considered explicitly [19]. Given the particles $\{\mathbf{x}_{k-1}^i, w_{k-1}^i\}_{i=1}^{N_p}$ drawn at the previous time step, a second set of particles $\{\mu_k^i, \widehat{w}_k^i\}_{i=1}^{N_p}$ is constructed in such a way that each μ_k^i is considered *highly likely* in view of $\mathbf{x}_{k-1}^i$ and the prior $p(\mathbf{x}_k|\mathbf{x}_{k-1}^i)$. For example, one may take $\mu_k^i = \mathrm{E}[\mathbf{x}_k|\mathbf{x}_{k-1}^i]$ or simply draw μ_k^i from $p(\mathbf{x}_k|\mathbf{x}_{k-1}^i)$. The *first-stage weights* $\widehat{w}_k^i$ are calculated by $\widehat{w}_k^i \propto w_{k-1}^i p(\mathbf{z}_k|\mu_k^i)$. The key of the APF is that a set of auxiliary indices m^i are then drawn such that $p(m^i = j) = \widehat{w}_k^j$ $(1 \leq i, j \leq N_p)$. Since the μ_k^i are deemed highly likely and both μ_k^i and $\widehat{w}_k^i$ incorporate the most recent observation, the auxiliary indices m^i are interpreted as indicating a set of highly likely samples $\mathbf{x}_{k-1}^{m^i}$ from among particles $\{\mathbf{x}_{k-1}^i, w_{k-1}^i\}_{i=1}^{N_p}$. The importance density for drawing new samples $\mathbf{x}_k^i$ in the APF is given by $p(\mathbf{x}_k|\mathbf{x}_{k-1}^{m^i})$. Finally, the updated *second-stage weights* are given by $w_k^i \propto p(\mathbf{z}_k|\mathbf{x}_k^i)/p(\mathbf{z}_k|\mu_k^{m^i})$. Relative to the SIR, the APF generally produces a set of particles with more uniform weights w_k^i, resulting in improved agreement between the particles and the true posterior.

18.3.3 Recent Improvements to Particle Filters

Several practical challenges arise when particle filters are used in real tracking applications. For example, the target motion may be so complex and unpredictable that it becomes difficult for the prediction stage to create a set of quality particles. Low SNRs may also deteriorate the observation model, leading to poor posterior estimation in the update stage. Consequently, many techniques have been proposed recently with the goal of improving the efficiency and quality of the basic particle-filtering approach. A modular approach to empirically evaluate and analyze recent improved particle filters was proposed in [13], in which particle-filtering algorithms were classified into two groups according their objectives. One group aims to construct an effective sample set, and the other is targeted on improving particle diversity. An optimal particle-filtering algorithm was developed in [13] for figure tracking that combines both techniques to improve particle quality. In this chapter, we use a different taxonomy that is focused specifically on differences between the most recently developed methods and compares various particle-filtering techniques with respect to their effects on particle generation and redistribution. Thus, the taxonomy used here is complementary to the one in [13], providing insight into how particle filters can be improved for a specific tracking application. As shown in Fig. 18.2, we classify methods for increasing the efficiency and quality of the particle sets as being either *top-down* or *bottom-up*; methods of both types can be incorporated in a single computational flow.

18.3.3.1 Top-Down Methods

The top-down methods focus on the dynamic model or try to develop better proposals for particle sampling. For example, a new importance function was introduced in [20] that is a combination of the posterior and the prior importance densities. In [21], a second particle filter was employed to estimate the optimal proposal

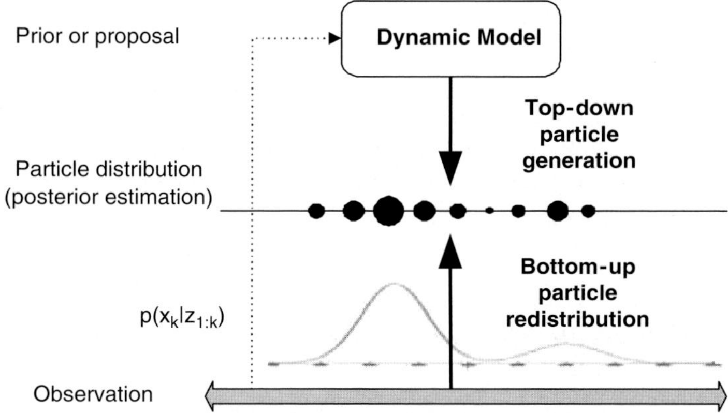

Fig. 18.2 Particle filtering and improvements to particle quality. The dotted line from the observation to the system dynamic model indicates that the current observations are used for proposal generation. The weight of a particle is indicated by its size.

distribution. An unscented particle filter incorporating the unscented Kalman filter to integrate the current observation for generating proposal distributions was introduced in [22]. In [23], the inference in a high-dimensional state space was factorized by several lower-dimensional ones in an iterative fashion. The commonality between these approaches is that the sampling scheme is improved at the system model level by involving the current observations.

18.3.3.2 Bottom-Up Methods

The bottom-up methods tend to directly control the particle distribution by moving or redistributing samples toward or near the modal positions in the state space. Typical algorithms falling in this category include the kernel particle filter in [24], which involves kernel-based density estimation of the posterior and uses the mean shift as a gradient estimation and mode-seeking procedure and the quasi-random sampling scheme [25], by which the sampling is done with a chosen set of quasi-random points that span the sample space so that they are maximally far away from one another. In this way, samples are efficiently redistributed and well spread in the state space.

18.3.3.3 Hybrid Methods

Some recent particle-filtering algorithms combine both top-down and bottom-up strategies to improve the sample set. For example, the hybrid particle filter in [26] involves both the mean shift (bottom-up) and an adaptive transition model (top-down). Specifically, the mean shift is used to move particles toward the local mode, and the adaptive transition model adjusts the system model according to the state prediction error. The algorithms in [27] and [28] can also be regarded as hybrid methods, by which particles are randomly resampled from a localized restriction of the observation density or a localized importance function after the prediction step. The boosted particle filter proposed in [10] is also a hybrid approach by which an independent detector (Adaboost) is combined with the motion model to produce a hybrid proposal for adding new tracks.

18.3.4 Challenges of Affine-Invariant Tracking

As we indicated in Section 18.2.2, there are two significant challenges associated with the affine-invariant target-tracking approach proposed here relative to the tracking problem studied in [7]. Because of these difficulties, both the standard APF and SIR algorithms fail to provide satisfactory tracking performance. The first challenge is that the affine-invariant tracking formulation is a combined parameter and state estimation problem with continuous aspect parameters by which we convert the parameter inference problem into an optimal filtering problem by augmenting new state variables. It is interesting to note in this regard that it was suggested in [29] that "due to the lack of ergodicity of the extended state process, such approaches are bound to fail." Moreover, the fact that the aspect parameters are continuous-valued random variables makes the simultaneous state and parameter estimation problem both significantly more difficult and more costly compared to related formulations in which the parameters are discrete. The second significant challenge arises from the highly peaked, low-entropy observation likelihood function associated with this problem. If the newly drawn samples fail to fall in close proximity to the modes of the likelihood function, the discrete representation of the posterior density is subject to rapid and profound degradation [27].

18.4 Boosted and Adaptive Particle Filters

Based on two recent techniques, we develop in this section two enhanced particle filters for the affine-invariant target-tracking problem. One is the adaptive auxiliary particle filter (AAPF), which improves particle quality by adjusting the system model (a top-down approach). The other is the boosted auxiliary particle filter (BAPF), which incorporates a local detector to add more promising particles (a bottom-up approach). We first review the main ideas of particle filter consistency checks and then discuss the two filters in detail. Hereafter, we use the terms *boosting* and *adaptation* specifically to refer to the way these general techniques are implemented in the AAPF and BAPF.

18.4.1 Track Quality Indicator

Particle filter consistency checks provide an important means of gaining insight into the confidence level associated with a Monte Carlo estimation scheme. Several test statistics were proposed in [3] for monitoring the consistency or tracking confidence of the particle filter. The main idea in [3] is that if the observation sequence $\mathbf{z}_{1:k}$ does not conform to the statistics prescribed by the sequence of densities $p(\mathbf{z_k}|\mathbf{z}_{1:k-1})$ given by

$$p(\mathbf{z}_k|\mathbf{z}_{1:k-1}) = \int p(\mathbf{z}_k|\mathbf{x_k})p(\mathbf{x_k}|\mathbf{z}_{1:k-1})d\mathbf{x}, \qquad (18.15)$$

then the particle filter is deemed inconsistent in the sense that the system model is inadequate for describing the observations in a statistically significant way. One may also interpret (18.15) as the degree of matching between the current observation and the current particle configuration. In other words, (18.15) indicates how well the particle distribution explains the current observation. However, the statistic proposed in [3] depends on the past observations, necessitating the definition of a window that specifies how many observations from the past should used for computing (18.15). Also, we think this test is most plausible in systems with fewer observation variables.

The general notion of a consistency check, however, motivates us to introduce a track quality indicator that estimates the tracking performance based on the data term (18.10) and weight w_k^i associated with each sample $\mathbf{x}_k^i$ in the current particle set. Here, it should be kept in mind that the weight w_k^i is proportional to the likelihood (18.8). The track quality indicator is given by

$$\phi_k = \sum_{i=1}^{N_p} w_k^i \lambda_k^i, \qquad (18.16)$$

which provides an estimate of the mean value of the data term for the current particle configuration.

The importance of monitoring how well the hypothesis associated with each particle matches with the observations was argued in [3]. Moreover, as we pointed out in Section 18.2.3, the data term λ_k^i may be regarded as the result of a matched filtering or correlation operation between the observation $\mathbf{z}_k$ and the clutter-free target frame $\mathbf{H}(\mathbf{x}_k^i)$, which motivates the use of λ_k^i in (18.16). Thus, ϕ_k provides an evaluation of the overall quality of the current particle configuration. Large values of ϕ_k indicate that the posterior state density represented by the particle set is consistent with the observations, giving us a high confidence that the tracker is working well. Similarly, small values of ϕ_k suggest that the target has been lost or that the consistency of the particle filter may be at issue. We choose to incorporate the data term in (18.16) as opposed to using the likelihood function alone for evaluating track quality because the shape of $\lambda(Z_k)$ shown in Fig. 18.1 is more like that of an ideal likelihood function as described in [17]. Furthermore, the high selectivity of $p(Z_k|\mathbf{x}_k)$ shown in Fig. 18.1 is undesirable in terms of obtaining a stable estimate of the tracking performance.

18.4.2 BAPF Algorithm

The main idea behind the BAPF algorithm is the introduction of a matched filter-based detector to boost the tracking performance. Unlike the approach in [10], in which the detector and tracker were two independent processes, here we want to encourage positive interaction between them. The nature of the interaction is that the matched filter kernel of the detector is derived from the most recently tracked target signature, whereas the incorporation of *boosting* particles from the detector can improve tracking performance by increasing the overlap between the particle distribution and the underlying posterior.

18.4.2.1 Initialization

Initialization plays an important role in most particle filters, especially when the state space is complicated. To initialize the state variables x_1, y_1, s_1, and θ_1 at the first frame, we generate an ensemble of target templates with different rotation and scaling parameters uniformly distributed in $[s_{min}, s_{max}]$ and $[\theta_{min}, \theta_{max}]$, respectively. Given a target template $\mathbf{G}(s, \theta)$, we compute a similarity map $\mathbf{M}(s, \theta)$ with values that are assumed proportional to the probability that a particular instance of the template is present in a given frame. This requires that we remove the structured clutter from the observed frame $\mathbf{z}_1$ by convolution with the template of the GMRF model:

$$\mathbf{K} = \begin{bmatrix} 0 & -\beta_v & 0 \\ -\beta_h & 1 & -\beta_h \\ 0 & -\beta_v & 0 \end{bmatrix}.$$

The similarity map $\mathbf{M}(s, \theta)$ is then defined by $\mathbf{M}(s, \theta) = \mathbf{z}_1 * \mathbf{K} * \mathbf{G}(s, \theta)$, where $*$ denotes convolution. A set of initial particles is selected from $\mathbf{z}_k$ corresponding to maxima in $\mathbf{M}(s, \theta)$. Each particle is assigned initial position and aspect parameters as specified by $\mathbf{M}(s, \theta)$. A similar process is used by the detector to produce boosted particles for the tracker.

18.4.2.2 Interaction between Detector and Tracker

For some threshold T_λ, our goal is to improve tracking performance by boosting with a local detector in frame k if $\phi_k < T_\lambda$. The detector derives its reference template from the estimated target state

$$\mathbf{x}_{k-1}^* = [x_{k-1}^* \ y_{k-1}^* \ s_{k-1}^* \ \theta_{k-1}^*]^T \tag{18.17}$$

in the frame before ϕ_k dropped below threshold. The detector is further aided by being constrained in the state space to a small search window about $\mathbf{x}_{k-1}^*$. Based on the dynamic target aspect model, we assume that $|s_k - s_{k-1}^*| \leq 2\Delta_s$ and $|\theta_k - \theta_{k-1}^*| \leq 2\Delta_\theta$. Therefore, we uniformly draw samples in the range $s_{k-1}^* \pm 2\Delta_s$ and $\theta_{k-1}^* \pm 2\Delta_\theta$

to initialize the detector template $\mathbf{G}(s,\theta)$, which is then convolved with a small region of interest (ROI) of width $2\Delta_x + 1$ and height $2\Delta_y + 1$ centered about the point (x^*_{k-1}, y^*_{k-1}) in the frame $\mathbf{z}_k$ where the ROI parameters Δ_x and Δ_y are determined by the state transition equation (18.1). Similar to the particle initialization process, a set of boosting particles can be selected from the matched filter results. The velocities of the boosting particles are assigned by applying block motion estimation between frames $k-1$ and k. The boosting particles are then mixed with the current particle set to perform Monte Carlo estimation in the BAPF.

18.4.2.3 BAPF Implementation

Pseudocode for the BAPF is given in Table 18.1, in which the samples μ_k^j are drawn from the transition prior. The threshold T_λ that triggers boosting can be obtained from the first few frames where tracking is usually stable and reliable, or it can be

Table 18.1 Pseudocode for the BAPF algorithm

---

1. Initialization:
    For $j = 1,\ldots,N_p$
        Draw $\mathbf{x}_0^j \sim \arg\max(\mathbf{M}(s,\theta))$ and set $w_0^j = 1/N_p$
    End
2. For $k = 1,2,\ldots$
    Draw samples for the first time:
    For $j = 1,\ldots,N_p$
        Draw $\mu_k^j \sim p(\mathbf{x}_k|\mathbf{x}_{k-1}^j)$ and compute $\widehat{w}_k^j \propto p(\mathbf{z}_k|\mu_k^j) \cdot w_{k-1}^j$
    End
    Normalize such that $\sum_{j=1}^{N_p} \widehat{w}_k^j = 1$
    Draw samples for the second time:
    For $j = 1,\ldots,N_p$
        Draw $m^j \sim \{1,2,\ldots,N_p\}$ such that $p(m^j = i) = \widehat{w}_k^i$, $i = 1,2,\ldots,N_p$
        Draw $\mathbf{x}_k^j \sim p(\mathbf{x}_k|\mathbf{x}_{k-1}^{m^j})$
        compute $w_k^j \propto \dfrac{p(\mathbf{z}_k|\mathbf{x}_k^j)}{p(\mathbf{z}_k|\mu_k^{m^j})}$ and λ_k^j according to (18.10)
    End
    Normalize such that $\sum_{j=1}^{N_p} w_k^j = 1$
    Evaluate the current tracking performance by computing ϕ_k according to (18.16)
    If $\phi_k < T_\lambda$ (poor tracking)
        $\diamond$ A local detector specified by $\mathbf{x}_{k-1}^*$ induces N_d boosting particles $\mathbf{x}_k^{(N_p+1)},\ldots,\mathbf{x}_k^{(N_p+N_d)}$
        For $j = 1,\ldots,N_p + N_d$ compute $w_k^j \propto p(\mathbf{z}_k|\mathbf{x}_k^{(j)})$
        End
        Normalize such that $\sum_{j=1}^{N_p+N_d} w_k^j = 1$
        Compute $\mathbf{x}_k^*$ and resample to keep N_p particles
    Else (good tracking)
        Compute $\mathbf{x}_k^*$
    End
End

---

predefined according to the target template and the sensor noise. The number of boosting particles N_d can be constant or a variable depending on ϕ_k. We normally take N_d in the range $0.02N_p \leq N_d \leq 0.05N_p$. Resampling is applied after the boosting operation to return the overall number of particles to N_p. The computational load required for boosting can be managed by restricting the size of the search area within the state space that is used to define the local detector and by adjusting the ROI parameters and the number of boosting particles.

18.4.2.4 Comments on BAPF

The BAPF algorithm is depicted schematically in Fig. 18.3, where a boosting operation is shown. The results of several simple simulations appear in Fig. 18.4. The particle distribution is shown with respect to the individual state variables. It is seen that the particle distribution after boosting and resampling is enhanced, and its mode is much closer to that of the ground truth posterior, which is shown as a Kronecker delta.

18.4.3 AAPF Algorithm

When the target is highly maneuverable and capable of accelerations that approach the limits of the white noise acceleration model (18.1), the standard SIR and APF

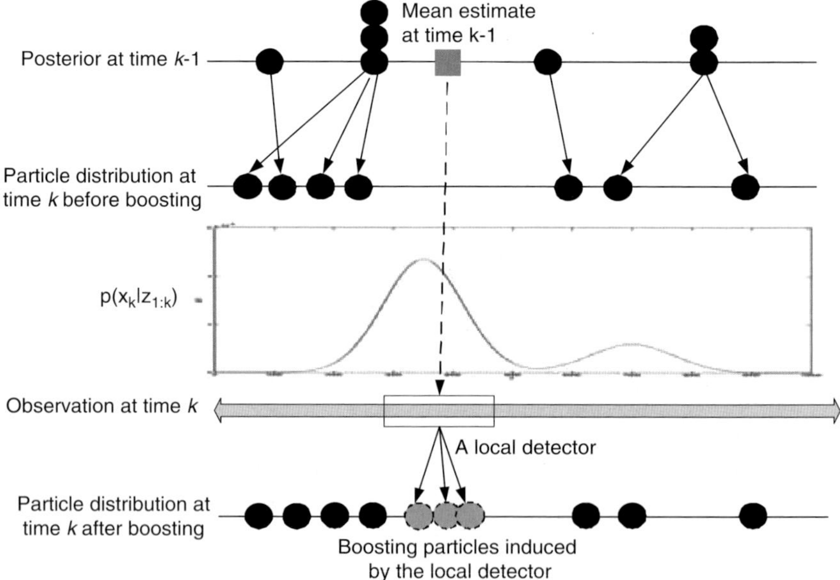

Fig. 18.3 Schematic depiction of the BAPF algorithm illustrating a boosting operation

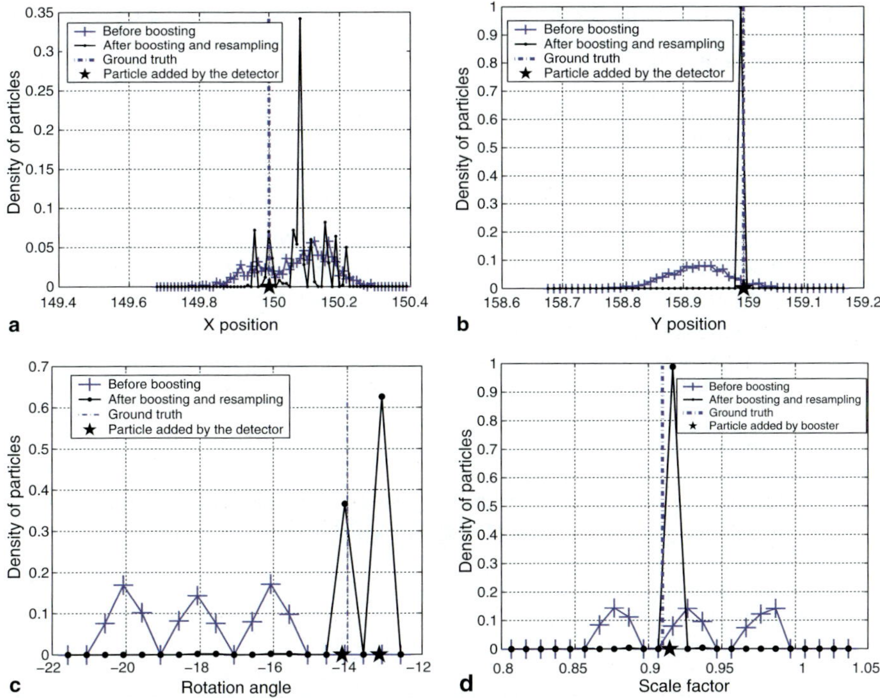

Fig. 18.4 Effect of BAPF on the particle distribution with respect to **a**, **b** position estimate, **c** rotation estimate, and **d** scale estimate

algorithms perform particularly poorly in the affine-invariant target-tracking problem. The situation is exacerbated by the high-dimensional state space and the high selectivity of the likelihood function shown in Fig. 18.1, which increases the chance that all of the particles may fall away from the modes of the posterior. When this occurs, the tracker fails and the target is lost. Boosting is unlikely to be helpful for this particular problem since the boosted particles will be concentrated near the last tracked position of the target and hence will also fail to coincide with the modes of the true posterior in general. In this section, we propose the AAPF as a solution The AAPF adjusts the target motion model by adaptively increasing the noise variances. The number of particles is also increased to reduce the chances of the particle distribution becoming too sparse [9].

18.4.3.1 AAPF Implementation

In the AAPF, the system model is adaptively adjusted based on ϕ_k. Although it is clear from Fig. 18.1 that the likelihood function is also highly selective with respect to the target aspect parameters, in the interest of simplicity we here consider only adjustment of the two variances associated with the velocity drift noise $\mathbf{w}_k$ in (18.1).

Table 18.2 Pseudocode for the AAPF algorithm

---

1. Initialization: (as in BAPF):
2. For $k = 1, 2, \ldots$

    Execute the normal APF and compute ϕ_k (as in BAPF)

    If $\phi_k < T_\lambda$ (poor tracking)

        $\diamond$ Set $\hat{N}_p = nN_p$

        $\diamond$ Make a set $\hat{\mathbf{x}}_{k-1}^l$ of $\hat{N}_p$ particles by making n copies of each $\mathbf{x}_{k-1}^j$ $(j = 1, \ldots, N_p)$ and

          assign corresponding weight W_{k-1}^l as $\frac{w_{k-1}^j}{n}$, $(l = 1, \ldots, \hat{N}_p)$

        $\diamond$ Draw μ_k^l from $p_{adp}(\mathbf{x}_k^l | \mathbf{x}_{k-1}^l)$ where the noise variances are scaled by n,

          assign weights as $\widehat{W}_k^l = p(\mathbf{z}_k | \mu_k^l) \cdot W_{k-1}^l$

        $\diamond$ Normalize such that $\sum_{l=1}^{\hat{N}_p} \widehat{W}_k^l = 1$

        $\diamond$ Use $\widehat{W}_k^l$ to obtain another set of particles $\widehat{\mathbf{x}}_k^l$ as in the second step of the APF using

          $p_{adp}(\mathbf{x}_k^l | \mathbf{x}_{k-1}^{n^l})$ and assign weights $\widetilde{W}_k^l \propto \frac{p(\mathbf{z}_k | \widehat{X}_k^l)}{p(\mathbf{z}_k | \mu_k^{m^l})}$

        $\diamond$ Normalize such that $\sum_{l=1}^{\hat{N}_p} \widetilde{W}_k^l = 1$

        $\diamond$ Compute $\mathbf{x}_k^*$

        $\diamond$ Reduce the particle set to N_p by choosing the top N_p weighted particles from $\widehat{\mathbf{x}}_k^l$;

          the selected particles are denoted as $\mathbf{x}_k^j$ with weights w_k^j, $j = 1, \ldots, N_p$

        $\diamond$ Normalize such that $\sum_{j=1}^{N_p} w_k^j = 1$

    Else   (good tracking)

        Compute $\mathbf{x}_k^*$

    End

  End

For a scaling factor n (distinct from the target aspect parameter s), the number of particles is increased to $\hat{N}_p = nN_p$ during the adaptation step to improve the overlap between the particle distribution and the underlying posterior. In general, n can be a variable dependent on ϕ_k or a constant. At the end of each adaptation step, the number of particles is reduced back to N_p by retaining the particles with the largest weights. Pseudocode for the AAPF algorithm is given in Table 18.2.

18.4.3.2 Comments on AAPF

The adaptation step of the AAPF algorithm is depicted schematically in Fig. 18.5. When the adaptive mode is triggered by ϕ_k, the motion model noise variances are increased, and the number of particles is increased to $\hat{N}_p$. Each particle from the posterior at time $k - 1$ is propagated n times according to the transition model with increased noise variances. The results of several simple simulations are shown in Fig. 18.6 with respect to the individual state variables, where the adaptation step has resulted in increased overlap between the particle distribution and the true posterior.

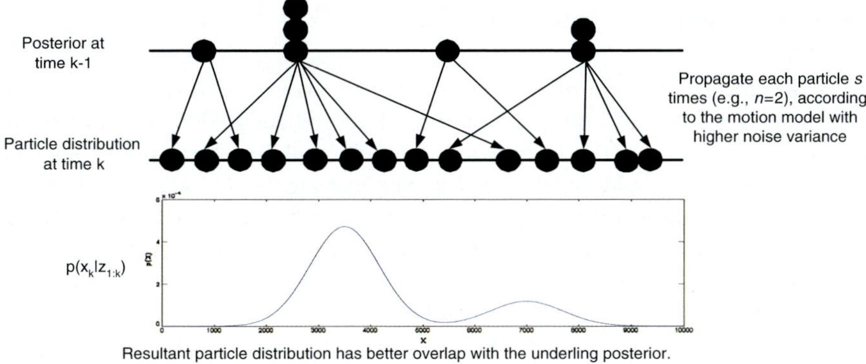

Fig. 18.5 Schematic depiction of an AAPF adaptation step

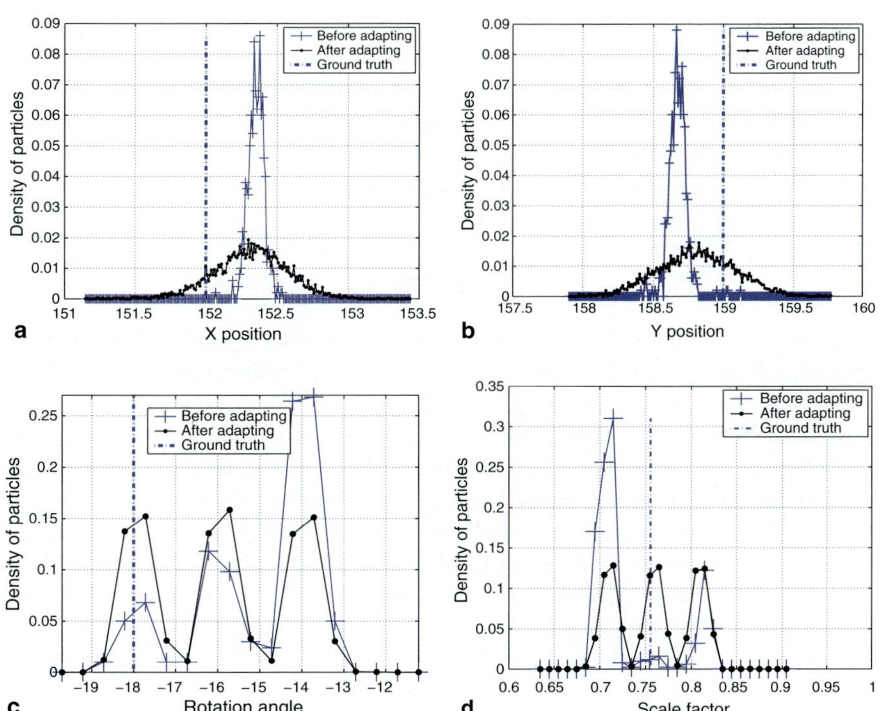

Fig. 18.6 Effect of the AAPF adaptation step on particle distributions with respect to **a**, **b** position estimate, **c** rotation estimate, and **d** scale estimate

18.4.4 Additional Remarks on BAPF and AAPF

In practice, the BAPF algorithm is more efficient than the AAPF algorithm due
to the smaller number of additional particles required for BAPF. In addition, the
computational overhead associated with the BAPF local detector is mild compared
to that of maintaining additional particles. The two algorithms were developed to
address different complementary problems related to the high-dimensional affine-
invariant target tracking state space. If the target still conforms to the state model
but is lost because of low SNR or simply because it fails to coincide with any of
the current particles, both algorithms can be helpful; however, BAPF will be both
more effective and more efficient. When the target is lost because it accelerates
beyond the limits of the state model, AAPF is expected to be more effective since
the boosting particles of BAPF will generally be too focused. Thus, it is of interest to
consider a hybrid top-down and bottom-up approach incorporating both BAPF and
AAPF. Implementation of such a scheme would require development of a means
for determining which algorithm to invoke when a degradation of the track quality
indicator ϕ_k is detected.

Balancing the diversity and focus of the particle configuration is essential to minimize the chance of particles becoming trapped in local minima. This issue is particularly important in higher-dimensional state spaces. The concept of *effective sample
size* was introduced to evaluate the quality of the particles with the goal of monitoring the degeneracy problem. The effective sample size is given by [18]

$$N_{eff} = \frac{1}{\sum_{j=1}^{N_p} (w_k^j)^2}. \tag{18.18}$$

A relatively small value N_{eff} indicates a diversified sample set in which the particle
weights have a large variance, and a relative large value implies a focused sample set
in which the particle weights have a small variance. Good tracking performance
requires a balanced particle distribution [2], which is reflected by a balanced value
N_{eff}. When $\phi_k < T_\lambda$, the tracking performance deteriorates. This can be explained
as a consequence of an unbalanced particle distribution, which corresponds to a
value N_{eff} that is too large or too small. We expect both the BAPF and the AAPF
will have positive effects on the balance of the particle distribution. In Section 18.5,
we compare the values N_{eff} as well as ϕ_k obtained before and after boosting and
adaptation.

18.5 Experimental Results

We tested the BAPF and APFF algorithms against a simulated infrared image sequence 30 frames long with interframe time interval $\Delta = 0.04\,$s. The sequence was
generated by adding GMRF noise fields to a real infrared background of 200×200
pixels. A synthetic target was injected with a base template size of 15×35 pixels

such that the peak target-to-clutter ratio (PTCR) was 5.6 dB. The motion of the target centroid was generated from the white noise acceleration model with initial velocity $\dot{x}_1 = 2.0$ pixels/frame and $\dot{y}_1 = 0.3$ pixels/frame. The ROI parameters for BAPF were set at $\Delta_x = \Delta_y = 4$. The target rotation angle was allowed to vary within a range of $[-30°, 30°]$ with $\Delta_\theta = 2°$, while the scale was allowed in the range $[0.5, 1.5]$ with $\Delta_s = 0.05$. Several sample frames are shown in the first column of Fig. 18.11, where the target is almost invisible to the human observer. The standard SIR and APF algorithms were also run against these sequences, but consistently failed to provide a track lock for any but the first few frames.

The correlation between the track quality indicator ϕ_k and the position error in the tracked centroid $||[x_k^* \; y_k^*]^T - [\hat{x}_k^* \; \hat{y}_k^*]^T||_{\ell^2}$ is shown in Fig. 18.7 for one run of the BAPF algorithm. Perfect correlation would be indicated by a straight-line plot with negative slope. The fact that the data are generally aligned in a negatively sloped band shows that there is correlation and demonstrates the validity of the track quality indicator as a quantification of track quality and consistency of the particle filter. However, the question of how to select an appropriate threshold for ϕ_k that indicates the necessity of boosting or adaptation remains open. In [9], a similar tracking indicator was directly used to control the system model and number of particles where no threshold is needed. A similar strategy could be implemented for BAPF and AAPF.

The BAPF algorithm used $N_p = 1{,}000$ particles, with an additional $N_d = 20$ particles being added from the local detector when boosting was triggered. For the AAPF algorithm, the number of particles was $N_p = 500$. The AAPF parameter n was set to 10, so that $\hat{N}_p = 5{,}000$ particles was used during adaptation. For the chosen T_λ, nearly half of the 30 frames triggered a boosting step in the BAPF or an adaptation step in the AAPF. Therefore, the average number of particles used for BAPF was

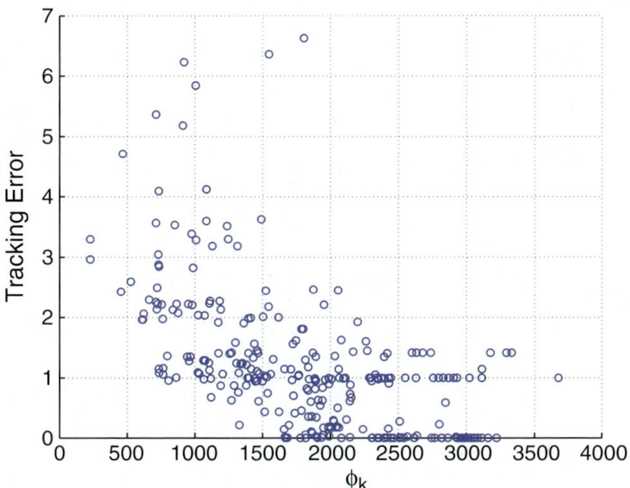

Fig. 18.7 Correlation between the position error in pixels and the track quality indicator ϕ_k

slightly more than 1,000, whereas in the AAPF it was approximately 2,500. In this case, the BAPF was clearly more efficient than the AAPF in the sense of using fewer particles (as expected). In addition to providing accurate and stable estimates of the continuous target aspect parameters, the tracking performance in terms of position error for the BAPF and AAPF was comparable to the techniques given in [7], in which a finite set of multiaspect target templates was employed.

The position error and tracked aspect parameters for both BAPF and AAPF are shown in Fig. 18.8 and Fig. 18.11, where it is seen that both algorithms were able

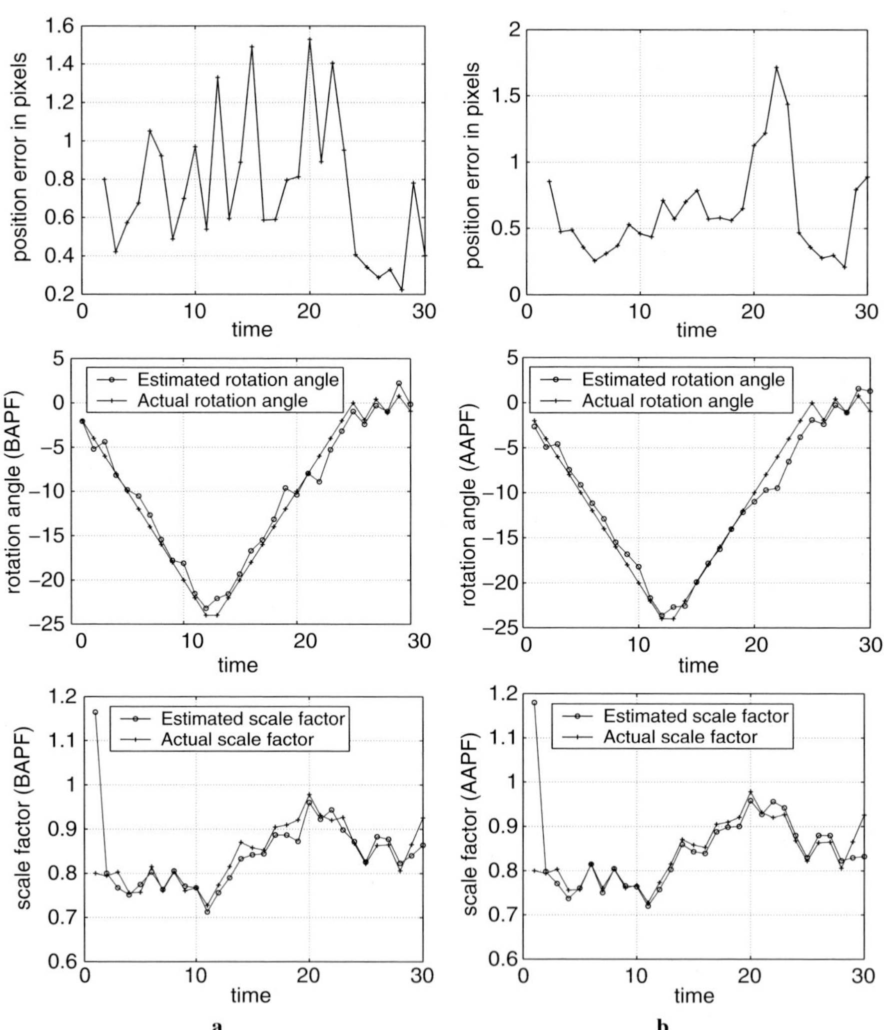

Fig. 18.8 Position error and tracked aspect parameters for **a** BAPF and **b** AAPF, averaged over 30 Monte Carlo runs. *Top row*: position error. The average error is 0.65 pixels for BAPF and 0.64 pixels for AAPF. *Middle row*: tracked rotation angle θ_k^*. *Bottom row*: tracked scale parameter s_k^*.

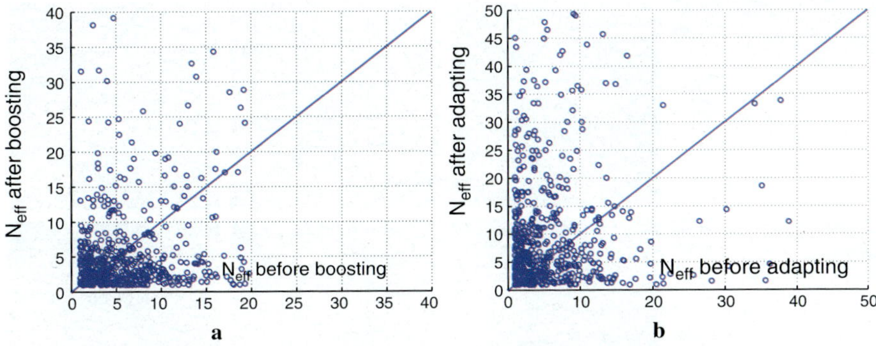

Fig. 18.9 Scatterplots showing N_{eff} before and after boosting and adaptation. **a** BAPF, **b** AAPF

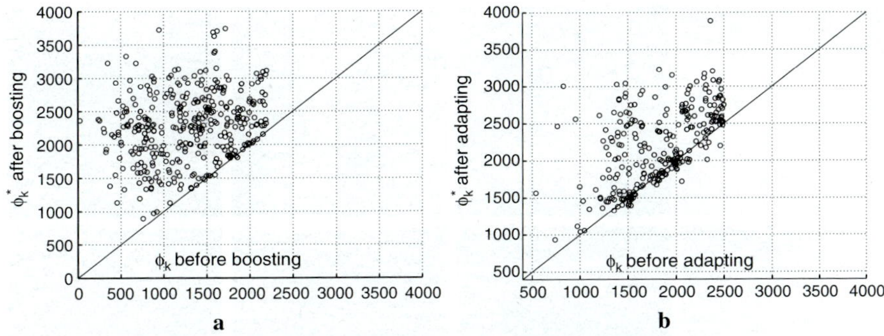

Fig. 18.10 Improvement in track quality ϕ_k due to boosting and adaptation. **a** BAPF, **b** AAPF

to handle the affine-invariant target-tracking problem with similar tracking performance. However, neither BAPF nor AAPF was able to track under the most severe SNR scenarios that were demonstrated in [7], probably as a result of the more complicated nature of the state space used here due to the continuous-valued aspect parameters.

To further examine the BAPF and AAPF in terms of their ability to adjust the balance of the particle distribution, we compare the values of N_{eff} before and after boosting and adaptation in Fig. 18.9. Here, we see that both the BAPF and AAPF can balance the diversity and the focus of the particle distribution by effectively adjusting N_{eff}, which is the key to good tracking performance. However, N_{eff} alone cannot be used to judge the tracking performance, as discussed in [3]. A better tracking indicator should measure the consistency between the predictions and observations directly, as in (18.15). However, the track quality indicator ϕ_k defined in (18.16) seems to provide a reasonable alternative to (18.15), which is difficult to estimate in practice. As shown in Fig. 18.10, the BAPF boosting steps and AAPF adaptation steps are both highly effective for improving the track quality indicator ϕ_k. Figure 18.11 provides the tracking results of the APF, BAPF, and AAPF algorithms.

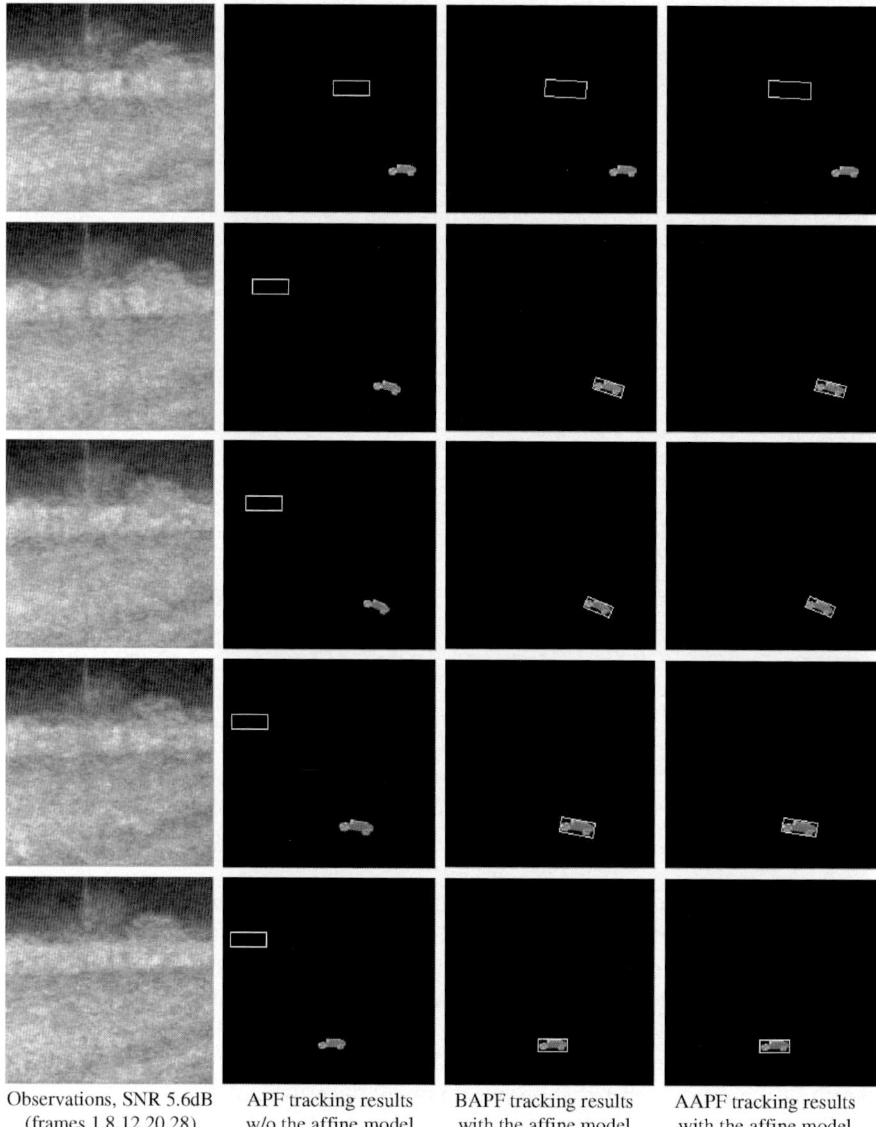

<table>
<tr><td>Observations, SNR 5.6dB
(frames 1,8,12,20,28)</td><td>APF tracking results
w/o the affine model</td><td>BAPF tracking results
with the affine model</td><td>AAPF tracking results
with the affine model</td></tr>
</table>

Fig. 18.11 Tracking results of APF, BAPF, and AAPF algorithms

18.6 Conclusions

In this chapter, we developed an affine-invariant target-tracking approach by which the motion of the target centroid is characterized by a white noise acceleration model, and the target aspect is modeled by an affine transformation applied to the base target template. Compared to other techniques that represent the target aspect by a fixed set of templates, the affine-invariant approach developed here enables us to effectively track the aspect parameters as continuous random variables. This problem formulation is characterized by a complex, high-dimensional state space that causes the standard SIR and APF particle-filtering algorithms to fail. By incorporating two recent particle-filtering enhancements, boosting and adaptation, into the APF, we developed the BAPF and AAPF tracking algorithms, both of which performed well against the affine-invariant tracking problem. Moreover, motivated by recently developed particle filter consistency checks, we introduced a new track quality indicator based on the likelihood function and a data term (matched filtering result) that quantifies agreement between the observed and expected target signatures. The track quality indicator was used to detect degradations in tracking performance and trigger a boosting step in BAPF or an adaptation step in AAPF. Both algorithms succeeded in effectively adjusting the balance between the diversity and focus of the particle distribution and improving the tracking confidence.

Acknowledgments This work was supported by the U.S. Army Research Laboratory and the U.S. Army Research Office under grant W911NF-04-1-0221.

Chapter's References

1. Gordon, N.J., Salmond, D.J., Smith, A.F.M. Novel approach to nonlinear/non-Gaussian Bayesian state estimation. IEE Proceedings-F (Radar, Signal Process.) 140(2) (1993) 107–113

2. Doucet, A., Freitas, J.F.G., Gordon, N.J.: Sequential Monte Carlo Methods in Practice. Springer-Verlag, New York (2001)

3. Heijden, F.V.D.: Consistency check for particle filters. IEEE Trans. Pattern Anal. Machine Intell. 28(1) (2006) 140–145

4. Chang, C., Ansari, R., Khokhar, A.: Multiple object tracking with kernel particle filter. In: Proc. IEEE Int. Conf. Comput. Vision, Pattern Recog. (2005) 566–573

5. Han, B., Zhu, Y., Comaniciu, D., Davis, L.: Kernel-based Bayesian filtering for object tracking. In: Proc. IEEE Int. Conf. Comput. Vision Pattern Recog. (2005) 227–234

6. Deutscher, J., Blake, A., Reid, I.D.: Articulated body motion capture by annealed particle filtering. In: Proc. IEEE Int. Conf. Comput. Vision Pattern Recog. (2000) 126–133

7. Bruno, M.G.S.: Sequential importance sampling filtering for target tracking in image sequences. IEEE Signal Process. Lett. 10(8) (2003) 246–249

8. Bruno, M.G.S., Moura, J.M.F.: Multiframe detection/tracking in clutter: Optimal performance. IEEE Trans. Aerosp. Electron. Syst. 37(3) (2001) 925–946

9. Zhou, S., Chellappa, R., Mogghaddam, B.: Adaptive visual tracking and recognition using appearance-adaptive models in particle filters. IEEE Trans. Image Process. 13(11) (2004) 1491–1505

10. Okuma, K., Taleghani, A., de Freitas, N., Little, J.J., Lowe, D.G.: A boosted particle filter: Multitarget detection and tracking. In: Proc. 8th Eur. Conf. Comput. Vision LNCS 3021 (2004) 28–39

11. Rathi, Y., Vaswani, N., Tannenbaum, A., Yezzi, A.: Particle filtering for geometric active contours with application to tracking moving and deforming objects. In: Proc. IEEE Int. Conf. Comput. Vision Pattern Recog. (2005) 2–9

12. Bruno, M.G.S.: Bayesian methods for multiaspect target tracking in image sequences. IEEE Trans. Signal Process. 52(7) (2004) 1848–1861

13. Wang, P., Rehg, J.M.: A modular approach to the analysis and evaluation of particle filters for figure tracking. In: Proc. IEEE Int. Conf. Comput. Vision Pattern Recog. (2006) 790–797
14. Moura, J.M.F., Balram, N.: Noncausal Gauss Markov random fields: Parameter structure and estimation. IEEE Trans. Inform. Theory 39(4) (1993) 1333–1355
15. Moura, J.M.F., Balram, N.: Recursive structure of noncousal Gauss Markov random fields. IEEE Trans. Inform. Theory 38(2) (1992) 334–354
16. Foley, J., van Dam, A., Feiner, S., Hughes, J.: Computer Graphics: Principles and Practice. 2nd edn. Addison-Wesley, Boston (1990)
17. Lichtenauer, J., Reinders, M., Hendriks, E.: Influence of the observation likelihood function on particle filtering performance in tracking applications. In: 6th IEEE Int. Conf. Automatic Face Gesture Recog. (2004) 767–772
18. Arulampalam, M.S., Maskell, S., Gordon, N., Clapp, T.: A tutorial on particle filters for online nonlinear/non-Gaussian Bayesian tracking. IEEE Trans. Signal Process. 50(2) (2002) 174–188
19. Pitt, M.K., Shephard, N.: Filtering via simulation: auxiliary particle filters. J. Am. Stat. Assoc. 94(446) (1999) 590–599
20. Huang, Y., Djuric, P.M.: A hybrid importance function for particle filtering. IEEE Signal Process. Lett. 11(3) (2004) 404–406
21. Shen, C., Brooks, M.J., van den Hengel, A.: Augmented particle filtering for efficient visual tracking. In: Proc. IEEE Int. Conf. Image Process. 3 (2005) 856–859
22. Rui, Y., Chen, Y.: Better proposal distributions: Object tracking using unscented particle filter. In: Proc. IEEE Int. Conf. Comput. Vision, Pattern Recog. 2 (2001) 786–793
23. Wu, Y., Huang, T.S.: A co-inference approach to robust visual tracking. In: Proc. IEEE Int. Conf. Comput. Vision 2 (2001) 26–33
24. Chang, C., Ansari, R.: Kernel particle filter for visual tracking. IEEE Signal Process. Lett. 12(3) (2005) 242–245
25. Philomin, V., Duraiswami, R., Davis, L.S.: Quasi-random sampling for condensation. In: Proc. Eur. Conf. Comput. Vision 2 (2000) 134–149
26. Maggio, E., Cavallaro, A.: Hybrid particle filter and mean shift tracker with adaptive transition model. In: Proc. IEEE Int. Conf. Acoust. Speech Signal Proc., Philadelphia, (2005)
27. Torma, P., Szepesvri, C.: Enhancing particle filters using local likelihood sampling. In: Proc. 8th Eur. Conf. Comput. Vision, LNCS 3021 (2004) 16–27
28. Torma, P., Szepesvari, C.: On using likelihood-adjusted proposals in particle filtering: Local importance sampling. In: Proc. 4th Int. Symp. Image, Signal Process. Anal. (2005) 58–63
29. Andrieu, C., Doucet, A., Tadic, V.: Particle methods for change detection, system identification and control. Proc. IEEE 92(3) (2004) 423–438

Index